S0-DIJ-656

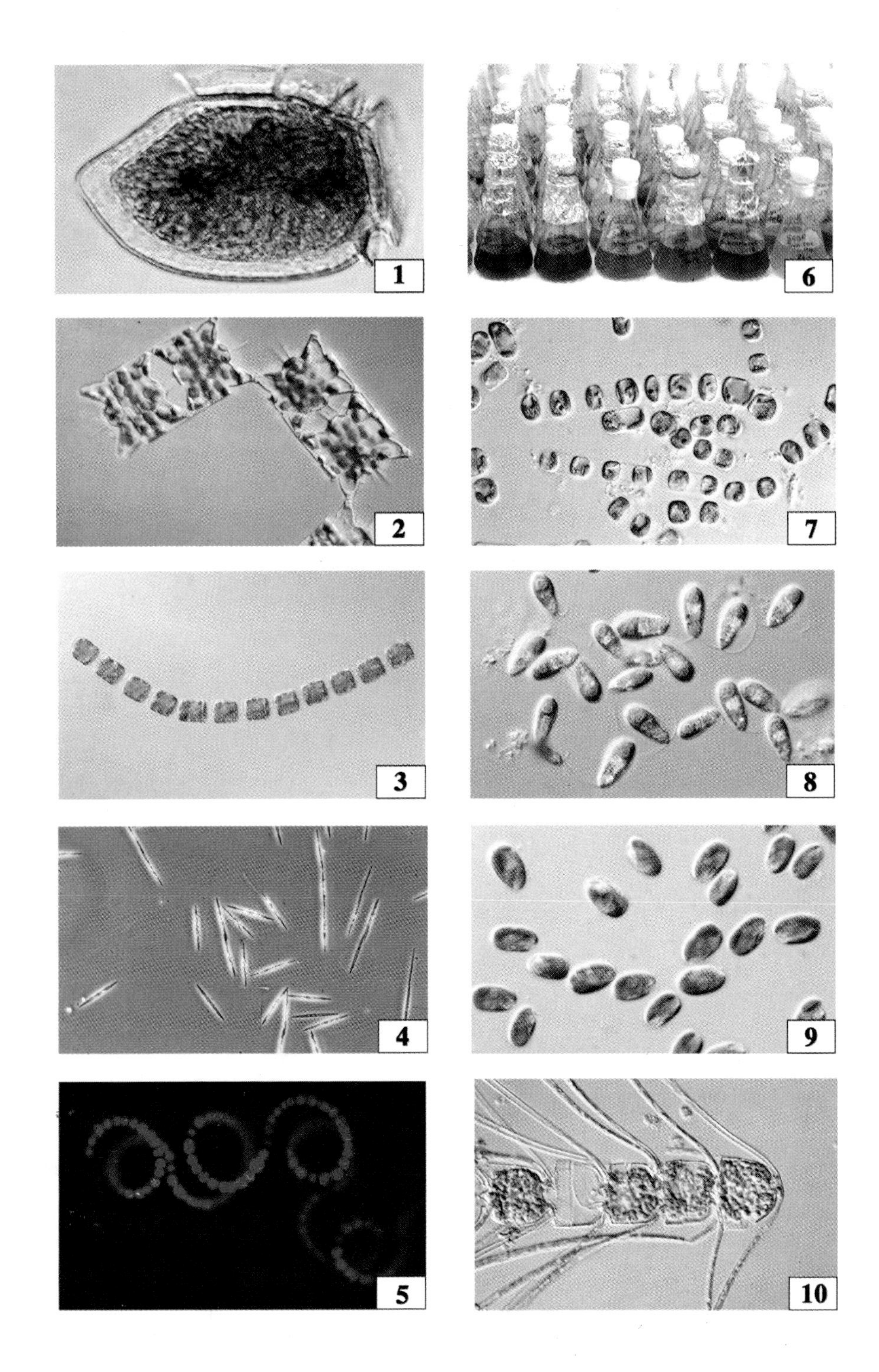
1
2
3
4
5
6
7
8
9
10

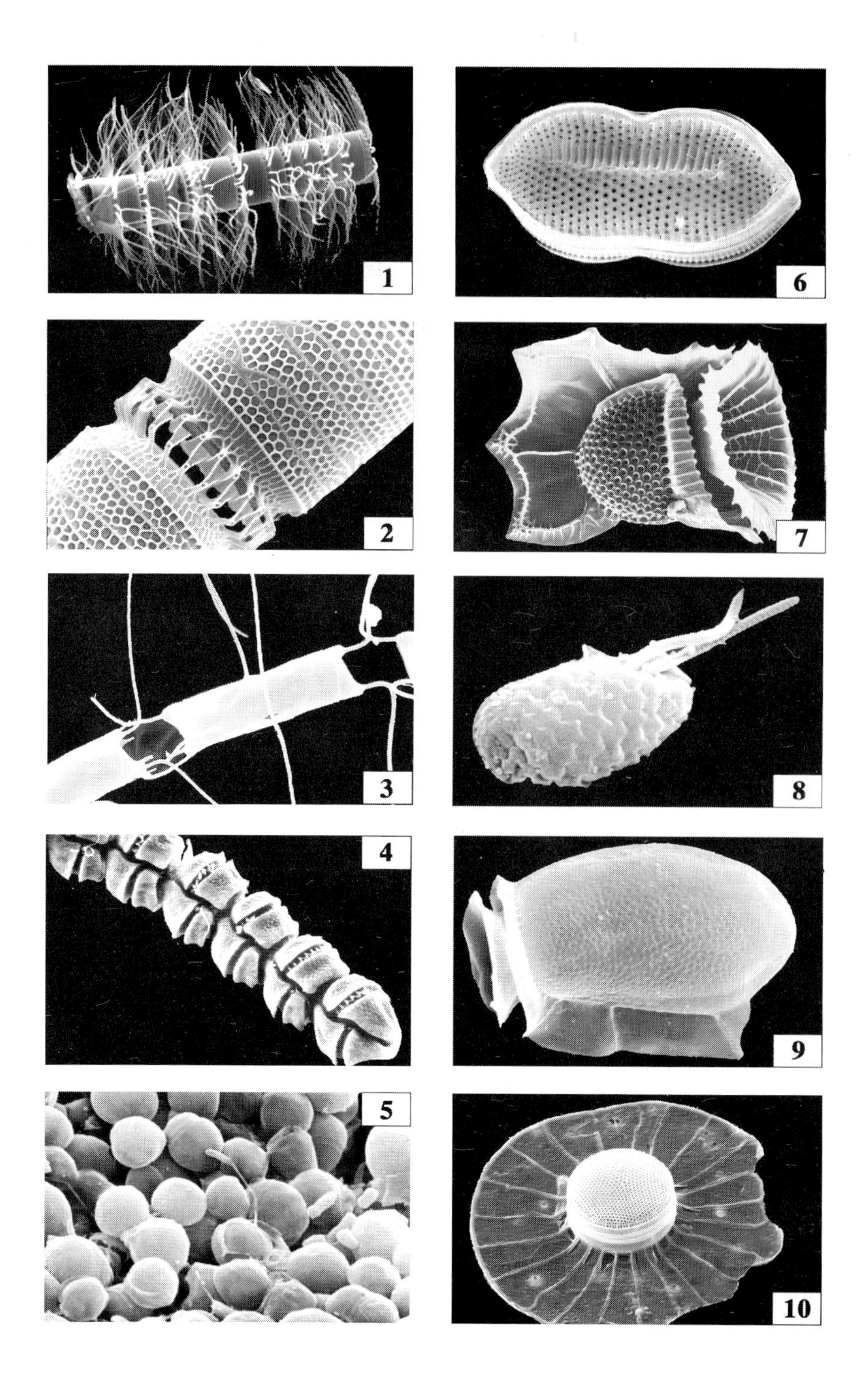
1
2
3
4
5
6
7
8
9
10

Algal Cultures, Analogues of Blooms and Applications

Volume 1

Editor

D.V. Subba Rao

Bedford Institute of Oceanography
Dartmouth, NS
Canada

Enfield (NH), USA   Plymouth, UK

Photos of microalgal species

Front cover: *Emiliania huxleyi* **a coccolithophore**

(Credit: Dr. S.W. Jeffrey, CSIRO, Australia)

Color plate as frontispiece

1. *Dinophysis norvegica* 2. *Biddulphia* sp. 3. *Thalassiosira* sp. 4. *Pseudo-nitzschia pungens f. multiseries* 5. *Anabaena circinalis* 6. Small volume cultures 7. *Skeletonema costatum* 8. *Dunaliella tertiolecta,* 9. *Rhodomonas salina* 10. *Chaetoceros* sp.

(Credits: #1. Dr. Rajashree Gouda # 2, 4. and 10. Subba Rao and # 3,5,6,7,8,9 Dr. S.W. Jeffrey)

Black and white plate on the reverse of frontispiece

1. *Bacteriastrum* sp. 2. *Paralia sulcata* 3. *Chaetoceros lascinosum* 4. *Gymnodinium catenatum* 5. Picoplankters 6. Naviculoid diatom 7. *Ornithocercus magnificus* 8. Cryptophyte 9. *Dinophysis fortii* 10. *Planktoniella sol.*

(Credits: #1, 2, 3, 4, 6, 7, 8, 9,10 Dr. S.W. Jeffrey and #5 Subba Rao)

SCIENCE PUBLISHERS
An imprint of Edenbridge Ltd., British Channel Islands.
Post Office Box 699
Enfield, New Hampshire 03748
United States of America

Internet site: *http://www.scipub.net*

sales@scipub.net (marketing department)
editor@scipub.net (editorial department)
info@scipub.net (for all other enquiries)

Library of Congress Cataloging-in-Publication Data

Algal cultures, analogues of blooms and applications / editors, D.V. Subba Rao.
p. cm.
Includes bibliographical references (p.).
ISBN 1-57808-393-1
1. Algae—Cultures and culture media. 2. Algal blooms. I. Subba Rao, D.V.

QK565.2.A438 2005
579.8--dc22 2005051701

ISBN (Set) 1-57808-393-1
ISBN (Vol. 1) 1-57808-392-3
ISBN (Vol. 2) 1-57808-394-X

Published by Science Publishers, Enfield, NH, USA
An imprint of Edenbridge Ltd.
Printed in India.

Dedication

In friendship to Dr. George F. Humphrey, Australia and Dr. James E. Stewart, Canada, and to the memory of my parents Sastri and Seshamma Durvasula.

SUBBA RAO
Editor

Preface

Marine phytoplankters, the free-floating photosynthetic life, play a crucial role in the production of oxygen and in food web dynamics in the seas. Their diversity in taxonomy, morphology, size, and nutritional requirements continue to fascinate biological oceanographers. The spatial and temporal variations of algae are enormous. To obtain a steady supply of algae for biochemical and physiological experimentation, it soon became necessary to culture the algae under defined laboratory conditions. Although studies on natural assemblages of marine phytoplankton and laboratory cultured algae were initiated about the same time (1893), unlike the former, culture studies did not progress till the 1960s as rapidly as one would have wished.

Due to the development of tracer carbon-14 technique in 1952, interest in studying marine micro algal cultures has been growing rapidly and many unexpected and exciting discoveries have already emerged. Four examples that involve marine micro algae may be cited: Discovery of photosynthetic picoplankton, UV light and climate, Geoengineering and climate and genetic engineering. New techniques have been developed in recent years and rapid advances have been made relating measurements of primary organic production of marine micro algae to their photosynthetic pigments, evident from the plethora of papers and reviews published. In general, principles of terrestrial plant physiology and biochemistry have been extended to study the physiological ecology of marine micro algae. Algal cultures are being used as excellent experimental materials to model growth, nutrient kinetics, physiological ecology, pollution research, phycotoxin research, remote sensing and climatic studies. However, most of the cultures are isolated from temperate seas and very few from other regions. It is to be noted that utility of algal cultures in tropics is mostly limited mariculture and to species isolated elsewhere but not native to their seas.

Despite some differences, several similarities exist between the data obtained on blooms and cultures. As a result of the ease with which some of the algae can be cultured, considerable interest is currently evinced to see if cultures could be used to gain insights to understand some of the ecological principles such as species succession, periodicities, physiological adaptations. Additionally, research is focused on the application of algae in mariculture operations, marine biotechnology, in space research for waste recycling systems, and as source of natural compounds such as antiviral and antifungal compounds and pharmaceuticals.

Of the 5000 confirmed taxa of marine micro algae, about 300 species contribute to blooms, both benign and toxigenic. More and more of novel nuisance phytoplankton blooms are recorded and the connection between their global expansion and human activities is actively sought. About 500 species, mostly from the temperate seas, are brought into culture and about 30 species are studied in considerable detail. A few important toxigenic dinoflagellate taxa still baffle any attempts to culture. Needless to add that research on algae will be actively pursued in the decades to come.

When Science Publishers, New Hampshire approached me to collate and edit a volume on marine micro algae, I thought it would be useful to broaden the range of topics and bring out a thematic volume with recent developments in micro algal research. This book contributed by colleagues from 14 nations encompasses numerous scientific disciplines. I tried to involve serious researchers who have made excellent contributions on marine micro algae. They were requested to incorporate the latest findings specifically to address how best algal cultures can be utilized as analogues of natural blooms, their utility in understanding the ecological principles and their applications in biotechnology. Each chapter is contributed by an expert or group of experts, reviewed internally by colleagues and by outside referees as well. I am grateful to each of the contributors for their high level of professional and scholarly efforts, cordial, and prompt cooperation extended to me. I have gained from their efforts but the omissions and commissions are mine. The scientific opinions expressed in this book are those of the authors and not that of any institution.

This book is not intended to be a compendium of everything worth knowing about marine micro algae given the fact that the knowledge base is constantly expanding.

It is hoped that this volume will be useful to our colleagues in biological oceanography as well as other scientists, advanced undergraduate and graduate students as a summary of current thoughts in physiological ecology.

Acknowledgements

Special thanks are due to Dr. Shirley Jeffrey for the generous assistance with most pictures of algae, and to Dr. Rajashree Gouda. My sincere thanks are extended to Mr. Arthur Cosgrove, Technographics, Bedford Institute of Oceanography, for his artistic skills in the design of the cover and the plates of algae. For her infinite patience, excellent help with the formatting, correspondence and unstinting support I am most grateful to my wife Bala T. Durvasula.

Contents

List of Contributors

Paulo Abreu
Departmento de Oceanografia
Fundação Universidade Federal do Rio Grande (FURG)
Caixa Postal 474, 96201-900 Rio Grande, RS, Brazil.

Nicholas G. Adams
National Marine Fisheries Service, NFSC
2725 Montlake Blvd., East
Seattle, Washington 98112, USA.

Geneviève Arzul
Ifremer, DEL-PC, BP 70
F-29280 Plouzané, France.

Elisa Berdalet
Institut de Ciències del Mar
Centre Mediterrani d'Investigacions Marines i Ambients (CSIC)
Pg. Marítim, 37-49
E-08003 Barcelona, Catalunya, Spain.

Bopaiah Biddanda
Annis Water Resources Institute and Lake Michigan Center
Grand Valley State University
740 W Shoreline Drive, Muskegon, MI 49441, USA.
Email: biddandb@gvsu.edu

Ilya Blum
Department of Mathematics
Mount Saint Vincent University
Halifax, N.S. Canada, B3M2J6

Deborah A. Bronk
Department of Physical Sciences
The College of William and Mary
Virginia Institute of Marine Science
Route 1208; Greate Rd.
Gloucester Point, VA 23062, USA.
Email: bronk@vims.edu

Malcolm R. Brown
CSIRO Marine Research and Aquafin CRC
GPO Box 1538
Hobart, Tasmania 7001
Australia.

Pascal Claquin
Laboratoire de Biologie et de Biotechnologies Marines
Université de Caen Basse-Normandie
Esplanade de la paix, 14032 Caen Cedex, France.

Andrew T. Davidson
Australian Antarctic Division
Channel Highway, Kingston
Tasmania 7050, Australia.

Pedro Duarte
CEMAS – University Fernando Pessoa
Praça 9 de Abril, 349, 4249-004 Porto
Portugal.

Ravi V. Durvasula
Department of Epidemiology and Public Health
Yale University School of Medicine, New Haven, CT, USA.
Email: ravi.durvasula@yale.edu

Sonya T. Dyhrman
Biology Department MS #32, Woods Hole Oceanographic Institution
Woods Hole,
Massachusetts 02543 USA.

Marta Estrada
Institut de Ciències del Mar
Centre Mediterrani d'Investigacions Marines i Ambients (CSIC)
Pg. Marítim, 37-49
E-08003 Barcelona, Catalunya, Spain.

Sabine Flöder
Institut für Botanik
Universität zu Köln
Gyrhofstr. 15 50931 Köln
Germany.
Email sabine.floeder@uni-koeln.de

Kevin J. Flynn
Institute of Environmental Sustainability
University of Wales
Swansea, Singleton Park
SA2 8PP, UK.

Patrick Gentien
Ifremer, CREMA,
F-17137 L'Houmeau, France.

Geir Johnsen
Biological institute
Norwegian University of Science and Technology
N-7491 Trondheim, Norway.

Johan U. Grobbelaar
Department of Plant Sciences
Botany, University of the Free State
Bloemfontein 9300, South Africa.

Tung-Yuan Ho
Department of Earth and Environmental Sciences
National Chung Cheng University,
Ming-Hsiung, 621,
Chia-Yi, Taiwan.
Email: tyho@ccu.edu.tw

S.W. Jeffrey
CSIRO Marine Research
GPO Box 1538, Hobart
Tasmania, 7001, Australia.

Oliver Kilian
Fachbereich Biologie
Universität Konstanz
78457 Konstanz, Germany.

Jacco C Kromkamp
Centre for Estuarine and Marine Ecology
Netherlands Institute of Ecology
PO box 140
4400 AC Yerseke, the Netherlands.
Email: j.kromkamp@nioo.knaw.nl

Peter G. Kroth
Fachbereich Biologie
Universität Konstanz
78457 Konstanz, Germany.

Thierry Lebeau,
Laboratoire de Biologie Marine (UPRES EA 2663),
Institut des Substances et des Organismes de la Mer (ISOmer)
Université de Nantes, 2, rue de la Houssinière,
BP 92208, 44322 Nantes cedex 3, France.

Jane Lewis
Research Centre Director
School of Biosciences
University of Westminster
115, New Cavendish Street
London W1W 6UW
UK.

Alan J. Lewitus
Belle W. Baruch Institute for Marine and Coastal Sciences
University of South Carolina, and
Marine Resources Research Institute

South Carolina Department of Natural Resources
P.O. Box 12559, Charleston, South Carolina, USA.
USA 29422-2559.

Scott K. Matthews
Department of Epidemiology and Public Health
Yale University School of Medicine, New Haven, CT, USA.

Marina Montresor
Stazione Zoologica 'A. Dohrn'
Villa Comunale
80121 - Naples, Italy.

Keizo Nagasaki
Harmful Algae Control Section, Harmful Algal Bloom Division,
National Research Institute of Fisheries and Environment of Inland Sea
2-17-5 Maruishi, Ohno, Saeki, Hiroshima 739-0452, Japan.
Email: nagasaki@affrc.go.jp

Clarisse Odebrecht
Departmento de Oceanografia
Fundação Universidade Federal do Rio Grande (FURG)
Caixa Postal 474, 96201-900 Rio Grande, RS, Brazil.

Youlian Pan
Institute for Information Technology
National Research Council of Canada
1200 Montreal Road, Building M50
Ottawa
Canada, KIA OR6

Louis Peperzak
National Institute for Coastal and Marine Management/RIKZ
P.O. Box 8039
NL-4330 EA Middelburg
The Netherlands.

Jean-Michel Robert
Laboratoire de Biologie Marine (UPRES EA 2663),
Institut des Substances et des Organismes de la Mer (ISOmer)
Université de Nantes, 2, rue de la Houssinière, BP 92208, 44322 Nantes cedex 3, France.

Egil Sakshaug
Biological institute
Norwegian University of Science and Technology
N-7491 Trondheim, Norway.

Ulrich Sommer
IFM-Geomar, Leibniz-Institute of Marine Science
Düsternbrooker Weg 20
24105 Kiel, Germany.

Pazhani Sundaram
Recombinant Technologies
LLC, Science Park
New Haven, CT, USA.

Ranjini K. Sundaram
Department of Epidemiology and Public Health
Yale University School of Medicine, New Haven, CT, USA.

Theodore J. Smayda
Graduate School of Oceanography University of Rhode Island
Kingston, RI 02881, USA.
Email: tsmayda@gso.uri.edu

D.V. Subba Rao
Bedford Institute of Oceanography
P.O. Box 1006, Dartmouth
NS, Canada, B2Y 4A2.

S. Swaminathan
Department of Mathematics and Statistics
Dalhousie University
Halifax
Nova Scotia, Canada, B3H 3J5.

Peter Thompson
Principal Research Scientist
CSIRO Marine Research, Hobart
Tasmania, 7001, Australia.

John K. Volkman
and Malcolm R. Brown
CSIRO Marine Research and Aquafin CRC
GPO Box 1538
Hobart, Tasmania 7001
Australia.
Email: john.volkman@csiro.au

S.W. Wright
Australian Antarctic Division, and Antarctic Climate and Ecosystems CRC
Channel Highway, Kingston, Australia

1

Why Study Algae in Culture?

D.V. Subba Rao[1]

[1]ERD, Bedford Institute of Oceanography, P.O. Box 1006, Dartmouth, N.S. Canada, B2Y 4A2.

Abstract

Marine micro algae play a central role in the biogeochemical cycles and sustain all forms of life in the oceans. Studies on marine phytoplankton and algal cultures were contemporaneous and aimed at increasing the harvestable food from the sea. Due to the ease with which samples can be collected, spatial and temporal studies of phytoplankton assemblages progressed rapidly. Geographical variations in the qualitative and quantitative abundances were related to the prevailing physical and chemical, particularly the nutrient conditions. Due to the non-availability of dependable methodology and defined culture media, process-oriented studies such as nutrient assimilation, photosynthesis and respiration rates did not progress for nearly 50 years as one would have wished.

Subsequent availability of sophisticated analytical techniques for nutrients and chlorophylls, carbon-14 tracer method to measure photosynthesis facilitated measurements of nutrient kinetics, division rates, carbon assimilation rates, adaptations to environmental variables, and biochemical variables both in natural and culture populations. Several principles of plant physiology were extended to cultures grown under monotonic conditions of temperature and light and the data demonstrated existence of similarities between natural populations, particularly the blooms, and culture populations.

The various chapters in this book present studies that facilitate our understanding of the structure and physiological functioning of the marine microalgae. Recent culture studies have shed light on life cycles of algae, biochemical and physiological changes associated with cell cycles, multi-nutrient kinetics, allelopathic interactions, fluorescence dynamics, and physiological adaptations of algae. As a result because of these developments it is now possible to understand some aspects of the ecological principles governing phytoplankton dynamics, and the structure and functioning of the marine microalgae and their production characteristics.

Algal cultures are increasingly applied in mariculture studies, and in those aimed at mechanisms of production of biotoxins, measurements of biomass and production using remote sensing techniques, remediation of marine pollution, as sources of natural biochemical and bioactive compounds, pharmaceuticals, in understanding the role of algae in climatic variations and their applications in genetic engineering. It is possible that studies based on cultures grown as single species and in mixed species, in dialysis bags, or diffusion chambers or in microcosms or mesocosms under simulated natural conditions in respect of temperature, spectral characteristics of light and at nutrient levels comparable to those in nature would be instructive and facilitate our understanding of these primary producers.

INTRODUCTION

Phytoplankton, a term coined in 1897, describes the free-floating pulsating photosynthetic microalgae in the aquatic environment. Marine phytoplankton cell volume in the oceans ranges between 6.10^{-20} and 4.10^{-9} m^3 (Raven, 2001) and account for nearly 50% of global primary production by plants on earth. Thus marine microalgae play a crucial role as a biological pump by consuming carbon dioxide at the surface and sequestering carbon as particulate and dissolved organic to deep oceans (Buesseler et al. 2004). Nearly 75% of the surface primary production is recycled in the euphotic zone through consumption by heterotrophs and about 20% of the surface production is transferred to the mesopelagic zone. Only 5% of the surface primary production makes it to the bottom (Coale, 2001). The main purpose of phytoplankton studies is to understand the production processes so that primary production, particularly in the nitrogen deficient oceanic waters, can be primed and food production from the oceans could be augmented to feed the anticipated nine billion human population. At present the annual global harvest from the oceans is approximately 103×10^6 tonnes which is about 6% of the human protein requirement (Schmitt, 1970) equivalent to 200×10^6 cattle.

As microalgae sustain either directly or indirectly on all marine life including the commercially important fisheries, studies on microalgae are crucial to our understanding of the structure and functioning of the aquatic ecosystems. Additionally, marine microalgae produce an array of bioactive compounds. To understand their diversity, size, metabolism and their role in the biogeochemical cycles, cultures of microalgae readily provide the necessary reference material for experimentation. Algal cultures are utilized to understand their life cycles and their role in microbial loops, energy transformation, and species succession. Although not enough is known, algal cultures are finding increasing applications in biotechnology, genetic engineering and space research.

In the marine environment there are about 17,000 photosynthetic species of which about 7000 are planktonic (Falkowski et al. 2003) of which about 5000 are confirmed (H. John Heinz III Center for Science, 2002; Sournia 1995, Sournia et al. 1991). These microalgae are diversified in morphology, size, metabolism and biochemistry. The simplest and the most abundant in the oceans are the cyanobacterial cells ($10^8 l^{-1}$) such as the coccoid picoplankter *Synecchocystis* (Furhman and Capone, 2001) and the most ornate are the dinoflagellates (see Frontispiece and inside cover page) such as *Ornithocercus magnificus* "with extarvagant development of bizzare excrescences on the thecae-spines flattened into wing-like expansions and so forth – that have usually been interpreted as organs of floatation" (Hart, 1963).

From an evolutionary point of view the cyanobacteria have existed for some 3500 million years (The Archaean Period) and the first algae began about 1200 million years ago in The Proterozoic Era (Bengston, 1994). The explosion of sea life began in the Paleozoic period (540-250 million years) and the first diatom occurred about 120 million years in the early Cretaceous (Gersonde and Harwood 1990). The planktonic ways of life, particularly of diatoms, evolved over geological time are the best adapted to their environmental conditions to out-compete the larger forms (Fogg, 1991).

CLASSIFICATION OF PHYTOPLANKTON

The phytoplankters range from the smallest unicellular picoplankton (< 3 μm diameter) to 8 cm long chains of 10,000,000 cells of the diatom *Navicula gravelliana* (Hendy, 1964). Based on their cell size, three major categories of phytoplankton are recognized i.e. the picoplankton (< 3 μm), the nanoplankton (< 20 μm) and the net or microplankton (> 20 μm). Of significant interest is the recent discovery of plankton life the SAR 11 cells (< 0.25 to 0.7 μm cells from Sargasso Sea) in the oceans (Morris, 2002). Their estimated microbial biomass of 2.4×10^{28} cells in the oceans with 50% in the euphotic zone will have a major impact on regional geochemistry and on the surrounding plankton. Although their exact physiological activities and roles in oceanic carbon cycling are not known, their abundance and presence in the northwestern Sargasso Sea waters to depths of 3000 m and in Oregon coastal surface waters suggests they are efficient competitors for resources.

Phytoplankton are represented by both Prokaryotes and Eukaryotes which differ in their characteristics as shown below (Table 1.1). While both groups provide interesting research material, the Eukaryotes are valuable tools in biotechnological industries and genetic engineering studies (see chapters: 21. Kilian and Kroth, 22. Lebeau and Robert, 24. Durvasula et al.).

Table 1.1 Characteristics of Prokaryotes and Eukaryotes

Characteristics	*Prokaryotes (Example-Blue greens)*	*Eukaryotes (Example-Other algae)*
Size	Small	Usually 10 × larger
Genetic make up	Haploid	Both haploid and diploid
Genetic system	Simple	Complex
Nuclei	Not membrane enclosed	Membrane enclosed
Chromosomes	Absent	Occur usually as homologous pairs
Division	Binary fission	Cell division by mitosis
Introns (non-coding regions in coding regions of the gene	Absent	Present
Organelles like Mitochondria and chloroplasts	Absent	Membrane enclosed; mitochondria extract energy, chloroplasts capture solar energy
Histones associated with DNA	Absent	Histones present
Cell contents	Peptidoglycan present	Absent
	Single RNA polymerase	Multiple RNA polymerase
Metabolic products	Long-chain poly unsaturated fatty acids	Saturated and monosaturated fatty acids, triglycerides extracellular polysacharides, D-galactose and L-galactose
Fe: P	High	Low
Siderophores	Present	Absent
Metabolism – carbon source	Oxygenic photosynthesis Aerobic respiration and Anaerobic respiration Anaerobic fermentation Anoxic photosynthesis Chemoautotrophy	Oxygenic photosynthesis Aerobic respiration

Algal Phylogeny and Pigments

There are 12 major taxonomic phyta or divisions representing marine phytoplankton. Common to all phytoplankters is the photosynthetic pigment chlorophyll *a*, which has become the touchstone for measuring algal biomass. The presence of additional characteristic auxiliary pigments imparts the characteristic color to specific algae (see chapter 2 Jeffrey and Wright). A strong correlation exists between the pigment types and phylogenetic trees of 37 species of algae (Zapata et al. 2004). Based on the utility of accessory pigments in their photosynthetic physiology, these algae can be separated into those with chlorophyll *b* – the green plastid group, that contrasted with the chlorophyllide *c* plastid-containing group. The latter group have been ecologically successful over the past 250,000,000 years (Grebyk et al. 2002). The only exception to this is the prokaryote *Acaryochloris marina* which has chlorophyll *d* as its major photosynthetic pigment (Miyashita et al. 1996, Kurano and Miyachi, 2004). Despite a few exceptions to pigment distributions, four lines of algae are recognized (Table 1.2).

Earlier studies

The Kiel school in Germany initiated phytoplankton studies in 1897. Until 1920 these studies were concerned with primary systematics, spatial and temporal variations with estimates of relative abundance, plankton dynamics, geographical patterns and seasonal changes (Hart 1963, Mills, 1989). Subsequently, between 1920 and 1935 physical and chemical events were implicated to explain formation of seasonal plankton blooms based on the nutrient cycles particularly of phosphate and nitrogen, related changes in phytoplankton to light and effective length of day (Mills 1989). Emphasis continued mostly on spatial and temporal distribution patterns of phytoplankton (Talling, 1984) in relation to environmental variables such as temperature, salinity, light and nutrient levels that could be routinely determined with ease. Culturing diatoms with a view to harvest more food from the sea however, had been initiated at Plymouth (Allen, 1910). Utilizing the "Dark and Light bottle technique" of Gaarder and Gran (1927), Marshall and Orr (1928) and Jenkin (1937) initiated studies on oxygen exchange by diatom cultures suspended at various depths in the sea.

In the seas, the only thing constant is the constant change of the qualitative and quantitative abundance of the populations. This variability is rather an inherent property of planktonic systems (Colijn et al. 1998). The qualitative and quantitative abundance, growth and production of phytoplankton progress over various scales of space and time (Harris, 1986) are governed by a suite of physical, chemical and biological variables. In the temporal scale these events range from < 1 min (molecular processes) to > 1 year (growth cycle of zooplankters) and in the spatial scale from 1 mm (< 0.25 to $0.7\ \mu m$ SAR 11 cells) to 10 km (zooplankton patches).

Nutrition

Their diverse modes of nutrition make the microalgae fascinating and explains the great plasticity in their end products (Fogg, 1989). Most

Table 1.2 Major lines and divisions of algae

Line of algae	*Characteristic pigments*	*Phyta or Division*	*Representatives*	*Total species**	*Example*
Golden-brown	*Chl a,c*	Bacillariophyta	Diatoms	12,000	*Coscinodiscus*
		Dinophytal, Chrysophyta	Dinoflagellates	4,000	*Dinophysis*
		Haptophyta	Golden algae	1,200	*Chrysosphaera*
		Xanthophyta	Prymnesiophytes	300	*Emiliana*
		Cryptophyta	Yellow greens	600	*Tribonema*
		Eumastigophyta	Cryptomonads	200	*Cryptomonas*
			Yellow-green algae	600	*Dinobryon*
Green	*Chl a,b*	Chlorophyta	Green flagellates	17,000	*Chalmydomonas*
		Euglenophyta	Euglenoids	900	*Euglena*
Red	*a and phycocyanin, phycoerythrin, carotene, allophycocyanin*	Rhodophyta	Red algae	6,000	*Porphyra*
Blue-green	*a and phycocyanin, phycobilin, phycoerythrin*	Cyanophyta	Blue greens	2,000	*Synecchocystis*
		Prochlorophyta	Prochlorophytes		*Prochloron*

*Includes fresh water and marine (See Graham and Wilcox 2000) microalgae are autotrophic and photosynthetic but contain exponents of all methods of nutrition. A few are capable of osmotrophy to assimilate dissolved organic substances, particularly the extracellular amino acid oxidases and proteolytic enzymes (see chapters: 9 Bronk and Flynn; 10, Lewitus). The 'new' dissolved organic nitrogen may even stimulate growth of many algae including those forming harmful algal blooms. Some of the unicellular, nonheterocystis, symbiotic nitrogen-fixing cyanobacteria exist as symbionts within the host cells of the coral (Lesser et al. 2004) or in the colonial diatom *Hemiaulus hauckii* (Carpenter et al. 1999). The heterocystous, N_2-fixing cyanobacterial *Richelia intracellularis* exists as an endosymbiont in the diatom and the symbiotic association adds an average of 45 mg Nm^2d^{-1} to the water column (Carpenter et al. 1999). While a few dinoflagellates are parasitic, some are holozoic as in *Gymnodinium fungiform*. Hundreds of this dinoflagellate attach to the surface of prey organism by a peduncle and ingest the cytoplasm or body fluids of its prey (Moree and Spero, 1981). The most bizzare mode of nutrition is in the toxic estuarine dinoflagellate *Pfiesteria piscicida* and *P. shumwayae* that ambush the prey. These are attracted to fish, produce toxins that cause stress, disease and death to estuarine fish (Burkholder et al. 2001).

Algal Blooms and Species Succession

As a review of the extensive literature on the qualitative and quantitative abundance of marine microalgae in the marine environment and the governing factors is beyond the scope of this chapter, some salient aspects where culture experiments contributed or would contribute further to our understanding of the physiological ecology of marine microalgae are briefly presented. In the temperate waters when the local conditions are favorable for algal growth several species grow rapidly. However, only a few species dominate leading to the formation of a bloom (> 20 μg chl *a* l^{-1}), usually a major one during spring and again a minor one during fall (Cushing, 1975). The spring bloom is attributed to seasonal mixing, spring thaw, availability of nutrients, more solar radiation essential for photosynthesis, and general warming of the environment. For a bloom to occur it is imperative that the water column be seeded with resting stages of diatoms with an ability to exploit favorable conditions; the vegetative form of the initial population results in a bloom (Ishikawa and Furuya, 2004). It may be noted that life histories of algae will have to be worked out with algal cultures (see chapter: 3 Montresor and Lewis). At low N: P ratios recruitment of resting stages induced blooms of *Microcystis* (Stahil-Delbango et al. 2003). Even in the oceanic waters niche segregation exists between *Prorochlorococcus* and *Synchococcus* in the central Atlantic depending on light (Augusti, 2004).

Seasonal blooms are ephemeral and may be caused by the complex life cycles associated with the physiological stages of the alga. Bloom formation may be regulated by phases in the life cycle of the alga which may be timed by an endogenous clock (Anderson and Kieafer, 1987) or by the lunar phase (Wyatt and Jenkinson, 1997). Usually the bloom lasts up to 21 days and during this time a few of the initial species are replaced by others, ecologically known as species succession. Species succession could be interpreted in terms of bacterial action (see chapter: 5 Biddanda et al.) or due to nutrient ratios and enrichment as shown with algal assemblages in a microcosm enrichment experiment (Estrada et al. 2003; see chapter 15: Flöder and Sommer). It is also possible that microalgae at the species level produce biologically active molecules the "allelochemicals" which cause a temporary dominance of the producer (the "donor") over the "receptor" species. This "chemical warfare" (Smetacek, 2001) influences species succession, biodiversity and food web structure (see chapter: 4 Arzul and Gentien). Smayda in this book reopened the Redfield-Braarud debate and advocated a functional group approach to understand the ecophysiology of species and their succession. He further argued about the need to combine autecological and synecological concepts, field and culture approaches to discern properties diagnostic of physiological status (see chapter: 7 Smayda).

In addition to local conditions, events on a greater scale also influence phytoplankton successions. Due to riverine fluxes, and increased usage of fertilizers the atomic ratios of nitrogen: phosphorus: silica in the coastal waters have changed globally from 16:1:16 to 26:1:16 causing species succession (see chapter: 8 Tung-Yuan Ho). Ratios below 1:1 Si: N, seem to replace diatoms by non-diatoms and P and Si limitation seems to lead to noxious blooms (Turner et al. 2003). During spring diatoms and dinoflagellates generally decreased in the Baltic proper but decreased in the Kattegat. In the Kattegat and Baltic cyanobacteria decreased (Wasmund and Uhlig, 2003). The North Atlantic Oscillation (NAO) seems to cause climatic variability in the northeast United States, as well as in the North Sea, Celtic Sea, Skagerrak and\Central North Atlantic. NAO resulted in a long-term increase in temperature ca 2.3°C to 3.0°C during the winter quarter from 1964 to 1996 (Smayda et al. 2004). This in turn decreased the occurrences of the benign winter-spring blooming diatom *Detonula confervacea, Skeletonema costatum*. Climatic changes due to NAO seem to effect the distribution of three toxic species of *Dinophysis* in the Gullmarg Fjord, Sweden (Belgrano et al. 1999).

The number of species contributing to a bloom varies and increases towards the tropics. In the Arctic approximately 15 species, mostly diatoms, contribute to the bloom (Subba Rao, et al. 1988) whereas in the temperate coastal seas about 30 species, mostly centric diatoms go through a succession pattern and contribute to the bloom (Platt and Subba Rao, 1970, Subba Rao and Smith, 1978). In the waters off Spain a few microflagellates and 10 diatoms contributed to the spring bloom (Casa et al. 1999) and in the Black Sea although 179 species were recorded, 7 diatoms and 22 dinoflagellates constituted the bloom (Túrkoğlu and Koray, 2002). For example, in Narragansett Bay, Rhode Island during a 38 year study 9 diatoms contributed to the bloom during 1959–1978 and during 1990–1996; during the intervening years 1979–1989 diatoms decreased, flagellates increased and six harmful algal bloom species occurred (Borkman and Smayda, Subba Rao et al. 2003). In the arid zone waters off Kuwait 28 species, mostly diatoms, constituted the bloom (Subba Rao, et al. 1999). In the coastal tropical waters of the Bay of Bengal, although about 100 species were abundant during spring, 36 species dominated the bloom and followed a succession pattern (Subba Rao, 1971, 1973). In the Arabian Sea on the west coast of India 74 species were abundant but 39 of them contributed to blooms (Subrahmanyan and Sarma, 1961). In experimental studies, small-scale turbulence of only a few millimeters could influence the extracellular micronutrient environment, leading to nitrogen assimilation, dominance of diatoms or dinoflagellates and appearance or disappearance of blooms (see chapter: 13 Berdalet and Estrada).

There are about 300 bloom forming species (H. John Heinz III Center for Science 2002) of which about 78 species are toxin producers including 60 dinoflagellates (Subba Rao, 2002). Atypical algal blooms are non-seasonal, usually monospecific and cause red-tide phenomenon with biomass levels as high as ~900 μg chl *a* l^{-1} (Subba Rao et al. 2003). Some of the red-tide species can be toxigenic leading to Paralytic Shell fish poisoning (PSP), Diarrhetic Shellfish poisoning (DSP), Neurotoxin Shellfish poisoning (NSP) and Ciguetera episodes with far reaching harmful effects on the human health and on commercially important fisheries. Some of the causative species of the monospecific blooms are not amenable for culturing thereby necessitating concentration of cells utilizing the right sieves for physiological studies as in *Gonyaulax digitale* (Amadi et al. 1992) or the toxigenic *Dinophysis norvegica* (Subba Rao and Pan, 1993). Such species when isolated and eventually cultured serve as research material for determining their growth, physiology (see chapter: 23 Peperzak and Dyhrman) and biochemistry to gain insights into their functioning.

Formation of blooms in offshore waters is quite rare and is mostly detected by satellite imagery. The most widespread species is the coccolithophore *Emiliania huxleyi* that contributes dimethylsulphide to the atmosphere. Blooms of this algal species have been observed in the southeast Bering Sea (Olson and Strom, 2002, Sukhanova et al. 2004), North Sea (Robinson 2002), and the Central North Atlantic (Balch et al. 1996). They were characterized by chlorophyll levels that ranged from 0.4 to 4.50 μg Chl *a* l^{-1} (Olson and Strom, 2002), low photosynthesis: respiration of 0.9 (Robinson et al. 2002), and the waters yielded dimethylsulphide (DMS) ranging 1.06 to 93.8 nmol dm^{-3} (Malin et al. 1993). Elsewhere blooms can be advected as in the toxic dinoflagellate *Gymnodinium breve* (= *Ptychodiscus brevis*); these originate in the Gulf of Mexico and are advected into the South Atlantic Bight by the Florida current and the Gulf Stream (Tester et al. 1993). Off the Western tropical Atlantic, Coles et al. (2004) located blooms of the diazotrophic cyanophyte *Trichodesmium* during summer.

Viruses

It is possible that the abundant viruses (upto 10^7 ml^{-1}) play a significant role in the species succession, in the mortality of phytoplankton and may terminate algal blooms through lysogeny (Bratbak et al. 1993, McDaniel et al. 2002; see chapter: 6 Nagasaki). Specific cyanophages seem to influence *Prorochlorococcus* or *Synechococcus* hosts differentially (Sullivan et al. 2003). Virus-like particles (VLPs) isolated from natural populations in the coastal bays of New Jersey and in culture populations of the brown tide bloom-forming *Aureococcus anophagefferens* were similar and potentially caused the termination of the bloom (Gastrich et al. 2004). Virus-mediated lysates are critical to the recycling of organically complexed iron in the coastal upwelling high-nutrient low-chlorophyll (HNLC) waters, that support up to 90% of primary production. Viruses seem to play an important role in the regeneration of bioavailable iron (Wilhelm et al. 2001).

Energy Transformation

Sun is the primary source of energy and ~1200 kilocalories reach a square meter surface area in a 12 day and ranges with the latitude. In the low latitudes i.e. 0–30° N and 0–30° S it is ~3000 kcal m^2 d^{-1} and at the mid latitudes 30–60° N and 30–60° S it is ~ 1500 kcal m^2 d^{-1} and much less in the higher latitudes. The total solar energy received by earth is roughly equivalent to 10,000 times the total energy consumed by humanity. In the oceans as light travels through the water column it is attenuated exponentially with increasing depth following the Beer-Lambert law expressed as:

$$\mathbf{A = \varepsilon bc}$$

Where A is absorbance (no units, since $\mathbf{A} = log_{10}\, P_0/P$)
ε is the molar absorbtivity with units of L mol^{-1} cm^{-1}
b is the path length expressed in centimeters.
c is the concentration of the compound in solution, expressed in mol $L^{-1.}$

Studies of the electromagnetic spectrum of visible light (350–750 nm) showed that water absorbs light differentially i.e. the longer the wavelength, the lower the energy and faster it gets absorbed as follows:

Table 1.3 Light transmission characteristics in the sea

Color	*Wavelength nm*	*Depth (m) where most is absorbed*
Red	600–750	5–10
Orange	600–575	10–15
Yellow	500–575	15–25
Green	525–475	30–50
Blue	475–500	60–100
Violet	450–400	10–30

Most (~50%) of the infra-red, is absorbed in the first one meter of the sea. Ultraviolet light (< 400 nm), the shortest wavelength with the highest energy is an exception; it is scattered by particles. Of the visible spectrum the red component gets absorbed first and the blue component penetrates deeper. In the oceanic waters 84% of the incoming radiation is absorbed in the top one meter and about 99% in the top 10 m. In the coastal waters which have more suspended particles than in the oceanic province, 99.5% of the light is absorbed in the top 10 m through absorption by algae and also through scattering by water molecules and suspended particles. Chlorophyll, because of its maximal light absorption coefficient between the wavelengths 430–670 nm, is the most important molecule on the earth and photosynthesis is the most important physiological process in the transformation of solar energy into chemical energy (see chapters: 19 Shakshug and Jensen; 18 Grobellar). The upper 200 m of the ocean is the illuminated zone also known as the "euphotic zone" characterized by active photosynthesis. The 200–1000 m is the "twilight zone" or the mesopelagic zone. The deeper waters with no light constitute the "aphotic zone". Members of each algal division have their characteristic photosynthetic pigments and harvest light energy in the visible spectrum (350–750 nm) and thus provide a spectral signature. Based on the spectral signatures using remote sensing it is possible to map detailed distributions of photosynthetic biomass in surface layers (see chapter: 2 Jeffrey and Wright).

Photosynthetic Production and Biogeochemical Cycles

The introduction of tracer carbon-14 technique (Steemann-Nielsen, 1952) opened up a new chapter in phytoplankton physiology. Phytoplankton being autotrophic convert inorganic carbon in the seawater to organic matter via photosynthesis and supply oxygen to the oceans. They are referred to as primary producers or organic producers, analogous to terrestrial pastures.

Phytoplankters could be unicellular or multicellular. In distribution they occur under extremes of climatic conditions between the tropics and the polar waters. Usually limited to the illuminated or euphotic waters, these biota act as floating oxygen farms, carry on photosynthesis during which inorganic carbon dioxide is converted into organic matter.

$$6CO_2 + 6H_2O + \text{light 8 photons} \Rightarrow C_6H_{12}O_6 + 6O_2$$

If we include the macronutrient (N, P, S) and the trace element (Fe, Zn, Mn,) uptake:

$$106CO_2 + 16NO_3 + PO_4 + SO_4 + 10^{-2}\,Fe + 4 \times 10^{-3}\,Zn + 4 \times 10^{-4}\,Mn \Rightarrow (C106\ H263\ O110\ N16\ PS) + 138O_2$$

In this transformation two products are most important: a) the organic matter produced constitutes the foundation of food web in the marine environment, and b) the waste product oxygen sustains all metabolic activity which involves the transfer of electrons between oxidized and reduced substances in reactions called Redox reactions.

Annually about 1.25×10^{24} cal of solar energy strikes the earth of which 10% is available to photosynthetic pigments for photosynthesis (Fig. 1.1). Assuming photosynthesis is 2% efficient, the annual potential primary productivity is 2.5×10^{21} cal or 2×10^{17} g C or 200×10^9 t C y^{-1} (Isaacs, 1969). The oceans receive nutrients and 750×10^{15} g C y^{-1} via the dust storms and other atmospheric activities (Adhiya and Chisholm, 2001). From the land and soil, anthropogenic activities contribute 2.05×10^{18} g C y^{-1}. In the oceans

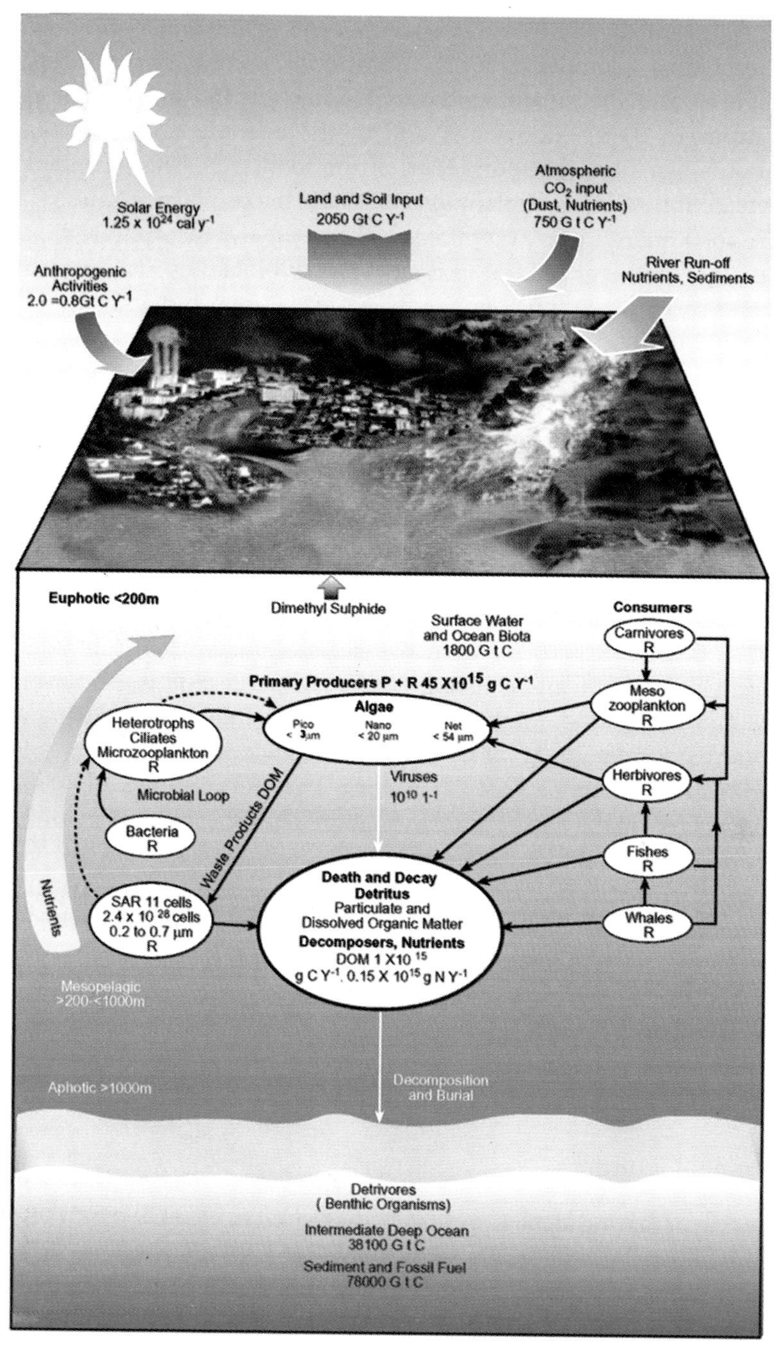

Fig. 1.1 Biogeochemical relationships in the ocean

there are essentially five interacting components a) the primary producers b) the herbivores – competing for the autotrophs, secondary producers, the carnivores and the supracarnivores feeding on the herbivores c) the decomposers d) the detrivores and e) the bacteria and viruses. The microorganisms are the foundation of the food web. The picoplankton, nanoplankton and the net plankton constitute the primary producers; they utilize solar energy, carry on photosynthesis (P) and incorporate 45×10^{15} g C y^{-1} (Falkowski et al. 2004). Like the rest of biota they also respire and consume oxygen. Carbon bound in the surface waters and oceans amounts to 18×10^{18} g C (Adhiya and Chisholm 2001). The consumers consisting of the mesozooplankton, herbivores, and the carnivores consume oxygen and release metabolic wastes into the environment. Upon their death and decay the primary producers and the consumers sink and contribute to the dissolved organic matter (DOM) pool; this DOM pool receives 1×10^{15} g C y^{-1} and 0.15×10^{15} g N y^{-1} and acts as a storehouse of nutrients (Moron and Zepp, 1997). The DOM pool also acts as a source of food for the detrivores – the benthic organisms; the unused DOM gets buried in the sediments and contributes to intermediate and the deep ocean reserve of 38 $\times 10^{18}$ g C and 78×10^{18} g C to the sediment and fossil fuel reserves respectively (Adhiya and Chisholm, 2001). Bacteria and viruses act on the DOM and release the nutrients most needed in the euphotic waters (see chapter: 5 Biddanda et al.). The microbial loop comprising of the heterotrophic bacteria, SAR 11 bacteria and microciliates play a very significant role in the energy transformation. These heterotrophs utilize the waste products from the primary producers and funnel the nutrient energy back to the primary producers. Assuming an ellipsoid bacterial cell of 0.4 μm × 0.4 μm and 0.4 pg C per/μm^3 of bacterial biomass, the SAR 11 bacteria (Morris et al. 2002) account for ~322×10^{11} g C. It is of interest to note that some of the microalgae release dimethylsulfide (DMSU) to the atmosphere. Although there are no precise estimates of the total DMSU released, Watson and Liss (1998) suggest it may be a powerful influence on the climate by changing the number of cloud condensate nuclei.

Microalgal cultures

Culturing of marine algae, specifically diatoms dates back to 1893 (Miquel, 1893), almost simultaneously with the coining the term phytoplankton. Initially *Thalassiosira gravida*, a diatom was cultured in artificial sea water to raise and support production of animal life in the sea (Allen and Nelson, 1910), but subsequently it was evident that addition of 1% of natural sea water or a minute trace of an organic substance such as aerobic soil bacteria upon peat acted as a powerful stimulant to algal growth (Allen 1914). Since then physiologists and biochemists around the world evinced a keen interest, a sort of academic curiosity, in utilizing algal cultures for various purposes (Appendix 1, Table 1.1). Although several 100 strains have been brought into culture only approximately 30 species have been studied in considerable detail as feed organisms for invertebrates and commercially important organisms (Appendix 1, Table 1.2). A variety of methods are available to culture marine micro algae (Fogg, 1975). The medium could be a semisolid such seawater agar slants or liquid medium; it is either precisely defined by enrichment of artificial sea-water (Appendix 1, Table 1.3) or natural seawater enriched with known quantities of nutrients and soil extract of an unknown composition (Appendix 1, Table 1.4). Algal cultures contaminated with bacteria can be purified by a variety of methods including most commonly, treatment with antibiotics. Usually they are monospecific but very few investigators have succeesfully grown two or more species simultaneously. While growing athecate dinoflagellates and a few armoured dinoflagellates such as *Alexandrium, Prorocentrum,* and a few *Ceratium* is possible, culturing *Dinophysis* is still a problem. At best *Dinophysis* could be maintained through a few generations for few months (Subba Rao, 1995). Only in recent years has it been possible to culture the Cyanophyte *Trichodesmium* (Appendix 1, Table 1.5). Culture volumes for the different algae range from a few milliliters to > 10 m^3 mesocosms. Batch cultures and continuous cultures are raised and utilized for various applications (Appendix 1, Table 1.6).

Although the importance of mixed species cultures in species succession was realized by Allen and Nelson (1910), obtaining reproducible

Table 1.4 Broad comparison of seasonal blooms, atypical blooms and batch cultures

Feature	*Seasonal bloom*	*Atypical bloom*	*Batch culture*
Growth medium	Moderately Nutrient rich sea water	Nutrient rich	~200 times
Bacteria	Present	Present	Absent
Number of species	~25	1–5	1
Duration	21 days	Few days	Not applicable
Cells 10^6 l^{-1}	<10	>10	>200
Chlorophyll *a* μg l^{-1}	20–30	>30–<570	~250
Divisions day^{-1}	2.6	1.5	3.4
Carbon assimilation ratios μg C h^{-1} μg chl *a*	<12	<12	23
Photosynthesis: Respiration	<10:1	<10:1	10:1

multispecies cultures is still a problem. Multispecies cultures are useful in discerning whether dominance of a species results from its environmental fitness or its reproductive capacity. For example in Prymnesiophyceae which are better competitors than the dinoflagellates these two factors are closely related and contribute to their dominance whereas in several dinoflagellates the reproduction cycle of the individual species is more important for its succession (Riegman et al. 1996). In a mixed culture of three dinoflagellates *Prorocentrum micans, Exuviella cordata* and *Glenodinium kovalevskii* and a chrysophyte *Olisthodiscus luteus*, the strong competitive ability of *G. kovalevskii* to grow despite limiting nutrients and its high growth rate contributed to its dominance over the other species (Fedorov and Il'Yash, 1992).

Application of Plant Physiology Principles

Although physiological ecology of terrestrial plants is still evolving as a separate discipline, phytoplanktologists with a bent towards reductionism, laws and analysis have borrowed and applied principles of plant physiology liberally. Such an application is complex and is not always either possible or expedient, particularly so when designing multifactorial experiments under near-*in situ* conditions. Caution should be applied while extrapolating small-scale laboratory single species test results to the large-scale natural populations (Englund and Cooper, 2003) and species sensitivity should be taken into consideration. Although several excellent experimental designs are available (Underwood, 1981 and 1997, Scheiner and Gurevitch, 2001), in some of the experiments the levels of variables were far from those encountered in nature and were misinterpreted far beyond the original context. A case in point is von Liebig's law of the minimum and the concept of iron limitation in phytoplankton (De Baar, 1994).

Studies aimed at understanding the mechanism of primary organic production by phytoplankton and physiological processes such as determination of photosynthetic activity in relation to temperature and nutrient uptake kinetics gained momentum. It should be emphasized that the distribution of phytoplankton is free-floating, and three dimensional in contrast to the stationary and two-dimensional distributions of terrestrial biomass. Several studies were carried to understand the growth and development of microalgal populations in relation to mineral nutrient (phosphate, silicate, nitrogen) uptake and utilization. Investigations have been carried out on carbon linked photosynthesis and its dependence on chlorophyll and light reactions. Based on their photosynthetic efficiency (μg

C µg Chl *a* h^{-1}) the concept of 'sun' and 'shade adaptation' was extended to marine algal populations; the former characterized by a higher carbon assimilation ratios (µg C µg Chl *a* h^{-1}), high light saturation (I_k) and less photoinhibition in contrast to the 'shade adapted populations' with low assimilation, low saturation and with strong photoinhibition (see chapter: 17 Duarte). New data on phytoplankton in the Arctic waters suggest that natural populations seem to accumulate more chlorophyll *a* in excess of their division leading to low carbon assimilation ratios (µg C µg h^{-1} chl *a*) comparable to senescent algae (Subba Rao, 1988).

Utility of Microalgal Cultures

Microalgal cultures grown under defined conditions of growth are particularly suitable as feed in mariculture operations because of their nutritional value (see chapter: 12 Volkman and Brown). Several cultures are utilized to verify some of the physiological ecology concepts on the structure and allometric functioning of algae which is not possible with the natural assemblages and bulk populations. An example illustrates this. With a ten-fold increase of radius of the alga, its sinking speed, and boundary layer thickness increased ten-fold; however the minimum energy cost of flagellar motility, average specific absorption coefficient for chlorophyll *a* and its intracellular UV-absorbing compounds and the area of membrane per unit volume available for solute fluxes decreased (Raven and Kübler, 1999). Research in this area has been motivated not only by their utility as teaching aids, but also as experimental organisms to test hypotheses about growth rates in relation to selected concentrations of trace elements. Growth of monospecific cultures ranged between 0.85 to 1.61 div $d^{-1}$and were comparable to those of enriched natural assemblages from the Bedford Basin (Subba Rao, 1981). Several studies were carried out to test the physiological response of microalgae to treatment with germanium (Subba Rao 1980), humic acids (Prakash et al. 1973); to test allelopathic effects (Subba Rao et al. 1994); production of domoic acid in relation to stress due to silicate (Pan et al. 1995a, 1995b), phosphate limitation (Pan et al. 1996), and enrichment with lithium (Subba Rao et al. 1998). In experimental studies mechanistic multi-nutrient (light, ammonia, nitrate, phosphorus, iron and silica) models can contribute to our understanding of the algal physiology (see chapters: 14 Flynn, 16 Thompson).

There is an increasing interest in the biotechnological applications of microalgae (Borowitzka and Borowitzka, 1992, Radmer and Parker, 1994).

Consequently this has led to a gold rush of discoveries (Azam and Worden, 2004). The following is a brief list of a few selected utilizations of microalgae:

- Teaching aids
- Determination of division rates of algae
- Proximate biochemical analyses of algae
- Investigating nutrient kinetics
- Determination of limiting factors for growth
- Algal toxins
- Physiological ecology of algae under a variety of conditions
- Microalgae in mariculture operations
- Algae as biomonitors of pollutants
- Marine biotechnology applications such as production of food additives
- Large-scale production of carbohydrates, proteins, lipids, enzymes and a host of useful compounds
- Algae in space research for waste recycling systems and as food
- Source of natural biochemical compounds such as antiviral and antifungal compounds and pharmaceuticals
- Genetic engineering

Not many species truly representative of marine phytoplankton, particularly the truly oceanic, are cultured and utilized to understand their physiological ecology. A literature search on marine microalgae (Appendix 1, Table 1.1) shows a maximum of 2100 strains of algae representing 475 species are held in culture. Of these only 126 species are utilized in experimental studies. Data based on ASFA search (Appendix 1, Table 1.2) showed detailed culture studies were limited to only 30 species. *Chlorella* was the most studied genus (889 studies), followed by *Isochrysis affinis galbana* (330), *Skeletonema costatum* (301) and *Phaeodactylum tricornutum* (287). The ease with which these can be isolated and cultured designated them as 'weed' or 'Cinderallas' of marine biologists.

Several factors dictate why only a few cultured species are studied; first and foremost being their capacity to grow rapidly. For example the 30 species include 12 flagellates, 9 diatoms, 8 dinoflagellates and one blue green (Appendix 1, Table 1.2). Most of the flagellates and diatoms are studied because of their utility as live feed in mariculture (see chapter: 12 Volkman and Brown), and biotechnological applications (see chapter: 22 Lebeau and Robert) and the dinoflagellates because of their economic and societal impact due to production of toxins causing Paralytic Shell fish poisoning (PSP), Diarrhetic Shellfish poisoning (DSP), Neurotoxin Shellfish poisoning (NSP) and Ciguetera episodes. These episodes have far reaching harmful effects on the human health and on commercially important fisheries. Blooms of the diatom *Pseudonitzschia* species resulted in Amnesic Shellfish poisoning (ASP) and has serious effects (see chapter: 25 Blum et al.). Millie et al. (1999) discussed the need for expanded collaborative research on the molecular, cellular, and ecophysiological bases of harmful algal blooms. The inability to culture species of the dinoflagellate genus *Dinophysis* implicated in Diarrhetic Shellfish poisoning (DSP) remains a major hindrance in phycotoxin studies.

Comparison of Algal Blooms and Cultures

Under laboratory conditions cultures are grown under constant temperature and light bacteria free; they are monospecific and grown in seawater enriched with up to 200 times more nutrients. Similarities between algal blooms and cultures include the pattern of growth curve between a bloom and a freshly seeded culture (Fig. 1.2). Both follow a set common pattern with four phases:

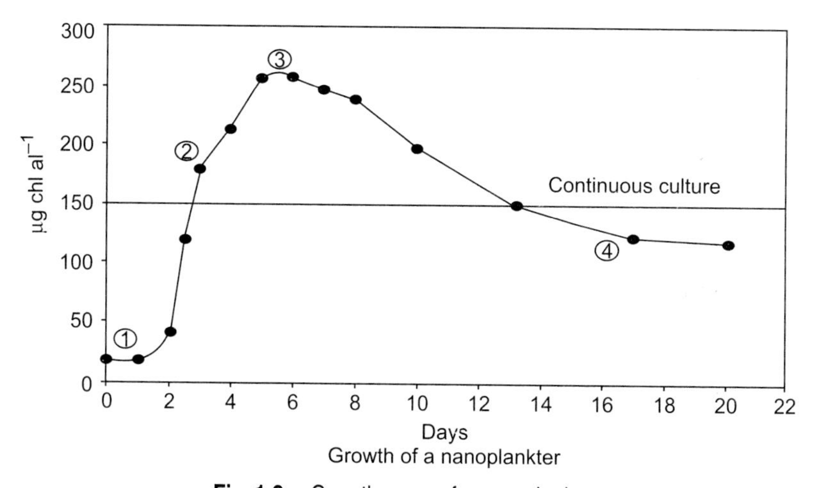

Fig. 1.2 Growth curve of a nanoplankter

1. An initial lag phase with no significant growth, 2. Followed by a log phase with an exponential growth leading to a maximum yield, 3. A stationary phase when the cell density attains a plateau and 4. A senescent phase with no further growth (Fogg, 1975). Log phase cultures yield higher biomass (> 200 cells 10^6 l^{-1} and chlorophyll levels > 30 Chl. *a* µg l^{-1}), high

division rates, high carbon assimilation rates, and photosynthesis: respiration ratios and compare favorably with data obtained on bulk phytoplankton bloom assemblages (Platt and Subba Rao, 1970). Division rates of a few cultured phytoplankters were similar to those obtained on blooms initiated through enrichment of Bedford basin water (Subba Rao, 1981).

Inadequacies in Comparative Field and Culture Studies

The sea as an environment imposes several experimental limitations which can be tackled only with culture experiments. Experimental studies utilizing microcosms and mesocosms (20–30 L) have played a significant role in understanding the physiological ecology of marine phytoplankton particularly while comparing field and laboratory studies (Estrada and Peters, 2002, Keller 2002). For example in simulating natural events and in studying the salinity effects on phytoplankton in estuarine habitats (Spies and Parsons, 1985) or in perturbation experiments (see chapter: 15 Flöder and Sommer) to study the response of assemblages of phytoplankton rather than on individual taxa to pulsed nutrient (with various Si:N:P ratios) enrichments (Estrada et al. 2003) microcosms were successfully utilized. Response of phytoplankton community (20 L) to environmental changes such as nutrients and irradiance is best understood by measuring instantaneous physiological indicators over short-term incubations and play an important role in validating models and in guiding new developments (Berges et al. 2004). Utilizing enriched mesocosms successive bloom events were observed with a diatom and a coccolithophorid *Emiliana huxleyi* that were preceded by a marked increase in the level of rbcLS mRNA (the large subunit of the Calvin cycle enzyme, RubisCO) gene expression (Wyman et al. 1998, 2000). Although comparative studies of marine microalgal response to natural and culture conditions would be instructive, only a few production studies have been made along these lines. Probably due to the logistics these studies progressed rather independently, and were not integrated.

Before the carbon-14 tracer technique was available, a few studies were carried out using oxygen exchange measurements to measure algal photosynthesis. Although this technique is laborious and replication becomes difficult it has the advantage of measuring net and gross production, the serious disadvantages are its imprecision, relatively small changes in oxygen relative to large background oxygen levels. Oxygen exchange measurements on diatom cultures suspended in Loch Striven established a quantitative relationship between photosynthesis and photometry and photosynthesis and respiration (Marshall and Orr, 1928). Similar measurements on cultures of the diatom *Coscinodiscus excentricus* in relation to submarine illumination (Jenkin, 1937) showed a close relationship between light energy and photosynthesis. Measurements on the diatom *Chaetoceros affinis* under the laboratory and field conditions showed similarities in photosynthetic characteristics in short exposures of 1–3 hours (Talling, 1960). Measurements of photosynthesis (P) and respiration (R) are very important in calculating net primary production but very few direct measurements are made because of the lower sensitivity of the oxygen exchange method. Both respiration and gross photosynthesis are dependent on the size of the alga (Banse, 1976). The few data based on batch cultures suggested existence of a linear relationship between P and R but R as a proportion of P can vary from 5% to 50% depending on the physiological age of the culture (Humphrey and Subba Rao, 1967, Humphrey, 1975, Falkowski and Owens, 1978). Comparative studies on *Skeletonema costatum* and natural populations indicated that diurnally fluctuating light could result in greater respiratory activity (Cosper, 1982; see chapter: 11 Kromkamp and Claquin).

Under natural conditions most microalgae experience energy fluctuations up to two orders magnitude as well as frequency variation of up to 10 orders. They constantly readjust their photosynthesis, photoinhibition and acclimation responses to these fluctuating conditions, a knowledge of which is essential for any predictions of their photosynthetic functioning at various spatial and temporal scales (see chapters: 18 Grobbelaar, and 17 Duarte). Under these conditions choosing the correct amplitude and frequency of these cycles is important (see Chapter: 16 Thompson). Of interest is a study on the spectral regimes related to algal absorption and to coastal, underwater light (Humphrey, 1983). Utilizing cultures of a diatom *Biddulphia aurita* and a dinoflagellate *Amphidinium carterae* Humphrey showed gross morphological changes, slower growth, crimping of thylakoid bands, increase in band number, changes in cellular pigment and carbon assimilation in response to stress. The spectral regime brought out the greatest differences in reaction; *Amphidinium* had decreased pigments, growth rate and lower assimilation numbers in contrast to *Biddulphia*.

Earlier field studies showed that diatoms subsist in iron-poor oceanic waters but using cultures of an oceanic diatom *Thalassiosira oceanica* and a coastal species *T. weissflogi*, Strzepek and Harrison (2004) demonstrated a fundamental difference in the photosynthetic architecture between these species. The oceanic species has a five-fold lower photosystem I and up to seven-fold lower cytochrome b_6f complex concentrations than the coastal species which is a photoacclimatory adaptation towards minimizing their iron requirement without compromising photosynthetic capacity (Strzepek and Harrison, 2004). Under various environmental conditions the $^{18}O/^{16}O$ and $^{15}N/^{14}N$ of internal nitrate of *T. weissflogii* were elevated relative to the nitrate in the medium by a proportion of ~1:1 but not in *T. oceanica, T. psuedonana* and the coccolithophorid *Emiliana huxleyi* (Granger et al. 2004).

Although important advances have been made through cross fertilization with techniques and methods from biochemistry and physiology, some ecological principles still remain to be adequately addressed with regard to marine microalgae. For example:

- Principles of limiting factors and minimum requirements of nutrients and response to ratios between nutrients
- Size, function and physiological scaling of metabolism-allometry
- Photosynthetic adaptation to light and temperature variation in photosynthetic pathways
- Physiological and biochemical adaptations to fluctuating light environments
- Ecophysiology of succession-interaction of species and principles of allelopathy
- Biochemistry and molecular biology of toxins (Plumley, 1997)
- Besides silicon metabolism to produce a characteristic frustule, diatoms may have several novel mechanisms that may be gene regulated (Armbrust et al. 2004) and include:
  (a) Acclimation, adaptation and natural selection in microalgae
  (b) Whether microalgae can perceive red/far-red light (Leblanc, 1999) but not green light
  (c) Protein transport into plastids of microalgae
  (d) Whether phylogenetically different algae such as the unicellular diatom *Thalassiosira weisflogii* and the multicellular brown sea weed *Ascophyllum nodosum* have functional similarities with respect to C_4 photosynthesis
  (e) Pathway of CO_2 delivery to RubisCo in diatoms (Reinfelder et al. 2001; Johnson et al. 2001), and
  (f) The high proportion and production of polyunsaturated fats in some microalgae and their oxidation to feed intermediate metabolism.

RECENT ADVANCES

(a) UV Light and Climate

Studies on the role of ultraviolet on phytoplankton physiology gained prominence because of its key role in carbon cycle and global climate regulation. Ultraviolet light splits the 'normal' oxygen molecules (O_2) into chemically active, single oxygen atoms (O). These single oxygen atoms combine with O_2 to form ozone, O_3:

$$O_2 \text{ —(ultraviolet light)} \longrightarrow O + O$$
$$O + O_2 \longrightarrow O_3$$

The ozone layer in the upper atmosphere acts as a shield, which filters out much of the injurious ultraviolet rays from sunlight, making the Earth's surface safe for living things. In the Antarctic because of a hole in the ozone layer, UV impinging upon the algae had deleterious effects on the algae. Phytoplankton population exposed to UV dropped by 6–12% compared to those not exposed to UV and the production decreased by almost 65% following one-exposure to UV (Dalton 2002; see chapter: 20 Davidson). Although there is no direct relationship between iron and UV light, it can affect organic compounds some of which may affect the solubility of iron (Waite and Szymczak, 1993, Matsunaga et al. 1998). Recent studies showed that two dinoflagellates synthesize micosporine-like amino acids (MAAs) that are highly packaged in intact cells; the MAAs absorb up to 80% of UV and increase photoprotection to cells (Laurion et al. 2004). Ultraviolet –ß radiation (280–320 nm) affects the marine algae, the atmosphere and therefore the climate (see Chapter: 20 Davidson).

(b) Geoengineering: Climate

With a view to cure global warming by reducing atmospheric levels of CO_2, 'geoengineering' is actively considered, particularly in the high-nutrient, low-chlorophyll (HNLC) regions. Pelagic waters have been artificially enriched with iron to prime production of algal blooms that would assimilate CO_2 and some of the organic carbon would sink into deep oceans (Chisholm et al. 2001). Since 1994 four small-scale enrichment experiments (~100 km^2) were conducted in the equatorial Pacific and Southern Ocean with little net transfer of CO_2. However, results of Southern Ocean Iron Fertilization Experiment (SOFeX) showed that iron fertilization on the ocean south of New Zealand resulted in a 200km long phytoplankton bloom, only 1,000 tonnes of carbon would sink below 100 meters contrary to earlier laboratory estimates of 100,000 tonnes which raises doubts about oceans acting as carbon sink.

(c) Genetic engineering studies

The eukaryotic algae, particularly the diatoms with their fast division rates are most suitable for genetic engineering manipulations of key biochemical pathways and hold a promising future (Stevens and Purton, 1997). In several planktonic diatoms physiological characteristics seem to be associated with genetically structured populations. In *Skeletonema costatum* in Narragansett Bay the genetic and physiological characteristics varied over a seasonal cycle (Gallagher, 1980). Three populations in the diatom *Ditylum brightwelli*, distinctly differentiated by their genetic and physiological characteristics, existed in the estuarine waters of the Strait of Juan de Fuca and Puget Sound (Rynearson and Armbrust, 2004). The most recent study (Armbrust, 2004) reported 34 million-base pair draft nuclear genome in the centric diatom *Thalassiosira pseudonana* and identified novel genes for silicic acid transport, with high affinity for iron uptake, nitrogenous compounds and biosynthetic enzymes for several types of polyunsaturated fatty acids. Massive genomic surveys of marine microalgae comparable to the human genome project are undertaken. The most significant advance in algal biotechnology is the heterotrophic cultivation of *Phaeodactylum tricornutum* (Zaslavskaia, 2001) for metabolic engineering. These authors engineered *P. tricornutum* with either a human (glut 1) or *Chlorella* (hup 1) glucose transporter gene thus converting a photosynthetic autotroph to a heterotroph capable of obtaining exogenous glucose in the absence of light energy. Similar to plate tectonics that has required a rethinking of marine geology, the advances being made in genetic engineering of marine microalgae would soon revolutionize and necessitate our rethinking about physiological ecology of marine phytoplankton (Knauss, 2002).

ACKNOWLEDGEMENTS

I am grateful to Dr. James E. Stewart for helpful discussions and constructive comments on this manuscript.

REFERENCES

Adhiya, J. and S.W. Chisholm. 2001. Is ocean fertilization a good carbon sequestration operation? Masschusetts Inst. Technology Publication No. LFEE 2001-001, 58 pp.

Allen, E.J. 1914. On the culture of the plankton diatom *Thalassiosira gravida* Cleve, in artificial seawater. J. Mar. Biol. Assoc. U.K. 10: 417-434.

Allen, E.J. and E.W. Nelson. 1910. On the artificial culture of marine plankton organisms. J. Mar. Biol. Assoc. U.K. 8: 421-474.

Amadi, I., D.V. Subba Rao and Y. Pan. 1992. Red water: *Gonyaulax digitale* bloom in the Bedford Basin, Nova Scotia, Canada. Botanica Marina 35: 451-458.

Anderson, D.M. and B.A. Kiefer. 1987.An endogenous annual clock in the toxic marine dinoflagellate *Gonyaulax tamarensis*. Nature 325: 616-617.

Armbrust, E.V. et al. 2004. The genome of the diatom *Thalassiosira psuedonana*: Ecology, evolution, and metabolism. Science 306: 79-86.

Augusti, S. 2004. Viability and niche segregation of *Proclorococcus* and *Synechococcus* cells across the Central Atlantic Ocean. Aq. Microbial. Ecol. 36: 53-59.

Azam, F., and A.Z. Worden. 2004. Microbes, molecules, and marine ecosystems. Science 303: 1622-1624.

Balch W.M., K.A. Kirkpatrick, P. Holligan, D. Harbour, and E. Fernandez. 1996. The 1991 cocolithophore bloom in the Central North Atlantic. 2. Relating optics to coccolith formation. Limnol. Oceanogr. 41: 1684-1696.

Banse, K. 1976. Rates of growth, respiration and photosynthesis of unicelulr algae as related to cell size–a review. J. Phycol. 12: 135-140.

Belgrano, A., O. Lindhal and B. Hernroth. 1999. North Atlantic Oscillation (NAO) primary productivity and toxic phytoplankton in the Gullmar Fjord, Sweden. (1985-1996). Proc. Roy. Soc. London, B. 266: 425-430.

Bengston, S. (ed.) 1994. Early life on Earth. Columbia Univ. Press, USA.

Berges, J.A., C.E. Gibson and B.M. Stewart. 2004. Physiological responses of phytoplankton communities in the Irish Sea to simulated upwelling. Hydrobiologia 517: 121-132.

Borkman. T. and T.J. Smayda. 2003. Long-term patterns of Narragansett Bay phytoplankton driven by decadal shifts in phytoplankton habitat, 2nd Symposium on Harmful Marine Algae in the U.S., Woods Hole, MA (USA), 8-13 Dec 2003. (World Meeting Number 000 7156).

Borowitzka, M.A. and L.J. Borowitzka. 1992. Microalgal biotechnology, Cambridge University Press, Cambridge, UK.

Bratbak, G., J.K. Egge, and M. Heldal. 1993. Viral mortality of the marine alga *Emiliania huxleyi* (Haptophyceae) and termination of algal blooms. Mar. Ecol. Prog. Ser. 93: 39-48.

Buesseler K.O., J.E. Andrews, S.M. Pike and M.A. Charette. 2004. The effects of iron fertilization on carbon sequestration in the Southern Ocean. Science 304: 414-417.

Burkholder, J.M., H.B. Glasgow and N. Deamer-Melia. 2001. Overview and present status of the toxic *Pfiesteria* complex (Dinophyceae). Phycologia 40: 186-214.

Carpenter, E.J., J.P. Montoya, J. Burns, M.R. Mulholland, A. Subramaniam and D.G. Capone. 1999. Extensive bloom of a N_2-fixing diatom/cyanobacterial association in the tropical Atlantic Ocean Mar. Ecol. Prog. Ser. 185: 273-283.

Casa, B., M. Varela and A. Bode. 1999. Seasonal succession of phytoplankton species on the coast of Coruna (Galacia, northwest Spain) Bol.Inst.Esp.Oceanogr. 15: 413-429.

Chisholm, S., P.G. Falkowski and J.C. Cullen, 2001. Oceans: Dis-crediting ocean fertilization. Science 294: 309-310.

Coale, K.H. 2001. Iron fertilization. Encyclopaedia of Ocean Sciences, 3: pp 1385-1397. *In:* (J.H.) Steele, S.A. Thorpe and KK Turekian (eds) Encyclopedia of Ocean Sciences. Academic Press, San Diego USA.

Coles, V.J., C. Wilson and R.R. Hood. 2004. Remote sensing of new production fuelled by nitrogen fixation. Geophys. Res. Lett.31: L06301, doi: 10.1029/2003GLO19018

Colijn, F., U. Tillmann and T. Smayda (eds.) 1998. The temporal variability of plankton and their physico-chemical environment. ICES J. Mar. Sci. 55: 557-824.

Cosper, E. 1982. Effects of diurnal fluctuations in light intensity on the efficiency of growth of *Skeletonema costatum* (Grev.) Cleve (Bacillariophyceae) in a cyclostat. J. Exp. Mar. Biol. Ecol. 65: 229-239.

Cushing, D.H. 1975. Marine Ecology and Fisheries. Cambridge University Press. Cambridge, UK.

Dalton, R. 2002. Ocean tests raise doubts over use of algae as carbon sink. Nature. 420: 722.

De Baar, H.J.W. 1994. von Liebig's law of the minimum and plankton ecology (1899-1991). Prog.Oceanog. 33: 347-386.

Englund, G. and S.D. Cooper. 2003. Scale effects and extrapolation in ecological experiments. Advances in Ecological Research 33: 161-213.

Estrada, M. and F. Peters. 2002. Microcosms: Applications in marine phytoplankton studies. pp 359-370 *In:* D.V. Subba Rao (ed) Pelagic Ecology Methodology, A.A. Balkema Publishers, Lisse, The Netherlands.

Estrada, M., E. Berdalet, M. Vila and C. Marassê. 2003. Effects of pulsed nutrient enrichment on enclosed phytoplankton: ecophysiological and successional responses. Aquat. Microb. Ecol. 32: 61-71.

Falkowski, P.G. and T.G. Owens. 1978. The effect of light intensity on photosynthesis and dark respiration in six species of marine phytoplankton. Mar. Biol. 45: 289-295.

Falkowski P.G., E.A. Laws, R.T. Barber and J.W. Murray. 2003. Phytoplankton and their role in primary, new and export production. Ch. 4 pp. 99-121. In : (ed.) M.J.R. Fasham, Ocean Biogeochemistry, Springer, Berlin, Germany.

Falkowski, P.G., M.E. Katz, A.E. Knoll, A. Quigg, J.A. Raven, O. Schofield and F.J.R. Taylor. 2004. The evolution of modern eukaryotic phytoplankton. Science 305: 354-360.

Fedorov, V.D. and L.V. Il'Yash. 1992. Adaptive mechanisms in life-history strategies of microalgae. Hydrobiol. J. 28: 38-45.

Fogg, G.E. 1975. Algal cultures and phytoplankton ecology. University of Wisconsin Press, Madison, WI (USA).

Fogg, G.E. 1989. The flexibility and variety of algal metabolism. Proc. Phytochem. Soc. Eur. 28: 3-12.

Fogg, G.E. 1991. The phytoplanktonic ways of life. New Phytol. 118: 191-132.

Furhman J.A. and D.G. Capone. 2001. Nifty nanoplankton. Nature 412: 593-594.

Furnas, M. 2002. Measuring the growth rates of phytoplankton in natural populations. Ch. 24. pp 221–250. In: (ed.) D.V. Subba Rao, Pelagic Ecology Methodology, A.A. Balkema, Lisse, The Netherlands.

Gaarder, T. and H.H. Gran. 1927. Investigations on the production of plankton in the Oslo Fjord. RaPports et Procès-Verbaux des Réunions, Conseil International pour l'Exploration de la Mer 42: 2-48.

Gallagher, J.C. 1980. Population genetics of *Skeletonema costatum* (Bacillariophyceae) in Narragansett Bay. J. Phycol. 16: 464-474.

Gastrich, M.D., J.A. Leigh-Bell, C. Gobler, O.R. Anderson, S.W. Wilhelm and M. Bryan. 2004. Viruses as potential regulators of regional brown tide blooms caused by the alga, *Aureococcus anophagefferens*. Estuaries 27: 112-119.

Gersonde, R. and D. M. Harwood. 1990. Lower cretaceous diatom from ODP Leg 113 site 693 (Weddell Sea). Part 1. Vegetative cells. Proc. Ocean Drilling Prog. Sci. Res. 113: 365-402.

Graham L.E. and L.W. Wilcox, 2000. Algae, Prenitice Hall, NJ USA.

Granger J., D.M. Sigman, J.A. Needoba and P.J. Harrison. 2004. Coupled nitrogen and oxygen isotope fractionation of nitrate during assimilation by cultures of marine phytoplankton. Limnol. Oceanogr. 49: 1763-1773.

Grebyk, D., O. Schofield, C. Vetriani, and P.G. Falkowski. 2002. The Mesozoic radiation of eukaryotic algae: The portable plastid hypothesis. J. Phycol. 39: 259-267

Harris, G.P. 1986. Phytoplankton ecology: Structure, Function and Fluctuations. Chapman and Hall Ltd., London, UK.

Hart, T.J. 1963. Speciation in marine phytoplankton. pp 145-155. *In:* J.P. Harding and N. Tebble. (eds.) Speciation in the Sea. Systematics Association Publication Number 5. The Systematics Association, London, U.K.

Hendey, N.I. 1964. An Introductory Account of the Smaller Algae of British Coastal Waters. Minsitry of Agriculture, Fisheries and Food, Fishery Investigations Series IV, 317 pp + XLV plates.

Humphrey, G.F. 1975. The photosynthesis: respiration ratio of some unicellular marine algae. J. Exp.Mar. Biol. Ecol. 18: 111-119.

Humphrey, G.F. 1983. The effect of the spectral composition of light on the growth, pigments, and photosynthesis rate of unicellular marine algae. J. Exp. Mar. Biol. Ecol. 66: 49-67.

Humphrey, G.F. and D.V. Subba Rao. 1967. Photosynthetic rate of *Cylindotheca closterium*. Aust. J. Mar. Freshw. Res. 18: 123 – 127.

Ishikawa, A. and K. Furuya. 2004. The role of diatom resting stages in the onset of the spring bloom in the East China Sea. Marine Biology, 145: 633-639.

Isaacs, J.D. 1969. The nature of oceanic life, Scientific American. 221: 146-162.

Jenkin, P.M. 1937. Oxygen production by the diatom *Coscinodioscus excentricus* Ehr. In relation to submarine illumination in the English Channel. Jour. Mar. Biol. Asso. U.K. 22: 301-342.

Johnson, A.M., J.A. Raven, J. Beardall and R.C. Leegood. 2001. Photosynthesis in a marine diatom. Nature 412: 40-41.

Keller, A.A., C.A. Oviatt and E. Klos. 2002. Mesocosms: Applications to phytoplankton ecology and production. pp 371-389. *In:* D.V. Subba Rao (ed.) Pelagic Ecology Methodology, A.A. Balkema Publishers, Lisse, The Netherlands.

Knauss, J. 2002. Oceanography: The next fifty years. pp. 351-355 *In:* KR Benson and P.F. Rehbock. (ed.) Oceanographic History, The Pacific and Beyond. University of Washington Press, Seattle, USA.

Kurano, N. and S. Miyachi. 2004. Microalgal studies for the 21 Century. Hydrobiologia 512: 27–32.

Laurion, I., F. Blouin, and S. Roy. 2004. Packaging of microsporin-like amino acids in dinoflagellates. Mar. Ecol. Prog. Ser. 279: 297-303.

Leblanc, C., A. Falciatore, W. Watanabe and C. Bowler. 1999. Semi-quantitative RT-PCR analysis of photoregulated gene expression in marine diatoms. Plant Mol. Biol 40: 1031-1044.

Lesser, M.P., C.H. Mazel, M.Y. Gorbunov, and P.G. Falkowski. 2004. Discovery of symbiotic nitrogen-fixing cyanobacteria in corals. Science 305: 997-1000.

Malin, G., S. Turner, P. Liss, P. Holligan, and D. Harbour. 1993. Dimethylsulphide and dimethylsulphoniopropianate in the Northeast Atlantic during the summer coccolithophore bloom. Deep-Sea Res. 40: 1487-1508.

Marshall, S.M. and A.P. Orr. 1928. The photosynthesis of diatom cultures in the sea. Jour. Mar. Biol.Asso. U.K. 15: 321-360.

Matsunaga, K., J. Nishioka, K. Kuma, K. Toya, and Y. Suzuki. 1998. Riverine input of bioavailable iron supporting phytoplankton growth in Kesennuma Bay (Japan). Water Res. 32: 3436-3442.

McDaniel, L., L.A. Houchin, S.J. Williamson, and J.H. Paul. 2002. Lysogeny in marine *Synechococcus*. Nature 415: 496.

Miquel, P. 1893. De la culture artificielle des diatomés. Introduction. Le Diatomiste 1: 73-75.

Millie, D.F., C.P. Dionigi, O. Schofield, G.J. Kirkpatrick and P.A. Tester. 1999. The importance of understanding the molecules, cellular, and ecophysiological bases of harmful algal blooms. J. Phycol. 35: 1353-1355.

Mills, E.L. 1989. Biological Oceanography. Cornell University Press. Itheca, USA.

Miyashita, H., H. Ikemoto, N. Kurano, K. Adachi, M. Chira and S. Miyachi. 1996. Chlorophyll *d* as a major pigment. Nature 383: 402.

Moree, M.D. and H.J. Spero. 1981. Phagotrophic feeding and its importance to the life cycle of the holozoic dinoflagellate, *Gymnodinium fungiforme*. J. Phycol. 17: 43-51.

Moran M.A. and R.G. Zepp. 1997. Role of photoreactions in the formation of biologically labile compounds from dissolved organic matter. Limnol. Oceanogr. 42: 1307-1316.

Morris, R.M., M.S. Rappe, S.A. Cannon, K.L. Vergin, W.A. Siebold, C.A. Carlson and S.J. Glovannoal. 2002. SAR 11 clade dominates ocean surface bacterioplankton communities. Nature 420: 806-810.

Olson M.B. and S.L. Strom. 2002. Phytoplankton growth, microzooplankton herbivory and community structure in the southeast Bering Sea: insight into the formation and temporal persistence of an *Emiliania huxleyi* bloom. Deep-Sea Res. 49: 5969-5990.

Pan, Y., D.V. Subba Rao, K.H. Mann, R.G. Brown, and R. Pocklington. 1995 a. Effects of Silicate Limitation on Production of Domoic Acid, a Neurotoxin, by the Diatom *Psuedonitzschia pungens* f. *multiseries* (Hasle). I. Batch Culture Studies. Marine Ecology Progress Series 131: 225-233.

Pan, Y., D.V. Subba Rao, K.H. Mann, W.K.W. Li, and W.G. Harrison. 1995b. Effects of Silicate Limitation on Production of Domoic Acid, a Neurotoxin, by the Diatom *Psuedonitzschia pungens* f. *multiseries* (Hasle). II. Continuous culture studies. Marine Ecology Progress Series 131: 235-243.

Pan, Y., D.V. Subba Rao, and K.H. Mann. 1996. Changes in domoic acid production and cellular chemical composition of the toxigenic diatom *Psuedonitzschia multiseries* under phosphate limitation. J. Phycology. 32: 371-381.

Platt, T. and D.V. Subba Rao. 1970. Primary production measurements on a natural bloom. J. Fish. Res. Bd. Canada 27: 887-889.

Plumley, F.G. 1997. Marine algal toxins: Biochemistry, genetics, and molecular biology. Limnol. Oceanogr. 42: 1252-1261.

Prakash, A., M.A. Rashid and D.V. Subba Rao. 1973. Influence of humic substances on the growth of marine phytoplankton: Diatoms. Limnol. Oceanogr. 18: 516-524.

Radmer R.J. and B.C. Parker. 1994. Commercial applications of algae: opportunities and constraints. J. Appl. Phycol. 6: 93-98.

Raven, J.A. 2001. Primary production processes. pp 2284-2288 *In:* J.H., Steele, S.A. Thorpe and KK Turekian (eds.). Encyclopedia of Ocean Sciences. Academic Press, San Diego, USA.

Raven, J.A. 2003. Cycling silica–the role of accumulation in plants. New Phytologist 158: 419-421.

Raven, J. and J.E. Kubler. 2002. New light on the scaling of metabolic rate with the size of algae. J. Phycol. 38: 1-16.

Reinfelder, J.R., A.M.L. Kraepiel and F.M.M. Morel. 2001. Unicellular $C_{(4)}$ photosynthesis in a marine diatom. Nature 407: 996-999.

Robinson, C., C.E. Widdicombe, M.V. Zubkov, G.A. Tarran, A.E.J. Miller and A.P. Rees. 2002. Plankton community respiration during a coccolithophore bloom. Deep-Sea Res. 49: 2929-2950.

Riegman, R., M. De Boer, and L. De Snerpont Domis. 1996. Growth of harmful algae in multispecies cultures. J. Plankton Res. 18: 1851-1866.

Rynearson, T.A. and E.V. Armbrust. 2004. Genetic differentiation among populations of the planktonic marine diatom *Ditylum brightwelli* (Bacillariophyceae). J. Phycol. 40: 34-43.

Schiner, S.M. and J. Gurevitch. 2001. Design and analysis of ecological experiments. Oxford University Press. London.

Schmitt, W.R. 1970. World food. California Mar. Res. Commission CalCOFI Rept. 14: 32-42.

Smayda, T.J., D.G. Borkman, G. Beaugrand and A. Belgrano. 2004. Responses of marine phytoplankton populations to fluctuations in marine climate. pp 49-58. *In:* Marine Ecosystems and Climatic Variation. N.C. Stenseth, G. Ottersen, J.W. Hurrell and A. Belgrano (eds.) Oxford University Press, Oxford, UK.

Spies, A. and T.R. Parsons. 1985. Estuarine microplankton: An experimenmtal approach in combination with field studies. J. Exp. Mar. Biol. Ecol. 92: 63-81.

Sournia, A. 1995. Red tide and toxic marine phytoplankton of the world ocean: an inquiry into biodiversity. pp 103-112. In. Ed. (Lassus, P., G. Arzul, E. Erard, P. Gentien and C. Marcaillou. Harmful Marine Algal Blooms. Technique et Documentation-Lavoisier, Intercept Ltd. 1995.

Sournia, A., M.J. Chrétiennot-Dinet and M. Ricard. 1991. Marine phytoplankton: how many species in the world ocean? J. Plankton Res. 13: 1093-1099.

Smetacek, V. 2001. A watery arm race. Nature 411: 745.

Stahil-Delbango, A., Lars-Anders Hansson, and M. Gyllström. 2003. Recruitment of resting stages may induce blooms of *Microcystis* at low N:P ratios. J. Plankton Res. 25: 1099-1106.

Steemann-Nielsen, E. 1952. The use of radio-active carbon (^{14}C) for measuring organic production in the sea. Journal du Conseil International pour Exploration de la Mer 18: 117-140.

Stevens, D.R. and S. Purton. 1997. Genetic engineeriung of eukaryotic algae: progress and prospects. J. Phycol. 33: 713-722.

Strzepek R.F. and P.J. Harrison. 2004. Photosynthetic architecture differs in coastal and oceanic diatoms. Nature 431: 689-692.

Subba Rao, D.V. 1971. The phytoplankton off Lawson's Bay, Waltair, Bay of Bengal. pp. 101-126. In Prof. P.N. Ganapati 60th birthday Commemoration Volume, Andhra University, Waltair, India.

Subba Rao, D.V. 1973. Effects of environmental perturbations on short-term phytoplankton production off Lawson's Bay, a tropical coastal embayment. Hydrobiologia 43: 73-91.

Subba Rao, D.V. 1980. Measurement of primary production in phytoplankton groups by size-fractionation and by Germanic acid techniques. Oceanologica Acta 3: 31-42.

Subba Rao, D.V. 1981. Growth response of marine phytoplankters to selected concentrations of trace metals. Botanica Marina 24: 369–379.

Subba Rao, D.V. 1988. Species specific primary production measurements of Arctic phytoplankton. Br. Phycol. J. 23: 273-282.

Subba Rao, D.V. 1995. Life cycle and reproduction of the dinoflagellate *Dinophysis norvegica*. Aquat. Microb. Ecol. 9: 199-201.

Subba Rao, D.V. 2002. Algal cultures. pp 425-439. *In:* Pelagic Ecology Methodology. D.V. Subba Rao. (ed.) A.A. Balkema Publishers, Lisse, The Netherlands.

Subba Rao, D.V. and S. Smith. 1978. Temporal variation of size fractionated primary production in Bedford basin during the spring bloom. Oceanol. Acta. 10: 101-109.

Subba Rao, D.V., P.M. Dickie, and P. Vass. 1988. Toxic phytoplankton blooms in the Eastern Canadian Atlantic embayments. International Council for the Exploration of the Sea. C.M. 1988/1: 28 1-16.

Subba Rao, D.V. and Y. Pan. 1993. Photosynthetic characteristics of *Dinophysis norvegica* Claparde & Lachmann, a red-tide dinoflagellate. J. Plankton Res. 15: 965-976.

Subba Rao, D.V., Y. Pan and K. Mukhida. 1998. Production of domoic acid by *Pseudo-nitzschia multiseries* Hasle, affected by lithium P.S.Z.N.I: Marine Ecology, 19: 31-36.

Subba Rao, D.V., Y. Pan and S.J. Smith.1994. Allelopathy Between *Rhizosolenia alata* (Brightwell) and the Toxigenic *Psuedonitzschia pungens* f. *multiseries* (Hasle). pp 681-686 In: *Proceedings of the 6th Conference on Toxic Phytoplankton.* Nantes, France, Oct. 18-22, 1993. eds. P. Lassus et al.Lavoisier, Intercept Ltd.

Subba Rao, D.V. Faiza Al-Yamani and C.V. Nageswara Rao. 1999. Eolian dust affects phytoplankton in the waters off Kuwait, the Arabian Gulf. Naturwissenschaften, 86: 525-529. Germany.

Subba Rao, D.V., J.M. Al-Hassan, F. Al-Yamani, K. Al-Rafaie, W. Ismail, C.V. Nageswara Rao and M. Al-Hassan. 2003. Elusive red tides in Kuwait coastal waters. Harmful Algae News 24: 10-13.

Subrahmanyan, R. and A.H. Viswanatha Sarma. 1961. Studies on the phytoplankton of the west coast of India. Ind. J. Fish. 7: 307-336.

Sukhanova, I.N., M.V. Flint, T.E. Whilledge and E.J. Lessard. 2004. Coccolithophorids in the phytoplankton of the Eastern Bering Sea after the anomalous bloom of 1997. Oceanology 44: 665-678.

Sullivan, M.B. Waterbury, J., and Chisholm, S.W. 2003. Cyanophages infecting the oceanic cyanobacterium *Prochlorococcus*. Nature 424: 1047-1051.

Talling, J.F. 1960. Comparative laboratory and field studies of photosynthesis by a marine planktonic diatom. Limnol Oceanogr. 5: 62-77.

Talling, J.F. 1984. Past and contemporary trends and attitudes in work on primary productivity. J. Plankton Res. 6: 203-217.

Tester, P.A., M.E. Geesey and F.M. Vukovich. 1993. *Gymnodinium breve* and global warming: What are the possibilities? Toxic phytoplankton blooms in the sea, pp 67-72. Elsevier, Amsterdam (the Netherlands)

Turner, R.E., N.N. Rabalais, D. Justice and Q. Dortcj. 2003. Future aquatic nutrient limitations. Mar. Poll. Bull. 46: 1032-1034.

Túrkoğlu, M. and T. Koray. 2002. Phytoplankton species succession and nutrients in the Southern Black Sea (Bay of Sinop) Turk. J. Bot. 26: 235-252.

Underwood, A.J. 1981. Techniques of anlaysis of variance in experimental marine biology and ecology. Oceanography and Marine Biology: An annual Review. 19: 13-605.

Underwood, A.J. 1997. Experiments in Ecology: Their logical design and interpretation using Analysis of Variance. Cambridge University Press, Cambridge, UK.

Waite, T.D., and R. Szymczak. 1993 Particulate iron formation dynamics in surface waters of the eastern Caribbean. J. Geophys.Res. (C-Oceans) 98: 2371-2383.

Wasmund, N. and S. Uhlig. 2003. Phytoplankton trends in Baltic Sea. ICES Journal of Marine science. 60: 177-186.

Watson, A.J. and P.S. Liss. 1998 Marine biological control on climate via the carbon and sulphur geochemical cycles. Phil. Trans. Biol. Sci. 353: 41-51.

Wilhelm, S.W., L. Poorvin and D.A. Hutchins. 2001. Viral regeneration of bioavailable iron in the coastal Californian upwelling. American Society of Limnology 2001 Aquatic Sciences Meeting, Albuquerque, NM (USA), 12-16 Feb 2001.

Wyatt, T. and R. Jenkinson. 1997. Notes on *Alexandrium* population dynamics. J. Plankton Res. 19: 551-557.

Wyman, M., J.T. Davies, K. Weston, D.W. Crawford and D.A. Purdie. 1998. Ribulose-1, 5-bisphosphate Carboxylase/oxygenase (RubisCO) gene expression and photosynthetic activity in nutrient-enriched mesocosm experiments. Estuar. Coast. Shelf Sci. 46: 22-33.

Wyman, M., J.T. Davies and D.A. Purdie. 2000. Molecular and physiological responses of two classes of marine chromophytic phytoplankton (diatoms and prymnesiophytes) during the development of nutrient-stimulated blooms. Appl. Environ. Microbiol. 66: 2349-2357.

Zaslavskaia, L.A., J.C. Lippmeier, C. Shih, D. Ehrhardt, A.R. Grossman and K.E. Apt. 2001. Trophic conversion of an obligate photoautotrophic organism through metabolic engineering. Science 292: 2073-2075.

Zapata, M., S.W. Jeffrey, S.W. Right, F. Rodriguez, J.L. Garrido and L. Clementson. 2004. Photosynthetic pigments in 37 species (65 strains) of Haptophyta; implications for oceanography and chemotaxonomy. Mar. Ecol. Prog. Ser. 270: 83-102.

2

Photosynthetic Pigments in Marine Microalgae: Insights from Cultures and the Sea

S.W. Jeffrey[1] and S.W. Wright[2]

[1]CSIRO Marine Research, GPO Box 1538, Hobart, Tasmania, 7001, Australia
[2]Australian Antarctic Division, and Antarctic Climate and Ecosystems CRC, Channel Highway, Kingston, Tasmania 7050, Australia

Abstract

Knowledge of pigment distributions across algal phyla has increased in the past decades due to the greater availability of uni-algal cultures and improved performance of pigment chromatography techniques. The diversity of pigments can now be understood due to developments in genetic analysis and the endosymbiotic theory of the origins of plastid diversity. In this chapter, we briefly describe the chemistry of major chlorophylls, carotenoids and biliproteins, then consider the present knowledge of pigment distribution from the best recent data. The pigment complements of each algal Division/Class are described in 18 separate tables with reference to the Cyanobacterial radiation, the Green Algal lineage and the Red Algal lineage (according to the Delwiche 1999 scheme). A comprehensive overview table shows the distribution of 16 chlorophylls, 37 carotenoids and three phycobiliproteins types across 32 pigment groups. Key new features are the recognition of five pigment Types in the Chloroxybacteria (Cyanophyta and Prochlorophyta), three in the Prasinophyta, three in the Chrysophyta, eight in the Haptophyta and five in the Dinophyta. We discuss applications of pigment data from cultures to phytoplankton in the sea – from the remote sensing of surface chlorophyll *a* from satellites to chemotaxonomic analysis of phytoplankton pigment distributions in the water column.

INTRODUCTION

Living banks of marine microalgae (Culture Collections) are of fundamental importance in providing experimental material for marine photosynthetic

A list of abbreviation appears on the last page of the text of this chapter.

pigment studies. Under standard growth conditions, microalgal cultures are consistent and reproducible and represent a subset of the diverse range of the thousands of microscopic species that make up the floating pastures of the world's oceans. Spectacular in color, they provide an immense variety of photosynthetic pigments for harvesting those restricted portions of the visible spectrum available to them in different water types and depths. The green chlorophylls, the yellow, orange and red carotenoids, and the red and blue-green biliproteins are capable of collectively harvesting the entire range of wavelengths of the visible spectrum (350-750 nm), and ensure the efficient capture of the limited light energy available for photosynthesis in the sea.

It was these vibrant colors that gave rise to the chromatographic nomenclature (*khromatos,* color; *graphia,* writing), when photosynthetic pigments were first separated from the leaves of higher plants by Tswett (1906). Later application of chromatographic techniques to algal cultures showed a much greater diversity of pigments than those in higher plants, and they proved to be valuable taxonomic criteria. Pigments also became excellent markers for algal types in aquatic field studies.

In this chapter, we present current understanding of the pigment composition of microalgae based on culture studies and insights from the sea. First, the development of culture collections is briefly described, followed by the chemistry of key pigments, and analytical methods for their separation and analysis. At the end, we give some recommendations for the future.

Microalgal Culture Collections

Until the early 1950s, the availability of microalgal cultures for pigment analysis was restricted to easy-to-grow inshore 'weed' species (Pringsheim 1946). However in the late 1950s, Dr Luigi Provasoli and co-workers found that many microalgae needed vitamins for growth (B_{12}, thiamine and biotin), and that addition of chelating agents to culture media successfully controlled metal availability (Provasoli et al. 1957, Provasoli 1958). The partial replacement of soil extract in culture media which provided various trace elements, vitamins and metal-chelating organic compounds, was also an advance. Two important publications (Stein 1973, Guillard 1975) provided technical guidance. Later refinements included improved algal isolation techniques, matching chemical and physical factors used in culturing regimes with those of the natural environment (e.g. light quality, intensity and day-length, temperature, salinity), improving seawater purification techniques, maintaining cultures in axenic condition and development of new culture media for nano- and pico-plankton from offshore environments (Brand 1986, Keller et al. 1987, Harrison et al. 1988, Partensky et al. 1993, Jeffrey and LeRoi 1997). Many marine laboratories had small specialist collections in the 1960s and 1970s which expanded greatly in the 1980s and 1990s, helped by increasing technical knowledge and generous exchange of species and expertise by Collection curators.

Two stimuli to algal cultivation in the 1980s were the proliferation of international research cruises studying the biological productivity of the oceans (e.g. Joint Global Ocean Flux Study, JGOFS; Jeffrey and Mantoura 1997) – which was an issue of central concern to a world facing global climate change; and a new awareness of harmful algal bloom events worldwide (Hallegraeff 1993). These new challenges increased the need for access to important strains for experimentation and study (Jeffrey and LeRoi 1997), and allowed the search for new oceanic microalgae to expand due to increased seagoing capability. Recent surveys of Collections (Jameson 2001, Andersen 2003) shows close to 150 microalgal research collections worldwide, and several thousand marine strains now available for study.

Features of the newer Collections include increased holdings of nano-plankton and pico-plankton, availability of multiple strains of single species for biochemical and genetic analysis (isolates often from very different geographic origins), and new species for aquaculture and biotechnology applications. One concern has always been whether laboratory cultivated strains differ significantly from those of natural populations, this topic will be discussed later.

CHEMICAL STRUCTURES OF KEY MICROALGAL PIGMENTS

Three types of conjugated molecules act as light-harvesting pigments in marine microalgae – the chorophylls, carotenoids and biliproteins. The chemical structures of these groups have been extensively reviewed, e.g. *Chlorophylls* (Scheer 1991), *Carotenoids in Photosynthesis* (Young and Britton 1993), biliproteins in *Photosynthetic Pigments of Algae* (Rowan 1989) and marine phytoplankton pigments (Jeffrey et al. 1997a). Detailed chemical structures for most of the important marine chlorophylls and carotenoids are given in Jeffrey et al. (1997b). Only the briefest introduction will be given here.

Chlorophylls

Chlorophylls *a* and *b*, together with various carotenoids, are found in the light-harvesting complexes of the plastids (chloroplasts) of higher plants and green algae, where they transfer excitation energy from the macromolecular antenna collectors to the reaction centres, the site of the initial chemical reactions of photosynthesis. Chlorophylls *a* and *b* (Vernon and Seely 1966, Scheer 1991) are conjugated tetrapyrroles (chlorins) with one ring (D) completely reduced, containing a single bond between C-17 and C-18 (Fig 2.1). They contain a cyclopentanone ring (E) conjoint with ring C, and have a propionic acid side chain at C-17, esterified to the C_{20} alcohol, phytol. A central magnesium atom binds to the nitrogen atoms of the pyrrole rings, while retaining the capacity to bind electron donors on either side of the plane of the chlorin ring. Chlorophyll can thus bind water, attach to proteins, and form self-aggregates by bonding between the magnesium of one ring and the 13-keto group of another.

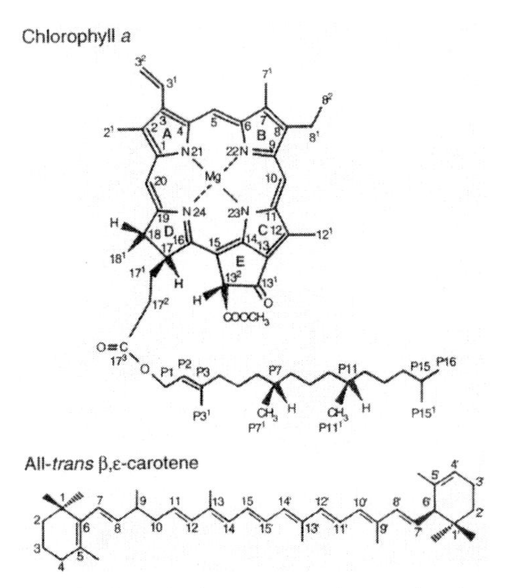

Fig. 2.1 The structure, IUPAC numbering system and stereochemistry of chlorophyll *a* and all-*trans* β,ε-carotene (from Jeffrey et al. 1997b)

Chlorophyll *b* differs from chlorophyll *a* by replacement of the methyl group at position C-7 of ring B with an aldehyde group, which changes its spectral properties and polarity, and increases its stability to photooxidation.

Divinyl Chls *a* and *b* replace Chls *a* and *b* as light-harvesting pigments in *Prochlorococcus* strains (Prochlorophyta; see Table 2.2, pigment Type 4). In both DV Chl *a* and DV Chl *b*, a vinyl group replaces the ethyl group of the parent chlorophyll at position C-8 on ring B (for structures see Jeffrey et al. 1997b).

Many chlorophyll derivatives are found both naturally and as artifacts of extraction. These may have lost the Mg atom (pheophytins), the phytol chain (chlorophyllides), both Mg and phytol (pheophorbides), they may rearrange (epimers) or oxidize (allomers) (Porra et al. 1997). These derivatives may complicate methods of analysis and it is essential that new HPLC techniques clearly differentiate between them (Wright et al. 1991, Mantoura et al. 1997). Jeffrey et al. (1997b) provide comprehensive Data and Graphics sheets for 47 of the most important chlorophylls and carotenoids found in marine algae.

Chl *d*, another chlorophyll of the chlorin type, has recently been found as the major chlorophyll in the symbiotic prochlorophyte *Acaryochloris marina* Miyashita et Chihara. Its chemical structure by current nomenclature is 3-desvinyl-3-formyl Chl *a* (Miyashita et al. 1997).

Members of the Chl *c* family are widely distributed accessory light-harvesting pigments in the golden-brown chromophyte algae, some of the prasinophytes and the prochlorophytes. Chls *c* differ from other chlorophylls in being porphyrin rather than chlorin derivatives, with a fully unsaturated tetrapyrrole macrocycle (i.e. ring D is unsaturated), but with a cyclopentanone ring and central magnesium atom like chlorophylls *a* and *b* (Fig. 2.1). An acrylic acid side chain with a very acidic carboxyl group (usually unesterified), replaces the C-17 (ring D) propionic acid side chain of chlorophylls *a* and *b*. However two Chl *c* pigments, MgDVP (also called DV-Pchlide) and Chl $c_{cs\text{-}170}$, retain the propionic acid. Structurally, polar Chls *c* are protochlorophyllides, but because they are functional light-harvesting pigments the term 'chlorophyll' is usually retained for both polar and non-polar Chls *c*.

Initially only two Chl *c* pigments (c_1 and c_2) were characterized (Jeffrey 1969, 1972). They differ in having an ethyl group (c_1) or a vinyl group (c_2) at C-8 of ring B of the macrocycle (Strain et al. 1971). Chl c_3 (Jeffrey and Wright 1987, Fookes and Jeffrey 1989) has a carbomethoxy group at position C-7 on ring B, and Chl $c_{cs\text{-}170}$ (Jeffrey 1989) on present evidence is thought to be the propionate derivative of Chl c_3 (Jeffrey et al. 1997b). In several non-polar Chl *c* pigments the Chl *c* moiety is covalently esterified at the C-17 acrylic acid to a massive galactolipid side chain (Chl c_2-MGDG; Garrido et al. 2000). These 'non-polar' Chl *c* pigments were discovered through application of the new advanced HPLC techniques to *Emiliania huxleyi* (Lohman) Hay et Mohler and other haptophyte microalgae (Zapata et al. 2004a).

The chemical structures of Chl *a*, *b* and *c* pigments and derivatives may be found in Jeffrey (1997a) and Zapata et al. (2004b). To date nine different Chl *c* pigments have been characterized.

Chls *a, b, c* and *d* have distinct spectral properties which are presented in the relevant papers cited above. Chls *a* and *b* have intense absorption bands in the red (640-700 nm) and blue-green (400-470 nm) spectral regions. The Chl *c* family is also spectrally distinct, with an oxorhodo-type visible spectrum. They have two low but predictable absorption bands at 580 and 630 nm in the orange spectral region, with strong absorption in the main Soret band in the blue-green region (420-450 nm). Chl *d* has two strongly red shifted absorption bands compared to Chl *a*, at 455 and 697 nm respectively (Miyashita et al. 1997). The absorption bands of chlorin and porphyrin-type Chls are dictated by differences in the conjugation pathway of their respective macrocycles.

Carotenoids

More than 800 carotenoids have been isolated and structurally characterized from the biosphere since the 1930s. Much of this work was done by an exceptional group of chemists (see Repeta and Bjørnland 1997 for references), and the field has been frequently reviewed (see regular *International Symposia on Carotenoids).* Approximately 60 different carotenoids have been found in marine microalgae in varying amounts and distributions, with about 29 useful as chemotaxonomic markers in oceanography (Bjørnland 1997a, b). Progress in structure elucidation of algal carotenoids, both the hydrocarbon carotenes and the oxygenated xanthophylls, has been an outstanding contribution of Professor S. Liaaen-Jensen and her colleagues. Most recent advances are reported in Bjørnland and Liaaen-Jensen (1989), Egeland and Liaaen-Jensen (1995) and Egeland et al. (1997).

A relatively small number of carotenoids are found in any one algal class, but most advanced HPLC methods are now revealing a greater diversity (e.g. in the haptophytes).

Carotenoids are usually yellow, orange or red isoprenoid, polyene pigments. The parent hydrocarbon, β,β-carotene, has a C_{40} skeleton with two 6-carbon β-ionone isocyclic rings at each end of the molecule. Fig. 2.1 shows another configuration: β,ε-carotene. Common modifications involve the degree of unsaturation, *cis – trans* isomers, double bond rearrangements including acetylenic and allenic units, glycosidic attachments, and oxygen functional groups (e.g. epoxides, ketones and hydroxyl derivatives). Loss of in-chain carbons may result in shortened skeletons e.g. the C_{37} skeleton of the abundant light-harvesting dinoflagellate carotenoid, peridinin.

In marine microalgae, carotenoids are mainly used as light-harvesting complexes with maximum absorbances from 420 to 550 nm. The violaxanthin $\rightleftarrows$ zeaxanthin and diadinoxanthin $\rightleftarrows$ diatoxanthin epoxide cycles provide essential light protection functions. The former plays a more dominant role in terrestrial plants, subjected to higher light environments than their marine counterparts. Carotenoids may also help to stabilize the photosynthetic apparatus (Scheer 2003). Detailed Data and Graphics sheets with chemical structures for thirty important marine algal carotenoids are given in Jeffrey et al. (1997b).

Phycobiliproteins

The third group of algal light-harvesting pigments are the phycobiliproteins of which there are three main types, phycoerythrobilins, phycocyanobilins and phycourobilins. These chromophores are open-chain tetrapyrroles and do not contain a central metal ligand like the chlorophylls. They form phycobiliproteins by covalent bonding through the cysteine residues of their *apo*protein with the 3-ethylidene and 18-vinyl groups of the chromophore (Porra et al. 1997). Covalent linkage may also occur by esterification of a chromophore propionic acid to a serine-hydroxyl of the *apo*protein. A recent summary of algal biliprotein structures, protein attachments, absorption spectra and biosynthesis is given by Scheer (2003).

Phycobiliproteins are found in light-harvesting macromolecular structures, the phycobilisomes, on the outer surface of the chloroplast membranes (the thylakoids) in the red algae and cyanobacteria. In the cryptomonads, which do not have phycobilisomes, the biliproteins are localized within the intra-thylakoid space, the lumen. Cryptomonad biliproteins have also been found within the thylakoid lumen of certain dinoflagellates.

The biliproteins absorb light in the green-yellow window of the visible light spectrum of the following wavelengths: 540 to 565 nm for the red phycoerythrobilins, 610 to 640 nm for the blue phycocyanobilins, and about 650 nm for the blue allophycocyanobilins.

Pigment Analytical Techniques

Parallel to the development of microalgal Culture Collections since the 1950s has been the increasing sophistication of pigment separation and analytical techniques (Jeffrey et al. 1997a, Jeffrey et al. 1999 [review], Garrido and Zapata 2004). These developments were driven by the recognition that algal pigments are valuable characters for taxonomy and phylogeny, are useful chemotaxonomic markers for phytoplankton populations in limnology and oceanography, and their role in algal photosynthesis and photoprotection is a challenging and important area for study.

In the early days of Culture Collections, the only pigment separation techniques were those of column chromatography (Strain et al. 1944, Strain 1958), while paper chromatography (Jeffrey 1961), and impregnated paper chromatography (Jensen 1966) provided micro-techniques. These methods separated the major pigments of algal classes and were used by Liaaen-Jensen and colleagues to facilitate the carotenoid structural analysis that has been the ongoing monumental contribution of the Trondheim group (Bjørnland and Liaaen-Jensen 1989). Thin-layer chromatography on organic and inorganic layers sharpened separations further, allowing Hager and Stransky (1970a, b) and the Liaaen-Jensen group to further characterize major pigments of many of the algal groups available in culture at the time. TLC has continued to aid carotenoid structural determinations (Repeta and Bjørnland 1997). Column and thin-layer chromatography using polyethylene powder allowed the polar Chl *c* pigments c_1 and c_2 to be separated (Jeffrey 1969, 1972), reverse-phase TLC (RP-TLC) revealed Chl c_3 (Vesk and Jeffrey 1987) and cellulose thin-layer chromatography applied to field oceanography showed for the first time the pigment composition of phytoplankton in the sea (Jeffrey 1974, Hallegraeff 1981, Jeffrey and Hallegraeff 1987). However, these techniques were time-consuming and were not suited to rapid, routine quantitation of field samples, and thus were not taken up by the oceanographic community at the time.

High-performance liquid chromatography (HPLC) revolutionized pigment analysis in the 1980s (Mantoura and Llewellyn 1983, Gieskes and Kraay 1983a, 1986, Wright and Shearer 1984, Zapata et al. 1987) permitting automated analysis and quantitation of pigments, with the possibility of online identification using diode-array detection. These methods greatly facilitated the routine analysis of hundreds of samples from ship-board field studies. Techniques were standardized in the 1988-89 SCOR intercalibration workshops, and the resulting monograph collated all available knowledge for oceanographers (Jeffrey et al. 1997a). An improved HPLC technique (Wright et al. 1991, Wright and Jeffrey 1997) was recommended which used a reverse-phase C_{18} monomeric column and ammonium acetate modifier, providing good resolution of 40 algal carotenoids and 12 chlorophylls and their derivatives, but lacking resolution of the polar chlorophyll *c* family, and the newly discovered DV-Chl *a* and DV-Chl *b* pigments (Goericke and Repeta 1992, 1993).

A wealth of new HPLC methods since the early 1990s has relied on new stationary phases - C_8, C_{30}, and polymeric C_{18} phases - (reviewed by Jeffrey et al. 1999, Garrido and Zapata 2004). These developments culminated in the superior methods of Garrido and Zapata (1997) and Zapata et al. (2000). Mobile phases incorporating pyridine modifiers were also introduced, achieving further resolution of five polar and three non-polar Chl *c* pigments, the Chl *a*/DV-Chl *a* and Chl *b*/DV-Chl *b* pairs, as well as several previously undiscovered carotenoids. Mass spectrometry coupled with online HPLC also provided new structural information from microgram quantities of pigments, thus expanding knowledge of the distribution of chlorophylls and carotenoids in particular microalgal groups (e.g. the Haptophyta, Zapata et al. 2004a; see Fig. 2.2.).

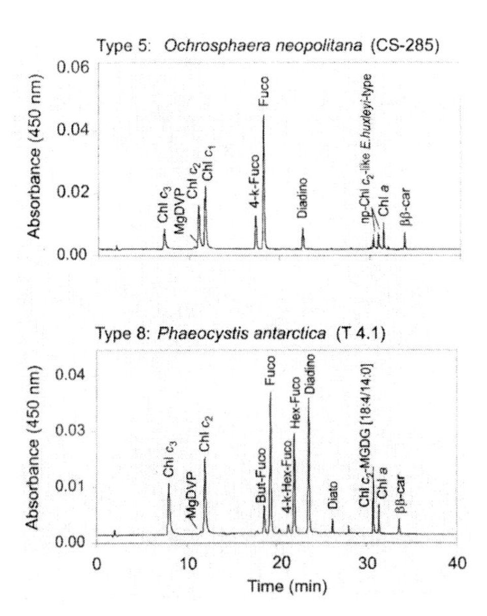

Fig. 2.2 HPLC separations of photosynthetic pigments from two haptophyte cultures, *Ochrosphaera neopolitana* Schussig and *Phaeocystis antarctica* Karsten by the method of Zapata et al. (2000). *O. neopolitana* is an example of a haptophyte with a Type 5 pigment composition, and *P. antarctica* is a representative of the Type 8 pigment suite. Absorbance detection at 436 nm (from Zapata et al. 2004a).

Unlike chlorophylls and carotenoids, analytical micromethods for biliproteins have been slow to develop within the marine community because those algal classes containing them (Cyanophyta, Rhodophyta, and Cryptophyta), are so easily recognized and counted by other techniques - epifluorescence microscopy (Johnson and Sieburth 1979, Waterbury et al. 1979) and flow cytometry (Olson et al. 1985, Li and Wood 1988, Chisholm et al. 1988). Since the biliproteins are water soluble, they cannot be analysed simultaneously with the lipophilic chlorophylls and carotenoids. Alloxanthin, zeaxanthin, and DV-Chl *a* are in any case more convenient HPLC markers for the biliprotein-containing cryptophytes, cyanobacteria and prochlorophytes respectively. Classical techniques for biliprotein analysis involve aqueous extraction and quantitation by spectrophometry (Siegleman and Kycia 1978) or spectrofluorimetry (Wyman 1992). Early results were extensively reviewed by Rowan (1989); later advances by Scheer (2003). Recently, a capillary electrophoretic method was developed (Viskari et al. 2002) but flow cytometry is likely to remain the method of choice for biliprotein-containing organisms in the field.

PIGMENT CHARACTERISTICS OF ALGAL DIVISIONS/CLASSES FROM STUDIES OF CULTURED MICROALGAE

Origins of Plastid Diversity

It is now generally accepted that plastids (chloroplasts) of eukaryotic algae are endosymbiotic organelles, originally derived from a previously free-living ancestral cyanobacterium (Bhattacharya 1997, Delwiche 1999, McFadden 2001, Palmer 2003). The host cell was a non-photosynthetic (heterotrophic) phagotrophic protist of unknown origin. Development of this ancient symbiosis eventually resulted in reduction of the size of the plastid genome, by gene transfer, loss and substitution, until the majority of the plastid proteins were encoded in the nuclear genome of the host. Further evolutionary development of this early endosymbiosis resulted in three major primary lineages, each clearly monophyletic: the Glaucocystophyta, the Chlorophyta and the Rhodophyta radiations (Moreira et al. 2000). The modern cyanobacterial radiation was derived directly from the ancestral cyanobacterium or its relatives without undergoing any symbioses (see Fig. 2.3 A adapted from Delwiche 1999).

Many other photosynthetic algae arose from secondary or even tertiary endosymbioses of cells from these lineages (see Fig. 2.3). In these cases an alga already equipped with a primary (or secondary) plastid, could be engulfed by another non-photosynthetic host cell, entering into a permanent or semi-permanent association with it. The history of these events can be seen in present day cells by the presence of vestigial nuclei (e.g. the nucleomorph), loss of cell compartments and organelles, the number of residual membranes surrounding the plastid (two, three or four indicating

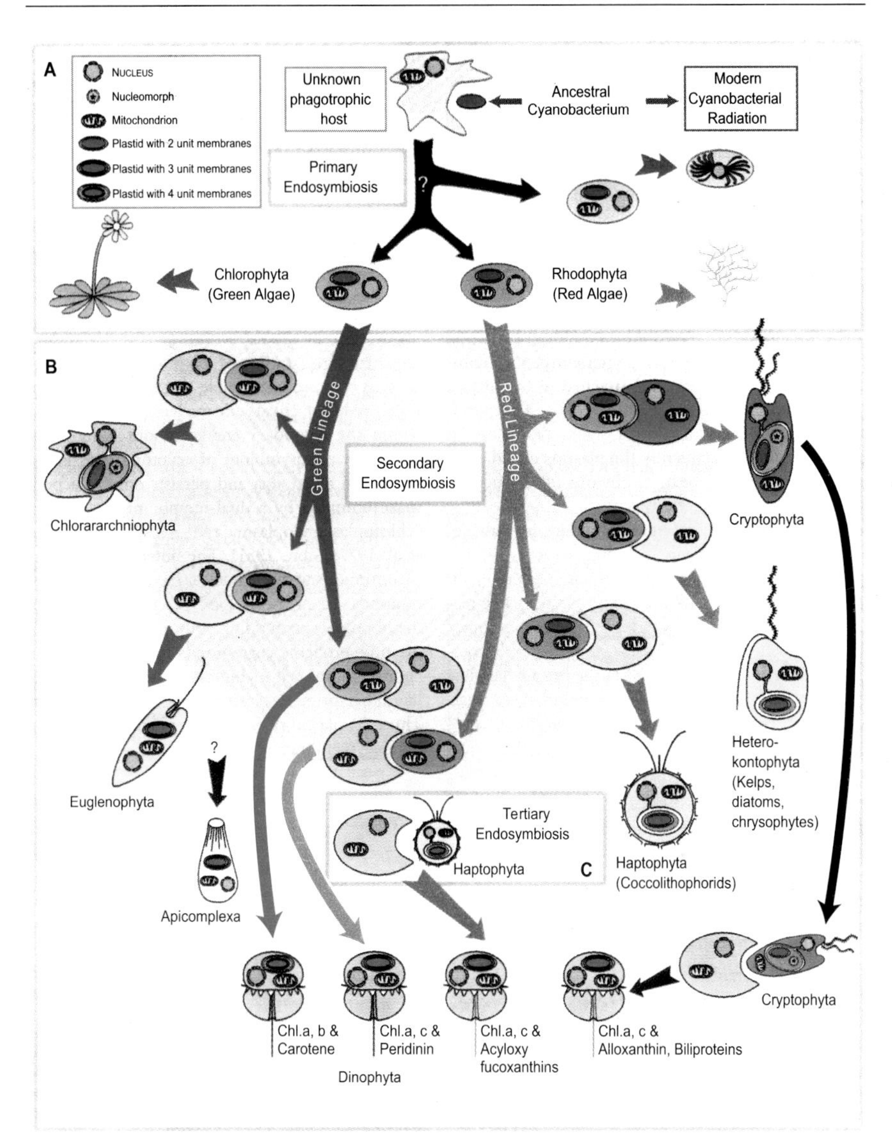

Fig. 2.3 Hypothetical evolution of plastid diversity via serial endosymbioses (slightly modified from Delwiche 1999, with permission)

primary, secondary or tertiary endosymbioses respectively – see Fig. 2.3) and analysis of nuclear and plastid genomes. Evidence from these sources supported the hypothesis (Cavalier Smith 2002) that plastids of heterokonts (diatoms, brown algae, chrysophytes etc), haptophytes, cryptophytes and dinoflagellates all arose from ancestral red algae by various secondary and tertiary endosymbioses (Fig. 2.3). By a similar process euglenophytes, chlorarachniophytes and green dinoflagellates acquired their plastids from the Chlorophyta. New evidence is also suggesting that in some groups (e.g. dinoflagellates) multiple plastid losses and replacements have occurred (Saldarriaga et al. 2001). Detailed description and discussion of these processes is outside the scope of this chapter but are presented here briefly to show evolutionary relationships between algal types. Important topical references are cited above.

Within the marine phytoplankton – the unicellular 'pastures' of the sea – may be found extensive populations of most microalgal lineages and their radiations which are even now undergoing changes in population balance. In the early primitive oceans, the fossil record shows clearly the dominance of the green superfamily, but in later evolutionary times, the balance has switched to ecological dominance of chromophyte algae from the red algal radiation. What forces have promoted these changes in the modern oceans is an active area of current research (Quigg et al. 2003, Grzebyk et al. 2003).

SCOR Recommended Cultures for Pigment Standards

One of the objectives of the SCOR Working Group 78, which ultimately produced the UNESCO Monograph (Jeffrey et al. 1997a) on *Photosynthetic Pigments in Oceanography*, was "to provide advice on the establishment and provision of reference chlorophylls, carotenoid and phycobiliprotein standards for calibration of HPLC, fluorescence and absorbance methods". Due to the high instability of isolated pigments and the commercial lack of all but a few pigment standards at the time, SCOR Working Group 78 recommended that the source of such standards should be well-characterized strains of microalgae, whose pigments had been securely identified by modern chemical methods. A small number of cultured species were selected (Table 2.1) from 12 algal classes that together contained most of the significant chlorophylls and carotenoids known to be found in the phytoplankton at that time. Simple directions for algal culturing were provided (Jeffrey and LeRoi 1997), plus pigment extraction, purification and storage techniques, and quantitative pigment analyses of the 12 cultures grown under standard conditions were reported (Jeffrey and Wright 1997). This recommendation of using cultured microalgae as a source of pure pigments has been adopted, and together with a new commercial resource, DHI Water and Environment, Horsholm, Denmark, provide authentic and easily available pigment standards for the research community. What is

Table 2.1 Microalgal species recommended as authentic sources of pigments by SCOR WG 78 (Jeffrey and Wright 1997). For details of growth conditions see Jeffrey and LeRoi (1997).

Species	*CSIRO Culture code*	*Class*	*Clonal designation and origin*
Amphidinium carterae Hulbert	CS-212	Dinophyceae	AMPHI, Plymouth 450
Rhodomonas salina[1] (Wislouch) Hill et Wetherbee	CS-174	Cryptophyceae	CCMP 56/00/00
Dunaliella tertiolecta Butcher	CS-175	Chlorophyceae	CCMP
Emiliania huxleyi (Lohman) Hay & Mohler	CS-57	Prymnesiophyceae	BT6
Euglena gracilis Klebs	CS-66	Euglenophyceae	Univ. NSW, strain Z
Micromonas pusilla (Butcher) Manton & Parke	CS-86	Prasinophyceae	UTEX LB991
Pavlova lutheri (Droop) Green	CS-182	Prymnesiophyceae	CCMP
Pelagococcus subviridis Norris	CS-99	Pelagophyceae	East. Aust. Current
Phaeodactylum tricornutum Bohlin	CS-29	Bacillariophyceae	Plymouth (1052/1)
Porphyridium purpureum[2] (Bory) Drew & Ross	CS-25	Rhodophyceae	Halifax, Canada
Pycnococcus provasolii Guillard	CS-185	Prasinophyceae	CCMP, 48-23AX
Synechococcus sp.	CS-197	Cyanophyceae	DC2

[1] Previously *Chroomonas salina*
[2] Previously *Porphyridium cruentum*

needed next are additional strains to supply the newly discovered pigments made available by the advanced HPLC techniques (e.g. DV-Chl *a*, DV-Chl *b*, the Chl *c* family (9 pigments), Chl *d*, new fucoxanthin derivatives etc.

Pigment Characteristics of Algal Divisions/Classes

An overview of major and taxonomically significant pigments from microalgal cultures across algal Divisions and Classes was previously carried out by Jeffrey and Vesk (1997). In that chapter, the focus was mainly on the biology of the major algal types found in marine phytoplankton with pigment distributions just one of the important characteristics. In this section we are focusing only on pigments from cultured microalgae (recent advances since 1997) and their distributions.

Jeffrey and Vesk (1997) listed exceptions to general pigment patterns in 15 footnotes to their Table 2.3. Subsequent research has shown that these species were representatives of sub-groups within the Division/Class, rather than being exceptions. We now recognize five pigment types within the Cyanophyta/Prochlorophyta, three within the Prasinophyta, three within the Chrysophyta, eight within the Haptophyta, and five within the Dinophyta.

These distributions will be considered here in the context of the Delwiche (1999) scheme (Fig. 2.3) listing the pigment characteristics of each Division/Class separately, together with a brief introduction to each Division/Class. We begin with members of the cyanobacterial radiation (Cyanophyta and Prochlorophyta; Table 2.2), followed by members of the green algal lineage (Chlorophyta, Prasinophyta and Euglenophyta; Tables 2.3 to 2.5) and finally members of the red algal lineage (Rhodophyta, Cryptophyta, Heterokontophyta and Dinophyta; Tables 2.6 to 2.19). Our focus will include only those groups commonly found in present-day phytoplankton. Finally, all significant pigments and pigment types currently known across algal Divisions/Classes are summarized in one overview table (Table 2.20).

The Cyanobacterial Radiation (Chloroxybacteria)

The chloroxybacteria are photosynthetic prokaryotes (Cyanophyta and Prochlorophyta) that range in size from pico-planktonic coccoid forms to filamentous or globular colonies several centimeters in diameter. They are widespread in freshwater and marine systems, and some species form endosymbiotic associations within tissues of marine invertebrates. They produce a great diversity of pigments including Chls *a, b*, MgDVP, the carotenoids β,β-carotene and zeaxanthin, and biliproteins as major components. Some chloroxybacteria produce DV-Chl *a*, DV-Chl *b*, or Chl *d* instead of Chl *a* – pigments unknown as major components in eukaryotes.

The present-day Chloroxybacteria can be separated into five pigment types, two in the Cyanophyta and three in the Prochlorophyta.

Cyanophyta

Pigment Types 1 and 2 are found in the Cyanophyta.

- Type 1 (comprising freshwater and colonial forms) contains Chl *a*, β,β-carotene, zeaxanthin, phycobiliproteins, plus a range of other carotenoids such as echinenone, myxoxanthophyll etc. (Hertzberg et al. 1971).
- Type 2 (comprising marine coccoid forms) contains a similar array of pigments but lacks the minor carotenoids of Type 1. The pigments of Type 2 resemble those of the extant unicellular Rhodophyta (Table 2.6).

Table 2.2 Pigment characteristics of the cyanobacterial radiation (Chloroxybacteria: Cyanophyta/Prochlorophyta)

Type 1	**Chlorophylls:**	**Chlorophyll a**
Freshwater and marine colonial forms	**Carotenoids:**	**β,β-carotene, zeaxanthin;** freshwater species may contain variable amounts of echinenone, canthaxanthin, myxoxanthophyll, oscillaxanthin, nostoxanthin, caloxanthin etc. (Hertzberg et al. 1971)
	Biliproteins:	Phycobiliproteins containing C-phycocyanin, allophycocyanin, C-phycoerythrin and/or R-phycoerythrin (Rowan 1989)
	Cultures used:	Freshwater species from families Oscillatoriaceae, Nostocaceae, Chroococcaceae etc (Hertzberg et al. 1971)
	Natural blooms:	*Trichodesmium* spp. (Carpenter et al. 1993)
	Culture Collections:	CCMP and other small Collections from the 1960s
Type 2	**Chlorophylls:**	**Chlorophyll *a***
Marine coccoid planktonic species	**Carotenoids:**	**β,β-carotene, zeaxanthin**
	Biliproteins:	Phycobiliproteins containing C-phycocyanin, allophycocyanin, C-phycoerythrin and/or R-phycoerythrin
	Cultures used:	*Synechococcus* spp., *Synechocystis* spp.
	Culture Collections:	CCMP, CSIRO Algal Culture Collection
Type 3	**Chlorophylls:**	**Chlorophylls *a, b,* MgDVP** (Larkum et al. 1994)
***Prochloron* strains**	**Carotenoids:**	**β,β-carotene, zeaxanthin, cryptoxanthin** and traces of β,β-carotene monoepoxide, mutachrome and echinenone (Foss et al. 1987)
	Biliproteins:	Absent
	Species used:	*Prochloron didemni* (symbiotic, and expressed from tropical ascidians); the free-living freshwater *Prochlorothrix* (Burger-Wiersma et al. 1986)
	Culture Collections:	*Prochloron* strains not available in culture
Type 4	**Chlorophylls:**	**DV-Chlorophyll *a,* DV-Chlorophyll *b,* MgDVP** (Helfrich et al. 1999); Chl *a, b* not detected
Marine *Prochlorococcus* strains containing DV-Chl *a* and DV- Chl *b*	**Carotenoids:**	**β,ε-carotene, zeaxanthin**
	Biliproteins:	Small amounts of novel phycoerythrin type III (Hess et al. 1996); no other biliproteins detected
	Cultures used:	*Prochlorococcus marinus* Chisholm
	Culture Collections:	Scripps Institution of Oceanography Collections, CCMP
	Pigment methods used:	HPLC (Goericke and Repeta 1992)

Contd.

Table 2.2 Type 4 Contd.

Type 5	**Chlorophylls:**	**Chlorophyll *d*;** minor concentrations of Chl *a* and a Chl *c*-like pigment
Symbiotic species:		
Acaryochloris marina	**Carotenoids:**	**β,ε-carotene, zeaxanthin**
containing Chl *d*	**Biliproteins:**	Traces of blue phycocyanin and allophycocyanin in biliprotein aggregates
	Species used:	*Acaryochloris marina* is found together with *Prochloron* in tropical ascidians and has been isolated in clonal culture
	Culture Collections:	Marine Biotechnology Institute, Kamaishi, Japan
	Pigment methods used:	HPLC (Miyashita et al. 1997)
	Comments:	*A. marina* is a *Prochloron*-like prokaryote within the cyanobacterial radiation (Chloroxybacteria). Chl *d* (max absorption 740 nm) benefits its host by harvesting far-red light outside the spectral range of other plants and cyanobacteria.

References for Types 1-5: Hertzberg et al. 1971, Bryant 1992, Carpenter et al. 1993, Larkum et al. 1994, Miyashita et al.1996, 1997, 2003, Marquardt et al. 1997, Jeffrey and Wright 1997, Hu et al. 1998, 1999, Helfrich et al. 1999, Goericke et al. 2000, Six et al. 2004

Prochlorophyta

The prochlorophytes are polyphyletic within the cyanobacterial radiation, and although some contain Chl *b*, they do not include the direct ancestor of chloroplasts (Palenik and Haselkorn 1992). Evidence also suggests that the green chloroplast ancestor and the prochlorophytes acquired their Chl *b* independently in convergent evolutionary events (Urbach et al. 1992). Pigment Types 3 to 5 are found in the Prochlorophyta.

- Type 3, known only in the ascidian symbiont, *Prochloron didemni* Lewin, and the free-living freshwater prochlorophyte *Prochlorothrix sp.*, contains Chl *a*, *b*, and MgDVP, β,β-carotene, zeaxanthin and cryptoxanthin. Type 3 lack biliproteins. Their pigments most closely resemble those of the green algal lineage.
- Type 4, found in the marine picoplanktonic prochlorophyte, *Prochlorococcus marinus* Chisholm, resembles the Type 3 pigmentation, but has DV-Chl *a* and DV-Chl *b*, instead of the normal monovinyl forms (Chls *a* and *b*) and small amounts of a novel phycoerythrin. Pigmentation of this type is unknown from eukaryotic lineages.
- Type 5, found in the ascidian symbiont, *Acaryochloris marina* Miyashita et Chihara, is unique in having Chl *d* as its major pigment, together with β,ε-carotene and zeaxanthin. It also contains some Chl *a*, a Chl *c*-like pigment and traces of biliprotein. This pigmentation is also unknown in eukaryotic lineages.

Table 2.2 summarizes the pigment characteristics of the Chloroxybacteria (Cyanophyta and Prochlorophyta). Pigments in bold font in Tables 2.2-2.19 represent typical major pigments.

The Green Algal Lineage

A primary endosymbiosis between a cyanobacterial ancestor and an unknown host initiated the green algal line, which comprises the Kingdom Viridiplantae, and all green algae and higher plants. Present-day planktonic representatives include chlorophytes, prasinophytes and (through secondary endosymbiosis) the euglenophytes.

Their characteristic pigments include Chl *a* and *b* as well as derivatives of β,β-carotene and β,ε-carotene. They do not produce biliproteins. Many species produce photoprotective pigments (notably violaxanthin, antheraxanthin and zeaxanthin – the 'violaxanthin cycle' pigments – as well as secondary carotenoids) that enable cells to withstand high light environments of coastal and inshore environments. The smallest known eukaryote, the prasinophyte *Ostreococcus tauri* Courties et Chrétiennot-Dinet, averages 0.97 μm in length (Courties et al. 1994), while their terrestrial descendents include the largest, the Californian Redwood.

Tables 2.3, 2.4 and 2.5 list the pigment characteristics of the Chlorophyta, Prasinophyta and Euglenophyta, respectively.

Table 2.3 Pigment characteristics of Chlorophytes (Chlorophyta)

Chlorophylls:	**Chlorophyll *a, b***
Carotenoids:	**Lutein, zeaxanthin, violaxanthin, antheraxanthin, 9'-*cis*-neoxanthin, β,β- and/or β,ε-carotene;** β,Ψ-carotene; canthaxanthin; loroxanthin is widely distributed but there is no obvious phylogenetic pattern, (Fawley 1991) *Secondary carotenoids* astaxanthin, β,β-carotene (promoted by culture age, high light and nitrogen limitation)
Cultures examined:	*Dunaliella tertiolecta* Butcher, *Scenedesmus obliquus* (Turpin) Kützing, *Chlorella pyrenoidosa* Chick, *Haematococcus pluvialis* Flotow; many species from early collections (Hager and Stransky 1970b).
Culture Collections used:	UTEX, CCMP, CSIRO Algal Culture Collection
Culture color:	Grass-green; *H. pluvialis* is red in cyst stages
Pigment methods used:	TLC; RP-HPLC (Wright et al. 1991, Zapata et al. 2000, Garrido and Zapata 2004)
Likely endosymbiosis:	Primary endosymbiosis of an unknown ancestral cyanobacterium with a heterotrophic eukaryotic host
References:	Hager and Stransky 1970b, Fawley 1991, Delwiche 1999, Zhekisheva et al. 2002

Table 2.4 Pigment characteristics of prasinophytes (Prasinophyta)

Chlorophylls:	**Chlorophylls *a, b,* MgDVP** present in all types; Chl $c_{cs\text{-}170}$ in one tropical strain of *Micromonas pusilla*
Carotenoids: **Type 1–common chlorophyte carotenoids**	**β,β-carotene, lutein, zeaxanthin, antheraxanthin, violaxanthin, *9'-cis*-neoxanthin**, β,ε-carotene; type species *Nephroselmis olivacea* Stein (Egeland et al. 1997)
Type 2–siphonaxanthin series	β,β-carotene, β,ε-carotene, zeaxanthin, antheraxanthin, violaxanthin, **neoxanthin** *plus* **siphonein (siphonaxanthin ester), 6'-hydroxy siphonein;** type species, *Pyramimonas amylifera* Conrad (Egeland et al. 1997)
Type 3–prasinoxanthin/uriolide series	Common chlorophyte carotenoids (see Type 1) *plus* ε,ε-carotene, **prasinoxanthin, uriolide, micromonal,** pre-prasinoxanthin, de-epoxyuriolide, dihydrolutein; sometimes traces of lutein: type species *Prasinococcus capsulatus* Miyashita et Chihara (Egeland et al. 1997)
Cultures examined:	Many classic prasinoxanthin spp. e.g. *Micromonas pusilla, Nephroselmis spp., Pyraminonas spp., Prasinococcus spp., Mantoniella sp., Bathycoccus prasinos* Eikrem et Throndsen; many new picoplanktonic isolates e.g. *Ostreococcus tauri* Courties et Chrétiennot-Dinet, *Crustomastix stigmatica* Zingone, *Dolichomastix tannilepis* Manton.
Culture Collections used:	CSIRO Algal Culture Collection, CCAP, Plymouth Culture Collection UK, CCMP, Sammlung von Algenkulturen, Pflanzen Physiologisches Institut, Universität Göttingen.
Culture color:	Pale-green to yellow-green.
Pigment methods used:	TLC (Ricketts 1966); HP-TLC (Jeffrey 1989); HPLC (Wright et al. 1991, Egeland et al. 1997, Jeffrey and Wright 1997)
Likely endosymbiosis:	Primary endosymbiosis of an ancestral cyanobacterium with a heterotrophic eukaryotic host. Prasinophytes are thought to pre-date the chlorophytes to yield the green algal lineage (Nakayama et al. 1998)
References:	Ricketts (1966, 1970), Courties et al.(1994), Egeland et al. (1995, 1997), Egeland and Liaaen-Jensen (1995), Jeffrey and Wright (1997), Delwiche (1999), Zingone et al. (2002)
Comments:	MgDVP is present in significant quantities in many prasinophytes as a light-harvesting pigment (Brown 1985). The paper by Egeland et al., (1997) on chemosystematic evaluation of carotenoids in prasinophytes is a classic, and is highly recommended.

Chlorophyta

The pigment complement of the Chlorophyta (Table 2.3) is widespread throughout the green lineage, including the higher plants. Typical pigments include Chls *a* and *b*, lutein, violaxanthin, antheraxanthin, zeaxanthin, 9'-*cis*-

Table 2.5 Pigment characteristics of euglenophytes (Euglenophyta)

Chlorophylls:	**Chlorophylls *a*, *b***
Carotenoids:	**9'-*cis*-neoxanthin, diadinoxanthin (major carotenoid),** diatoxanthin, **β,β-carotene, β,ε-carotene**
Cultures examined:	*Euglena gracilis* Klebs (freshwater), *Eutreptiella* Throndsen (marine)
Culture Collections used:	CCMP, CSIRO Algal Culture Collection
Culture color:	Grass-green
Pigment methods used:	HPLC (Wright et al. 1991)
Likely endosymbiosis:	Secondary endosymbiosis of an ancestral green alga with a heterotrophic eukaryotic flagellate
References:	Bjørnland 1982, Jeffrey and Wright 1997, Delwiche 1999
Comments:	More species need examination with new HPLC methods

neoxanthin, β,β-carotene and often β,ε-carotene. The molecular organization and function of Chls *a*, *b* and certain carotenoids has been brilliantly demonstrated in the crystal structure of the major light-harvesting complex of a higher plant (spinach), with the geometric arrangement of pigment molecules being deduced at atomic resolution (Liu et al. 2004). The chlorophyll binding sites and orientation of carotenoids for membrane stabilization and excess energy quenching are also reported. This work is relevant to the Chlorophyta since they contain the same pigment complement.

Prasinophyta

The prasinophytes are generally accepted as the most primitive members of the green algal lineage. They were initially recognized as scaly green flagellates, although many naked forms are now included. They are often important components of the nanoplankton and picoplankton.

Prasinophytes contain a diverse range of pigments (Table 2.4). In addition to Chl *a* and *b* (and often MgDVP as a major light-harvesting component), thirty carotenoids have been identified, of which fourteen are unique to the prasinophytes (Egeland et al. 1997).

Three carotenoid patterns were identified within the class by Egeland et al. (1997). Type 1 contain carotenoids typical of the chlorophytes, with lutein and lutein epoxide, Type 2 contain chlorophyte pigments (minus lutein) plus those of the siphonaxanthin series, while Type 3 contain chlorophyte carotenoids plus those of the prasinoxanthin series and the uriolide series, but with very little lutein. The distribution of the pigment types does not strictly conform with phylogenetic trees based on 16S rDNA (c.f. Nakayama et al. 1998, Fawley et al. 2000, Zingone et al. 2002).

Euglenophyta

Euglenophytes (Table 2.5) differ from other members of the green algal lineage representing a secondary endosymbiosis of an ancestral (eukaryotic) green alga with a heterotrophic flagellate. They are common in marine and freshwater systems, particularly if there is organic pollution (Jeffrey and Vesk 1997). Their pigment characteristics include Chls *a* and *b* but differ from other green algae with the chromophyte pigment diadinoxanthin as the major carotenoid, as well as some β,β-carotene and neoxanthin.

The Red Algal Lineage

The red algal lineage constitutes a second primary algal radiation (Delwiche 1999; Fig. 2.3). The red algae per se (Rhodophyta), which have plastids derived from a primary endosymbiotic event with an ancestral cyanobacterium, have very few modern unicellular representatives, but over 4000 species of macrophytes (the red seaweeds, Kraft 1981). Most present-day eukaryotic algae that contain plastids originally derived from the red algal ancestor are products of secondary or tertiary endosymbioses (Fig. 2.3 and Tables 2.6-2.19). These include the Kingdom Chromista (chromophytes), comprising cryptomonads, heterokonts (diatoms, bolidophytes, chrysophytes, raphidophytes, eustigmatophytes and brown seaweeds) and the Dinophyta (dinoflagellates). Only the cryptomonads retain the phycobiliproteins of the ancestral rhodophytes, albeit it in a different molecular orientation within the thylakoid lumen of the plastid rather than in phycobilisomes on the external thylakoid face. The chromophyte algae, while diverse in cell morphologies, ultrastructure and molecular phylogeny, share similar photosynthetic pigments – the Chl *c* family and unique carotenoids as described for each Division/Class (see below).

The Dinophyta have their own unique carotenoid, peridinin, derived and modified from the ancestral red algal plastid and found in about 50% of extant dinoflagellates. Due to their unusual ability to take in endosymbionts (Morden and Sherwood 2002), they may also show pigment suites originally derived from diatom, haptophyte, cryptophyte or prasinophyte prey species (see below). These symbioses can be permanent or semi-permanent associations.

Grzebyk et al. (2003) consider the diversity and ecological success of the red plastid phyla in the modern ocean may be due to their retention of "a complementary set of genes that potentially confer more capacity to autonomously express proteins regulating oxygenic photosynthetic and energy transduction pathways".

Rhodophyta

The extant unicellular rhodophytes are represented in present-day oceans by only a few genera (e.g *Rhodosorus*, *Rhodella* and *Porphyridium*). These nanoplanktonic forms are usually spherical (5-10 μm), non-flagellated, either solitary or found in colonies held together by mucilage. The red, water-soluble phycoerythrobilin, with lesser amounts of phycocyanobilin and allophycocyanin, are found in macromolecular complexes, the phycobilisomes, on the outer face of the thylakoids. Chl *a*, β,β-carotene, and zeaxanthin complete the pigment suite. Table 2.6 gives the pigment characteristics of the planktonic Rhodophyta, as currently known.

Table 2.6 Pigment characteristics of planktonic rhodophytes (Rhodophyta)

Chlorophylls:	**Chlorophyll *a***
Carotenoids:	**β,β-carotene, zeaxanthin**
Biliproteins:	Phycoerythrobilin, phycocyanobilin, allophycocyanin
Cultures examined:	*Porphyridium purpureum* (Bory) Drew & Ross
Culture Collections used:	CSIRO Algal Culture Collection
Culture color:	Deep red
Pigment methods used:	HPLC (Wright et al. 1991, Zapata et al. 2000)
Likely endosymbiosis:	Primary endosymbiosis of an ancestral cyanobacterium and a eukaryotic heterotrophic host
References:	Rowan 1989, Wright et al.1991, Delwiche 1999, Kopecky et al. 2002, Scheer 2003
Comments:	More unicellular rhodophytes should be examined by advanced HPLC techniques to identify possible chlorophyll *c* derivatives in 'hypothesized' rhodophyte precursors of the chromophyte lineage

Cryptophyta

The Cryptophyta are a well-defined group of mostly photosynthetic nanoplanktonic flagellates (Clay et al. 1999) that have affinities with both the Rhodophyta per se (phycobiliprotein accessory pigments) and other members of the red lineage, the Chl *c*-containing 'chromophyte' algae (Fig. 2.3). They are common in marine, estuarine and freshwater habitats. Due to their susceptibility to preservatives they may often be missed in field collections, but they can be detected in pigment samples by the specific marker carotenoid, alloxanthin (c.f. Gieskes and Kraay 1983a).

The biology and ultrastructure of these unique biflagellate cells are described in Jeffrey and Vesk (1997). They are immediately recognizable under the light microscope by their proteinaceous pellicle, subapical gullet or furrow, and ovoid asymmetric shape. Electron microscopy reveals an electron dense thylakoid lumen, ejectile organelles and a vestigial nucleus in the cryptophyte cytoplasm that indicates the secondary endosymbiotic event.

Cryptomonads are also found in symbiotic associations within dinoflagellates (e.g. *Gymnodinium acidotum* Nygaard and *Dinophysis* spp., see Table 2.18; Meyer-Harms and Pollehne 1998), and the red ciliate *Mesodinium rubrum* (Hibberd 1977). The pigment characteristics of the Cryptophyta are listed in Table 2.7.

Table 2.7 Pigment characteristics of cryptomonads (Cryptophyta)

Chlorophylls:	**Chlorophylls *a*, c_2,** MgDVP; Chlc_1 present in one strain of *Chroomonas* (Schimek et al. (1994)
Carotenoids:	**Alloxanthin, crocoxanthin, monadoxanthin, β,ε-carotene,** ε,ε-carotene, Ψ,Ψ-carotene
Biliproteins:	Cr-Phycoerythrins 545 or 555 (red); or Cr-Phycocyanins 630, 645 or 569 (blue) sited within the thylakoid lumen
Cultures examined:	Species of *Chroomonas, Cryptomonas, Rhodomonas, Geminigera*
Culture Collections used:	CSIRO Algal Culture Collection; University of Göttingen Culture Collection
Culture color:	Red, reddish brown or blue-green, depending on the biliprotein type present.
Pigment methods used:	TLC, HPLC (Wright et al.1991, Zapata et al. 2000, Schimek et al. 1994)
Likely endosymbiosis:	Secondary endosymbiosis of an ancestral red algal precursor and a heterotrophic protozoan host
References:	Jeffrey 1976a, Schimek et al. 1994, Jeffrey and Wright 1997, Clay et al. 1999, Jeffrey and Zapata (unpubl.)

Bacillariophyta

Diatoms are ubiquitous in the world's oceans and occur in all kinds of aquatic habitats - from the subsurface diatom communities of the polar sea ice to the spectacular centric diatom communities of tropical waters. The rapid growth of diatoms generally ensures that they are the first to appear in the water column after nutrient (nitrate) enrichment (e.g. the classic spring diatom blooms of northern temperate waters; Round 1981). These nutrient enrichments result from convective overturn of the water column after winter, as well as the increase in springtime light and temperature. Spring diatom blooms also occur in the coastal waters of eastern Australia, but are initiated by intrusions of nutrient-rich slope waters on to the shelf after winter (Hallegraeff and Jeffrey 1993). Species successions of both northern and southern hemisphere diatom blooms are amazingly similar (Round 1981, Jeffrey and Carpenter 1974).

Diatoms are the best known of all the unicellular planktonic algae, due to the exquisite architecture of their silica walls, visible in the light microscope, on which classical diatom taxonomy is based. More than 10,000 diatom taxa have been described, although the number of living species is probably closer to 5,000 (Jeffrey and Vesk 1997). Diatoms (5 to 200 μm) occur either as single cells or as chain-like colonies, and their overall symmetry has been used to separate two orders: the Centrales (centric diatoms), and the Pennales (pennate diatoms). Centric diatoms are usually found in the water column, whereas pennate diatoms, which have a gliding motility, are usually found on surfaces or in sediments.

The most comprehensive survey of diatom pigments used 51 species (71 strains) isolated from a spring diatom bloom (Stauber and Jeffrey 1988). TLC and RP-TLC pigment methods showed simple patterns across 13 families - Chls a, c_1 and c_2, fucoxanthin, diadinoxanthin and β,β-carotene. Chl c_1 was replaced by Chl c_3 in a few species. Oceanographers would benefit from a new diatom survey using advanced HPLC techniques, which might show new Chl *c* pigments and fucoxanthin derivatives across diatom taxa (Rodriguez et al. 2001).

The pigment charcteristics of the Bacillariophyta are listed in Table 2.8.

Bolidophyceae

Two new picoplanktonic flagellates, *Bolidomonas pacifica* Guillou et Chrétiennot-Dinet and *B. mediterranea* (Guillou et al. 1999a) have been placed within a recently erected class, the *Bolidophyceae* Guillou et Chrétiennot-Dinet. These two species from the eukaryotic picoplankton are biflagellates, approximately 1.2 μm in diameter, with no cell walls or external siliceous structures, and a cellular ultrastructure related to the Heterokonta. Analysis of the SSU rDNA gene place the two strains as a sister group to the diatoms. Pigment analyses support this relationship, with Chls a, c_1, c_2, c_3, fucoxanthin and diadinoxanthin being the major components.

Table 2.9 lists the pigment characteristsics of the Bolidophyceae.

Chrysophyta

Marine members of the Chrysophyta are the well-known silicoflagellates (Dictyophyceae), the picoplanktonic pelagophytes (Pelagophyceae; Andersen et al. 1993), the nanoplanktonic Parmales (Booth and Marchant 1987) and the Synurophyceae and Chrysophyceae (Andersen et al. 1993).

The silicoflagellates are unique unicells, the spongy cytoplasm of which is contained in a basket-like siliceous skeleton composed of a network of

Table 2.8 Pigment characteristics of diatoms (Bacillariophyta)

Chlorophylls:	**Chlorophylls *a*, c_1, c_2,** and MgDVP; c_3 replaces c_1 in 7 strains of *Thalassionema, Thalassiothrix* and *Nitzschia;* Chl c_2-P. *gyrans*-type is present in three species: *Minutocellus polymorphus* (Hargraves and Guillard) Hasle, Von Stosch & Syvertsen, *Stephanopyxis turris* (Greville) Ralfs in Pritchard and *Nitzschia bilobata* W. Smith (Jeffrey and Zapata, unpubl)
Carotenoids:	**Fucoxanthin, diadinoxanthin, diatoxanthin, β,β-carotene;** 19' – butanoyloxy-fucoxanthin present in one strain of *Thalassiothrix*
Cultures examined:	51 species (71 strains) – Stauber and Jeffrey (1988) Order Centrales: 41 species from 15 genera and 13 families Order Pennales: 29 species from 13 genera and 9 families *Pseudo-nitzschia multiseries* (Hasle) Hasle studied from recent Collections (Rodriguez et al. 2001)
Culture Collections used:	CSIRO Algal Culture Collection, CCMP
Culture color:	Gold, orange or brown, depending on culture age and light intensity used for growth
Pigment methods used:	Cellulose and polyethylene TLC; Merck RP-8 HPTLC; RP-HPLC (refs. below)
Likely endosymbiosis:	Secondary endosymbiosis (Chromophyte radiation) developed from an ancestral red alga and a heterotrophic eukaryotic host
References:	Stauber and Jeffrey (1988), Jeffrey (1997b), Delwiche (1999), Zapata et al. (1998), Rodriguez et al. (2001), Jeffrey and Zapata (unpubl.)
Comments:	Diatoms need further examination by the latest HPLC methods to test for new Chl *c* and fucoxanthin derivatives (e.g. Zapata et al. 2000)

Table 2.9 Pigment characteristics of the bolidophytes (Bolidophyceae)

Chlorophylls:	**Chlorophyll *a*, c_1, c_2, c_3**
Carotenoids:	**Fucoxanthin, diadinoxanthin, diatoxanthin,** β,β-carotene (trace)
Cultures examined:	*Bolidomonas pacifica, B. mediterranea*
Culture Collections used:	Roscoff Culture Collection
Culture color:	Pale brown
Pigment methods used:	HPLC (Vidussi et al. 1996)
Likely endosymbiosis:	Secondary endosymbiosis (chromophyte radiation) developed from unknown ancestral red alga
References:	Guillou et al. (1999a,b)
Comments:	This recently described class of picoplanktonic flagellates is a likely sister group to the diatoms, on the basis of rDNA analyses

tubular elements. They possess a single flagellum, with cell sizes from 20 to 100 μm. Although not dominant in present-day oceans, they form a significant component of the phytoplankton biomass in temperate and polar regions, favouring temperatures below 15 ºC. Their rich fossil record makes them valuable indicators in biostratigraphy and dating of marine sediments.

The pico-planktonic marine chrysophytes (Pelagophyceae) are widespread in the world oceans, first observed in the North Pacific Ocean (Lewin et al. 1977), and later in Norwegian and Australian waters and coastal waters of the USA. They have a unique pigment composition (with Chl c_3 and But-fucoxanthin but not Hex-fucoxanthin; Vesk and Jeffrey 1987).

The nanoplanktonic Parmales (2 to 5 μm spheres) are covered with ornamented siliceous plates, originally found in North Pacific and Antarctic waters, but now known worldwide. Seven species are known, but attempts to culture these cells have so far proved unsuccessful, and nothing is yet known of their lifecycles or division processes. Similarly, their pigment composition is unknown from culture, but analysis of a field sample dominated by Parmales suggests that they probably contain Chls *a*, c_2 and c_3, fucoxanthin, 19'-hexanoyloxyfucoxanthin, 19'-butanoyloxyfucoxanthin, diadinoxanthin, and β,β-carotene (Wright and van den Enden 2000a). If this pigment composition is verified by culture studies the Parmales would present a unique pigment complement to any other class within the Chrysophyta. However, the presence of 19′-hexanoyloxyfucoxanthin in the field sample may indicate a contaminant.

Table 2.10 summarizes the pigment characteristics of an unpublished study of 21 species of cultured chrysophytes from four classes (Jeffrey and Zapata, in preparation). Three pigment types were identified, but pigment

Table 2.10 Pigment characteristics of chrysophytes (Chrysophyta)

Type 1 (Chrysophyceae, Synurophyceae)	**Chlorophylls:**	**Chlorophyll *a*, c_1, Chl c_2-*P. gyrans*-type** (present in 10 of 10 species); Chl c_2 variable; Chl c_3 absent
	Carotenoids:	**Violaxanthin, zeaxanthin, antheraxanthin, fucoxanthin,** and **β,β-carotene** (always present), **Vaucheriaxanthin (9 of 10 species)**
	Cultures examined:	Chrysophyceae (5 spp.): *Chrysochaete, Dinobryon, Hibberdia, Ochromonas, Phaeoplaca spp.* Synurophyceae (5 spp.): *Mallomonas* (2 spp.), *Synura* (2 spp.), *Tessellaria*
	Culture Collections used:	CCMP, Vigo Culture Centre
	Culture color:	Pale-gold to gold
	Pigment methods used:	RP-HPLC (Zapata et al. 2000)
	Likely endosymbiont:	Secondary endosymbiosis between ancestral red alga and a heterotrophic eukaryotic host
	References:	Andersen and Mulkey (1983), Andersen et al. (1999), Jeffrey and Zapata (in preparation)

Contd.

Table 2.10 Contd.

Type 2 (Dictyophyceae)	**Chlorophylls:**	**Chlorophyll *a* and c_2** (always present); Chls c_1 and c_3 variable
	Carotenoids:	**Fucoxanthin, diadinoxanthin, β,β-carotene** (always present); diatoxanthin, zeaxanthin (variable); 19'-butanoyloxyfucoxanthin present in one species.
	Cultures examined:	*Dictyophyceae* (5 spp.): *Apedinella, Dictyocha, Mesopedinella, Pseudopedinella, Rhizochromulina*
	Culture Collections used:	CCMP
	Culture color:	Pale-gold to gold
	Pigment methods used:	RP-HPLC (Zapata et al. 2000)
	Likely endosymbiont:	Secondary endosymbiosis between ancestral red alga and a heterotrophic eukaryotic host
	References:	Andersen et al. 1999, Jeffrey and Zapata (in preparation)
Type 3 (Pelagophyceae)	**Chlorophylls:**	**Chlorophyll *a*, c_2** (always present); Chl c_3 (5 out of 6 species); Chl c_1 variable;
	Carotenoids:	**19'-butanoyloxyfucoxanthin, fucoxanthin, diadinoxanthin, diatoxanthin, zeaxanthin, β,β-carotene** (always present). Gyroxanthin-like and ε,ε-carotene variable
	Cultures examined:	Pelagophyceae (6 species): *Aureococcus, Aureomba, Pelagomonas, Pulvinaria, Sarcinochrysis, Pelagococcus*
	Culture Collections used:	CCMP
	Culture color:	Pale-gold to gold
	Pigment methods used:	RP-HPLC (Zapata et al. 2000, Wright et al. 1991)
	Likely endosymbiont:	Secondary endosymbiosis between ancestral red alga and a heterotrophic eukaryotic host
	References:	Vesk and Jeffrey (1987), Jeffrey and Vesk (1997), Jeffrey and Wright (1997), Andersen et al. (1999), Jeffrey and Zapata (in preparation)
	Comments:	For discussion of the Parmales, see text.

distributions were not always clear-cut within classes, and some variability was encountered. Whether this was due to poor health of some cultures, inaccurate taxonomic identification of some of the strains, or their inherent variability, cannot yet be determined. The pigments encountered were various combinations of Chls a, c_1, c_2, c_2-*P. gyrans*- type, c_3, with combinations of the carotenoids violaxanthin, antheraxanthin, zeaxanthin, fucoxanthin, 19'-butanoyloxyfucoxanthin, diadinoxanthin, and β,β-carotene. Vaucheriaxanthin, gyroxanthin, ε,ε-carotene, and a novel acyloxyfucoxanthin were also tentatively identified in some strains.

Raphidophyta

Raphidophytes are a small group of micro-flagellates (30 to 100 μm), previously regarded as freshwater forms, but becoming increasingly common as blooms in estuarine and coastal regions. Their importance stems from the potential damage they can cause to the aquaculture industry. The definitive pigment study was that of Fiksdahl et al. (1984), who identified fucoxanthin, violaxanthin, diadinoxanthin, zeaxanthin and β,β-carotene in the group, but more recently Mostaert et al. (1998) have found evidence of fucoxanthinol and 19'-butanoyloxyfucoxanthin in some genera. Marshall and Newman (2002) studied xanthophyll cycle protective mechanisms in these organisms under different environmental conditions. Raphidophytes now need to be re-examined by advanced HPLC methods. Table 2.11 lists the pigment characteristics of the Raphidophyta.

Table 2.11 Pigment characteristics of raphidophytes (Raphidophyta)

Chlorophylls:	**Chlorophyll *a*, c_1, c_2**
Carotenoids:	**Fucoxanthin, violaxanthin, diadinoxanthin, zeaxanthin, β,β-carotene;** fucoxanthinol and 19'-butanoyloxyfucoxanthin in some genera
Cultures examined:	Marine genera: *Chattonella, Fibrocapsa, Heterosigma, Haramonas*
Culture Collections used:	CSIRO Algal Culture Collection, University of Tasmania Plant Science Culture Collection, NIES, Tsukuba, Japan
Culture color:	Gold to golden-brown
Pigment methods used:	HPLC; TLC
Likely endosymbiosis:	Secondary endosymbiont derived from an ancestral red algal and a eukaryotic heterotrophic host
References:	Fiksdahl et al.(1984), Mostaert et al. (1998), Marshall and Newman (2002)
Comments:	Raphidophytes need to be examined by advanced HPLC methods for evidence of possible new Chl *c* and fucoxanthin derivatives. On present knowledge raphidophyte pigments resemble those of Chrysophytes Types 1 and 2

Eustigmatophyta

The eustigmatophytes are a small group of coccoid microalgae (2 to 4 μm in diameter), most of which produce characteristic motile cells (zoospores), containing a prominent red eye-spot (stigma) from which the Division gets its name (Hibberd 1990). Although they have been isolated from coastal waters, tide pools and the ocean's surface, their ecological role remains uncertain. Their importance stems from their high nutritive value and ease of culture, which enables them to be highly successful as live feeds in the aquaculture industry (Volkman et al. 1993).

The eustigmatophytes are usually grouped with the Chromophyta (Gentile and Blanch, 2001), although they have a unique suite of photosynthetic pigments not found in any other algal Division. Chl *a* is the only chlorophyll, with no accessory Chls *b* or *c*, and high concentrations of the light-harvesting and photo-protective carotenoid, violaxanthin. The unique carotenoid vaucheriaxanthin and its esters, together with minor amounts of antheraxanthin, zeaxanthin, canthaxanthin and β,β-carotene complete the pigment suite. Table 2.12 summarizes the pigment characteristics of the Eustigmatophyta.

Table 2.12 Pigment characteristics of yellow-green algae (Eustigmatophyta)

Chlorophylls:	**Chlorophyll *a***
Carotenoids:	**Vaucheriaxanthin esters, violaxanthin, antheraxanthin, zeaxanthin,** canthaxanthin (variable), β,β-carotene; secondary carotenoids appear under adverse conditions
Cultures examined:	*Nannochloropsis oculata* (Droop) Hibberd, *N. gaditana* Lubian
Culture Collections used:	CSIRO Algal Culture Collection, CCMP
Culture color:	yellow-green
Pigment methods used:	HPLC (Wright et al. 1991)
Likely endosymbiosis:	Unknown to these authors
References:	Norgård et al.(1974), Bjørnland and Liaaen-Jensen (1989), Volkman et al. (1993), Gentile and Blanch (2001)

Haptophyta

The golden haptophyte microalgae are widely distributed in the nanoplankton of the world's oceans, blooming seasonally at polar, equatorial and subtropical latitudes. The most well known are the coccolithophorids with beautiful calcite scales (Jeffrey and Vesk 1997); least known are the fragile species covered with organic scales, since they are readily destroyed by standard preservatives. Photosynthetic pigment suites have therefore become important markers for detecting field populations of haptophytes (Jeffrey and Wright 1994).

Jeffrey and Wright (1994) distinguished four pigment types within the class using the best HPLC method of the time (Wright et al. 1991). Zapata et al. (2004a) recently distinguished eight pigment types using the advanced HPLC method of Zapata et al. (2000) in 65 strains across seven families. While Chls *a*, c_2 and MgDVP and the carotenoids fucoxanthin, diadinoxanthin and β,β-carotene were found in all strains examined, the eight pigment types were observed in unique distributions of new Chl *c* and fucoxanthin pigments (Tables 2.13, 2.14). Some pigment suites were restricted to haptophyte families, some to particular species. Van Lenning et al. (2003)

found three pigment types in the family Pavlovophyceae. It is hoped that the new HPLC techniques will resolve haptophyte field populations to greater accuracy than was previously possible. The paper by Zapata et al. (2004a) is the most comprehensive survey of an algal Division using the advanced HPLC methods.

Table 2.13 lists the pigment characteristics of the Haptophyta as currently known, and Table 2.14 summarizes the distributions of the eight pigment types across seven haptophyte families.

Table 2.13 Pigment characteristics of haptophytes (Haptophyta)

Chlorophylls:	All species contained **Chlorophylls *a*, c_2**, and MgDVP(trace); other chlorophyll *c* pigments identified were **Chl c_1, Chl c_2-P. *gyrans*-type,** Chl c_2- MGDG [18:4/14:0], non-polar Chl c_1*-like;* Chl c_2 -MGDG [14:0/14:0]; **Chl c_3;** MV-Chl c_3; For chlorophyll *c* distributions across species see Table 14. No phytylated chl *c* derivative was found (however see Nelson and Wakeham 1989)
Carotenoids:	All species contained **fucoxanthin, diadinoxanthin and β,β-carotene.** Other carotenoids identified were 4-keto-fucoxanthin, **19'-hexanoyloxyfucoxanthin, 4-keto-19'-hexanoyloxyfucoxanthin, 19'-butanoyloxyfucoxanthin.** For carotenoid distributions across species see Table 14.
Cultures examined:	37 species (65 strains) from seven families and two classes (Zapata et al. 2004a); Pavlovophyceae (7 strains) (Van Lenning et al. 2003)
Culture Collections used:	CCMP, CSIRO Algal Culture Collection, Australian Antarctic Division Culture Collection, Algobank, University of Caen, France
Culture color:	Pale-gold to deep orange, depending on culture age and light intensity for growth
Pigment methods used:	RP-HPLC (Garrido and Zapata 1997, Zapata et al. 2000, 2004a), mass spectrometry (Garrido et al. 2000)
Likely endosymbiosis:	Secondary endosymbiosis of an unknown red algal ancestor and a heterotrophic eukaryotic host
References:	Fawley (1989), Jeffrey and Wright (1994), Garrido et al. (2000), Palmer (2003), Van Lenning et al. (2003), Zapata et al. (2004a)

Dinophyta

Dinoflagellates are a unique group of unicellular flagellates (5 to 200 μm, Taylor 1987) comprising approximately 130 genera and 1200 living species. They are widely distributed in the world's oceans, forming specific species assemblages in tropical, sub-tropical and temperate environments. Inshore toxic dinoflagellate blooms are increasing worldwide and are of immense economic importance (Hallegraeff 1993).

Only about half the extant dinoflagellates are photosynthetic with plastids (surrounded by three membranes), containing Chls *a*, c_2, and the carotenoid

Table 2.14 Distribution of chlorophyll *c* pigments and fucoxanthin derivatives in eight pigment types across haptophyte families. n: no of species (strains) examined; tr: trace. Identification of pigments (code 1 to 14), see below, and abbreviations section.

Pigment type	Type species (examples)		*Chlorophyll C* pigments									Fucoxanthin derivatives				
Haptophyte family		n	1	2	3	4	5	6	7	8	9	10	11	12	13	14
Type 1		6 (7)														
Pavlovophyceae	*Pavlova lutheri* (Droop) Green															
Isochrysidaceae	*Chrysotila lamellosa* Anand. Emend. Green & Parke					•	•	•					•			
Pleurochrysidaceae	*Pleurochrysis roscoffensis* Chadfaud & Feldman															
Type 2		5(7)														
Pavlovophyceae	*Pavlova gyrans* Butcher		•			•	•	•					•			
Type 3		5(7)														
Isochrysidaceae	*Isochrysis galbana* Parke					•	•	•	•				•			
Type 4		4(6)														
Prymnesiaceae	*Prymnesium parvum* Carter			•		•	•	•	•	•			•			
Type 5		3(3)														
Hymenomonadaceae	*Ochrosphaera neopolitana* Schussnig			•		•	•	•	•				•			•
Type 6		2(12)														
Noëlaerhabdaceae	*Emiliana huxleyi* (Lohman) Hay & Mohler			•	•	•	•		•			tr	•	•	•	
Type 7		7(8)														
Prymnesiaceae	*Chrysochromulina polylepis* Manton & Parke			•		•	•		•		•	tr	•	•	•	
Type 8		5(15)														
Phaeocystaceae	*Phaeocystis antarctica* Karsten															
Prymnesiaceae	*Imantonia rotunda* Reynolds			•		•	•		•			•	•	•	•	
Isochrysidaceae	*Dicrateria inornata* Parke															

Pigment code - (Chlorophyll *c* pigments 1-9; fucoxanthin derivatives 10-14): 1. Chl c_2-*P. gyrans* type, **2.** Chl c_3, **3.** Monovinyl Chl c_3, **4.** MgDVP, **5**. Chl c_2, **6**. Chl c_1, **7.** Chl c_2-MGDG [18:4/14:0], **8.** Non-polar-Chl c_1–like, **9.** Chl c_2-MGDG [14:0/14:0], **10**. 19'-Butanoyloxyfucoxanthin, **11**. Fucoxanthin, **12**. 4-Keto-19'-hexanoyloxyfucoxanthin, **13**. 19'-Hexanoyloxyfucoxanthin, **14.** 4-Keto-fucoxanthin.

peridinin. These plastids originated early in evolutionary history from an ancestral red alga by a process of secondary endosymbiosis (Delwiche 1999, Ishida and Green 2002). The non-photosynthetic dinoflagellates have a wide range of nutritional strategies (Schnepf and Elbrächter 1992), and some novel modes of food acquisition have been detected.

Dinoflagellates clearly have an exceptional ability for forming symbiotic associations, and the history of their uptake, loss and possible replacement is an active area of study (Saldarriaga et al. 2001). These authors have shown that at least eight independent plastid losses have occurred in the evolution of dinoflagellates, and at least three or four instances of plastid replacement.

Types of symbioses known include

- ingesting prey species and 'enslaving' the devoured plastid for a permanent or semi-permanent association (e.g. haptophyte, cryptomonad, diatom and prasinophyte plastids within dinoflagellates; see Tables 2.16 to 2.19);
- capturing cyanophytes (e.g. *Synechocystis* spp.and *Synechococcus* spp.) and harboring them in chambers within the cells of heterotrophic tropical Dinophysiales (Gordon et al.1994); and
- forming extensive symbioses within tropical invertebrate hosts. The best known are the zooxanthellae (*Symbiodinium* spp.), which are widespread within corals, clams, radiolarians, acantharians and foraminiferans of coral reefs. These algal 'gardens' within tropical animals contribute significantly to the metabolic survival of their hosts (Trench et al. 1981).

The pigment characteristics of the Dinophyta are shown in Tables 2.15 to 2.19. Table 2.15 lists the pigment characteristics of the peridinin-containing dinoflagellates; Table 2.16 shows the characteristics of haptophyte-containing dinoflagellates, with 19'-hexanoyoxyfucoanthin as a major pigment; Table 2.17 gives the characteristics of diatom-containing dinoflagellates, with fucoxanthin as the major pigment; Table 2.18 shows the characteristics of the cryptomonad - containing dinoflagellates with alloxanthin and biliproteins; and Table 2.19 lists the characteristics of prasinophyte-containing dinoflagellates, with Chl *b* as the major diagnostic pigment.

Summary of Pigment Distributions

Advances in HPLC pigment technology and the availability of many more cultures for examination has expanded the number of pigment types identified within microalgal Divisions (Tables 2.2- 2.19). Table 2.20 presents

Table 2.15 Pigment characteristics of Type 1 dinoflagellates (Dinophyta) with peridinin

Chlorophylls:	**Chlorophylls *a*, c_2**, MgDVP
Carotenoids:	**Peridinin, diadinoxanthin, diatoxanthin,** dinoxanthin, peridininol, pyrroxanthin, P-457 (7',8'-dihydroneoxanthin-20'-al-3'-β-lactoside), β,β-carotene
Cultures examined:	21 species (22 strains) from 7 families (Jeffrey et al. 1975); *Gyrodinium aureolum* Hulbert
Culture Collections used:	Scripps Institution of Oceanography Dinoflagellate collection (Professor F.T. Haxo), CSIRO Algal Culture Collection, CCMP, IEO Vigo, Spain
Culture color:	Orange-red to red-brown
Pigment methods used:	TLC (cellulose; sucrose); RP-HPLC (Wright et al. 1991)
Likely endosymbiosis:	Secondary endosymbioses developed from an ancestral red algal and a eukaryotic heterotrophic host
References:	Jeffrey et al. (1975), Wright et al. (1991), Delwiche (1999), Zapata et al. (1998), Hansen et al. (2000)
Comments:	Dinoflagellates need further examination by latest HPLC methods (e.g. Zapata et al. 2000) to detect possible new chlorophyll *c* derivatives

Table 2.16 Pigment characteristics of Type 2 dinoflagellates (Dinophyta) with 19'-hexanoyloxyfucoxanthin

Chlorophylls:	**Chlorophylls *a*, c_2, c_3**, MgDVP; traces of c_1 and Chl c_2-MGDG [14:0/14:0] present in *Gymnodinium breve*
Carotenoids:	**19'-hexanoyloxyfucoxanthin, fucoxanthin, diadinoxanthin,** 19'-butanoyloxyfucoxanthin, diatoxanthin, **gyroxanthin diester**, β,β-carotene, β,ε-carotene. No peridinin, C_{37}-xanthophylls or the glycoside P457 detected
Cultures examined:	*Gymnodinium galatheanum* (Braarud) Taylor = *Karlodinium micrum* (Leadbeater & Dodge) Larsen, *Gymnodinium breve* Davis = *Karenia brevis* (Davis) Hansen and Moestrup, *Gymnodinium mikimotoi* Miyake et Kominami ex Oda
Culture Collections used:	CCMP; Instituto Espanol de Oceanografia, Vigo, Spain; original isolates from Bjørnland and Tangen (1979)
Culture color:	Golden brown
Pigment methods used:	TLC; RP-HPLC (Zapata et al. 2000)
Likely endosymbiosis:	Tertiary endosymbiosis of a haptophyte established within a dinoflagellate host (Tengs et al. 2000)
References:	Bjørnland and Tangen (1979), Bjørnland et al. (1987), Bjørnland and Liaaen-Jensen (1989), Bjørnland (1990), Zapata et al. (1998), Tengs et al. (2000), Bjørnland et al. (2000), Hansen et al. (2000), de Salas et al. (2003), Jeffrey and Zapata (unpubl.)

Table 2.17 Pigment characteristics of Type 3 dinoflagellates (Dinophyta) with fucoxanthin (no acyloxyfucoxanthins)

Chlorophylls:	**Chlorophylls** ***a***, c_1, c_2
Carotenoids:	**Fucoxanthin, diadinoxanthin, diatoxanthin, β,β-carotene;** no acyloxy fucoxanthins detected
Cultures examined:	*Kryptoperidinium foliaceum* (Stein) Lindemann (previously known as *Peridinium foliaceum*), *Peridinium balticum* (Lev.) Lemm
Culture Collections used:	Scripps Institution of Oceanography Dinoflagellate Collection; UTEX; CSIRO Algal Culture Collection
Culture color:	Gold
Pigment methods used:	TLC (sucrose; polyethylene); RP-HPLC (Zapata et al. 2000)
Likely endosymbiosis:	Tertiary endosymbiosis of a diatom within a dinoflagellate host; supported by ultrastructural studies (Tomas and Cox 1973, Jeffrey and Vesk 1976) and molecular phylogeny (Chesnick et al. 1997). However, see comments below.
References:	Jeffrey et al. (1975), Withers et al. (1977), Chesnick et al. (1997), Jeffrey and Zapata (unpub)
Comments:	Recent evidence Yoon et al. (2002), Morden and Sherwood (2002) suggests that dinoflagellate plastids containing Chls c_1, c_2 and fucoxanthin are ancestral to dinoflagellates with peridinin.

Table 2.18 Pigment characteristics of Type 4 dinoflagellates (Dinophyta) with phycobilins and/or alloxanthin

Chlorophylls:	**Chlorophyll** *a; c_2* present in *Dinophysis norvegica* Ehrenberg
Carotenoids:	**Alloxanthin** (*D. norvegica)*
Phycobilins	Cryptomonad-type phycoerythrin (red), Cryptomonad-type phycocyanin (blue-green)
Cultures examined:	*Gymnodinium acidotum* Nygaard*; Amphidinium wigrense.* No cultures available of *Dinophysis* spp.
Dinoflagellates from coastal blooms:	*Dinophysis* spp. – *Dinophysis norvegica, D. caudata* Saville-Kent*, D. fortii* Pavillard*, D. acuminata* Claparède et Lachmann
Species color	Red-brown to colorless (*Dinophysis* spp.*);* blue-green to colorless (*G. acidotum; A. wigrense)*
Pigment methods used:	Absorbance and fluorescence emission spectra of individual cells of *Dinophysis* spp. frozen before analysis (Hewes et al. 1998); Gold beads coated with phycoerythrin antibodies applied to sections of thylakoids of *Dinophysis* spp. (Vesk et al.1996); HPLC (Kraay et al. 1992)
Likely endosymbiosis:	Tertiary endosymbiosis of a cryptomonad in various stages of establishment within a dinoflagellate host
References:	Wilcox and Wedemeyer (1984), Hallegraeff and Lucas (1988), Schnepf and Elbrächter (1988, 1992), Lucas and Vesk (1990), Fields and Rhodes (1991), Vesk et al. (1996), Hewes et al.(1998), Meyer-Harms and Pollehne (1998), Hackett et al. (2003)
Comments:	Inspiration for culturing *Dinophysis* spp. required (see Hackett et al. (2003) for clues that *Dinophysis* may be feeding on red algae)

Table 2.19 Pigment characteristics of Type 5 dinoflagellates (Dinophyta) with chlorophyll *b*

Chlorophylls:	**Chlorophylls *a, b;*** no Chl *c* detected
Carotenoids:	β,β-carotene, neoxanthin, violaxanthin, and zeaxanthin. Prasinoxanthin was found in *Gymnodinium chlorophorum* (W.W.C. Gieskes, personal communication to Elbrächter and Schnepf 1996). However, prasinoxanthin was not detected in a *G. chlorophorum* field sample which had an unknown major carotenoid (P. Henriksen, personal communication to present authors). No peridinin or fucoxanthin detected
Cultures examined:	*Lepidodinium viride* Watanabe et al., *Gymnodinium chlorophorum* Elbrächter and Schnepf (1996)
Culture Collections used:	Oceanography Laboratory, NIES, Tsukuba, Japan; Sammlung von Algal Kulturen, Göttingen, Germany
Culture color:	Green
Pigment methods used:	Paper chromatography (Watanabe et al. 1987); HPLC (Dr. W.W.C. Gieskes, personal communication to Elbrächter and Schnepf 1996; details not published); P. Henriksen (personal communication to authors)
Likely endosymbiosis:	Secondary endosymbiosis of prasinophyte (green algae) within a dinoflagellate host (supported by ultrastructural and genetic analysis).
References:	Watanabe et al. (1987, 1990), Watanabe and Sasa (1991), Elbrächter and Schnepf (1996)
Comments:	Needs re-examination with latest HPLC methods, if only to determine the full pigment complement of the prasinophyte endosymbiont

our latest overview of the subject: the distribution of major and taxonomically significant pigments across microalgal Divisions/Classes. Only those pigments securely identified have been included; pigments 'expected' but not reported, such as β,ε-carotene in Type 4 dinoflagellates, are obviously not included. These pigment patterns are based on examination of significantly more species than were available in the 1997 compilation, but will surely change as more species are cultured and examined by the best methods, particularly in the little-known pico-eukaryote group (Potter et al. 1997).

PIGMENT INSIGHTS FROM NATURAL PHYTOPLANKTON POPULATIONS

Early Understandings

The first extensive records of phytoplankton biomass in the sea were from research cruises in the 1950s and 1960s. Data were based on a simple spectrophotometric method for Chls *a, b* and *c* (Richards and Thompson 1952), which was applied to summer and winter sections/depth profiles across large areas of the Indian and Pacific Oceans (Humphrey 1966, 1970).

Table 2.20 Distribution of major and taxonomically significant pigments across microalgal Divisions/Classes. ◆ = present; ◇ = significant, but not present in all species tested; t = trace. For pigment code, see Abbreviations.

	Algal Division/Class																														
	Cyanobacterial radiation					Green lineage					Red Lineage																				
Pigments	Cyanophyta		Prochlorophyta			Chlorophyta	Prasinophyta			Euglenophyta	Rhodophyta	Cryptophyta	Bacillariophyta	Bolidophyta	Chrysophyta			Raphidophyta	Eustigmatophyta	Haptophyta								Dinophyta				
Pigment types	1	2	3	4	5	1	1	2	3	1	1	1	1	1	1	2	3	1	1	1	2	3	4	5	6	7	8	1	2	3	4	5
Chlorophylls																																
Chl *a*	◆	◆	◆		◆	◆	◆	◆	◆	◆	◆	◆	◆	◆	◆	◆	◆	◆	◆	◆	◆	◆	◆	◆	◆	◆	◆	◆	◆	◆	◆	◆
DV-Chl *a*				◆																												
Chl *b*			◆			◆	◆	◆	◆	◆																						◆
DV-Chl *b*				◆																												
Chl *d*					◆																											
Chl *c* series																																
MgDVP			◆	◆			◆	◆	◆			◇	◇				t			t	t	t	t	t	t	t	t	◆	◆			
Chl c_1												◇	◆	◆	◆	◇	◇	◆		◆	◆	◆	◆	◆					t	◆		
Chl c_2												◆	◆	◆	◇	◆	◆	◆		◆	◆	◆	◆	◆	◆	◆	◆	◆	◆	◆	◆	
Chl c_3													◇	◆			◆						◆	◆	◆	◆	◆		◆			
Chl c_2 *P. gyrans* type													◇		◆						◆											
Chl c_2 MGDG [18/14]																						◆	◆	◆	◆	◆	◆					
Chl c_2 MGDG [14/14]																										◆			◇			
np Chl c_1 - like																							◆									
MV-Chl c_3																									◆							

Table 2.20 Contd.	1	2	3	4	5	1	1	2	3	1	1	1	1	1	1	2	3	1	1	1	2	3	4	5	6	7	8	1	2	3	4	5
Unknown Chl *c*-like					◆																											
Chl *c* cs-170									◇																							
Biliproteins																																
Allophycocyanin	◆	◆			t						◆																					
Phycocyanobilin	◆	◆			t						◆	◆																			◆	
Phycoerythrobilin	◆	◆		t							◆	◆																			◆	
Carotenes																																
Former IUPAC terminology																																
α β,ε				◆	◆	◆	◇	◇		◆		◆																	◆			
β β,β	◆	◆	◆			◆	◆	◆	◆	◆	◆		◆	t	◆	◆	◆	◆	◆	◆	◆	◆	◆	◆	◆	◆	◆	◆	◆	◆		◆
γ β,ψ						◇																										
ε ε,ε									◇			◆					◇															
Lycopene ψ,ψ												◆																				
Xanthophylls																																
Alloxanthin												◆																			◆	
Antheraxanthin						◆	◆	◆	◆						◆				◆													
Astaxanthin						◇																										
But-fucoxanthin													◇			◇	◆	◇							t	t	◆		◆			
Canthaxanthin	◇					◇													◇													
Cryptoxanthin			◆																													
Crocoxanthin												◆																				
Diadinoxanthin										◆			◆	◆		◆	◆	◆		◆	◆	◆	◆	◆	◆	◆	◆	◆	◆	◆		
Diatoxanthin										◆			◆	◆		◇	◆			◆	◆	◆	◆	◆	◆	◆	◆	◆	◆	◆		

Table 2.20 Contd.	1	2	3	4	5	1	1	2	3	1	1	1	1	1	1	2	3	1	1	1	2	3	4	5	6	7	8	1	2	3	4	5
Dinoxanthin																												◆				
Echinenone	◇		t																													
Fucoxanthin													◆	◆	◆	◆	◆	◆		◆	◆	◆	◆	◆	◆	◆	◆		◆	◆		
4-keto-fucoxanthin																								◆								
Gyroxanthin diester																	◇												◆			
Hex-fucoxanthin																									◆	◆	◆		◆			
4-keto-Hex-fucox																									◆	◆	◆					
Loroxanthin						◇																										
Lutein						◆	◆		t																							◇
Micromonal									◆																							
Monadoxanthin												◆																				
9'-*cis* neoxanthin						◆	◆	◆	◆	◆																						◆
P-457 + P-468																												◆				
Peridinin																												◆				
Peridininol																												◆				
Prasinoxanthin									◆																							◇
Pyrrhoxanthin																												◆				
Siphonaxanthin esters								◆																								
6-Hydroxy siphonein								◆																								
Uriolide									◆																							
Vaucheriax. esters															◆				◆													
Violaxanthin						◆	◆	◆	◆						◆			◆	◆													◆
Zeaxanthin	◆	◆	◆	◆	◆	◆	◆	◆	◆		◆				◆	◆	◆	◆	◆													◆

Although the methods over-estimated both Chls *a* and *c* because early extinction coefficients were too low (discussed by Jeffrey and Welschmeyer 1997, Humphrey and Jeffrey 1997), the program resulted in a comprehensive atlas of data from the Indian Ocean giving both chlorophyll and productivity profiles (Krey and Babenerd 1976).

TLC and early HPLC methods were used for field samples in the 1970s and 1980s and several new pigment insights from the sea were gained. These included:

- Chlorophyte pigments (Chl *b*) in the oligotrophic central North Pacific Ocean (Jeffrey 1976b), now in retrospect thought probably due to DV-Chl *b* from the then unknown *Prochlorococcus* populations;
- Alloxanthin in HPLC samples from the North Sea which probably originated from microscopically unrecognizable cryptomonads, first highlighted by Gieskes and Kraay (1983a);
- Unknown Chl *a* derivatives found in North Sea and tropical Atlantic phytoplankton HPLC samples by Gieskes and Kraay (1983b), probably the first separation of DV-Chl *a* from then unknown *Prochlorococcus marinus*;
- A new Chl *c*-like pigment (probably Chl c_3) separated by Gieskes and Kraay (1986) by HPLC from a North Sea bloom of the haptophyte *Corymbellus aureus*. All these observations foreshadowed later advances.

The subsequent sophistication of pigment separation techniques that came in the 1980s and 1990s, allowed far more accurate estimates of Chl *a* phytoplankton 'biomass', both from satellites, *in situ* fluorescence detection, and HPLC analysis.

Wider perceptions

Remote sensing of ocean colour by sensors on aircraft or satellites (e.g. NASA's SeaWiFS) has become a powerful tool for estimating surface algal chlorophyll concentrations in the world oceans from space. Spectral variations in light leaving the sea surface (differential absorption and backscatter of irradiance), corrected for atmospheric distortion, are detected by satellite-mounted spectroradiometers at wavelengths appropriate to the detection of phytoplankton pigments *in vivo* (Sathyendranath 1986). Remote sensing of Chl *a* now provides detailed images of surface phytoplankton distributions from all oceans on a daily basis. As an example, Fig. 2.4 shows exquisite patterns of surface Chl *a* concentrations around south-eastern

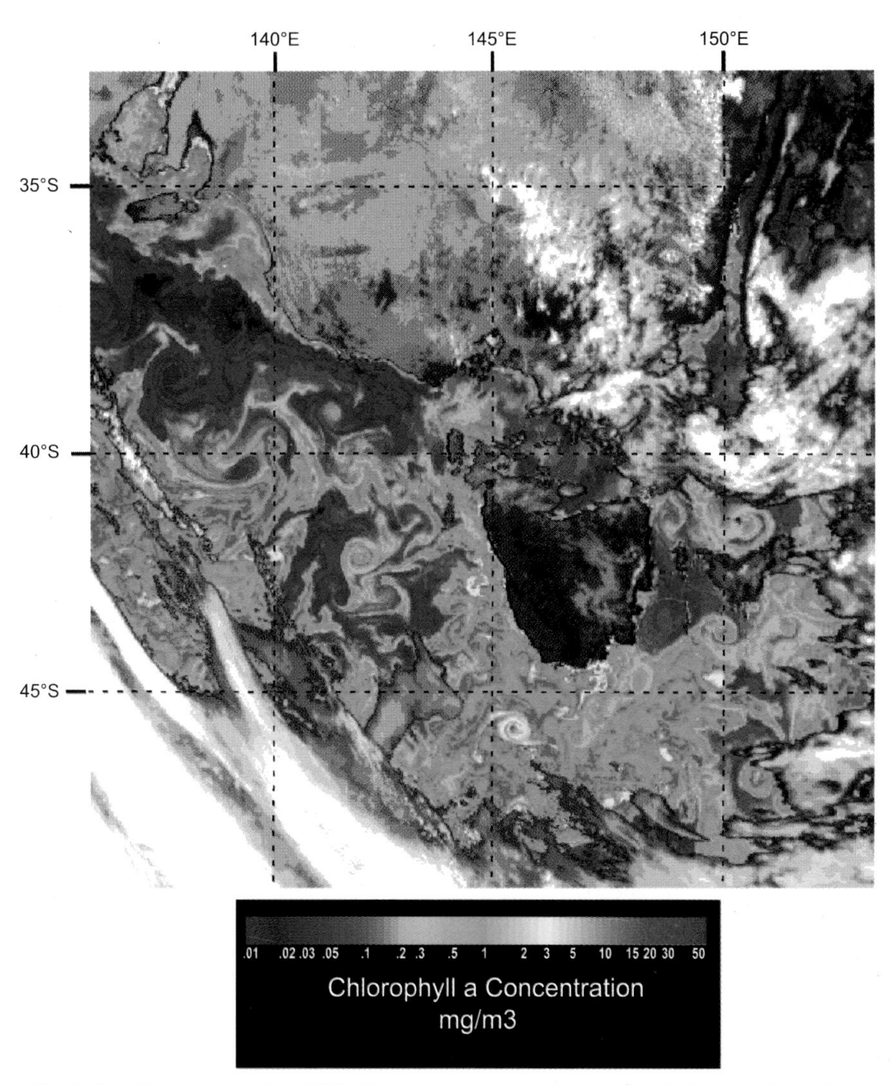

Fig. 2.4 Single scene, SeaWiFS false color image on 6 January 2003 of the South-East corner of Australia with Tasmania the large island in the center of the image (between 40°- 45° S and 145°- 150°E). The surface chlorophyll concentration is clearly shown and reveals mesoscale oceanographic features including upwelling along the NSW coast south from Sydney, and near Cape Jaffa (37 °S, 140 °E), plus many eddy features in the low-nutrient Subtropical Convergence region east and west of Tasmania, interacting with the higher nutrient Subantarctic Zone further south. A chlorophyll color scale is shown. This image has been provided by the SeaWiFS project, NASA/Goddard Space Flight Center and ORBIMAGE. We are indebted to Mr. Brian Griffiths for providing this figure.

Australia in the summer of 2003, revealing the response of phytoplankton to mesoscale oceanographic features such as eddies, fronts and coastal upwellings.

New higher resolution satellite sensors (e.g. MODIS with 36 wavelength bands) are now able to improve detection of spectral characteristics of water-leaving radiances from algal populations. Near-surface populations of diatoms and the cyanobacterium *Trichodesmium* can now be discriminated from mixed phytoplankton by ocean colour data, allowing these populations to be mapped even in optically complex waters (Subramaniam et al. 2002, Sathyendranath et al. 2004). Reliable estimates of Chl *a* concentrations from ocean colour data require algorithms that are appropriate to local algal populations (Morel 1997), highlighting the continued need for ground-truthing pigment samples of the study region.

While the satellite sensors can detect only near-surface Chl *a*, ground-truthing of chlorophyll in the water column from ships by fluorescence detection, can find and measure sub-surface chlorophyll maxima. HPLC pigment analysis can also locate the distribution of key accessory pigments or pigment suites that signal discrete populations of algal types in the water column.

Deeper visions

Use of individual pigments as chemotaxonomic markers is a powerful aid to studying field populations of phytoplankton, but only if they are unambiguous markers for a single algal source (such as peridinin or DV-Chl *a*, for Type 1 dinoflagellates or Type 4 prochlorophytes, respectively), or where a pigment is derived largely from a dominant taxon in a given phytoplankton population. This is often the case for zeaxanthin from cyanobacteria, or fucoxanthin from diatoms, when haptophytes and other non-diatom sources are insignificant, as determined by microscopy.

The variability of pigment composition in response to environmental conditions means that reliable quantitative taxonomic analyses of field populations can not simply be derived by applying the results of cultures grown under standard conditions. Environmental factors that strongly influence pigment composition include light intensity, quality and daylength (Wood 1985, Bidigare et al. 1989, Partensky et al. 1993, Johnsen et al. 1994, Moore et al. 1995), nutrient status, notably iron concentration (Goericke and Montoya 1998, van Leeuwe et al. 1998, Schlüter et al. 2000, Henriksen et al. 2002), growth phase, and strain differences (Stolte et al. 2000, Zapata, et al. 2004a). These factors need to be more rigorously tested experimentally with algal cultures under environmental conditions specific to particular aquatic habitats in order to obtain more reliable baseline pigment ratios for application to field studies.

Due to the uncertainty of pigment composition in the field, marker pigment: Chl *a* ratios must be computed from the data by use of simultaneous equations (Letelier et al. 1993) or matrix factorization (CHEMTAX software, Mackey et al. 1996). These techniques are now frequently used for determining pigment ratios in field samples (reviewed by Jeffrey et al. 1999).

For example Fig. 2.5 shows the application of CHEMTAX software to pigment data from a transect crossing the retreating Antarctic ice edge along longitude 150 ºE (Wright and van den Enden 2000b). This transect encompassed a region of complex oceanography that included the Antarctic Slope Front, a large cold-core eddy, the Antarctic Divergence (where relatively warm (+1.8 ºC) upwelling water brings nutrients into the surface layers), and retreating sea ice that produced a shallow layer of meltwater overlying the mixed layer. Fig. 2.5b shows the total Chl *a* concentrations along the transect. Subsequent panels show calculated contributions to Chl *a* by algal groups, apportioned by CHEMTAX on the basis of the concentrations of various marker pigments. CHEMTAX was able to separate the components of the phytoplankton populations and clearly showed that they were differently placed in relation to oceanographic features (Fig. 2.5c-j). The algal group names in Fig. 2.5 refer to pigment categories representing algal Divisions/Classes rather than the actual taxa themselves.

Clonal isolates from blooms

There has always been a concern that cultures from clonal isolates may not be truly representative of 'natural' species in field populations. We discuss here several examples from the literature in which pigment characteristics from cultures were found to match closely those of the blooms from which they were derived.

- Carreto et al. (2001) reported a massive outbreak of phytoplankton (99% *Gymnodinium* sp.) in southern Chile, which caused high mortality to fish and shellfish, and had a concentration of 8 – 9 x 10^6 cells L^{-1} in high density patches. Cells from the bloom were cultured, and using the advanced HPLC methods of Garrido and Zapata (1997) and Zapata et al. (2000), full suites of pigments, typical of Type 2 dinoflagellates, were found. Major pigments included 19′-hexanoyloxyfucoxanthin, fucoxanthin, 19′-butanoyloxyfucoxanthin, Chls *a*, c_2, c_3, Chl c_2 – MGDG (14:0/14:0) and gyroxanthin diester. The match between cultures from this bloom and other icthyotoxic

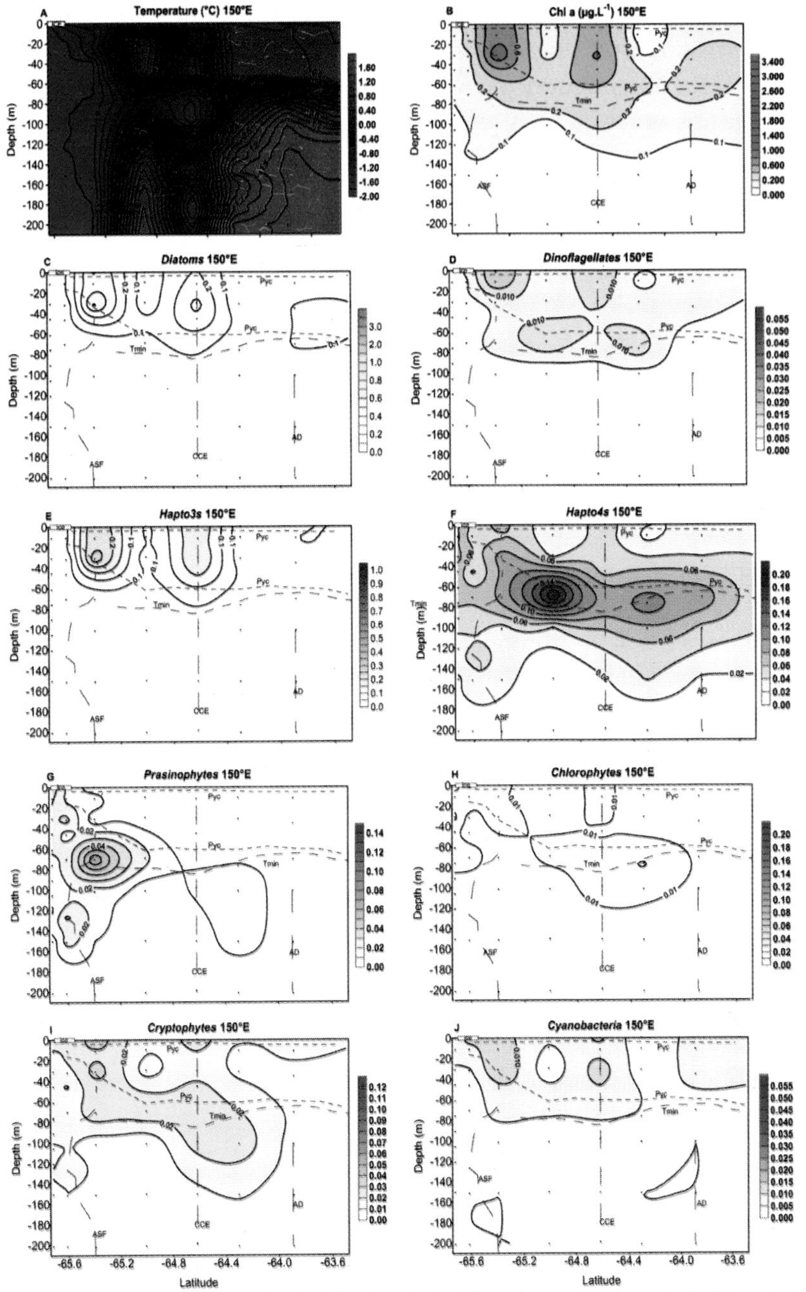

Fig. 2.5 Oceanographic section (20-23 March, 1996) and calculated phytoplankton stocks for the upper 200 m of the water column near the retreating Antarctic ice edge along longitude

Contd.

Gymnodinium species was excellent, thanks to the enhanced selectivity of the HPLC methods used.

- Virtually monoalgal populations of *Prochlorococcus* strains (up to 97% of the autotrophic biomass) were found at depth (80 to 140 m) in low-light, suboxic environments of the Arabian Sea and the eastern tropical North Pacific (Goericke et al. 2000). The carotenoid suites of these field populations were similar to those of other cultured *Prochlorococcus* strains, except for high concentrations of a dihydro-derivative of zeaxanthin, suggesting acclimation to low irradiances. It is suggested that *Prochlorococcus* is uniquely adapted to exploit this low-light, low-oxygen niche.
- *Prochlorococcus* is found at a range of depths over which light intensities can vary by up to four orders of magnitude. Moore et al. (1998) hypothesized that coexistence of genetically different populations adapted for growth at high- and low-light intensities might explain this wide adaptability. Multiple *Prochlorococcus* populations from two locations in the North Atlantic Ocean, sorted by flow cytometry into high-and low- fluorescence cells, were brought into culture, and compared with similar cultures from the Gulf Stream and the Sargasso Sea. Following phylogenetic diversity studies and pigment responses to light climate in these *Prochlorococcus* strains, it was concluded that multiple *Prochlorococcus* ecotypes could occur in a particular environment (possibly a general microbial characteristic), allowing greater survival to environmental stress than could be achieved by physiologically and genetically homogeneous populations (Moore et al. 1998).
- Blooms of diatoms in the East Australian Current off Sydney in 1980 allowed the isolation of over 50 diatom species in culture (Dr J. Stauber). These showed almost universal simple diatom pigment complements, with Chl a, c_1, c_2, fucoxanthin, diadinoxanthin and β,β-carotene (Stauber and Jeffrey 1988; and Table 2.8). These pigment

Fig. 2.5 Contd.

150 °E. Plots show A: Temperature (°C), B: Chl *a* ($\mu g.L^{-1}$). Panels C to J show contributions to the total Chl *a* by various algal taxa, calculated by CHEMTAX software as described in the text. C: Diatoms, D: Dinoflagellates; E: Haptophytes-3; F: Haptophytes-4; G: Prasinophytes; H: Chlorophytes; I: Cryptophytes; J: Cyanobacteria. Haptophyte types (3 & 4) were as defined by Jeffrey and Wright (1994), now Type 6 (cocco lithophorids) and Type 8 (Phaeocystis spp.), respectively, in Table 2.14. Sea ice is shown as a small rectangle at top left of each panel. Other oceanographic features include the Antarctic Slope Front (ASF), a cold-core eddy (CCE), Antarctic Divergence (AD), two pycnoclines (pyc), and the temperature minimum layer (Tmin). Dots indicate sample locations (from Wright and van den Enden 2000b).

suites matched those found in spring diatom bloom field samples in the same location (Hallegraeff 1981).

These examples comparing the pigment composition of cultures and the blooms from which they were isolated, give confidence that the physiology of cells from both environments are not dissimilar. A closer match will be gained when cultures can be grown under identical environmental conditions to those experienced by field populations – a challenge for the future!

CONCLUSIONS

Developments in genetic technology and formulation of the endosymbiotic theory have given a framework within which to understand the diverse array of pigments found in phytoplankton. These developments highlight the complexity of distributions, particularly in the dinoflagellates.

As more cultures are isolated and analysed, new pigments are being identified, giving increased power to discriminate taxa in mixed phytoplankton populations. Convenient sources of such pigments are required. Twelve algal cultures were recommended by SCOR Working Group 78 as sources of pure pigments (Table 2.1). Several additions should now be made to that list to include new pigments of taxonomic value found by the advanced HPLC methods. Possible candidates include: *Prochlorococcus marinus* (DV-Chl *a*, DV-Chl *b*), *Karenia brevis* (gyroxanthin diester), *Pyramimonas amylifera* (siphonaxanthin esters, 6-hydroxysiphonein), *Nannochloropsis oculata* (vaucheriaxanthin esters), *Chrysochromulina polylepis* (Chl c_2 MGDG [14:0/14:0]), and *Ochrosphaera neopolitana* (4-keto-fucoxanthin). Clonal designations and ease of culture of appropriate strains should be identified.

While pigment data from cultured algal species has already given great insights into phytoplankton populations in the sea, much more information is required to facilitate quantitative interpretation of oceanographic pigment data. In particular:

- The pigment composition of some classes remain unclear or inconsistent (e.g. Chrysophyta), a problem that needs isolation and genetic characterization of further strains.
- Most analyses of pigment composition in cultures have used standard conditions of light, nutrients etc. so that genetic differences between strains can be distinguished (Zapata et al. 2004a). These measurements must be extended to real underwater conditions, to determine how the accessory pigment : Chl *a* ratios respond to varying environments.

- More cultures need to be isolated and characterized, particularly the eukaryotic pico-plankton (López-García et al. 2001, Moon-van der Staay et al. 2001). For instance, Diez et al. (2001) found 255 clones of eukaryotic pico-plankton by DNA analysis, of which 53 were abundant but unknown and uncultured stramenopiles. Other well-known species, e.g *Dinophysis* spp. and members of the Parmales, have not yet been successfully cultured.
- More model marine chromophyte algae are needed for photosynthetic and light-harvesting research – "organisms that could be the 'spinach' of the algal light-harvesting world" (Macpherson and Hiller 2003). The present volume offers the terrestrial plant scientist further opportunities to select model organisms from our rich resource of cultured microalgae.

Clearly, cultures represent the key to understanding phytoplankton in the sea. There is a need to "re-connect" them with their phytoplankton cousins under conditions closer to those they would normally experience in their natural habitat.

ACKNOWLEDGEMENTS

We thank Mr. Brian Griffiths, CSIRO Division of Marine Research for helpful comments on the text and for provision of Fig 2.4, Dr. Peter Henriksen for unpublished pigment data, Ms. Louise Bell for graphics assistance, Ms. Denise Schilling for excellent and cheerful word-processing talent, and Dr. Subba Rao V. Durvasula for kind patience and encouragement. S.W.J. especially thanks Dr. Tony Haymet, Chief of the CSIRO Division of Marine Research for continued support.

Dedication

This chapter is dedicated to our friend and colleague, Professor Dr Fauzi Mantoura, Director of the International Atomic Energy Agency Marine Environment Laboratory, Monaco, with whom we had our first great adventure in defining, testing and writing the 1997 UNESCO monograph "Phytoplankton Pigments in Oceanography: Guidelines to Modern Methods".

Abbreviations

But-fucoxanthin	19′-butanoyloxyfucoxanthin
CCAP	The Culture Collection of Algae and Protozoa, UK

CCMP	Provasoli-Guillard Centre for Culture of Marine Phytoplankton
Chl	Chlorophyll
Chl c_2-MGDG [18:4/14:0]	Chlorophyll c_2-monogalactosyl diacylglyceride ester. [18:4/14:0] denotes the chain lengths (18, 14) and number of double bonds (4, 0) of the two esterified fatty acids, respectively
Chl c_2-*P. gyrans* type	Chlorophyll c_2 *Pavlova gyrans* type
Chl $c_{CS\text{-}170}$	Chlorophyll *c* from *Micromonas pusilla*, CS-170
DV-Chl	Divinyl chlorophyll
DV-Pchlide	Divinyl protochlorophyllide (synonym for MgDVP)
Hex-fucoxanthin	19′-hexanoyloxyfucoxanthin
HPLC	High performance liquid chromatography
HPTLC	High performance thin-layer chromatography
MgDVP	Magnesium divinyl pheoporphyrin a_5 mono-methyl ester
MODIS	Moderate Resolution Imaging Spectrometer
MV-Chl c_3	Monovinyl Chlorophyll c_3
NIES	National Institute of Environmental Studies, Tsubuka, Japan
np Chl c_1-like	Non polar Chlorophyll c_1-like
RP	Reverse phase
SCOR	Scientific Committee on Oceanographic Research
SSU rDNA	Small subunit ribosomal DNA
TLC	Thin-layer chromatography
UTEX	University of Texas Algal Culture Collection
Vaucheriax. esters	Vaucheriaxanthin esters

REFERENCES

Andersen, R.A. 2003. A world list of algal culture collections. pp. 753-766. *In* G.M. Hallegraeff, D.M. Anderson, and A.D. Cembella [eds.]. Manual on Harmful Marine Microalgae. UNESCO Monographs on Oceanographic Methodology, Vol. 11. UNESCO, Paris, France.

Andersen, R.A. and T.J. Mulkey. 1983. The occurrence of chlorophyll c_1 and c_2 in the Chrysophyceae. J. Phycol. 19: 289-294.

Andersen, R.A., G.W. Saunders, M.P. Paskind and J.P. Sexton. 1993. Ultrastructure and 18S RNA gene sequence for *Pelagomonas calceolata* gen. et sp. nov. and the description of a new algal class, the Pelagophyceae classis nov. J. Phycol. 29: 701-715.

Andersen, R.A., Y. Van de Peer, D. Potter, J.P. Sexton, M. Kawachi and T. LaJeunesse. 1999. Phylogenetic analysis of the SSUrRNA from members of the Chrysophyceae. Protist 150: 71-84.

Bhattacharya, D. 1997. Origins of algae and their plastids. Springer-Verlag, Vienna, Austria.

Bidigare, R.R., O. Schofield and B.B. Prezelin. 1989. Influence of zeaxanthin on quantum yield of photosynthesis of *Synechococcus* clone WH7803 (DC2). Mar. Ecol. Prog. Ser. 56: 177-188.

Bjørnland, T. 1982. Chlorophylls and carotenoids of the marine alga *Eutreptiella gymnastica*. Phytochem. 21: 1715.

Bjørnland, T. 1990. Carotenoid structures and lower plant phylogeny. pp. 21-37. *In* M.M. Mathews-Roth and R.F. Taylor. [eds.]. Carotenoids: Chemistry and Biology. Plenum Press, New York, USA.

Bjørnland, T. 1997a. Structural relationships between algal carotenoids. pp. 572-577. *In* S.W. Jeffrey, R.F.C. Mantoura and S.W. Wright. [eds.]. Phytoplankton Pigments in Oceanography: Guidelines to Modern Methods, UNESCO monographs on oceanographic methodology, Vol 10. UNESCO, Paris, France.

Bjørnland, T. 1997b. UV-vis spectroscopy of carotenoids. pp. 578-594. *In* S.W. Jeffrey, R.F.C. Mantoura and S.W. Wright. [eds.]. Phytoplankton Pigments in Oceanography: Guidelines to Modern Methods, UNESCO monographs on oceanographic methodology, Vol 10. UNESCO, Paris, France.

Bjørnland, T. and S. Liaaen-Jensen. 1989. Distribution patterns of carotenoids in relation to chromophyte phylogeny and systematics. pp. 37-60. *In* J.C. Green, B.S.C. Leadbeater and W.L. Diver [eds.] The Chromophyte Algae: Problems and Perspectives, Clarendon Press, Oxford, UK.

Bjørnland, T. and K. Tangen. 1979. Pigmentation and morphology of a marine *Gyrodinium* (Dinophyceae) with a major carotenoid different from peridinin and fucoxanthin. J. Phycol. 15: 457-463.

Bjørnland, T., A. Fiksdahl, T. Skjetne, J. Krane and S. Liaaen-Jensen. 1987. Gyroxanthin diester – an acetylenic allenic carotenoid. 8th International Symposium on Carotenoids, 21-31 July 1987. Boston, USA.

Bjørnland, T., A. Fiksdahl, T. Skjetne, J. Krane and S. Liaaen-Jensen. 2000. Gyroxanthin – the first allenic acetylenic carotenoid. Tetrahedron 56: 9047-9056.

Booth, B.C. and H.J. Marchant. 1987. Parmales, a new order of marine chrysophytes, with descriptions of three new genera and seven new species. J. Phycol. 23: 245-260.

Brand, L.E. 1986. Nutrition and culture of autotrophic ultraplankton and picoplankton. *In* T. Platt and W.E. Li. [eds.] Photosynthetic Picoplankton. Canad. Bull. Fish. Aquat. Sci., 214: 205-233.

Brown, J.S. 1985. Three photosynthetic antenna porphyrins in a primitive green alga. Biochim. Biophys. Acta 807: 143-146.

Bryant, D.A. 1992. Puzzles of chloroplast ancestry. Current Biology 2: 240-242.

Burger-Wiersma, T., M. Veenhuis, H.J. Korthals, C.C.M. Van de Wiel, and L.R. Muir. 1986. A new prokaryote containing chlorophylls *a* and *b*. Nature 320: 262-263.

Carreto, J.L., M. Seguel, N.G. Montoya, A. Clément and M.J. Carignan. 2001. Pigment profile of the ichthyotoxic dinoflagellate *Gymnodinium* sp. from a massive bloom in southern Chile. J. Plankton Res. 23: 1171-1175.

Carpenter, E.J., J.M. O'Neil, R. Dawson, D.G. Capone, P.J.A. Siddiqui, T. Roenneberg and B. Bergman. 1993. The tropical diazotrophic phytoplankter *Trichodesmium:* biological characteristics of two common species. Mar. Ecol. Progr. Ser. 95: 295-304.

Cavalier-Smith, T. 2002. Chloroplast evolution: secondary symbiogenesis and multiple losses. Curr. Biol. 12: R62-64.

Chesnick, J.M., W.H.C.F. Kooistra, U. Wellbrock and L. Medlin. 1997. Ribosomal RNA analysis indicates a benthic pennate diatom ancestry for the endosymbionts of the dinoflagellates *Peridinium foliaceum* and *Peridinium balticum* (Pyrrophyta). J. Eukaryot. Microbiol. 44: 314-320.

Chisholm, S.W., R.J. Olson, E.R. Zettler, R. Goericke, J.B. Waterbury and N.A. Welschmeyer. 1988. A novel free-living prochlorophyte abundant in the oceanic euphotic zone. Nature 334: 340-343.

Clay, B.L., P. Kugrens and R.E. Lee. 1999. A revised classification of the Cryptophyta. Bot. J. Linn. Soc. 131: 131-151.

Courties, C., A. Vaquer, M. Troussellier, J. Lautler, M.J. Chrétiennot, J. Neveux, C. Machado and H. Claustre. 1994. Smallest eukaryotic organism. Nature 370: 255.

Delwiche, C.F. 1999. Tracing the thread of plastid diversity through the tapestry of life. Amer. Nat. 154: S164-S177.

De Salas, M.F., C.J.S. Bolch, L. Botes, G. Nash, S.W. Wright and G.M. Hallegraeff. 2003. *Takayama* gen.nov. (Gymnodiniales, Dinophyceae), a new genus of unarmored dinoflagellates with sigmoid apical grooves, including the description of two new species. J. Phycol. 39: 1233-1246.

Díez, B., C. Pedrós-Alió and R. Massana. 2001. Study of genetic diversity of eukaryotic picoplankton in different oceanic regions by small-subunit rRNA gene cloning and sequencing. Appl. Environ. Microbiol. 67: 2932–2941

Egeland, E.S. and S. Liaaen-Jensen. 1995. Ten minor carotenoids from Prasinophyceae (Chlorophyta). Phytochem 40: 515-520.

Egeland, E.S., W. Eikrem, J. Throndsen, C. Wilhelm, M. Zapata and S. Liaaen-Jensen. 1995. Carotenoids from further prasinophytes. Biochem. Syst. Ecol. 23: 747-755.

Egeland, E.S., R.R.L. Guillard and S. Liaaen-Jensen. 1997. Additional carotenoid prototype representatives and a general chemosystematic evaluation of carotenoids in Prasinophyceae (Chlorophyta). Phytochem. 44: 1087-1097.

Elbrächter, M. and E. Schnepf. 1996. *Gymnodinium chlorophorum,* a new green bloom-forming dinoflagellate (Gymnodiniales, Dinophyceae) with a vestigial prasinophyte endosymbiont. Phycologia 35: 381-393.

Fawley, M.W. 1989. A new form of Chlorophyll *c* involved in light-harvesting. Plant Physiol. 91: 727-732.

Fawley, M.W. 1991. Disjunct distribution of the xanthophyll loroxanthin in the green algae. J. Phycol. 27: 544-548.

Fawley, M.W., Y. Yun and M. Qin. 2000. Phylogenetic analyses of 18S rDNA sequences reveal a new coccoid lineage of the Prasinophyceae (Chlorophyta). J. Phycol. 36: 387-393.

Fields, S.D. and R.G. Rhodes. 1991. Ingestion and retention of *Chroomonas* spp. (Cryptophyceae) by *Gymnodinium acidotum* (Dinophyceae). J. Phycol. 27: 525-529.

Fiksdahl, A., N. Withers, R.R.L. Guillard and S. Liaaen-Jensen. 1984. Carotenoids in the Raphidophyceae – a chemosystematic contribution. Comp. Biochem. Physiol. 78B: 265-271.

Fookes, C.J.R. and S.W. Jeffrey. 1989. The structure of chlorophyll c_3, a novel marine photosynthetic pigment. J. Chem. Soc. Chem. Comm. 23: 1827-1828.

Foss, P., R.A. Lewin and S. Liaaen-Jensen. 1987. The carotenoids of *Prochloron* sp. (Prochlorophyta). Phycologia 26: 142-144.

Garrido, J.L. and M. Zapata. 1997. Reversed-phase high-performance liquid chromatography of mono- and divinyl-chlorophyll forms using pyridine-containing mobile phases and polymeric octadecylisilica. Chromatographia 44: 43-49.

Garrido, J.L. and M. Zapata. 2004 Chlorophyll analysis by new high performance liquid chromatography methods. *In* B. Grimm, R.J. Porra, W. Rüdiger and H. Scheer. [eds.]. Chlorophylls and Bacteriochlorophylls: Biochemistry, Biophysics and Biological Function. Kluwer Academic Publishers, the Netherlands (in press).

Garrido, J.L., J. Otero, M.A. Maestro and M. Zapata. 2000. The main non-polar chlorophyll *c* from *Emiliania huxleyi* (Prymnesiophyceae) is a chlorophyll c_2-monogalactosyldiacylglyceride ester: a mass spectrometry study. J. Phycol. 36: 497-505.

Gentile, M.P. and H.W. Blanch. 2001. Physiology and xanthophyll cycle activity of *Nannochloropsis gaditana.* Biotechnol. Bioeng. 75: 1-12.

Gieskes, W.W.C. and G.W. Kraay. 1983a. Dominance of Cryptophyceae during the phytoplankton spring bloom in the central North Sea detected by HPLC analysis of pigments. Mar. Biol. 75: 179-185.

Gieskes, W.W.C. and G.W. Kraay. 1983b. Unknown chlorophyll *a* derivatives in the North Sea and the tropical Atlantic Ocean revealed by HPLC analysis. Limnol. Oceanogr. 28: 757-766.

Gieskes, W.W.C. and G.W. Kraay. 1986. Analysis of phytoplankton pigments by HPLC before, during and after mass occurrence of the microflagellate *Corymbellus aureus* during the spring bloom in the open northern North Sea in 1983. Mar. Biol. 92: 45-52.

Goericke, R. and J.P. Montoya. 1998. Estimating the contribution of microalgal taxa to chlorophyll *a* in the field – variations of pigment ratios under nutrient- and light-limited growth. Mar. Ecol. Prog. Ser. 169: 97-112.

Goericke, R. and D.J. Repeta. 1992. The pigments of *Prochlorococcus marinus:* the presence of divinyl chlorophyll *a* and *b* in a marine prokaryote. Limnol. Oceanogr. 37: 425-433.

Goericke, R. and D.J. Repeta. 1993. Chlorophylls *a* and *b* and divinyl chlorophylls *a* and *b* in the open sub-tropical North Atlantic Ocean. Mar. Ecol. Progr. Ser. 101: 307-313.

Goericke, R., R.J. Olson and A. Shalapyonok. 2000. A novel niche for *Prochlorococcus* sp. in low light sub-oxic environments in the Arabian Sea and the Eastern Tropical North Pacific. Deep-Sea Res. 47: 1183-1205.

Gordon, N., D.L. Angel, A. Neori, N. Kress and B. Kimor. 1994. Heterotrophic dinoflagellates with symbiotic cyanobacteria and nitrogen limitation in the Gulf of Aqaba. Mar. Ecol. Progr. Ser. 107: 83-88.

Grzebyk, D., O. Schofield, V. Vetriani and P.G. Falkowski. 2003. The Mesozoic radiation of eukaryotic algae: the portable plastid hypothesis. J. Phycol. 39: 259-267.

Guillard, R.R.L. 1975. Culture of phytoplankton for feeding marine invertebrates. pp. 29-60 *In* W.L. Smith and M. Chanley, [eds.]. Culture of Marine Invertebrate Animals. Plenum Press, New York, USA.

Guillou, L., M.-J. Chrétiennot-Dinet, L.K. Medlin, H. Claustre, S. Loiseaux-de Goer and D. Vaulot. 1999a. *Bolidomonas:* a new genus with two species belonging to a new algal class, the Bolidophyceae (Heterokonta). J. Phycol. 35: 368-381.

Guillou, L., S.-Y. Moon-van der Staay, H. Claustre, F. Partensky and D. Vaulot. 1999b. Diversity and abundance of Bolidophyceae (Heterokonta) in two oceanic regions. Appl. Environ. Microbiol. 65: 4528-4536.

Hackett, J.D., L. Maranda, H. Su Yoon and D. Bhattacharya. 2003. Phylogenetic evidence for the cryptophyte origin of the plastid of *Dinophysis* (Dinophysiales, Dinophyceae). J. Phycol. 39: 440-448.

Hager, A. and H. Stransky. 1970a. Das Carotinoidmuster und die Verbreitung des lichtinduzierten Xanthophyllcyclus in verschiedenen Algenklassen. V. Einzelne Vertreter der Cryptophyceae, Euglenophyceae, Bacillariophyceae, Chrysophyceae und Phaeophyceae. Arch. Mikrobiol. 73: 77-89.

Hager, A. and H. Stransky. 1970b. Das Carotinoidmuster und die Verbreitung des lichtinduzierten Xanthophyllcyclus in verschiedenen Algenklassen. III. Grünalgen. Arch. Mikrobiol. 72: 68-83.

Hallegraeff, G.M. 1981. Seasonal study of phytoplankton pigments and species at a coastal station off Sydney: importance of diatoms and the nanoplankton. Mar. Biol. 61: 107-118.

Hallegraeff, G.M. 1993. A review of harmful algal blooms and their apparent global increase. Phycologia 32: 79-99.

Hallegraeff, G.M. and S.W. Jeffrey. 1993. Annually recurrent diatom blooms in spring along the New South Wales coast of Australia. Aust. J. Mar. Freshw. Res. 44: 325-334.

Hallegraeff, G.M. and I.A.N. Lucas. 1988. The marine dinoflagellate genus *Dinophysis* (Dinophyceae): photosynthetic, neritic and non-photosynthetic, ocean species. Phycologia 27: 25-42.

Hansen, G., N. Daugbjerg and P. Henriksen. 2000. Comparative study of *Gymnodinium mikimotoi* and *Gymnodinium aureolum*, comb. nov. (= *Gyrodinium aureolum*) based on morphology, pigment composition, and molecular data. J. Phycol. 36: 394-410.

Harrison, P.J., P.W. Yu, P.A. Thompson, N.M. Price and D.J. Phillips. 1988. Survey of selenium requirements in marine phytoplankton. Mar. Ecol. Prog. Ser. 47: 89-96.

Helfrich, M., A. Ross, G.C. King, A.G. Turner and A.W.D. Larkum. 1999. Identification of [8-vinyl]-protochlorophyllide *a* in phototrophic prokaryotes and algae: chemical and spectroscopic properties. Biochim. Biophys. Acta. 1410: 262-272.

Henriksen, P., B. Riemann, H. Kaas, H.M. Sørensen and H.L. Sørensen. 2002. Effects of nutrient-limitation and irradiance on marine phytoplankton pigments. J. Plankton Res. 24: 835-858.

Hertzberg, S., S. Liaaen-Jensen and H.W. Siegelman. 1971. The carotenoids of blue-green algae. Phytochem 10: 3121-3127.

Hess, W.R., F. Partensky, G.W.M. van der Staay, J. M. Garcia-Fernandez, T. Börner and D. Vaulot. 1996. Coexistence of phycoerythrin and a chlorophyll *a/b* antenna in a marine prokaryote. Proc. Natl. Acad. Sci. USA 93: 11126-11130.

Hewes, C.D., B.G. Mitchell, T.A. Moisan, M. Vernet and F.M.H. Reid. 1998. The phycobilin signatures of chloroplasts from three dinoflagellate species: a microanalytical study of *Dinophysis caudata, D. fortii,* and *D. acuminata* (Dinophysiales, Dinophyceae). J. Phycol. 34: 945-951.

Hibberd, D.J. 1977. Observations on the ultrastructure of the cryptomonad endosymbiont of the red-water ciliate *Mesodinium rubrum*. J. Mar. Biol. Assoc. U.K. 57: 45-61.

Hibberd, D.J. 1990 Phylum Eustigmatophyta. pp. 326-333. *In* L. Margulis, J.O. Corliss, M. Melkonian, and D.J. Chapman [eds.]. Handbook of Protoctista. Jones and Bartlett, Boston, USA.

Hu, Q., H. Miyashita, I. Iwasaki, N. Kurano, S. Miyachi, M. Iwaki, and S. Itoh. 1998. A photosystem 1 reaction centre driven by chlorophyll *d* in oxygenic photosynthesis. Proc. Natl. Acad. Sci. USA. 95: 13319-13323.

Humphrey, G.F. 1966. The concentrations of chlorophylls *a* and *c* in the south-eastern Indian Ocean. Aust. J. Mar. Freshwat. Res. 17: 135-145.

Humphrey, G.F. 1970. The concentration of chlorophylls *a* and *c* in the south-west Pacific Ocean. Aust. J. Mar. Freshwat. Res. 21: 1-10.

Humphrey, G.F. and S.W. Jeffrey. 1997. Tests of accuracy of spectrophotometric equations for the simultaneous determination of chlorophylls *a, b,* c_1 *and* c_2 pp. 616-

621. *In* S.W. Jeffrey, R.F.C. Mantoura and S.W. Wright. [eds.]. Phytoplankton Pigments in Oceanography: Guidelines to Modern Methods, UNESCO monographs on oceanographic methodology, Vol 10. UNESCO, Paris, France.

Ishida, K. and B.R. Green. 2002. Second- and third-hand chloroplasts in dinoflagellates: Phylogeny of oxygen-evolving enhancer 1 (PsbO) protein reveals replacement of a nuclear-encoded plastid gene by that of a haptophyte tertiary endosymbiont. Proc. Natl. Acad. Sci. USA 99: 9294-9299.

Jameson, I. 2001. Microalgae Collections from internet and other sources. CSIRO Marine Research (unpublished report).

Jeffrey, S.W. 1961. Paper-chromatographic separation of chlorophylls and carotenoids from marine algae. Biochem. J. 80: 336-342.

Jeffrey, S.W. 1969. Properties of two spectrally different components in chlorophyll *c* preparations. Biochim. Biophys. Acta 177: 456-467.

Jeffrey, S.W. 1972. Preparation and some properties of crystalline chlorophyll c_1 and c_2 from marine algae. Biochim. Biophys. Acta 279: 15-33.

Jeffrey, S.W. 1974. Profiles of photosynthetic pigments in the ocean using thin-layer chromatography. Mar. Biol. 26: 101-110.

Jeffrey, S.W. 1976a. The occurrence of chlorophyll c_1 and c_2 in algae. J. Phycol. 12: 349-354.

Jeffrey, S.W. 1976b. A report of green algal pigments in the Central North Pacific Ocean. Mar. Biol. 37: 33-37.

Jeffrey, S.W. 1989. Chlorophyll *c* pigments and their distribution in the chromophyte algae. pp. 13-36. *In* Green, J.C., Leadbeater, B.S.C., and Diver W.L. [eds.]. The Chromophyte Algae: Problems and Perspectives. Clarendon Press, Oxford, UK.

Jeffrey, S.W. 1997a. Structural relationships between algal chlorophylls. pp. 566-571. *In* S.W. Jeffrey, R.F.C. Mantoura and S.W. Wright. [eds.] 1997a. Phytoplankton Pigments in Oceanography: Guidelines to Modern Methods, UNESCO monographs on oceanographic methodology, Vol 10. UNESCO, Paris, France.

Jeffrey, S.W. 1997b. Preparation of chlorophyll standards. pp. 207-238. *In* S.W. Jeffrey, R.F.C. Mantoura and S.W. Wright, [eds.]. Phytoplankton Pigments in Oceanography: Guidelines to Modern Methods. UNESCO Monographs on oceanographic methodology, Vol 10. UNESCO, Paris, France.

Jeffrey, S.W. and S.M. Carpenter. 1974. Seasonal succession of phytoplankton at a coastal station off Sydney. Aust. J. Mar. Freshw. Res. 25: 361-369.

Jeffrey, S.W. and M. Vesk. 1976. Further evidence for a membrane-bound endosymbiont within the dinoflagellate *Peridinium foliaceum*. J. Phycol. 12: 450-455.

Jeffrey, S.W. and G.M. Hallegraeff. 1987. Phytoplankton pigments, species and light climate in a complex warm-core eddy of the East Australian Current. Deep-Sea Res. 34: 649-673.

Jeffrey, S.W. and S.W. Wright. 1987. A new spectrally distinct component in preparations of chlorophyll *c* from the microalga *Emiliania huxleyi* (Prymnesiophyceae). Biochim. Biophys. Acta 894: 180-188.

Jeffrey, S.W. and S.W. Wright. 1994. Photosynthetic pigments in the Haptophyta. pp. 111-132. *In* J.C. Green and B.S.C. Leadbeater. [eds.]. The Haptophyte Algae. Clarendon Press, Oxford, UK.

Jeffrey, S.W. and J.M. LeRoi. 1997. Simple procedures for growing SCOR reference microalgal cultures. pp. 181-205. *In* S.W. Jeffrey, R.F.C. Mantoura, and S.W. Wright. [eds.]. Phytoplankton Pigments in Oceanography: Guidelines to Modern Methods.

UNESCO Monographs on oceanographic methodology, Vol 10. UNESCO, Paris, France.

Jeffrey, S.W. and R.F.C. Mantoura. 1997. Development of pigment methods for oceanography: SCOR-supported Working Groups and objectives. pp. 19-36. *In* S.W. Jeffrey, R.F.C. Mantoura and S.W. Wright. [eds.]. Phytoplankton Pigments in Oceanography: Guidelines to Modern Methods. UNESCO Monographs on oceanographic methodology, Vol 10. UNESCO, Paris, France.

Jeffrey, S.W. and M. Vesk. 1997. Introduction to marine phytoplankton and their pigment signatures. pp. 37-84. *In* S.W. Jeffrey, R.F.C. Mantoura and S.W. Wright. [eds.]. Phytoplankton Pigments in Oceanography: Guidelines to Modern Methods, UNESCO Monographs on Oceanographic Methodology, Vol 10. UNESCO, Paris, France.

Jeffrey, S.W. and N.A. Welschmeyer. 1997. Spectrophotometric and fluorometric equations in common use in oceanography. pp. 597-615. *In* S.W. Jeffrey, R.F.C. Mantoura and S.W. Wright. [eds.]. Phytoplankton Pigments in Oceanography: Guidelines to Modern Methods, UNESCO Monographs on Oceanographic Methodology, Vol 10. UNESCO, Paris, France.

Jeffrey, S.W. and S.W. Wright. 1997. Qualitative and quantitative HPLC analysis of SCOR reference algal cultures. pp. 343-360. *In* S.W. Jeffrey, R.F.C. Mantoura and S.W. Wright. [eds.]. Phytoplankton Pigments in Oceanography: Guidelines to Modern Methods. UNESCO monographs on oceanographic methodology, Vol 10. UNESCO, Paris, France.

Jeffrey, S.W., M. Sielicki, and F.T. Haxo. 1975. Chloroplast pigment patterns in dinoflagellates. J. Phycol. 11: 374-384.

Jeffrey, S.W., R.F.C. Mantoura and S.W. Wright. 1997a. [eds.] Phytoplankton Pigments in Oceanography: Guidelines to Modern Methods, UNESCO Monographs in Oceanographic Methodology, Vol. 10. UNESCO, Paris, France.

Jeffrey, S.W., R.F.C. Mantoura, and T. Bjørnland. 1997b. Data for the identification of 47 key phytoplankton pigments. pp 449-559. *In* S.W. Jeffrey, R.F.C. Mantoura and S.W. Wright. [eds.]. Phytoplankton Pigments in Oceanography : Guidelines to Modern Methods. UNESCO Monographs on oceanographic methodology, Vol 10. UNESCO, Paris, France.

Jeffrey, S.W., S.W. Wright and M. Zapata. 1999. Recent advances in HPLC pigment analysis of phytoplankton. Mar. Freshw. Res. 50: 879-896.

Jensen, A. 1966. Algal carotenoids V. Iso-fucoxanthin - a rearrangement product of fucoxanthin. Acta. Chem. Scand. 20: 1728-1730.

Johnsen, G., O. Samset, L. Granskog and E. Sakshaug. 1994. *In vivo* absorption characteristics in 10 classes of phytoplankton: Taxonomic characteristics and responses to photoadaption by means of discriminant and HPLC analysis. Mar. Ecol. Progr. Ser. 105: 149-157.

Johnson, P.W. and J. McN. Sieburth. 1979. Chroococcoid cyanobacteria in the sea: A ubiquitous and diverse phototrophic biomass. Limnol. Oceanogr. 24: 928-935.

Keller, M.D., R.C. Selvin, W. Claus and R.R.L. Guillard. 1987. Media for the culture of oceanic ultraphytoplankton. J. Phycol. 23: 633-638.

Kraay, G.W., M. Zapata and M.J. Veldhius. 1992. Separation of chlorophylls c_1, c_2 and c_3 of marine phytoplankton by reversed-phase-C18-high-performance liquid chromatography. J. Phycol. 28: 708-712.

Kraft, G.T. 1981. Rhodophyta: Morphology and Classification. pp 6-51. *In* C.S. Lobban and M.J. Wynne [eds.]. *The Biology of Seaweeds.* University of California Press, Berkeley, USA.

Krey, J. and B. Babenerd. 1976. Phytoplankton production: Atlas of the International Indian Ocean Expedition. Univ. Kiel, Germany.

Kopecký, J., M. Riederer and E. Pfündel. 2002. *Porphyridium purpureum* (formerly *P. cruentum*) contains β-carotene but no α-carotene. Algological Studies 104: 189-195.

Larkum, A.W.D., C. Scaramuzzi, G.C. Cox, R.G. Hiller and A.G. Turner. 1994. Light-harvesting chlorophyll *c*-like pigment in *Prochloron*. Proc. Natl. Acad. Sci. USA. 91: 679-683.

Letelier, R.M., R.R. Bidigare, D.V. Hebel, M. Ondrusek, C.D. Winn and D.M. Karl. 1993. Temporal variability of phytoplankton community structure based on pigment analysis. Limnol. Oceanogr. 38: 1420-1437.

Lewin, J., R.E. Norris, S.W. Jeffrey and B. Pearson. 1977. An aberrant chrysophycean alga, *Pelagococcus subviridis* Norris gen. nov., sp. nov., from the North Pacific. J. Phycol. 13: 259-266.

Li, W.K.W. and A.M. Wood. 1988. Vertical distribution of North Atlantic ultraphytoplankton: analysis by flow cytometry and epifluorescence microscopy. Deep-Sea Res. 35: 1615-1638.

Liu, Z., H. Yan, K. Wang, T. Kuang, J. Zhang, L. Gui, X. An and W. Chang. 2004. Crystal structure of spinach major light-harvesting complex at 2.72 Å resolution. Nature 428: 287-292.

López-García, P., F. Rodríguez-Valera, C. Pedrós-Alló and D. Moreira. 2001. Unexpected diversity of small eukaryotes in deep-sea Antarctic plankton. Nature 409: 603-607.

Lucas, I.A.N. and M. Vesk. 1990. The fine structure of two photosynthetic species of *Dinophysis* (Dinophysiales, Dinophyceae). J. Phycol. 26: 345-357.

Mackey, M.D., D.J. Mackey, H.W. Higgins and S.W. Wright. 1996. CHEMTAX- A program for estimating class abundances from chemical markers: application to HPLC measurements of phytoplankton pigments. Mar. Ecol. Progr. Ser. 144: 265-283.

Macpherson, A.N. and R.G. Hiller. 2003. Light-harvesting systems in chlorophyll *c*-containing algae. pp. 323-352. *In* B.R. Green and W.W. Parson [eds.]. Light Harvesting Antennas in Photosynthesis. Advances in Photosynthesis and Respiration, Vol. 13, Kluwer Academic Publishers, The Netherlands.

Mantoura, R.F.C and C.A. Llewellyn. 1983. The rapid determination of algal chlorophyll and carotenoid pigments and their breakdown products in natural waters by reverse-phase high-performance liquid chromatography. Anal. Chim. Acta. 151: 297-314.

Mantoura, R.F.C., R.G. Barlow and E.J.H. Head. 1997. Simple isocratic HPLC methods for chlorophylls and their degradation products. pp. 307-326. *In* S.W. Jeffrey, R.F.C. Mantoura and Wright. [eds.]. Phytoplankton Pigments in Oceanography : Guidelines to Modern Methods. UNESCO Monographs on Oceanographic Methodology, Vol 10. UNESCO, Paris, France.

Marquardt, J., H. Senger, H. Miyashita, S. Miyachi and E. Mörschel. 1997. Isolation and characterization of biliprotein aggregates from *Acaryochloris marina*, a *Prochloron*-like prokaryote containing mainly chlorophyll *d*. FEBS Letters 410: 428-432.

Marshall, J.A. and S. Newman. 2002. Differences in photoprotective pigment production between Japanese and Australian strains of *Chattonella marina* (Raphidophyceae). J. Exp. Mar. Biol. Ecol. 272: 13-27.

McFadden, G.I. 2001. Primary and secondary endosymbiosis and the origin of plastids. J. Phycol. 37: 951-959.

Miyashita, H., H. Ikemoto, N. Kurano, K. Adachi, M. Chihara and S. Miyachi. 1996. Chlorophyll *d* as a major pigment. Nature 383: 402-403.

Miyashita, H., K. Adachi, N. Kurano, H. Ikemoto, M. Chihara and S. Miyachi. 1997. Pigment composition of a novel oxygenic photosynthetic prokaryote containing chlorophyll *d* as the major chlorophyll. Plant Cell Physiol. 38: 274-281.

Miyashita, H., H. Ikemoto, N. Kurano , S. Miyachi and M. Chihara. 2003. *Acaryochloris marina* gen. et. sp. nov. (Cyanobacteria) an oxygenic photosynthetic prokaryote containing Chl *d* as a major pigment. J. Phycol. 39: 1247-1253.

Moon-van der Staay, S., R. De Wachter and D. Vaulot. 2001. Oceanic 18S rDNA sequences from picoplankton reveal unsuspected eukaryotic diversity. Nature 409: 607-610.

Moore, L.R., R. Goericke and S.W. Chisholm. 1995. Comparative physiology of *Synechococcus* and *Prochlorococcus:* influence of light and temperature on growth, pigments, fluorescence and absorptive properties. Mar. Ecol. Prog. Ser. 116: 259-275.

Moore, L.R., G. Rocap and S. W. Chisholm. 1998. Physiology and molecular phylogeny of co-existing *Prochlorococcus* ecotypes. Nature 393: 464-467.

Morden, C.W. and A.R. Sherwood. 2002. Continued evolutionary surprises among dinoflagellates. Proc. Nat. Acad. Sci. USA 99: 11558-11560.

Moreira, D., H. LeGuyader and H. Phillipè. 2000. The origin of red algae and the evolution of chloroplasts. Nature 405: 69-72.

Morel, A. 1997. Consequences of a *Synechococcus* bloom upon the optical properties of oceanic (case 1) waters. Limnol. Oceanogr. 42: 1746-1754.

Mostaert, A.S., U. Karsten, Y. Hara and M.M. Watanabe. 1998. Pigments and fatty acids of marine raphidophytes: a taxonomic re-evaluation. Phycol. Res. 46: 213-220.

Meyer-Harms, B. and F. Pollehne. 1998. Alloxanthin in *Dinophysis norwegica* (Dinophysiales, Dinophyceae) from the Baltic Sea. J. Phycol. 34: 280-285.

Nakayama, T., B. Marin, H.D. Kranz, B. Surek, V.A.R. Huss, I. Inouye and M. Melkonian. 1998. The basal position of scaly green flagellates among the green algae (Chlorophyta) is revealed by analysis of nuclear-encoded SSU rRNA sequences. Protist. 149: 367-380.

Nelson, J.R. and S.G. Wakeham. 1989. A phytol-substituted chlorophyll *c* from *Emiliania huxleyi;* (Prymnesiophyceae). J. Phycol. 25: 761-766.

Norgård, S., W.A. Svec, S. Liaaen-Jensen, A. Jensen and R.R.L. Guillard. 1974. Chloroplast pigments and algal systematics. Biochem. Syst. and Ecol. 2: 3-6.

Olson, R.J., D. Vaulot and S.W. Chisholm. 1985. Marine phytoplankton distributions measured using shipboard flow cytometry. Deep-Sea Res. 32: 1273-1280.

Palenik, B. and R. Haselkorn. 1992. Multiple evolutionary origins of prochlorophytes, the chlorophyll *b*-containing prokaryotes. Nature 355: 265-267.

Palmer, J.D. 2003. The symbiotic birth and spread of plastids: how many times and whodunit? J. Phycol. 39: 4-11.

Partensky, F., N. Hoepffner, W.K.W. Li, O. Ulloa and D. Vaulot. 1993. Photo-acclimation of *Prochlorococcus* sp. (Prochlorophyta) strains isolated from the north Atlantic and the Mediterranean Sea. Plant Physiol. 101: 285-296.

Porra, R.J., E.E. Pfündel and N. Engel. 1997. Metabolism and function of photosynthetic pigments. pp. 85-126. *In* S.W. Jeffrey, R.F.C. Mantoura and S.W. Wright. [eds.]. Phytoplankton Pigments in Oceanography: Guidelines to Modern Methods, UNESCO monographs on oceanographic methodology, Vol 10. UNESCO, Paris, France.

Potter, D., T.C. La Jeunesse, G.W. Saunders and R.A. Andersen. 1997. Convergent evolution masks extensive biodiversity among marine coccoid picoplankton. Biodiv. Conserv. 6: 99-107.

Pringsheim, E.G. 1946. Pure Cultures of Algae. Cambridge University Press, London, UK.

Provasoli, L. 1958. Nutrition and ecology of protozoa and algae. Ann. Rev. Microbiol. 12: 279-308.

Provasoli, L., J.J.A. McLaughlin and M.R. Droop. 1957. The development of artificial media for marine algae. Arch. Mikrobiol. 25: 392-428.

Quigg, A., Z.V. Finkel, A.J. Irwin, Y. Rosenthal, T-Y. Ho, J.R. Reinfelder, O. Schofield, F.M.M. Morel and P.G. Falkowski. 2003. The evolutionary inheritance of elemental stoichiometry in marine phytoplankton. Nature 425: 291-294.

Repeta, D.J. and T. Bjørnland. 1997. Preparation of carotenoid standards. pp. 239-260. *In* S.W. Jeffrey, R.F.C. Mantoura and S.W. Wright. [eds.]. Phytoplankton Pigments in Oceanography: Guidelines to Modern Methods, UNESCO monographs on oceanographic methodology, Vol 10. UNESCO, Paris, France.

Richards, F.A. and T.G. Thompson. 1952. The estimation and characterization of plankton populations by pigment analyses. II. A spectrophotometric method for the estimation of plankton pigments. J. Mar. Res. 11: 156-172.

Ricketts, T.R. 1966. Magnesium 2,4-divinylphaeoporphyrin a_5 monomethyl ester, a protochlorophyll-like pigment present in some unicellular flagellates. Phytochem. 5: 223-229.

Ricketts, T.R. 1970. The pigments of the Prasinophyceae and related organisms. Phytochem. 9: 1835-1842.

Rodriguez, F., Y. Pazos, J. Maneiro, S. Fraga, and M. Zapata. 2001. HPLC pigment composition of phytoplankton populations during the development of *Pseudo-nitzschia* spp. blooms. pp 199-201. *In* G.M. Hallegraeff, S.I. Blackburn, C.J. Bolch and R.J. Lewis, [eds.]. Harmful Algal Blooms 2000. UNESCO Paris, France.

Round, F.E. 1981. The Ecology of Algae, Cambridge Univ. Press, Cambridge, UK.

Rowan, K.S. 1989. Photosynthetic Pigments of Algae. Cambridge University Press, Cambridge, U.K.

Saldarriaga, J.F., F.J.R. Taylor, P.J. Keeling and T. Cavalier-Smith. 2001. Dinoflagellate nuclear SSU rRNA phylogeny suggests multiple plastid losses and replacements. J. Mol. Evol. 53: 204-213.

Sathyendranath, S. 1986. Remote sensing of phytoplankton: a review, with special reference to picoplankton. pp. 561-583. *In* T. Platt and W.K.W. Li [eds.]. Phytosynthetic Picoplankton. Can. Bull. Fish Aquat. Sci.

Sathyendranath, S., L. Watts, E. Devred, T. Platt, C. Caverhill and H. Maass. 2004. Discrimination of diatoms from other phytoplankton using ocean-colour data. Mar. Ecol. Prog. Ser. 272: 59-68.

Scheer, H. [ed.] 1991. *Chlorophylls*. CRC Press, Boca Raton, Florida, USA.

Scheer, H. 2003. The pigments. pp. 29-81. *In* B.R. Green and W.E. Parsons [eds.]. Light-harvesting Antennas in Photosynthesis. Advances in Photosynthesis and Respiration, Vol 13. Kluwer Academic Publishers, the Netherlands.

Schimek, C., I.N. Stadnichuk, R. Knaust and W. Wehrmeyer. 1994. Detection of chlorophyll c_1 and magnesium-2,4-divinylpheoporphyrin a_5 monomethyl ester in cryptophytes. J. Phycol. 30: 621-627.

Schlüter, L., F. Møhlenberg, H. Havskum, and S. Larsen. 2000. The use of phytoplankton pigments for identifying and quantifying phytoplankton groups in coastal areas: testing the influence of light and nutrients on pigment/chlorophyll *a* ratios. Mar. Ecol. Prog. Ser. 192: 49-63.

Schnepf, E. and M. Elbrächter. 1988. Cryptophycean-like double membrane bound chloroplast in the dinoflagellate *Dinophysis* Ehrenb.: evolutionary, phylogenetic and toxicological implications. Bot. Acta. 101: 196-203.

Schnepf, E. and M. Elbrächter. 1992. Nutritional strategies in dinoflagellates. Europ. J. Protistol. 28: 3-24.

Siegelman, H.W. and J.H. Kycia. 1978. Algal biliproteins pp. 71-79. *In* J.A. Hellebust and J.S. Craigie. [eds.]. Handbook of Phycological Methods. Physiological and Biochemical Methods. Cambridge University Press, Cambridge, UK.

Six, C., J.C. Thomas, B. Brahamsha, Y. Lemoine and F. Partensky. 2004. Photophysiology of the marine cyanobacterium *Synechococcus* sp. WH8102, a new model organism. Aquat. Microb. Ecol. 35: 17-29.

Stauber, J.L. and S.W. Jeffrey. 1988. Photosynthetic pigments in fifty-one species of marine diatoms. J. Phycol. 24: 158-172.

Stein, J.R. 1973. Handbook of Phycological Methods: Culture Methods and Growth Measurements, Cambridge University Press, Cambridge, U.K.

Stolte, W., G.W. Kraay, A.A.M. Noordeloos and R. Riegman. 2000. Genetic and physiological variation in pigment composition of *Emiliania huxleyi* (Prymnesiophyceae) and the potential use of its pigment ratios as a quantitative physiological marker. J. Phycol. 36: 529-539.

Strain, H.H. 1958. Chloroplast pigments and chromatographic analysis. 32nd Annual Priestly Lecture, Pennsylvania State University Press, USA.

Strain, H.H., B.T. Cope, G.N. McDonald, W.A. Svec and J.J. Katz. 1971. Chlorophylls c_1 and c_2. Phytochem 10: 1109-1114.

Strain, H.H., W.M. Manning and G. Hardin. 1944. Xanthophylls and carotenes of diatoms, brown algae, dinoflagellates and sea anemones. Biol. Bull. 86: 169-191.

Subramaniam, A., C.W. Brown, R.R. Hood, E.J. Carpenter and D.G. Capone. 2002. Detecting *Trichodesmium* blooms in SeaWiFS imagery. Deep-Sea Res. II 49: 107-121.

Taylor, F.J.R. [ed.] 1987. The biology of dinoflagellates. Blackwell, Oxford, UK.

Tengs, T., O.J. Dahlberg, K. Shalchian-Tabrizi, D. Klaveness, K. Rudi, C.F. Delwiche and K.S. Jakobsen. 2000. Phylogenetic analyses indicate that the 19'-hexanoyloxy-fucoxanthin-containing dinoflagellates have tertiary plastids of haptophyte origin. Mol. Biol. Evol. 17: 718-729.

Tomas, R.N. and E.R. Cox. 1973. Observations on the symbiosis of *Peridinium balticum* and its intracellular alga. I. Ultrastructure. J. Phycol. 9: 304-323.

Trench, R.K., D.S. Wethey and J.W. Porter. 1981. Observations on the symbiosis with zooxanthellae among the Tridacnidae (Mollusca, Bivalvia). Biol. Bull. 161: 180-198.

Tswett, M. 1906. Adsorptionanalyse und chromatographische methode. Anwendung auf die chemie des chlorophylls. Bert. Bot. Ges. 24: 384-393.

Urbach, E., D.L. Robertson and S.W. Chisholm. 1992. Multiple evolutionary origins of prochlorophytes within the cyanobacterial radiation. Nature 355: 267-270.

van Leeuwe, M.A., K.R. Timmermans, H.J. Witte, G.W. Kraay, M.J.W. Veldhuis and H.H.W. de Baar. 1998. Effects of iron stress on chromatic adaptation by natural phytoplankton communities in the Southern Ocean. Mar. Ecol. Prog. Ser. 166: 43-52.

Van Lenning, K., M. Latasa, M. Estrada, A.G. Sáez, L. Medlin, I. Probert, B. Véron and B. Young. 2003. Pigment signatures and phylogenetic relationships of the Pavlovophyceae (Haptophyta). J. Phycol. 39: 379-389.

Vernon, L.P. and G.R. Seely. [eds.] 1966. The Chlorophylls. Academic Press, New York, USA.

Vesk, M. and Jeffrey S.W. 1987. Ultrastructure and pigments of two strains of the picoplanktonic alga *Pelagococcus subviridis* (Chrysophyceae). J. Phycol. 23: 322-336.

Vesk, M., T.P. Dibbayawan and P.A. Vesk. 1996. Immunogold localization of phycoerythrin in chloroplasts of *Dinophysis acuminata* and *D. fortii* (Dinophysiales, Dinophyta). Phycologia 35: 234-238.

Vidussi, F., H. Claustre, J. Bustillos-Guzman, C. Cailliau and J. Marty. 1996. Determination of chlorophylls and carotenoids of marine phytoplankton: separation of chlorophyll *a* from divinyl-chlorophyll *a* and zeaxanthin from lutein. J. Plankton Res., 18: 2377-2382.

Viskari, P.J., C.S. Kinkade and C.L. Colyer. 2002. Determination of phycobiliproteins by capillary electrophoresis with laser-induced fluorescence detection. Electrophoresis 22: 2327-2335.

Volkman, J.K., M.R. Brown, G.A. Dunstan and S.W. Jeffrey. 1993. The biochemical composition of marine microalgae from the class Eustigmatophyceae. J. Phycol. 29: 69-78.

Watanabe, M.M. and T. Sasa. 1991. Major carotenoid composition of an endosymbiont in a green dinoflagellate *Lepidodinium viride*. J. Phycol. 27 (Sup): 75.

Watanabe, M.M., Y. Takeda, T. Sasa, I. Inouye, S. Suda, T. Sawaguchi and M. Chihara. 1987. A green dinoflagellate with chlorophylls *a* and *b*: Morphology, fine structure of the chloroplast and chlorophyll composition. J. Phycol. 23: 382-389.

Watanabe, M.M., S. Suda, I. Inouye, T. Sawaguchi and M. Chihara. 1990. *Lepidodinium viride* gen. et sp. nov. (Gymnodiniales, Dinophyta), a green dinoflagellate with a chlorophyll *a* and *b*-containing endosymbiont. J. Phycol. 26: 741-751.

Waterbury, J.B., S.W. Watson, R.R.L. Guillard and L.E. Brand. 1979. Widespread occurrence of a unicellular marine, planktonic, cyanobacterium. Nature 277: 293-294.

Wilcox, L.W. and G.J. Wedemeyer. 1984. *Gymnodinium acidotum* Nygaard (Pyrrophyta), a dinoflagellate with an endosymbiotic cryptomonad. J. Phycol. 20: 236-242.

Withers, N.W., E.R. Cox, T. Tomas, and F.T. Haxo. 1977. Pigments of the dinoflagellate *Peridinium balticum* and its photosynthetic endosymbiont. J. Phycol. 13: 354-358.

Wood, M.A. 1985. Adaptation of photosynthetic apparatus of marine ultraplankton to natural light fields. Nature 316: 253-255.

Wright, S.W. and S.W. Jeffrey. 1997. High resolution HPLC system for chlorophylls and carotenoids of marine phytoplankton. pp. 327-341. *In* S.W. Jeffrey, R.F.C. Mantoura and S.W. Wright. [eds.]. Phytoplankton Pigments in Oceanography: Guidelines to Modern Methods, UNESCO monographs on oceanographic methodology, Vol 10. UNESCO, Paris, France.

Wright, S.W. and J.D. Shearer. 1984. Rapid extraction and high-performance liquid chromatography of chlorophylls and carotenoids from marine phytoplankton. J. Chromatog. 294: 281-295.

Wright, S.W. and R.L. van den Enden. 2000a. Phytoplankton community structure and stocks in the East Antarctic marginal ice zone (BROKE survey, Jan – Mar 1996) determined by CHEMTAX analysis of HPLC pigment signatures. Deep-Sea Res. 47: 2363-2400.

Wright, S.W. and R.L. van den Enden. 2000b. Phytoplankton distribution and abundance off East Antarctica, as determined by CHEMTAX analysis of HPLC pigment data (BROKE survey, January-March 1996). ANARE Research Note 103, Australian Antarctic Division, Hobart (ISBN 0642.253).

Wright, S.W., S.W. Jeffrey, R.F.C. Mantoura, C.A. Llewellyn, T. Bjørnland, D. Repeta and N.A. Welschmeyer. 1991. Improved HPLC method for the analysis of chlorophylls and carotenoids from marine phytoplankton. Mar. Ecol. Prog. Ser. 77: 183-196.

Wyman, M. 1992. An in vivo method for the estimation of phycoerythrin concentrations in marine cyanobacteria (*Synechococcus* spp.). Limnol. Oceanogr. 37: 1300-1306.

Yoon, H.S., J.D. Hackett and D. Bhattacharya. 2002. A single origin of the peridinin and fucoxanthin containing plastids in dinoflagellates through tertiary endosymbiosis. Proc. Nat. Acad. Sci. USA 99: 11724-11729.

Young, A. and G. Britton [eds.] 1993. Carotenoids in Photosynthesis. Chapman & Hall, London, UK.

Zapata M., F. Rodríguez and J.L. Garrido. 2000. Separation of chlorophylls and carotenoids from marine phytoplankton: a new HPLC method using a reversed phase C_8 column and pyridine-containing mobile phases. Mar. Ecol. Prog. Ser. 195: 29-45.

Zapata, M., A.M. Ayala, J.M. Franco and J.L. Garrido. 1987. Separation of chlorophylls and their degradation products in marine phytoplankton by reversed-phase high-performance liquid chromatography. Chromatographia 23: 26-30.

Zapata, M., J. Freire, and J.L. Garrido. 1998. Pigment composition of several harmful algae as determined by HPLC using pyridine-containing mobile phases and a polymeric octadecylsilica column. pp 304-307. *In* B. Reguera, J. Blanco, M.L. Fernández and T. Wyatt. [eds.]. Harmful Algae. UNESCO, Paris, France.

Zapata, M., S.W. Jeffrey, S.W. Wright, F. Rodríguez, J.L. Garrido and L. Clementston. 2004a. Photosynthetic pigments in 37 species (65 strains) of Haptophyta: implications for oceanography and chemotaxonomy. Mar. Ecol. Prog. Ser. 270: 83-102.

Zapata, M. and J.L. Garrido, and S.W. Jeffrey. (in press). Chlorophyll *c* pigments: current status. *In* B. Grimm, R.J. Porra, W. Rüdiger and H. Scheer [eds.]. Chlorophylls and Bacteriochlorophylls: Biochemistry, Biophysics and Biological Function. Kluwer Academic Publishers. The Netherlands (in press).

Zhekisheva, M., S. Boussiba, I. Khozin-Goldberg, A. Zarka and Z. Cohen. 2002. Accumulation of oleic acid in *Haematococcus pluvialis* (Chlorophyceae) under nitrogen starvation or high light is correlated with that of astaxanthin esters. J. Phycol. 38: 325-331.

Zingone, A., M. Borra, C. Brunet, G. Forlani, W.H.C.F. Kooistra and G. Procaccini. 2002. Phylogenetic position of *Crustomastix stigmatica* sp. nov. and *Dolichomastix tenuilepis* in relation to the Mamiellales (Prasinophyceae, Chlorophyta). J. Phycol. 38: 1024-1039.

3

Phases, Stages and Shifts in the Life Cycles of Marine Phytoplankton

Marina Montresor[1] *and Jane Lewis*[2]

[1]Stazione Zoologica 'A. Dohrn' Villa Comunale 80121 - Naples, Italy
[2]School of Biosciences University of Westminster 115, New Cavendish Street London W1W 6UW United Kingdom

Abstract

Marine phytoplankton includes a wide variety of species, however, information on their life histories is still extremely scanty, and is limited to a few classes. The vast majority of phytoplanktonic species are characterized by complex life cycles that comprise morphologically and physiologically distinct stages. The development of culture media and culturing techniques permits the establishment of clonal strains, and the manipulation of culture conditions allows the different life stages to be produced and the possibility of furthering our understanding of the factors that induce and modulate shifts among them. Phylogenetic lineages have distinct life cycle patterns; nevertheless, they share the occurrence of a sexual cycle and the capability of forming resting stages. Diatoms are characterized by a peculiar division modality that causes the progressive reduction of population cell size; in several species, this sets the dimensional window within which sexual reproduction occurs. Dinoflagellates are an extremely heterogeneous lineage including phototrophic, mixotrophic and heterotrophic taxa. Their wide diversity is mirrored by a great diversity in their life cycles, however we are just beginning to unravel this complexity. Different kinds of resting stages can be produced by phytoplankton species, some capable of withstanding short-term disturbances, others to remain dormant and/or quiescent for years. Laboratory investigations with cultured strains coupled with *in situ* studies provide information on the factors inducing shifts among the different phases and elucidate their role in species ecology.

INTRODUCTION

Phytoplankton (mostly unicellular species) increase their number and consequently their biomass by vegetative division. However, the vast majority of phytoplankton organisms have complex, heteromorphic life histories, including stages that differ in their morphology, size, ultrastructure, ploidy, and physiology. Distinctive stages, which do not reproduce vegetatively but undergo resting periods of variable lengths, are known among groups spreading over the whole phylogenetic diversity, e.g. dinoflagellates, diatoms, haptophytes, raphidophyceans and prasinophyceans. Moreover, single cells, multicellular colonies and/or cells with markedly different size ranges can all be included in the life cycle of the same species. A sexual cycle including syngamy and meiosis is reported for an increasingly large number of species. The sexual phase can be linked to the production of distinctive life stages, such as resting cysts in dinoflagellates, auxospores in diatoms, and stages with heteromorphic coccoliths in haptophytes. Different species have different life history patterns, reflecting adaptations to distinct environmental conditions gained through their evolutionary history, and we are just beginning to unravel the intricate patterns of this diversity. Shifts among life stages, induced by endogenous rhythms or by the interaction with the surrounding chemical-physical and biological environment, can have profound implications for population dynamics, such as bloom development and decay, and species succession (Garcés et al. 2002).

Life cycle observations of microalgae have been made since the very beginning of phytoplankton studies (Pouchet 1883). Such observations were, however, sporadic; early studies relied on the use of field samples and were dependent on frequent sampling in the field, or survival of mixed cultures in the laboratory to follow through life histories. The existence of resting stages and their potential role in explaining different timing of diatom appearance during the annual cycle was postulated by Hensen (1887), who observed the presence of heavily silicified spores inside diatom frustules in phytoplankton samples collected in the Kiel Bight. Some of the first evidence that diatoms have complex life histories, including a sexual cycle, dates back to Klebahn (1896) who described auxospore formation in the freshwater species *Rhopalodia gibba* (Ehrenberg) O. Müller. However, the finding that fossil remnants of dinoflagellates were resting stages produced within non-fossilizable vegetative cells is relatively recent (Evitt 1961). It was during the second half of the past century that germination experiments and studies carried out on cultured species started providing evidence for the link between dinoflagellate motile cells and their non-motile benthic stages, several of which show morphological similarity with fossil stages (e.g. Wall et al. 1967, Fensome et al. 1993).

Isolation and maintenance of unialgal cultures allows direct and repeated observation of life cycle changes to be made until all aspects of the behavior of the organism under study are known and a coherent picture is established. Latterly the advent of ultra-clean, readily available tissue culture technology has greatly improved this endeavor. However, not all microalgae are amenable to culture and the detailed analysis of life cycles requires an unusual degree of dedication from the observer, so after more than a century of effort, relatively few species have been examined in a thorough fashion. Furthermore, our understanding of the role that phytoplankton life histories have in species ecology is dramatically hampered by the limited possibilities of our current technology to carry out extensive observations *in situ*. We can all see when a tree starts producing leaves, flowers, fruits and seeds, and how they are dispersed by animals or wind, how insects transport pollen from one flower to the other. However, the vast majority of the corresponding events in phytoplankton is hidden from our perception and is only captured almost by chance by the standard sampling techniques used in biological oceanography. Manipulation of culture conditions and experimental settings is often needed to obtain shifts among stages or to induce the onset of specific processes. All that can be tricky: experimental conditions lend themselves to testing of a single or a few parameters at a time, oversimplifying the complex interactive networking of factors acting at sea. However, culture studies do allow us to describe the backbone of the multifaceted life histories of phytoplankters vital to our improved understanding of their ecology and evolution.

In this chapter we aim to illustrate our assertion that heterogeneous life cycles are common in a variety of microalgae and are driven by broadly similar mechanisms over a variety of groups. We aim to establish that cultures provide an excellent means of exploring these life cycles and the drivers of life cycle phases, improving our interpretational skills and providing hypotheses and methodologies for testing in field situations. We do not aim to provide an extensive review of all the achievements obtained in life history studies for phytoplankton organisms. Several review papers dealing with different aspects of marine phytoplankton life histories have been published in the last decades and the reader should refer to them to gain a more complete overview (Fryxell 1983, Steidinger and Walker 1984, Pfiester and Anderson 1987, Pfiester 1989, Round et al. 1990, Billard 1994, McQuoid and Hobson 1996, Edlund and Stoermer 1997, Garcés et al. 2002). Rather we seek to illustrate selected aspects of the life cycle of marine phytoplankton organisms, mainly focusing on diatoms and dinoflagellates, with the aim of providing a picture of the diversity and complexity of their life cycles, on the achievements obtained by culture studies and on a selection of the aspects that, in our opinion, would be worthwhile pursuing in the future.

CULTURING TO ELUCIDATE ELEMENTS OF LIFE HISTORIES

When naming a phytoplankton species, we visualize an image that synthesizes the main morphological features of the vegetative cell. Rarely do we think of a composite image constituted of morphologically or dimensionally different stages, although this would be a more accurate picture for many species. Being able to play different games i.e. having a heteromorphic life cycle, can be an advantage and allows a species to spread its genotype into distinct forms, sizes and stages, and thus expand the range of environmental conditions under which it can survive. Only by following a single species (often a single cell) in culture are we able to reconcile the distinct stages of the life cycle into a single picture. This does not only provide basic information on the biology of the different taxa, but also contributes to a better circumscription of species identity, and to achieve further elements for tracing their evolutionary histories. In fact, the phylogenetic framework of phytoplanktonic taxa is still largely based on the morphological features of only one stage in the life cycle, very seldom considering its whole diversity.

Diatoms

Small, Medium, Large, X-large

Diatoms are encased in a rigid box-shaped silica frustule constituted by one larger (the lid) and one smaller (the box) valve. Upon division, the two daughter cells keep the parental valves as the outer larger ones and synthesize the smaller ones. As vegetative reproduction proceeds, this peculiar mechanism of cell division causes a progressive cell size reduction of the population. For most diatom species size enlargement is only possible following sexual reproduction and the formation of the auxospore, within which the maximum-sized cell is reconstituted. However, the notable size range a diatom can experience during the life cycle merits some consideration per se. Cell volume can span over one order of magnitude: from 2900 to 100 μm^3 for *Chaetoceros curvisetus* Cleve, from the larger-sized initial cell to the minimum dimension at which cell death occurs (Furnas

1978). A considerable variation in cell volume also occurs in the pennate diatom *Pseudo-nitzschia delicatissima* (Cleve) Heiden, where larger and intermediate cells are capable of forming chains, whereas the smallest size fraction is present only as single cells (M.M. unpubl). Thus, the same species can experience different morphologies within its vegetative life cycle, presumably with implications for survival success. For example, different predators can graze upon specific size classes, and long colonies formed only in distinct size ranges would deter some of them. Interestingly, also parasites seem to have preferences for distinct size classes, as demonstrated in a number of freshwater species (Holfeld 2000). There are few species for which the whole size range has been determined in culture conditions and we must be cautious in applying this information to the field, as it seems that the smallest sizes recorded in culture may not be found in the natural environment (Paasche 1973). A puzzling exception is represented by *Pseudo-nitzschia galaxiae* Lundholm and Moestrup, a species that shows considerable morphological variability along its wide size range (from 82 to 10 μm). Cells of the smallest size (≤ 20 μm) regularly bloom during the late winter in the Gulf of Naples (Mediterranean Sea) (Cerino et al. 2005). This species also represents a good example of how culture studies can help in linking different morphotypes. Cells in the smallest size range, in fact, would never have been identified as members of the genus *Pseudo-nitzschia* when observed by light microscopy: they are single-celled and can be easily mistaken for the fusiform morphotype of *Phaeodactylum tricornutum* Bohlin. However, observations of size reduction in culture, ultrastructural features and genetic analyses proved them to belong to *Pseudo-nitzschia galaxiae*.

Based on a general allometric rule, larger cells have lower division rates due to the large energy requirements for duplicating cell biomass. An inverse relationship between cell size and division rates was indeed reported for several diatom species in which growth was estimated over different size classes (Findlay 1972, Paasche 1973, Durbin 1977). However, higher growth rates were reported for populations with the larger cell volume in *Thalassiosira weissflogii* (Grunow) Fryxell and Hasle (= *T. fluviatilis* Hustedt), where size distribution within the population was also observed to influence size and timing of division peaks (Chisholm and Costello 1980, Costello and Chisholm 1981).

Sexual Cycle

Diatoms have a diplontic life cycle, where gamete fusion rapidly follows meiosis and the dominant stage is diploid. Centric and pennate diatoms have different reproductive modalities: the first group is characterized by oogamy i.e. motile uniflagellate sperm cells fuse with non motile egg cells (Fig. 3.1a), while the second group produces non-motile, morphologically undifferentiated but functionally distinct, gametes (Round et al. 1990). The diploid zygote produced after syngamy is not surrounded by the silica frustule and enlarges forming the auxospore, within which a large-sized vegetative cell is formed (Figs. 3.1a, 3.2). This life stage should not to be confused with the spore that is a resting stage, generally unrelated to the sexual phase. The sexual cycle in diatoms thus links the meiotic process and consequent genetic recombination to the re-establishment of the maximum size of the population.

Studies on the life cycle of diatoms have been numerous in the first half of the past century, when the intricate conjugation modalities of pennate, mainly freshwater species, were carefully depicted using natural samples and culture material (reviewed in Drebes 1977, Round et al. 1990, Edlund and Stoermer 1997). However, it was only relatively recently that definitive evidence for oogamus reproduction in centric diatoms was provided by culture studies (von Stosch 1950). Until recently, diatoms were considered homothallic (Drebes 1977). This still holds true for centric species, where auxospore formation occurs in cultures established from the isolation of a single cell, but there is increasing evidence that pennate diatoms are basically heterothallic, i.e. conjugation only occurs when compatible strains are mixed. This evidence mainly stems from a number of culture studies carried out on freshwater species (e.g. Mann 1989, Mann et al. 1999, Chepurnov et al. 2002) but a heretothallic life cycle has been also described for species of the marine genus *Pseudo-nitzschia* (Davidovich and Bates 1998). Extremely complex breeding systems exist among diatoms. Monoecious (capable of self-reproduction within a strain), unisexual (of different mating type, they have to meet a cell of complementary mating type to produce the auxospore) and bisexual (they can act as different mating types, both + and -, when crossed with other strains) strains exist, for example, in *Achnanthes longipes* C.A. Agardh. The results of crosses between different reproductive types and the effects of inbreeding have been unraveled in a series of elegant laboratory studies (Chepurnov and Mann 1997, Chepurnov and Mann 1999, Chepurnov 2000).

Vegetative Enlargement

There are exceptions to the general rule that diatoms have to go through sexual reproduction to regain size, but very little is known about the occurrence and relevance of this process. Cells can extrude the cytoplasm

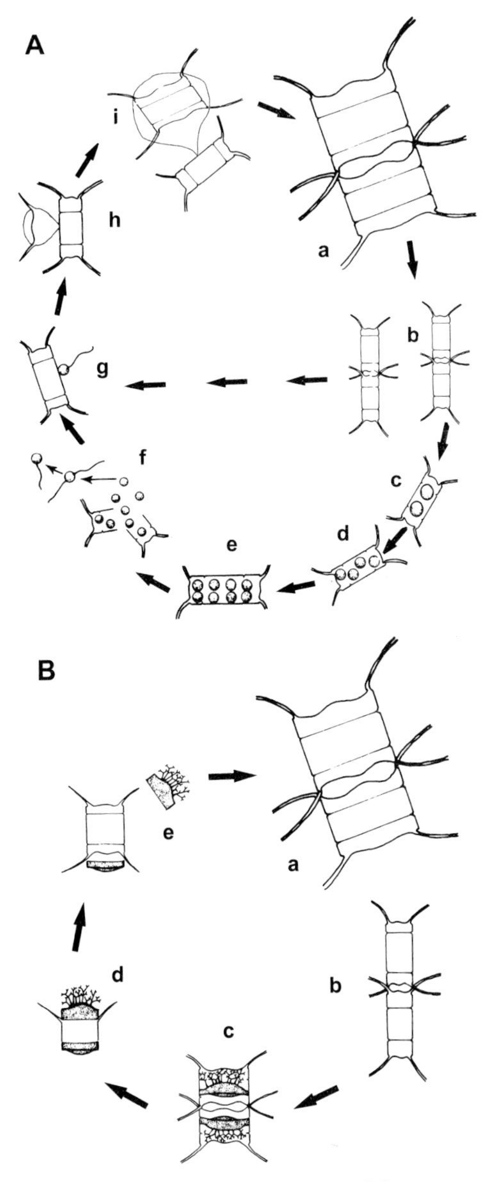

Fig. 3.1 Schematic drawing representing the life cycle of *Chaetoceros diadema* (modified from French III and Hargraves 1985). A) Vegetative cells diminish their diameter (a, b); motile sperm cells are produced within male gametangia (c-f) and fertilize the female gametangium containing the egg cell (g); the initial vegetative cell forms within the auxospore envelope (h, i). B) Spores can be formed within cells of any size (a-c); spore valves separate and setae are formed following germination (d), spore valves cast off (e).

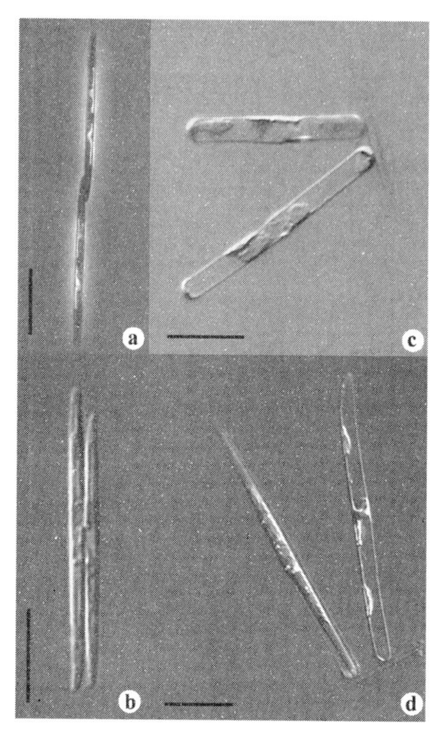

Fig. 3.2 Light micrographs of different life stages of *Pseudo-nitzschia delicatissima.* Chain-forming vegetative cells (a), paired gametangia (b), two early-stage auxospores (c), two mature auxospores, note the almost complete large initial cell inside the left auxospore (d). Scale bars = 20 μm (Courtesy of Alberto Amato, Naples).

from the silica frustule and build an auxospore-like structure within which a larger vegetative cell is built. This process, called 'vegetative cell enlargement' was first described by von Stosch (1965a) as a manipulation tool for maintaining cultures without inducing sexual reproduction. A number of centric diatoms have been shown to undergo this process in culture (*Skeletonema costatum* (Greville) Cleve, Gallagher 1983; *Leptocylindrus danicus* Cleve, French III and Hargraves 1986; *Coscinodiscus wailesii* Gran and Angst, Nagai et al. 1995) but also the pennate diatom *Achnanthes longipes* has been shown to regain size apparently without undergoing a sexual phase (Chepurnov and Mann 1999). Species capable of regaining size bypassing the sexual process could have a selective advantage in avoiding the cost (in terms of energy requirements for producing gametes) and risks (in terms of the possible failure of finding a mate, or failure of the mating process) of sexual reproduction.

Resting Stages

Diatoms, as many other phytoplankters produce resting stages, either spores or resting cells, which have been reported for about 120 marine species (McQuoid and Hobson 1996). Spores are identified by their heavily silicified frustules, whose morphology can be relatively similar or drastically different from that of the corresponding vegetative cells (Hargraves 1976) (Figs 3.1b, 3.3). Unlike the formation of dinoflagellate resting cysts, spore formation in diatoms is not linked to the sexual phase but it is the result of two subsequent division processes in which the modified valves are synthesized (Fig. 3.1b). Resting cells are apparently undifferentiated from the normal vegetative cells and only their ultrastructure reveals reduced cytoplasmic organelles and a high amount of reserve material that characterize all stages deputed to a resting phase (Anderson 1975).

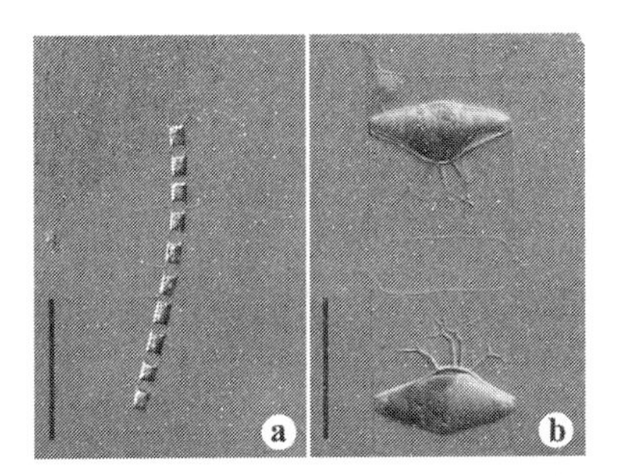

Fig. 3.3 *Chaetoceros diadema*, light micrographs of vegetative cells (a) and resting spores (b). Scale bars = 200 μm (a) and 20 μm (b) (Courtesy of Ugo Sacchi, Naples).

Evidence for the capability to produce resting stages came from laboratory observations showing that cells were able to stop growing when stored in the dark and resume growth when re-exposed to the light (Antia and Cheng 1970). Not only spore-forming diatoms are able to survive prolonged dark conditions, demonstrating that morphologically undifferentiated resting cells could represent a widespread survival strategy among diatoms (Peters 1996, Peters and Thomas 1996, Lewis et al. 1999). Both spores and resting cells are able to retain intact their photosynthetic apparatus, maintain low respiration rates, and are able to withstand prolonged darkness. Laboratory investigations showed higher C:N and C:Chl *a* ratios in resting stages, providing evidence for an enhanced storage of reserve material, low photosynthetic and respiratory activities (Anderson 1976, Hargraves and French 1983, Kuwata et al. 1993). Dark survival seems to be markedly longer for species living at high latitudes, which are able to regain successful growth after four to nine months of storage, as compared to temperate species, whose survival is shorter (Peters 1996, Peters and Thomas 1996). This might represent a species-specific adaptation to the long-lasting adverse conditions for vegetative growth (prolonged darkness) that are found at higher latitudes. A recharging mechanism by which spores would be able to perform short-term photosynthesis, so as to take advantage of occasional re-suspension in the water column and prolong their survival capability, was also suggested (French and Hargraves 1980).

Spores are easily identified in phytoplankton or sediment trap samples and their role in population dynamics has been addressed in different environments, such as coastal seas and upwelling areas. Spores have higher sinking rates and can accumulate on the bottom sediments or at the pycnocline in the water column, from where they are resuspended following upwelling events or the seasonal mixing of the water column (Garrison 1981, Pitcher 1986, Eilertsen and Wyatt 2000). Spores do not have an obligate dormancy, but are quiescent stages that can rapidly resume vegetative growth, thus inoculating a new population in the water column. By contrast, the role of resting cells in population dynamics is largely unknown, with the exception of a few reports from freshwater systems, where resting cells have been recorded in sediments and were able to rejuvenate when exposed to the light under normal oxygen concentrations (Sicko-Goad et al. 1989).

The life cycle of the diatom *Leptocylindrus danicus* appears to be unique since the resting spore is formed within the auxospore, the product of the sexual cycle (French III and Hargraves 1985). The maximum size cell is produced following the germination of the spore and this led to investigations aimed at following the occurrence of sexual events at sea by monitoring size distribution of the natural population (French III and Hargraves 1986). However, only for one out of 70 strains brought into culture, was it possible to observe sexual reproduction and spore formation, whereas in the other strains cell enlargement occurred through vegetative enlargement. Upon EM examination, these asexual strains were shown to lack a sub-central pore in the valve and were designated as *L. danicus* Cleve var. *apora* French III and Hargraves. Laboratory investigations thus showed that apparently identical strains of the same species turned out to have minor morphological differences but substantial differences in their life histories, with considerable implications for population dynamics.

Dinoflagellates

Heteromorphic vegetative stages

The formation of heteromorphic stages during the vegetative cycle is also known for dinoflagellates. One of the distinctive features of some species (e.g. several species of the genus *Alexandrium*, *Gymnodinium catenatum* Graham, *Pyrodinium bahamense* Plate) is the capability of forming chains, where cells are joined in a head-to-tail fashion. Depending on different temperature and salinity conditions, cultures of *Gymnodinium catenatum* are dominated by either single cells or colonies of different lengths, suggesting a notable level of plasticity (Blackburn et al. 1989). Chain formation has been suggested as an adaptation to favor bloom formation by increasing cell motility, thus allowing chain-formers to exploit effectively stratified water columns (Margalef 1998). Sub-populations with different size and growth rates have been recorded in *Karenia mikimotoi* (Miyake and Kominami *ex* Oda) G. Hansen and Moestrup (= *Gymnodinium* cf. *nagasakiense* Takayama and Adachi) (Partensky and Vaulot 1989). Small-sized cells are often interpreted as gametes in the dinoflagellate life cycles, but in *K. mikimotoi* small cells were able to maintain vegetative division at high rates, thus probably gaining a competitive advantage of building up biomass during bloom development. Small cells, previously considered as distinct species, have been recorded also for a number of species of the genus *Dinophysis* both in natural samples and in laboratory incubations (Reguera and González-Gil 2001). These small cells are engulfed by normal-sized cells originating a planozygote with two trailing flagella; it is thus likely that they are a stage in the sexual cycle.

Dinoflagellates can also switch between motile and non-motile stages during their vegetative phase. These non-motile stages, known as temporary or pellicle cysts, have been often reported in cultures and have been interpreted as an adaptation to withstand short-term adverse environmental conditions (Garcés 2002). Cells shed flagella and thecal plates (ecdysis) and appear as featureless globular bodies. *Alexandrium taylorii* Balech produces temporary cysts in culture and in the natural environment, where it shifts from motile stage to temporary cyst over a circadian rhythm. Temporary cysts produced during the night phase revert to the flagellate stage during the light cycle, thus alternatively colonizing water column and surface sediments, possibly enhancing their survival capabilities (Garcés et al. 1998). Temporary cysts are also known for *Lingulodinium polyedrum* (Stein) Dodge (Fig. 3.4) where this process has been reported in early observations

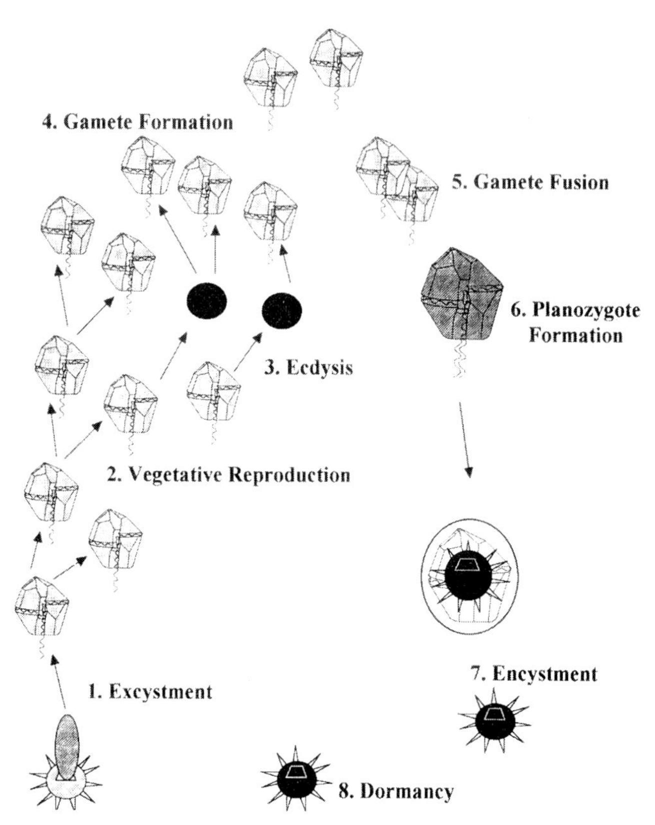

Fig. 3.4 Schematic drawing representing the life cycle of *Lingulodinium polyedrum* (reproduced, with permission, from Lewis and Hallett 1977). 1. Hypnozygote (cyst), in the sediment (diploid condition). When suitable conditions prevail excystment may take place to form a naked cell (planomeiocyte), which subsequently divides to form vegetative cells (haploid condition). Excystment takes around 15 min and thecal formation thereafter takes several h. 2. Vegetative reproduction occurs by binary fission – either by ecdysis or by thecal sharing. 3. Ecdysis may take place in response to environmental cues or as part of cell division. 4. Gamete formation. 5. Gamete fusion, after cells come together, fusion takes some 45 min. 6. Planozygote formation. The planozygote has distinctive 'ski track' flagella and is enlarged with respect to vegetative cells, often with broad intercalary bands. 7. Cyst (hypnozygote) formation takes place within the theca and an outer expanding membrane. Cyst formation is rapid (within 20 min) and is complete with the rupture of the outer membrane. The cyst subsequently falls to the sediment and outer thecal material is lost or decays. 8. Cyst undergoes a mandatory dormancy period when excystment cannot take place, thereafter the cyst remains quiescent until suitable conditions for excystment prevail.

of bloom material (Torrey 1902). In this species ecdysis can be induced by a variety of unfavorable conditions such as cooling, reduced photoperiod, oxygen depletion, pH decrease, and mechanical stress (Lewis and Hallett 1997).

In some species temporary cysts can also act as division cysts, giving rise to a variable number of motile cells upon germination (Montresor 1995, Garcés et al. 1998, Parrow and Burkholder 2003a, b). Indeed, mechanisms of vegetative division are notably diverse among dinoflagellates, with species undergoing binary fission in their planktonic stages, keeping a half or completely discharging the parental cell wall and building new thecae, or forming division cysts from which a variable number of flagellate stages emerge (Elbrächter 2003). Careful studies with culture material are required to sort out the different division strategies. These studies, besides providing additional characters for building phylogenetic relationships among species, will also provide a better characterization of the species lifestyle.

Sexual Cycle

Dinoflagellates have a haplontic life cycle, where meiosis rapidly follows syngamy, and the dominant stage is haploid. The general scheme of the sexual cycle (Fig. 3.4) involves the conjugation of isomorphic or anisomorphic (i.e. of similar or different morphology, respectively) gametes. Contrary to reports for diatoms, gametes are generally indistinguishable from vegetative cells, and gametogenesis may be affected by an internal switch within vegetative cells that does not affect their external morphology (Xiaoping et al. 1989, Probert et al. 2002). In some species gamete formation may be preceded by a reduction division that provides cells with smaller size and reduced pigmentation (von Stosch 1973). This differentiation is remarkable in species of the genus *Ceratium*, where smaller and morphologically different male gametes are engulfed by normal-shaped female gametes (von Stosch 1964, von Stosch 1965b), or in *Pyrophacus* where a small cell with a distinct shape and plate pattern conjugate with a larger cell (Pholpunthin et al. 1999). Once formed, gametes must locate each other. There has been little experimental evidence of possible mechanisms in dinoflagellates and it has been speculated that toxins might play a role acting as pheromones (Wyatt and Jenkinson 1997). Destombe and Cembella (1990) investigated mating type determination and gametic recognition in *Alexandrium* and concluded that efficiency of mating depended on compatibility between clones. However, gametic recognition did not guarantee mating success and it has been observed that gamete fusion is not necessarily followed by formation of viable zygotes (Probert et al. 2002). The product of gamete fusion is a planozygote, a diploid large motile cell, characterized by two longitudinal flagella. In some species, this stage undergoes dramatic morphological and physiological transformations and turns into a resting cyst. Cysts have notably thick and resistant walls (Kokinos et al. 1998) and internally have a high proportion of storage material. The meiotic division that reestablishes the haploid vegetative stage occurs in the planozygote and is evidenced by the pairing of chromosomes within the nucleus, a process called nuclear cyclosis (Pfiester 1989). When the planozygote encysts, meiosis can occur either within the cyst or in the motile stage that emerges after cyst dormancy is over. As a result of culturing experiments, species have been defined as homothallic (self-compatible) or heterothallic (two mating types, usually denoted + and -). Experiments are carried out following incubation of pairs of strains and cyst production is evidence for sexual reproduction and strain compatibility. However, while positive results offer evidence for sexual compatibility, negative ones can be biased by non-optimal experimental conditions for encystment or a long-term culture history, which often impairs encystment success. In such mating experiments Destombe and Cembella (1990) showed that *Alexandrium tamarense* (Lebour) Balech (= *Alexandrium excavatum* (Braarud) Balech and Tangen) was not simply heterothallic and noted that incompatibility between clones relaxed over time in culture. Also *Gymnodinium catenatum* has a multiple group mating system with varying levels of compatibility, planozygote viability, and encystment success between groups (Blackburn et al. 2001). This calls for thorough investigations in order to properly evaluate the strength of isolation barriers among dinoflagellates. *Scrippsiella trochoidea* (Stein) Loeblich III constitutes an example of homothallic life cycle: here the production of peculiar cysts surrounded by calcareous spines occurs within a culture established by the isolation of a single cell, without the need for crossing strains with distinct mating types (Montresor et al. 2003). Multiple strains should be always tested when assessing dinoflagellate mating behavior. In fact, in the isolation of a single cell from which a culture is derived it would be possible to isolate a planozygote without being aware that this was the case giving rise to a notionally clonal culture (i.e. a culture established by the isolation of a single vegetative cell) that is actually mixed (i.e. a culture containing two mating types). The same is true of cultures derived from single cysts.

The vast majority of information on dinoflagellate life cycles stems from phototrophic species, due to the greater feasibility of maintaining them in culture. However, a large number of dinoflagellates are heterotropic (Jacobson 1999) or mixotrophic (Stoecker 1999), which implies that an optimal food source has to be identified in order to grow them in the laboratory. *Pfiesteria piscicida* Steidinger and Burkholder and *P. shumwayae*

Glasgow and Burkholder are heterotrophic dinoflagellates that can cause massive fish kills and induce neurological problems in humans. Early work on *P. piscicida* described a multi-phase, complex life cycle including different kinds of cysts and amoeboid stages (Burkholder and Glasgow jr. 1997). Recently, there has been a debate in the scientific community concerning the veracity of this complexity in the life cycle of these species. Detailed laboratory investigations focused on asexual and sexual reproduction in algal-fed clonal cultures provided evidence that asexual division occurs within division cysts and sexual reproduction is characterized by fusion of gametes, formation of planozygotes, zygotic cyst and finally release of flagellate cells (Litaker et al. 2002, Parrow et al. 2002, Parrow and Burkholder 2003b, 2004). There is some evidence that sexual stages (fusing gametes and planozygotes) were more abundant in mixed clones of *Pfiesteria shumwayae* but further investigations are required to assess the mating behavior of heterotrophic dinoflagellates (Parrow and Burkholder 2003b). The discrepancies between the more classical, haplobiontic life cycle and the previous reports including the high number of diversified life stages may reside in the observation that toxic *Pfiesteria* strains, where toxicity is induced by the presence of fish prey, have a more complex life cycle (Burkholder et al. 2001, Burkholder and Glasgow 2002).

Haptophytes

Haptophytes include flagellate species surrounded by organic scales as well as coccolithophorids, in which the cell is covered by calcium carbonate platelets of biogenic origin: the coccoliths. Haptophytes have a haplo-diplobiontic life cycle, where meiosis and syngamy are separated in space and time, implying that both diploid and haploid stages can perform vegetative division. Alternation of morphologically different life stages, at times characterized by different ploidy levels, is widely known for haptophytes, although detailed information on the modalities and occurrence of the sexual phase are unknown for most species (Billard 1994). Recent investigations confirm the ability of haploid and diploid phases to reproduce indefinitely vegetatively and provide evidence of sexual fusion and meiosis (Houdan et al. 2004).

Coccoliths can be distinguished on the basis of their crystallographic ultrastructure: heterococcoliths are a mixture of crystals of variable shape whereas holococcoliths are made by one single type of calcium carbonate crystal. The finding that coccolithophorids can have a life cycle including morphologically distinct forms, including both coccolith and non-coccolith bearing stages or coccolith with different morphological and crystallographic ultrastructure, dates back to the 1960s (von Stosch 1955, Parke and Adams 1960). Parke and Adams (1960) succeeded in culturing the motile holococcolith-bearing *Crystallolithus hyalinus* Gaarder and Markali. A non-motile stage developed in old cultures and, within a few days, large-sized coccoliths started to be produced underneath the hyaline layer containing the small holococcoliths. That coccoid stage fit the description of the heterococcolith-bearing *Coccolithus pelagicus* (Wallich) Schiller, thus providing evidence that the two species were in fact distinct morphotypes within the life cycle of the same species. The life cycle of *Emiliania huxleyi* (Lohmann) Hay and Mohler, one of the key-species of the modern oceans, also encompasses different morphotypes: a flagellate stage (S-cell) surrounded by organic scales, a coccoid stage surrounded by heterococcoliths (C-cell), and a naked coccoid stage (N-cells) (Klaveness 1972). All three stages are capable of reproducing vegetatively by binary fission but occasionally N-cells appear in cultures of C-cells, and S-cells can appear in cultures of N-cells; S-cells can switch to C-cells and *vice versa*. Paasche and Klaveness (1970) first showed that C- and N-cells have the same amount of DNA, while S-cells have a smaller chromosome number. Flow cytometric analysis definitively confirmed that S-cells have half the DNA of C-cells, thus representing the haploid and diploid stage, respectively, most probably linked by a sexual phase (Green et al. 1996).

In the last few years, intensive sampling efforts have allowed the observation of several holococcolith-heterococcolith combinations, i.e. cells surrounded by two layers of coccoliths normally considered to belong to different taxa (Cros et al. 2000, Geisen et al. 2002, Young and Henriksen 2003). Analogous to that known for *Crystallolithus hyalinus*/*Coccolithus pelagicus* and *E. huxleyi*, the holococcolith-heterococcolith combinations have been interpreted as the product of syngamy between two haploid holococcolith stages that give rise to a transition stage in which the heterococcoliths of the diploid stage are produced on top of. This fits into a haplo-diplobiontic life cycle reported for other haptophyceans (*Chrysochromulina* Edvardsen and Vaulot 1996, Paasche et al. 1990; *Prymnesium* Larsen and Edvardsen 1998). Phase shifts have been studied in culture only for a few species (Billard, 1994, Geisen et al. 2002) and, recently, phase transitions have been obtained through the manipulation of experimental condition for *Calyptrosphaera sphaeroidea* Schiller (Noël et al. 2004). Improved culturing techniques, coupled with ploidy analysis and mating tests will undoubtedly elucidate the characteristics of these complex life cycles and the factors that induce shifts among the stages. Besides considerable implications for coccolithophorid taxonomy, the existence of distinct life stages could imply different niche partitioning in the natural environment.

The genus *Phaeocystis* includes species whose life cycle includes motile and non-motile stages and, in some cases (*P. antarctica* Karsten, *P. globosa* Scherffel, *P. pouchetii* (Hariot) Lagerheim), the formation of large colonies. However, species within the genus differ in the number and ultrastructural features of the different stages and there are species for which colony formation is not known, or is constituted by an aggregation of a few non-motile stages within a mucus matrix (*P. jahnii* Zingone, Zingone et al. 1999). Cells differ in size and relative DNA content and are linked into a complex life pattern where, however, syngamy has not been located yet (Rousseau et al. 1994, Peperzak et al. 2000).

DRIVERS FOR SHIFTS AMONG LIFE CYCLE STAGES

We have illustrated how cultures are the primary means of establishing the variety and nature of life cycle stage in marine microalgae. But cultures are also a unique tool to investigate timing and triggers inducing shifts among stages, which is essential information to understand the role of heteromorphic life cycles in species ecology.

When does Sex Occur?

Locating the timing and occurrence of sexual phases such as gametic fusion (syngamy) and meiotic events is challenging both in the field and when following a single species in culture. These events can be ephemeral processes in e.g. haptophytes, where stages of different ploidy suddenly appear in culture but the cytological process originating them has not yet been observed (Larsen and Edvardsen 1998). Culturing has been used to investigate drivers for the sexual phase in dinoflagellates. Attention has been almost exclusively focused on cyst-forming species in which the planozygote can transform into a resting stage (see below). However, the formation of cyst cannot be considered an unequivocal indicator of the occurrence of sexual reproduction: apparently not all planozygotes perform this transformation and a better estimate of the onset and relevance of sexual reproduction would be provided by gamete formation (Anderson 1998). Unfortunately, gametes are often indistinguishable from vegetative cells, fusing pairs can be mistaken for dividing cells and even unequivocal identification of planozygotes can be difficult in some species. All this makes our appreciation of the occurrence and importance of the sexual phase in dinoflagellates in the field and also in culture extremely limited.

Diatoms are somehow less prudish and provide morphological evidence for gamete formation and conjugation process. The attainment of a species-specific size window (usually 30-40% of the maximum cell size, Drebes 1977) seems to be a prerequisite for the induction of gametogenesis and thus sexual reproduction in diatoms (Fig. 3.5). However, within this permissive window, an environmental stimulus is required to operate the switch from asexual to sexual phase. Light (Furnas 1985, Vaulot and Chisholm 1987, Armbrust et al. 1990), nutrients (French III and Hargraves 1985), temperature shifts (Drebes 1966), salinity (Schmid 1995), but most often a combination of several of these are effective in inducing gametogenesis and auxospore formation or modulating their occurrence. At times an increase and at times a decrease of the environmental factors are effective. Gametogenesis in *Thalassiosira weissflogii* can be induced by shifting the culture from saturating light to darkness, and again to continuous light, but only the fraction of cells in the G1 phase of the cell cycle respond to the trigger, showing a possible link between gametogenesis and cell cycle (Vaulot and Chisholm 1987). In the pennate diatom *Pseudo-nitzschia multiseries* (Hasle) Hasle sexual reproduction occurs only when exponentially growing strains are mixed without any apparent need for special manipulations (Davidovich and Bates 1998). Daylength has seldom been tested as a factor inducing gametogenesis and both positive, negative and lack of effect of different light:dark cycles on the sexual phase have been detected (Holmes 1966, Hiltz et al. 2000). This diversity of effective triggers for sexual reproduction is not surprising considering the notable phylogenetic and ecological diversity of the species that have been investigated.

In recent years, the general assumption that sexual reproduction occurs in the smallest size range has been challenged by observations on both centric and pennate diatoms. In *Thalassiosira weissflogii* spermatogenesis took place regardless of the size range of the population (Armbrust et al. 1990); an overlap in size ranges of initial cells and gametangia has been reported for *Coscinodiscus granii* Gough (Schmid 1995). Also in the pennate *Pseudo-nitzschia multiseries* the size range for sexual induction seems to be rather wide, ranging from 23 to 70% of the maximal cell size (Davidovich and Bates 1998); in *P. delicatissima*, cells can undergo sexualization over almost their entire size range (Amato et al. 2005) (Fig. 3.5). The information is still too scanty to draw general conclusions, however, it is likely that species living in different habitats (planktonic vs. benthic), or with different mating behaviors, have distinct constraints on the occurrence of sexual events. A restricted narrow size range could be effective in preventing too frequent sexual events in benthic diatoms, where distances among cells are generally

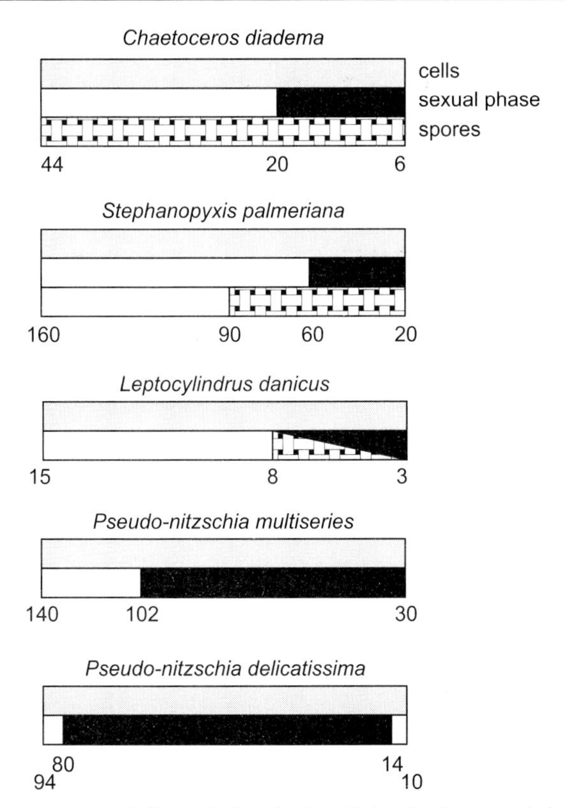

Fig. 3.5 Size range of vegetative cells (gray), size window for the sexual phase (black) and for spore formation (dashed) in centric (*C. diadema, S. palmeriana* and *L. danicus*) and pennate (*P. multiseries* and *P. delicatissima*) diatoms (modified from Drebes 1966, French III and Hargraves 1985, Hiltz et al. 2000 and Amato et al. 2005).

smaller and gamete encounters are facilitated by sliding movements. A too narrow size range could instead represent a serious limit for planktonic species, in which cell encounter rate is less frequent. This might be especially true for pennate heterothallic species, which have the further challenge of mating with a cell of opposite polarity and, in contrast to centric diatoms, lack multiple sperm cells that enhance fertilization success.

Unfortunately, there are almost no records for the occurrence of sexual events at sea, possibly due to the difficulty of identifying fragile sexual stages in preserved samples, or to the fact that they represent a very small proportion of the total cell number. A notable exception is represented by the record of a bloom of *Corethron pennatum* (Grunow) Ostenfeld (= *C. criophilum* Castracane) undergoing a massive sexual phase in the Southern Ocean (Crawford 1995). The examination of cell size distribution over the years allows insights into the timing of sexuality. In fact, the sexual event is manifested by the appearance of large cells, the initial cells produced in the auxospore, in the population. This meticulous work has been carried out only for a few freshwater species (*Cocconeis scutellum* Ehrenberg var. *ornata* Grunow, *Asterionella formosa* Hassall, *Nitzschia sigmoidea* (Nitzsch) W. Smith, *Aulacoseira subarctica* (O. Müller) Hawroth) and results show that the life cycle in the different species takes variable time periods to complete, spanning from 1 to 40 years (Mizuno and Okuda 1985, Mann 1988, Jewson 1992).

Centric diatoms are homothallic, i.e. sexual reproduction occurs within a monoclonal culture. However, mechanisms could have evolved to avoid or limit inbreeding in the natural environment, by analogy with what is known for plants. In *Skeletonema costatum*, high irradiance values induce the formation of female gametangia, whereas at low irradiance male gametangia are produced (Migita 1967). In *Coscinodiscus concinnus* Wm. Smith, male gamete production seems to occur over a wide range of temperature and photoperiod conditions, with a preference for short daylengths (Holmes 1966). Female gametangia were not distinguishable, but auxospore formation showed a restricted range of occurrence, with preference for short daylengths and higher temperatures. These results suggest that gametangia of the two sexes might form at different environmental conditions. Different timing in gamete formation might represent another possible mechanism to limit self-fertilization. In *Chaetoceros curvisetus* a daily timing in male gametogenesis has been shown, which started after 12 to 16 h from the onset of the light cycle, regardless of daylength (Furnas 1985); unfortunately, no corresponding information is available for the female gametangia.

From Growing to Resting and Back

Diatoms

Experiments with batch cultures showed that spore formation takes place as a response to either nitrogen or phosphorous depletion (Durbin 1978, Davis et al. 1980, Hargraves and French 1983, Oku and Kamatani 1995). The presence of silica for building up the heavy silicified wall is essential for spore formation. Kuwata and Takahashi (1990) and Kuwata et al. (1993) showed that *Chaetoceros pseudocurvisetus* Mangin forms resting spores when nitrogen is depleted but silica is available, while resting cells were produced under nitrogen limitation but in the absence of silica. This species thus shows two complementary strategies to withstand different levels of nutrient depletion. Spore formation in *Chaetoceros pseudocurvisetus* takes place also in growth media where nutrients were not depleted (Oku and Kamatani 1995) and this parallels numerous findings in the natural environment where, indeed, spore formation is recorded towards the end of the bloom but not necessarily under nutrient-limiting conditions (Garrison 1981, Pitcher 1986). The two cold-water species *Thalassiosira nordenskioeldii* Cleve and *Detonula confervacea* (Cleve) Gran produced spores under nitrogen limitation only at the lower range (0-10 °C) of the temperature window for vegetative growth (up to 15 °C) and the survival time after dark storage was higher at lower temperatures (Durbin 1978). Hence temperature sets a latitudinal barrier for the distribution of this species, and resting spore formation acts as a survival strategy for cold and dark conditions recorded during the winter at high latitudes.

There is a puzzling and poorly explored link between spores and diatom size range. At least in some species, spore production does not take place over the whole size spectrum but is restricted or more common in certain size windows. In *Leptocylindrus danicus* and *Stephanopyxis palmeriana* (Greville) Grunow spore formation only occurs in cells whose cell diameter is toward the lower size range, while *Chaetoceros diadema* (Ehrenberg) Gran produces spores within cells of any size (Drebes 1966, French III and Hargraves 1985) and *Stephanopyxis turris* (Arnott in Greville) Ralfs in Pritchard only in cells within the wider size range (Fig. 3.5). Smaller strains of *Ditylum brightwellii* (West) Grunow showed lower commitment to spore formation, whereas larger strains produced more resting spores (Hargraves 1982). Thus, for this species a strain-specific capability for spore formation has been recorded: not all strains could be induced to produce resting stages, and strains not committed to spore formation when in the smaller size range did not form spores even after vegetative cell enlargement.

Dinoflagellates

The most frequently cited factors capable of inducing encystment are nutrient (nitrogen or phosphorous) deficiency (Pfiester and Anderson 1987) (Table 3.1). However, there is conflicting evidence from laboratory and field studies where sexuality, as judged by cyst formation, appears to have occurred in times of nutrient sufficiency (Anderson and Morel 1979, Anderson et al. 1983, Montresor et al. 1998). This probably reflects methodological problems centered on the scale of nutrient measurements, the difficulty of determining the nutrient status of cells, and the identification of gametes. Nutrient concentration measurements in the field and in culture are made on water samples that integrate seawater volumes in the order of tens of milliliters, while cells *in situ* experience a wide variety of nutrient

Table 3.1 Factors inducing cyst formation in culture

Species	*Factors*	*References*
Alexandrium catenella	Bacterial promotion	Adachi et al. 1999
Alexandrium lusitanicum	P, N or Fe depletion, turbulence	Blanco 1995
Alexandrium pseudogonyaulax	N or P depletion	Montresor and Marino 1996
Alexandrium tamarense	N or P depletion, light, temperature, medium purity, Fe stress	Turpin et al. 1978, Anderson and Lindquist 1985, Anderson et al. 1984, Doucette et al. 1989
Cryptoperidiniopsoids	Prey availability, spontaneous	Parrow and Burkholder 2003a
Ensiculifera sp.	P or N depletion	Blanco 1995
Gymnodinium catenatum	Nutrient depletion	Blackburn et al. 1989
Gymnodinium nolleri	N or P depletion, temperature	Ellegaard et al. 1998
Gymnodinium fungiforme	Prey reduction	Spero and Moree 1981
Gymnodinium pseudopalustre	Short daylength	von Stosch 1973
Gyrodinium instriatum	Cell contact	Uchida 2001
Gyrodinium uncatenum	N or P depletion, temperature	Anderson et al. 1985
Lingulodinium polyedrum	P or N depletion, turbulence, conditioning by other species	Blanco 1995
Pentapharsodinium tyrrhenicum	Short daylength, temperature, nutrient depletion	Sgrosso et al. 2001
Pfiesteria piscicida	Prey availability, darkness	Anderson et al. 2003
Pfiesteria shumwayae	Prey availability, spontaneous	Parrow and Burkholder 2003b
Polykrikos kofoidii	spontaneous, prey starvation	Morey-Gaines and Ruse 1980, Nagai et al. 2002
Scrippsiella lachrymose	Nutrient depletion	Olli and Anderson 2002
Scrippsiella operosa (=*Calciodinellum operosum*)	Daylength, temperature, nutrient depletion	Sgrosso et al. 2001
Scrippsiella rotunda	Day length, temperature, nutrient depletion	Sgrosso et al. 2001
Scrippsiella trochoidea	N or P depletion, temperature, salinity, cell contact, vitamin addition	Wall et al. 1970, Watanabe et al. 1982, Binder and Anderson 1987, Uchida 1991, Uchida 1991, Uchida 2001
S. trochoidea var. *aciculifera*	Daylength, temperature, nutrient depletion	Sgrosso et al. 2001

concentrations in the scale of microliters. The switch to gamete formation in an individual cell must be driven by the intracellular status of that particular cell, which could be markedly different within the population. Laboratory investigations have suggested that gamete formation takes place when intracellular nutrient concentrations fall to a threshold value (Anderson et al. 1983, Anderson and Lindquist 1985, Anderson et al. 1985). However, is one nutrient pathway critical over another? Interpretation of field and laboratory data in order to answer this question is complicated by the fact that the different pathways governing the internal physiology of the cell are interdependent. For example, P-stressed cells may be forced to shut down an N-uptake pathway because of a lack of ATP, hence they also become N-stressed (Syrett 1981, Davies and Sleep 1989, Flynn et al. 1996). Detailed experiments following intracellular amino acid pools in *Alexandrium minutum* Halim suggest that it is the N-pathway that is critical, and that when cells reach a critical intracellular N-status, which may be driven in the first instance by P-limitation, cells act as gametes (Probert 1999). It has been noted in this and other laboratory studies that gamete formation was reversible if intracellular status changed.

The environmental conditions (e.g. temperature and salinity) under which encystment occurs may be more restricted than the conditions suitable for vegetative growth. Factors that have been documented to influence the process are: temperature (Ellegaard et al. 2002), day length (Sgrosso et al. 2001), density of cells (Uchida 2001), and iron depletion (Doucette et al. 1989, Blanco 1995). One can hypothesize that nutrient stress triggers gamete formation and, once formed, gamete survival, mating success, and the subsequent steps are modulated by other environmental factors. Bioassays carried out with bacterial isolates from natural populations were shown to promote encystment in *Alexandrium catenella* (Whedon and Kofoid) Balech and strains of Proteobacteria were also detected, which showed an inhibitory effect on the encystment process (Adachi et al. 1999, Adachi et al. 2001). However, care has to be taken in such experiments to separate the effects of bacterial presence per se and any effects they may be having on the physical environment of the culture (for example nutrient limitation). A few records of encystment exist for heterotrophic species: *Polykikos kofoidii* Chatton is reported to produce cysts either spontaneously (Morey-Gaines and Ruse 1980) or following starvation (Nagai et al. 2002), while *Gymnodinium fungiforme* Anissimova encysted in response to prey deprivation (Spero and Moree 1981). Other records link sexuality in heterotrophic species with dense cultures and swarming behavior around senescent prey (Parrow et al. 2002). Non-toxic strains of *Pfiesteria piscicida* are capable of producing long-term resting cysts in food-limited cultures, apparently without undergoing the sexual phase, while short-lasting hypnozygotes were produced following gametic fusion (Litaker et al. 2002).

The formation of temporary cysts is not linked to the sexual phase and these cysts are considered as short-term resting stages produced as a response to environmental disturbance. Evidence has been presented showing that *Alexandrium ostenfeldii* (Paulsen) Balech et Tangen can shift from the motile stage to temporary cyst to avoid infection by the parasite *Parvilucifera infectans* Norén et Moestrup (Toth et al. 2004). Motile cells are very sensitive to the parasite infection, while temporary cysts are more resistant possibly because sealed by a continuous cyst wall. *A. ostenfeldii* shifts to the encysted stage also when exposed to the culture medium where the parasite was growing thus suggesting the presence of a water-soluble chemical signal 'perceived' by the dinoflagellate.

After encystment, the cyst undergoes dormancy or a mandatory resting phase, during which no germination can occur because of an active endogenous inhibition of growth. This is followed by a quiescent period when excystment may occur under favorable conditions and it is thus modulated by an exogenous control (Anderson 1998). Of the possible factors that might affect the dormancy period, temperature has been the most frequently cited. In the studies where this has been investigated in the laboratory, dormancy period was reduced at higher storage temperatures, as might be predicted from physiological considerations (Anderson 1980, Montresor and Marino 1996). Intriguingly, it can be seen from pooling dormancy results that the length of dormancy of one species varies with the geographic origin of the population under consideration (Hallegraeff et al. 1998). When stored at around 5°C, dormancy varies from four months (*Alexandrium tamarense* strains from the Gulf of Maine, USA) to 12 months (*A. tamarense* strains from St. Lawrence estuary, Canada) (Anderson 1980, Castell Perez et al. 1998). It is tempting to speculate on a latitudinal trend here but this is perhaps unwise on the basis of so few studies.

Sporadic records are available concerning cyst survival (reviewed in Lewis et al. 1999). From these data, heterotrophic species (e.g. *Protoperidinium*) appear to have shorter survival times, up to two years, than autotrophic species (e.g. *Scrippsiella, Lingulodinium*) of up to around 10 years. These survival times have to be regarded as minima, as they often reflect the time period of the experiment. Most data are from laboratory-held material and it is not clear that these survival times will be the same in the environment. Indeed, survival times recorded for *Ceratium hirundinella* (O.F. Müller) Schrank from varved field samples (6.5 and 12.5 years) are among the highest known (Huber and Nipkow 1923). Indirect estimates from sedimentological data support field survival times of at least these values (Keafer et al. 1992) and field data from Sweden implies even greater survival times of up to 37 years for both autotrophic (*Gymnodinum nolleri* Ellegaard and Moestrup, *Protoceratium reticulatum* (Claparède and Lachmann) Bütschli) and heterotrophic (*Oblea rotunda* (Lebour) Balech ex Sournia and *Diplosalis* sp.) species (McQuoid et al. 2002).

Germination conditions have been investigated for a limited range of species. Attempts have been made to assess cyst germination *in situ* (Ishikawa et al. 1995) but the majority of data have been gained from laboratory studies. Where there has been detailed investigation, it seems that oxygen is required for cyst germination (Anderson et al. 1987, Rengefors and Anderson 1998, Kremp and Anderson 2000). It has also been shown that light may be obligate for germination for some although not all species (Binder and Anderson 1986, Park and Hayashi 1993, Nuzzo and Montresor 1999) and that it may also influence the rate of germination (Anderson et al. 1987). Further than this, temperature plays a key role setting a species-specific window in which germination can take place and within that window increasing temperatures tend to accelerate germination (Binder and Anderson 1987, Lewis and Hallett 1997). Presence of predators (Rengefors et al. 1998), endogenous rhythms (Anderson and Keafer 1987), and growth factors (Costas et al. 1993) can also affect excystment. We can illustrate this interplay of different factors with an example borrowed from a freshwater system that provides a simplified scenario with marked seasonal forcing, as compared to coastal temperate areas (Rengefors and Anderson 1998, Rengefors et al. 1998). *Ceratium hirundinella* is a summer species, whereas *Peridinium aciculiferum* Lemmermann is recorded in the water column under the ice in winter. In the laboratory, *C. hirundinella* was found to germinate between 6 and 21°C after a maturation period of 4.5 months, whereas *P. aciculiferum* germinated in a narrow temperature window between 3 and 7.5 °C, after a dormancy period of 2.5 months. Maximum germination in laboratory was achieved at the time of year the species would normally occur in the field, suggestive (although not unequivocal proof) of control by an endogenous clock. Experiments with exudates from zooplankton cultures showed that germination was, however, reduced in the presence of exudates at all temperatures in the case of *P. aciculiferum* and at intermediate temperatures in the case of *C. hirundinella*. As a large robust species, *C. hirundinella* suffers little grazing pressure and germination at the most suitable temperatures is not affected by grazing cues. In contrast, the winter species (*P. aciculiferum*), more vulnerable to grazers, germinates at low temperatures when grazing pressure would be low in the lake; furthermore germination is reduced when cues for grazers are present.

The proportion of cysts that germinate in laboratory studies is variable but, when the mandatory dormancy period is over and optimal conditions are met, high final excystment rates (80-100%) are usually reported. However, in the field these rates will be tempered by the age of the cysts, and hence their viability, and their environmental opportunity (e.g. if they are buried in anoxic layers of the sediment they will not germinate). Very few studies have considered the conditions affecting the survival or success of the planomeiocyte, the motile stage produced from cyst germination. Kremp (2001) showed that for *Scrippsiella hangoei* (Schiller) Larsen and *Peridiniella catenata* (Levander) Balech, light was important in survival and, counter intuitively, turbulence improved survival of *P. catenata*. This is an important but neglected area. In our experience in the laboratory, germination is no guarantee of success; cysts will excyst under conditions where subsequently the planomeiocyte will perish.

LIFE CYCLES AND BLOOM DYNAMICS

The challenging endeavor is to integrate the knowledge of life histories gained from culturing species in the laboratory to the sea in order to understand crucial aspects of species ecology, such as bloom dynamics, species succession and recurrence. Models are starting to be developed in which population dynamics are examined not only in the light of physical and chemical constraints, but also considering different aspects of species-specific life history traits (Eilertsen and Wyatt 2000, Gentien 2002, Wyatt 2002). However, there is a desperate need for data to feed these models. Blooms are the result of a balance between accumulation and losses, represented by cell death caused by grazing, parasitic or viral arracks. Biomass accumulation is enhanced by favorable physical and chemical conditions that increase growth rates and limit dispersal. However, defense capabilities should not be disregarded as factors improving fitness and success of blooming species (Smetacek 2001). *Phaeocystis globosa* forms large, multicellular balloon-like colonies surrounded by a tough wall that confers resistance to viral attacks and grazing. This might explain the notable ecological success of this species and its capability of producing extensive blooms (Hamm et al. 1999). Species-specific allelochemicals could also play a significant role in reducing predator pressure, and thus bloom development and maintenance (Matsuoka et al. 2000, Tillmann and John 2002). Life cycle strategies have important implications for the onset, maintenance, and termination of blooms, as demonstrated by studies coupling laboratory experiments with field observations (Imai et al. 1998, Smayda 1998, Anderson 1998, Rengefors and Anderson 1998, Kremp and Heiskanen 1999, Kremp and Anderson 2000). Production of resting stages can be responsible for bloom termination (Kremp and Heiskanen 1999) and laboratory experiments support field observations demonstrating that massive encystment can occur in some species (Sgrosso et al. 2001, Olli and Anderson 2002). *In situ* studies are presently hampered by our limited observational capabilities for locating and sampling thin layers where cells might accumulate, assessing germination rates of resting stages in the sediments, identifying different life stages in the water column. Laboratory studies might fail in considering key-triggering factors, or interactions among them, and caution must be taken in extrapolating the results obtained with cultures to species dynamics in the natural environment (Caceres and Schwalbach 2001). However, combining the two approaches and taking advantage of new molecular and technological tools, represent the best possible strategy and will act as cross-validation of working hypotheses.

One of the most puzzling features of phytoplankton blooms is their temporal occurrence: some species show surprisingly regular seasonal patterns over the years (Zingone et al. 2002), while others bloom with apparently longer periodicity (Gjøsæter et al. 2000). Research efforts have been allocated in defining environmental windows for an optimal growth of vegetative cells, but those investigations failed in providing a convincing framework to explain species occurrence and recurrence. Specific timing in the transition among life cycle phases might provide significant insights. An endogenous clock has been advocated to explain regular germination patterns of *Alexandrium tamarense* cysts (Anderson and Keafer 1987), and a possible control by lunar phases on the germination process has been also hypothesized (Wyatt and Jenkinson 1997). Regulation of diatom spore germination and growth by photoperiod has been proposed to support the synchronized timing of *Chaetoceros* blooms over a wide latitudinal range (Eilertsen et al. 1995). However, recurrent blooms are produced by many species for which benthic resting stages are not known and alternative mechanisms need to be researched to explain their punctual appearances (Noji et al. 1986).

CONCLUSIONS

Considerable diversity has been recorded in life cycle patterns, morphology and physiology among the different species. However, even a basic pool of information is far from being complete. We have to consider that important achievements concerning different aspects of phytoplankton life cycles have been attained only in recent years. As an example, the peculiar microreticulate cyst of the harmful dinoflagellate *Gymnodinium catenatum* was described less than 20 years ago (Bravo 1986). By now, we know that there are at least two other species that produce microreticulate cysts (Bolch et al. 1999). We also know the main features of the life cycle of *G. catenatum* and its complex mating behavior, but we still know very little about the closely related species. The genus *Alexandrium* includes more than 20 species, several of which are harmful, but solid information concerning life strategies is available only for a few. Similar statements can be made for diatoms, where the life cycle of the ASP-producing *Pseudo-nitzschia* was elucidated only a few years ago (Davidovich and Bates 1998). And what pitiful knowledge of the life cycle of other diatoms or coccolithophorids, key-players in different biogeochemical cycles, do we have? As a result of our deliberations we would like to highlight a few areas that we believe merit further consideration.

One Strain is not Enough

Phytoplankton species are characterized by variable levels of intraspecific diversity in their morphological features, physiological performances (Brand 1989), and life cycle traits. Strain-specific variability in resting cell formation has been detected for the raphidophycean *Heterosigma akashiwo* (Hada) Hada (Han et al. 2002), diversity in encystment success was reported for the dinoflagellate *Scrippsiella trochoidea* (Montresor et al. 2003), and a strain-specific commitment in spore formation has been reported for the diatom *Ditylum brightwellii* (Hargraves 1982). Extrapolating any information we gain from the study of a single strain to the whole species or population can be misleading and multiple, recently isolated, strains should be considered to get a proper picture of life cycle traits.

Culture Maintenance

A wide variety of culture media have been developed in recent years for phototrophic phytoplankton, however, there are species whose growth in culture is still problematic (Nishitani et al. 2003). The setting of specific diets is crucial for maintaining in culture mixotrophic and heterotrophic species where prey quality and availability might influence phase transitions (Anderson et al. 2003, Nagai et al. 2002, Parrow and Burkholder 2003a).

The Role of Interactions Among Species

Establishing a clonal culture implies, by definition, the isolation of a single cell, from which a clonal population is obtained. Shifts among life cycle stages can be induced by changes in physical or nutritional conditions, but the interaction with other species mediated by direct cell contact or by the action of chemical compounds should be also considered as potential triggers for changes. The dinoflagellate *Heretocapsa circularisquama* has been found to kill *Karenia mikimotoi* by cell contact but it transforms into a temporary cysts when cell concentration of *K. mikimotoi* exceeds a threshold concentration (Uchida et al. 1999, Uchida 2001). A density-dependent mechanism mediated by cell contact was also postulated to induce the transformation of planozygote into cyst (Uchida 2001). Infochemicals produced by grazers seem to inhibit cyst germination in dinoflagellates (Rengefors et al. 1998) and increase colony size of *Phaeocystis globosa* (Jakobsen and Tang 2002).

Experimental Settings

Phytoplankton cultures are generally kept under constant environmental (temperature, irradiance, photoperiod, turbulence) conditions, while in the field cells experience continuous variability, both at micro- and at a seasonal scale. Small culture vessels impair cell migrations, and extremely high cell concentrations attained in batch cultures might cause unnatural pH values or nutrient concentrations. All these aspects need more consideration when designing experimental settings for testing different aspects of species life cycles and behaviors.

Considerable attention has been devoted in the last decades to investigate the role of abiotic factors in inducing and modulating phytoplankton life cycle strategies. This avenue has provided important and interesting results and it is worthy of pursuit. However, it is now time to explore also the potential of biotic signals and interactions in activating shifts among life stages. By analogy to what is known for the terrestrial environment, phytoplankton organisms evolved in a fluctuating physical-chemical environment but also in a biological context, represented by interactions, with other species, with grazers and pathogens. Life histories are the product of this intricate evolutionary history and it is our challenge to unravel this complexity improving our observational capabilities in the environment and our experimental skills in the laboratory.

REFERENCES

Adachi, M., T. Kanno, T. Matsubara, T. Nishijima, S. Itakura and M. Yamaguchi. 1999. Promotion of cyst formation in the toxic dinoflagellate *Alexandrium* (Dinophyceae) by natural bacterial assemblages from Hiroshima Bay, Japan. Mar. Ecol. Prog. Ser. 191: 175-185.

Adachi, M., T. Matsubara, R. Okamoto, T. Nishijima, S. Itakura and M. Yamaguchi. 2001. Inhibition of cyst formation in the toxic dinoflagellate *Alexandrium* (Dinophyceae) by bacteria from Hiroshima Bay, Japan. Aquat. Microb. Ecol. 26: 223-233.

Amato, A., L. Orsini, D. D'Alelio and M. Montresor. 2005. Life cycle, size reduction patterns and ultrastructure of the planktonic diatom *Pseudo-nitzschia delicatissima*. J. Phycol. 41: 542-556.

Anderson, D.M. 1980. Effects of temperature conditioning on development and germination of *Gonyaulax tamarensis* (Dinophyceae) hypnozygotes. J. Phycol. 16: 166-172.

Anderson, D.M. 1998. Physiology and bloom dynamics of toxic *Alexandrium* species, with emphasis on life cycle transitions. pp. 29-48. *In* D.M. Anderson, A.D. Cembella and G.M. Hallegraeff [eds]. 1998. Physiological Ecology of Harmful Algal Blooms. Springer-Verlag, Berlin, Heidelberg, Germany.

Anderson, D.M. and B.A. Keafer. 1987. An endogenous annual clock in the toxic marine dinoflagellate *Gonyaulax tamarensis*. Nature 325: 616-617.

Anderson, D.M. and N.L. Lindquist. 1985. Time-course measurements of phosphorus depletion and cyst formation in the dinoflagellate *Gonyaulax tamarensis* Lebour. J. Exp. Mar. Biol. Ecol. 86: 1-13.

Anderson, D.M. and F.M.M. Morel. 1979. The seeding of two red tide blooms by the germination of benthic *Gonyaulax tamarensis* hypnocysts. Estuar. Coastal Mar. Sci. 8: 279-293.

Anderson, D.M., S.W. Chisholm and C. Watras, C.J. 1983. Importance of life cycle events in the population dynamics of *Gonyaulax tamarensis*. Mar. Biol. 76: 179-189.

Anderson, D.M., D.W. Coats and M.A. Tyler. 1985. Encystment of the dinoflagellate *Gyrodinium uncatenum*, temperature and nutrient effects. J. Phycol. 21: 200-206.

Anderson, D.M., D.M. Kulis and B.J. Binder. 1984. Sexuality and cyst formation in the dinoflagellate *Gonyaulax tamarensis*, cyst yield in batch cultures. J. Phycol. 20: 418-425.

Anderson, D.M., C.D. Taylor and E.V. Armbrust. 1987. The effects of darkness and anaerobiosis on dinoflagellate cyst germination. Limnol. Oceanogr. 32: 340-351.

Anderson, J.T., D.K. Stoecker and R.R. Hood. 2003. Formation of two types of cysts by a mixotrophic dinoflagellate, *Pfiesteria piscicida*. Mar. Ecol. Prog. Ser. 246: 95-104.

Anderson, O.R. 1975. The ultrastructure and cytochemistry of resting cell formation in *Amphora coffaeformis* (Bacillariophyceae). J. Phycol. 11: 272-281.

Anderson, O.R. 1976. Respiration and photosynthesis during resting cell formation in *Amphora coffaeformis* (Ag.) Kütz. Limnol. Oceanogr. 21: 452-456.

Antia, N.J. and J.Y. Cheng. 1970. The survival of axenic cultures of marine planktonic algae from prolonged exposure to darkness at 20 °C. Phycologia 9: 179-184.

Armbrust, E.V., S.W. Chisholm and R.J. Olson. 1990. Role of light and the cell cycle on the induction of spermatogenesis in a centric diatom. J. Phycol. 26: 470-478.

Billard, C. 1994. Life cycles. pp. 167-186. *In* J.C. Green and B.S.C. Leadbeater [eds]. The Haptophyte Algae. Systematics Association, Clarendon Press, Oxford, UK.

Binder, B.J. and D.M. Anderson. 1986. Green light-mediated photomorphogenesis in a dinoflagellate resting cyst. Nature 322: 659-661.

Binder, B.J. and D.M. Anderson. 1987. Physiological and environmental control of germination in *Scrippsiella trochoidea* (Dinophyceae) resting cyst. J. Phycol. 23: 99-107.

Blackburn, S., G.A. Hallegraeff and C.J. Bolch. 1989. Vegetative reproduction and sexual life cycle of the toxic dinoflagellate *Gymnodinium catenatum* from Tasmania, Australia. J. Phycol. 25: 577-590.

Blackburn, S.I., C.J.S. Bolch, K.A. Haskard and G.M. Hallegraeff. 2001. Reproductive compatibility among four global populations of the toxic dinoflagellate *Gymnodinium catenatum* (Dinophyceae). Phycologia 40: 78-87.

Blanco, J. 1995. Cyst production in four species of neritic dinoflagellates. J. Plankton Res. 17: 165-182.

Bolch, C.J.S., A.P. Negri and G.M. Hallegraeff. 1999. *Gymnodinium microreticulatum* sp. nov. (Dinophyceae): a naked, microreticulate cyst-producing dinoflagellate, distinct from *Gymnodinium catenatum* and *Gymnodinium nolleri*. Phycologia 38: 301-313.

Brand, L.E. 1989. Review of genetic variation in marine phytoplankton species and the ecological implications. Biological Oceanography 6: 397-409.

Bravo, I. 1986. Germinacion de quistes, cultivo y enquistamiento de *Gymnodinium catenatum* Graham. Invest. Pesq. 50: 313-321.

Burkholder, J.M. and H.B. Glasgow. 2002. The life cycle and toxicity of *Pfiesteria piscicida* revisited. J. Phycol. 38: 1261-1267.

Burkholder, J.M., H.B. Glasgow and N. Deamer-Melia. 2001. Overview and present status of the toxic *Pfiesteria* complex (Dinophyceae). Phycologia 40: 186-214.

Burkholder, J.M. and H.B. Glasgow J.R. 1997. *Pfiesteria piscicida* and other *Pfiesteria*-like dinoflagellates: behavior, impacts and environmental control. Limnol. Oceanogr. 42: 1052-1075.

Caceres, C.E. and M.S. Schwalbach. 2001. How well do laboratory experiments explain field patterns of zooplankton emergence? Freshwater Biology 46: 1179-1189.

Castell Perez, C., S. Roy, M. Levasseur and D.M. Anderson. 1998. Control of germination of *Alexandrium tamarense* (Dinophyceae) cysts from the lower St. Lawrence estuary (Canada). J. Phycol. 34: 242-249.

Cerino, F., L. Orsini, D. Sarno, C. Dell'Aversano, L. Tartaglione and A. Zingone. 2005. The alternation of different morphotypes in the seasonal cycle of the toxic diatom *Pseudo-nitzschia galaxiae*. Harmful Algae, 4: 33-48.

Chepurnov, V.A. 2000. Variation in the sexual behaviour of *Achnanthes longipes* (Bacillariophyta). III. Progeny of crosses between monoecious and unisexual clones. Eur. J. Phycol. 35: 213-223.

Chepurnov, V.A. and D.G. Mann. 1997. Variation in the sexual behaviour of natural clones of *Achnanthes longipes* (Bacillariophyta). Eur. J. Phycol. 32: 147-154.

Chepurnov, V.A. and D.G. Mann. 1999. Variation in the sexual behaviour of *Achnanthes longipes* (Bacillariophyta). II. Inbreed monoecious lineages. Eur. J. Phycol. 34: 1-11.

Chepurnov, V.A., D.G. Mann, W. Vyverman, S. K. and D.B. Danielidis. 2002. Sexual reproduction, mating system, and protoplast dynamics of *Seminavis* (Bacillariophyceae). J. Phycol. 38: 1004-1019.

Chisholm, S.W. and J.C. Costello. 1980. Influence of environmental factors and population composition on the timing of cell division in *Thalassiosira fluviatilis* (Bacillariophyceae) grown on light/dark cycles. J. Phycol. 16: 375-383.

Costas, E., S. Gonzalez-Gil, A. Aguilera and V.L. Rodas. 1993. An apparent growth factor modulation of marine dinoflagellate excystment. J. Exp. Mar. Biol. Ecol. 166: 241-249.

Costello, J.C. and S.W. Chisholm. 1981. The influence of cell size on the growth rate of *Thalassiosira weissflogii*. J. Plankton Res. 3: 415-419.

Crawford, R.M. 1995. The role of sex in the sedimentation of a marine diatom bloom. Limnol. Oceanogr. 40: 200-204.

Cros, L., A. Kleijne, A. Zeltner, C. Billard and J.R. Young. 2000. New example of holococcolith-heterococcolith combination coccospheres and their implication for coccolithophorid biology. Mar. Micropaleontol. 39: 1-34.

Davidovich, N.A. and S. Bates. 1998. Sexual reproduction in the pennate diatom *Pseudo-nitzschia multiseries* and *P. pseudodelicatissima* (Bacillariophyceae). J. Phycol. 34: 126-137.

Davies, A.G. and J.A. Sleep. 1989. The photosynthetic response of nutrient-depleted cultures of *Skeletonema costatum* to pulses of ammonium and nitrate; the importance of phosphate. J. Plankton Res. 11: 141-164.

Davis, C.O., J.T. Hollibaugh, D.L.R. Seibert, W.H. Thomas and P.J. Harrison. 1980. Formation of resting spores by *Leptocylindrus danicus* (Bacillariophyceae) in a controlled experimental ecosystem. J. Phycol. 16: 296-302.

Destombe, C. and A. Cembella. 1990. Mating-type determination, gametic recognition and reproductive success in *Alexandrium excavatum* (Gonyaulacales, Dinophyta), a toxic red-tide dinoflagellate. Phycologia 29: 316-325.

Doucette, G.J., A.D. Cembella and G.L. Boyer. 1989. Cyst formation in the red tide dinoflagellate *Alexandrium tamarense* (Dinophyceae), effects of iron stress. J. Phycol. 25: 721-731.

Drebes, G. 1966. On the life history of the marine plankton diatom *Stephanopyxis palmeriana*. Helgol. wiss. Meeresunters. 13: 104-114.

Drebes, G. 1977. Sexuality. pp. 250-283. *In* D. Werner [ed.]. The Biology of Diatoms. Blackwell Scientific Publications, Oxford.

Durbin, E.G. 1977. Studies on the autecology of the marine diatom *Thalssiosira nordenskioeldii*. II. The influence of cell size on growth rate, and carbon, nitrogen, chlorophyll *a* and silica content. J. Phycol. 13: 150-155.

Durbin, E.G. 1978. Aspects of the biology of spores of *Thalassiosisra nordenskioeldii* and *Detonula confervacea*. Mar. Biol. 45: 31-37.

Edlund, M.B. and E.F. Stoermer. 1997. Ecological, evolutionary, and systematic significance of diatom life histories. J. Phycol. 33: 897-918.

Edvardsen, B. and D. Vaulot. 1996. Ploidy analysis of the motile forms of *Chrysochromulina polylepis* (Prymnesiophyceae). J. Phycol. 32: 94-102.

Eilertsen, H.C. and T. Wyatt. 2000. Phytoplankton models and life history strategies. S. Afr. J. mar. Sci. 22: 323-338.

Eilertsen, H.C.H.R., S. Sandberg and H. Tollefsen. 1995. Photoperiodic control of diatom spore growth: a theory to explain the onset of phytoplankton blooms. Mar. Ecol. Prog. Ser. 116: 303-307.

Elbrächter, M. 2003. Dinophyte reproduction: progress and conflicts. J. Phycol. 39: 629-632.

Ellegaard, M., D.M. Kulis and D.M. Anderson. 1998. Cysts of the Danish *Gymnodinium nolleri* Ellegaard et Moestrup sp. ined. (Dinophyceae): studies on encystment, excystment and toxicity. J. Plankton Res. 20: 1743-1755.

Ellegaard, M., J. Lewis and I. Harding. 2002. Cyst-theca relationship, life cycle, and effects of temperature and salinity on the cyst morphology of *Gonyaulax baltica* sp. nov. (Dinophyceae) from the Baltic Sea area. J. Phycol. 38: 775-789.

Evitt, W.R. 1961. Observations on the morphology of fossil dinoflagellates. Micropaleontology 7: 385-420.

Fensome, R.A., F.J.R. Taylor, G. Norris, W.A.S. Sarjeant, D.J. Wharton and G.L. Williams. 1993. A classification of living and fossil dinoflagellates. Micropaleontology, Special Publication Number 7. Sheridan Press, Hanover, Pennsylvania, USA.

Findlay, I.W.O. 1972. Effects of external factors and cell size on the cell division rate of a marine diatom, *Coscinodiscus pavillardii* Forti. Int. Revue ges. Hydrobiol. 57: 523-533.

Flynn, K.J., K. Flynn, E.H. John, B. Reguera, M.I. Reyero and J.M. Franco. 1996. Changes in toxins, intracellular and dissolved free amino acids of the toxic dinoflagellate *Gymnodinium catenatum* in response to changes in inorganic nutrients and salinity. J. Plankton Res. 18: 2093-2111.

French, F. and P.E. Hargraves. 1980. Physiological characteristics of plankton diatom resting spores. Marine Biology Letters 1: 185-195.

French III, F.W. and P.E. Hargraves. 1986. Population dynamics of the spore-forming diatom *Leptocylindrus danicus* in Narragansett Bay, Rhode Island. J. Phycol. 22: 411-420.

Fryxell, G.A. 1983. Survival Strategies of the Algae. Cambridge University Press, Cambridge, UK.

Furnas, M. 1978. Influence of temperature and cell size on the division rate and chemical content of the diatom *Chaetoceros curvisetum* Cleve. J. Exp. Mar. Biol. Ecol. 34: 97-109.

Furnas, M.J. 1985. Diel synchronization of sperm formation in the diatom *Chaetoceros curvisetus* Cleve. J. Phycol. 21: 667-671.

Gallagher, J.C. 1983. Cell enlargement in *Skeletonema costatum* (Bacillariophyceae). J. Phycol. 19: 539-542.

Garcés, E. 2002. Temporary cysts in dinoflagellates. pp. 46-48. *In* E. Garcés, A. Zingone, M. Montresor, B. Reguera and B. Dale [eds]. LIFEHAB: Life History of Microalgal Species Causing Harmful Blooms. European Commission, Brussels, Belgium. http://www.icm.csic.es/bio/projects/lifehab/.

Garcés, E., M. Delgado, M. Masò and J. Camp. 1998. Life history and *in situ* growth rates of *Alexandrium taylori* (Dinophyceae, Pyrrhophyta). J. Phycol. 34: 880-887.

Garcés, E., A. Zingone, M. Montresor, B. Reguera and B. Dale. 2002. LIFEHAB: Life History of Microalgal Species Causing Harmful Blooms. European Commission, Brussels, Belgium. http://www.icm.csic.es/bio/projects/lifehab/.

Garrison, D.L. 1981. Monterey Bay phytoplankton. II. Resting spore cycles in coastal diatom populations. J. Plankton Res. 3: 137-156.

Geisen, M., C. Billard, A.T.C. Broerse, L. Cros, I. Probert and J.R. Young. 2002. Life-cicle associations involving pairs of holococcolithophorid species: intraspecific variation or cryptic speciation? Eur. J. Phycol. 37: 531-550.

Gentien, P. 2002. Models of bloom dynamics: what is needed to incorporate life cycles and life stages? pp. 103-108. *In* E. Garcés, A. Zingone, M. Montresor, B. Reguera and B. Dale [eds]. LIFEHAB: Life History of Microalgal Species Causing Harmful Blooms. European Commission, Brussels, Belgium. http://www.icm.csic.es/bio/projects/lifehab/.

Gjosaeter, J., K. Lekve, N.C. Stenseth, H.P. Leinaas, H. Christie, E. Dahl, D.S. Danielssen, B. Edvardsen, F. Olsgard, E. Oug and E. Paasche. 2000. A long-term perspective on the *Chrysochromulina* bloom on the Norwegian Skagerrak coast 1988: a catastrophe or an innocent incident? Mar. Ecol. Prog. Ser. 207: 201-218.

Green, J.C., P.A. Course and G.A. Tarran. 1996. The life-cycle of *Emiliania huxleyi*: a brief review and a study of relative ploidy levels analysed by flow cytometry. J. mar. Syst. 9: 33-44.

Hallegraeff, G.M., J.A. Marshall, J. Valentine and S. Hardiman. 1998. Short cyst-dormancy period of an Australian isolate of the toxic dinoflagellate *Alexandrium catenella*. Marine Freshwater Research 49: 415-420.

Hamm, C.E., D.A. Simson, R. Merkel and V. Smetacek. 1999. Colonies of *Phaeocystis globosa* are protected by a thin but tough skin. Mar. Ecol. Prog. Ser. 187: 101-111.

Han, M.-S., Y.-P. Kim and R.A. Cattolico. 2002. *Heterosigma akashiwo* (Raphidophyceae) resting cell formation in bacht culture: strain identity versus physiological response. J. Phycol. 38: 304-317.

Hargraves, P.E. 1976. Studies on marine plankton diatoms. II. Resting spores morphology. J. Phycol. 12: 118-128.

Hargraves, P.E. 1982. Resting spores formation in the marine diatom *Ditylum brightwellii* (West) Grun. ex V.H. pp. 33-46. *In* D.G. Mann [ed.]. Proceedings of the Seventh Diatom Symposium. Otto Koeltz Sci. Publ., Koenigstein, Germany.

Hargraves, P.E. and F.W. French. 1983. Diatom resting spores: significance and strategies. pp. 49-68. *In* G.A. Fryxell [ed.]. Survival Strategies of the Algae. Cambridge University Press, Cambridge, UK.

Hensen, V. 1887. Über die Bestimmung des Planktons oder des im Meere treibenden Material an Pflanzen und Tiere. Kommisssion zur wissenschaftlichen Untersuchungen der deutschen Meere in Kiel, 1882-1886. V. Bericht, Jahrgang 12-16: 1-107.

Hiltz, M., S.S. Bates and I. Kaczmarska. 2000. Effect of light:dark cycles and cell apical length on the sexual reproduction of the pennate diatom *Pseudo-nitzschia multiseries* (Bacillariophyceae) in culture. Phycologia 39: 59-66.

Holfeld, H. 2000. Infection of the single-celled diatom *Stephanodiscus alpinus* by the cytrid *Zygorhizidium*: parasite distribution within host population, changes in host cell size and host-parasite size relationship. Limnol. Oceanogr. 45: 1440-1444.

Holmes, R.W. 1966. Short-term temperature and light conditions associated with auxospore formation in the marine centric diatom *Coscinodiscus concinnus* W. Smith. Nature 209: 217-218.

Houdan, A., C. Billard, D. Marie, F. Not, A.G. Sáez, J.R. Young, and I. Probert, 2004. Flow cytometric analysis of relative ploidy levels in holococcolithophore-heterocooclithophore (Haptophyta) life cycles. Systematics and Biodiversity 1: 453-465.

Huber, G. and F. Nipkow 1923. Experimentelle Untersuchungen über die Entwicklung und Formbildung von *Ceratium hirundinella* O.F.M. Flora New Series 16: 114-215.

Imai, I., M. Yamaguchi and M. Watanabe. 1998. Ecophysiology, life cycle, and bloom dynamics of *Chattonella* in the Seto Inland Sea, Japan. pp. 95-112. *In* D.M. Anderson, A.D. Cembella and G.M. Hallegraeff [eds]. Physiological Ecology of Harmful Algal Blooms. Springer-Verlag, Berlin Heidelberg, Germany.

Ishikawa, A., N. Fujita and A. Taniguchi. 1995. A sampling device to measure *in situ* germination rates of dinoflagellate cysts in surface sediments. J. Plankton Res. 17: 617-651.

Jacobson, D.M. 1999. A brief history of dinoflagellate feeding research. J. Eukaryot. Microbiol. 46: 376-381.

Jakobsen, H.H. and K.W. Tang. 2002. Effects of protozoan grazing on colony formation in *Phaeocystis globosa* (Prymnesiophyceae) and the potential costs and benefits. Aquat. Microb. Ecol. 27: 261-273.

Jewson, D.H. 1992. Size reduction, reproductive strategy and the life strategy of a centric diatom. Phyl. Trans. R. Soc. Lond. B 336: 191-213.

Keafer, B.A., K.O. Buesseler and D.M. Anderson. 1992. Burial of living dinoflagellate cysts in estuarine and nearshore sediments. Mar. Micropaleontol. 20: 147-161.

Klaveness, D. 1972. *Coccolituhs huxleyi* (Lohm.) Kamptner. II. The flagellate cell, aberrant cell types, vegetative propagation and life cycles. Br. Phycol. J. 7: 309-318.

Klebahn, H. 1896. Beiträge zur Kenntnis der Auxosporenbildung. I. *Rhopalodia gibba*. Jahrbuch für wissenschaftliche Botanik 29: 595-654.

Kokinos, J.K., T.I. Eglington, M.A. Goni, J.J. Boon, P.A. Martoglio and D.M. Anderson. 1998. Characterization of a highly resistant biomacromolecular material in the cell wall of a marine dinoflagellate resting cyst. Org. Geochem. 28: 265-288.

Kremp, A. 2001. Effects of cyst resuspension on germination and seeding of two bloom-forming dinoflagellates in the Baltic Sea. Mar. Ecol. Prog. Ser. 216: 57-66.

Kremp, A. and D.M. Anderson. 2000. Factors regulating germination of resting cysts of the spring bloom dinoflagellate *Scrippsiella hangoei* from the northern Baltic Sea. J. Plankton Res. 22: 1311-1327.

Kremp, A. and A.-S. Heiskanen. 1999. Sexuality and cyst formation of the spring-bloom dinoflagellate *Scrippsiella hangoei* in the coastal northern Baltic Sea. Mar. Biol. 134: 771-777.

Kuwata, A., T. Hama and M. Takahashi. 1993. Ecophysiological characterization of two life forms, resting spores and resting cells, of a marine planktonic diatom, *Chaetoceros pseudocurvisetus*, formed under nutrient depletion. Mar. Ecol. Prog. Ser. 102: 245-255.

Kuwata, A. and M. Takahashi. 1990. Life-form population responses of a marine planktonic diatom, *Chaetoceros pseudocurvisetus*, to oligotrophication in regionally upwelled waters. Mar. Biol. 107: 503-512.

Larsen, A. and B. Edvardsen. 1998. Relative ploidy levels in *Prymnesium parvum* and *P. patelliferum* (Haptophyta) analyzed by flow cytometry. Phycologia 37: 412-424.

Lewis, J. and R. Hallett. 1997. *Lingulodinium polyedrum* (*Gonyaulax polyedra*) a blooming dinoflagellate. Oceanogr. Mar. Biol. Annu. Rev. 35: 97-161.

Lewis, J., A.S.D. Harris, K.J. Jones and R.L. Edmonds. 1999. Long-term survival of marine planktonic diatoms and dinoflagellates in stored sediment samples. J. Plankton Res. 21: 343-354.

Litaker, R.W., M.W. Vandersea, S.R. Kibler, V.J. Madden, E.J. Noga and P.A. Tester. 2002. Life cycle of the heterotrophic dinoflagellate *Pfiesteria piscicida* (Dinophyceae). J. Phycol. 38: 442-463.

Mann, D.G. 1988. Why didn't Lund see sex in *Asterionella*? A discussion of the diatom life cycle in nature. pp. 385-412. *In* F.E. Round [ed.]. Algae and the Aquatic Environment. Biopress, Bristol, UK.

Mann, D.G. 1989. The species concept in diatoms: evidence for morphologically distinct, sympatric gamodemes in four epipelic species. Pl. Syst. Evol. 164: 215-237.

Mann, D.G., V.A. Chepurnov and S.J.M. Droop. 1999. Sexuality, incompatibility, size variation and preferential polyandry in natural populations and clones of *Sellaphora pupula* (Bacillariophyceae). J. Phycol. 35: 152-170.

Margalef, R. 1998. Red tide and ciguatera as successful ways in the evolution and survival of an admirable old phylum. pp. 3-7. *In* B. Reguera, J. Blanco, M.L. Fernandez and T. Wyatt [eds]. Harmful Algae. Xunta de Galicia and IOC, Santiago De Compostela, Spain.

Matsuoka, K., H.-J. Cho and D.M. Jacobson. 2000. Observations on the feeding behavior and growth rates of the heterotrophic dinoflagellate *Polykrikos kofoidii* (Polykrikaceae, Dinophyceae). Phycologia 39: 82-86.

McQuoid, M.R., A. Godhe and K. Nordberg. 2002. Viability of phytoplankton resting stages in the sediments of a coastal Swedish fjord. Eur. J. Phycol. 37: 191-201.

McQuoid, M.R. and L.A. Hobson. 1996. Diatom resting stages. J. Phycol. 32: 889-902.

Migita, S. 1967. Sexual reproduction of centric diatom *Skeletonema costatum*. Bull. Jap. Soc. Sci. Fish. 33: 392-398.

Mizuno, M. and K. Okuda. 1985. Seasonal change in the distribution of cell size of *Cocconeis scutellum* var. *ornata* (Bacillariophyceae) in relation to growth and sexual reproduction. J. Phycol. 21: 547-553.

Montresor, M. 1995. The life history of *Alexandrium pseudogonyaulax* (Gonyaulacales, Dinophyceae). Phycologia 34: 444-448.

Montresor, M. and D. Marino. 1996. Modulating effect of cold-dark storage on excystment in *Alexandrium pseudogonyaulax* (Dinophyceae). Mar. Biol. 127: 55-60.

Montresor, M., S. Sgrosso, G. Procaccini and W.H.C.F. Kooistra. 2003. Intraspecific diversity in *Scrippsiella trochoidea* (Dinophyceae): evidence for cryptic species. Phycologia 42: 56-70.

Montresor, M., A. Zingone and D. Sarno. 1998. Dinoflagellate cyst production at a coastal Mediterranean site. J. Plankton Res. 20: 2291-2312.

Morey-Gaines, G. and R.H. Ruse. 1980. Encystment and reproduction of the predatory dinoflagellate *Polykrikos kofoidi* Chatton (Gymnodiniales). Phycologia 19: 230-236.

Nagai, S., Y. Hori, T. Manabe and I. Imai. 1995. Restoration of cell size by vegetative cell enlargement in *Coscinodiscus wailesii* (Bacillariophyceae). Phycologia 34: 533-535.

Nagai, S., Y. Matsuyama, H. Takayama and Y. Kotani. 2002. Morphology for *Polykrikos kofoidii* and *P. schwartzii* (Dinophyceae, Polykrikaceae) cysts obtained in culture. Phycologia 41: 319-327.

Nishitani, G., K. Miyamura and I. Imai. 2003. Trying to cultivation of *Dinophysis caudata* (Dinophyceae) and the appearance of small cells. Plankton Biol. Ecol. 50: 31-36.

Noël, M.-H., M. Kawachi and I. Inouye. 2004. Induced dimorphic life cycle of a coccolithophorid, *Calyptrosphaera sphaeroidea* (Prymnesiophyceae, Haptophyta). J. Phycol. 40: 112-129.

Noji, T., U. Passow and V. Smetacek. 1986. Interaction between pelagial and benthal during autumn in Kiel Bight. I. Development and sedimentation of phytoplankton blooms. Ophelia 26: 333-349.

Nuzzo, L. and M. Montresor. 1999. Different encystment patterns in two calcareous cyst-producing species of the dinoflagellate genus *Scrippsiella*. J. Plankton Res. 21: 2009-2018.

Oku, O. and A. Kamatani. 1995. Resting spore formation and phosphorus composition of the marine diatom *Chaetoceros pseudocurvisetus* under various nutrient conditions. Mar. Biol. 123: 393-399.

Olli, K. and D.M. Anderson. 2002. High encystment success of the dinoflagellate *Scrippsiella* cf. *lachrymosa* in culture experiments. J. Phycol. 38: 145-156.

Paasche, E. 1973. The influence of cell size of growth rate, silica content, and some other properties of four marine diatom species. Nowegian Journal of Botany 20: 151-162.

Paasche, E., B. Edvardsen and W. Eikrem. 1990. A possible alternate stage in the life cycle of *Chrysochromulina polylepis* Manton et Parke (Prymnesiophyceae). Nova Hedwigia Beih. 100: 91-99.

Paasche, E. and D. Klaveness. 1970. A physiological comparison of coccolith-forming and naked cells of *Coccolithus huxleyi*. Arch. Microbiol. 73: 143-152.

Park, H.D. and H. Hayashi. 1993. Role of encystment and excystment of *Peridinium bipes* f. *oculatum* (Dinophyceae) in freshwater red tides in Lake Kizaki, Japan. J. Phycol. 29: 435-441.

Parke, M. and I. Adams. 1960. The motile (*Crystallolithus hyalinus* Gaarder and Markali) and non-motile phases in the life-history of *Coccolithus pelagicus* (Wallich) Schiller. J. Mar. Biol. Assoc. U.K. 39: 263-274.

Parrow, M.W. and J.A.M. Burkholder. 2003a. Estuarine heterotrophic cryptoperid-iniopsoids (Dinophyceae): life cycle and culture studies. J. Phycol. 39: 679-696.

Parrow, M.W. and J.A.M. Burkholder. 2003b. Reproduction and sexuality in *Pfiesteria shumwayae* (Dinophyceae). J. Phycol. 39: 697-711.

Parrow, M.W. and J.A.M. Burkholder. 2004. The sexual life cycles of *Pfiesteria piscicida* and cryptoperidiniopsoids (Dinophyceae). J. Phycol. 40: 664-673.

Parrow, M., J.-A.M. Burkholder, N.J. Deamer and C. Zhang. 2002. Vegetative and sexual reproduction in *Pfiesteria* spp. (Dinophyceae) cultured with algal prey, and inferences for their classification. Harmful Algae 1: 5-33.

Partensky, F. and D. Vaulot. 1989. Cell size differentiation in the bloom-forming dinoflagellate *Gymnodinium* cf. *nagasakiense*. J. Phycol. 25: 741-750.

Peperzak, L., F. Colijn, E.G. Vrieling, W.W.C. Gieskes and J.C.H. Peeters. 2000. Observations of flagellates in colonies of *Phaeocystis globosa* (Prymnesiophyceae); a hypothesis for their position in the life cycle. J. Plankton Res. 22: 2181-2203.

Peters, E. 1996. Prolonged darkness and diatom mortality. II. Marine temperate species. J. Exp. Mar. Biol. Ecol. 207: 43-58.

Peters, E. and D.N. Thomas. 1996. Prolonged darkness and diatom mortality. I. Marine Antarctic species. J. Exp. Mar. Biol. Ecol. 207: 25-41.

Pfiester, L.A. 1989. Dinoflagellate sexuality. Int. Rev. Cytol. 114: 249-272.

Pfiester, L.A. and D.M. Anderson. 1987. Dinoflagellate reproduction. pp. 611-648. *In* F.J.R. Taylor [eds]. The Biology of Dinoflagellates. Blackwell Sci. Publ., Oxford, UK.

Pholpunthin, P., Y. Fukuyo, K. Matsuoka and Y. Nimura. 1999. Life history of a marine dinoflagellate *Pyrophacus steinii* (Schiller) Wall and Dale. Bot. Mar. 42: 189-197.

Pitcher, G.C. 1986. Sedimentary flux and the formation of resting spores of selected *Chaetoceros* species at two sites in the southern Benguela system. S. Afr. J. mar. Sci. 4: 231-244.

Pouchet, G. 1883. L'histoire des cilio-flagelles. Journal l'Anatomie et Physiologie Paris 19: 399-455.

Probert, I. 1999. Sexual reproduction and ecophysiology of the marine dinoflagellate *Alexandrium minutum* Halim. PhD thesis, University of Westminster, UK.

Probert, I., J. Lewis and E. Erard-Le Denn. 2002. Morphological details of the life history of *Alexandrium minutum* (Dinophyceae). Cryptogam. Algol. 23: 343-355.

Reguera, B. and S. González-Gil. 2001. Small cell and intermediate cell formation in species of *Dinophysis* (Dinophyceae, Dinophysiales). J. Phycol. 37: 318-333.

Rengefors, K. and D.M. Anderson. 1998. Environmental and endogenous regulation of cyst germination in two freshwater dinoflagellates. J. Phycol. 34: 568-577.

Rengefors, K., I. Karlsson and L.-A. Hansson. 1998. Algal cyst dormancy: a temporal escape from herbivory. Proc. Royal Soc. Lond. B 265: 1353-1358.

Round, F.E., R.M. Crawford and D.G. Mann. 1990. The diatoms. Biology and morphology of the genera. Cambridge University Press, Cambridge.

Rousseau, V., D. Vaulot, R. Casotti, V. Cariou, J. Lenz, J. Gunkel and M. Bauman. 1994. The life cycle of *Phaeocystis* (Prymnesiophyceae): evidence and hypotheses. J. mar. Syst. 5: 23-39.

Schmid, A.M.M. 1995. Sexual reproduction in *Coscinodiscus granii* Gough in culture: a preliminary report. pp. 139-159. *In* D. Marino and M. Montresor [eds]. Proceedings of the 13th International Diatom Symposium 1994. Biopress, Bristol, UK.

Sgrosso, S., F. Esposito and M. Montresor. 2001. Temperature and daylength regulate encystment in calcareous cyst-forming dinoflagellates. Mar. Ecol. Prog. Ser. 211: 77-87.

Sicko-Goad, L., E.F. Stoermer and J.P. Kociolek. 1989. Diatom resting cell rejuvenation and formation: time course, species records and distribution. J. Plankton Res. 11: 375-389.

Smayda, T.J. 1998. Ecophysiology and bloom dynamics of *Heterosigma akashiwo* (Raphidophyceae). pp. 113-132. *In* D.M. Anderson, A.D. Cembella and G.M. Hallegraeff [eds]. Physiological Ecology of Harmful Algal Blooms. Springer-Verlag, Berlin, Heidelberg, Germany.

Smetacek, V. 2001. A watery arms race. Nature 411: 745.

Spero, H.J. and M.D. Moree. 1981. Phagotrophic feeding and its importance to the life cycle of the holozoic dinoflagellate *Gymnodinium fungiforme*. J. Phycol. 17: 43-51.

Steidinger, K.A. and L.M. Walker. 1984. Marine plankton life cycle strategies. CRC Press, Inc., Boca Raton, Florida, USA.

Stoecker, D.K. 1999. Mixotrophy among dinoflagellates. J. Eukaryot. Microbiol. 46: 397-401.

Syrett, P.J. 1981. Nitrogen metabolism of microalgae. pp. 182-210. *In* T. Platt [eds]. Physiological bases of phytoplankton ecology. Canadian Bulletin of Fisheries and Aquatic Sciences,

Tillmann, U. and U. John. 2002. Toxic effects of *Alexandrium* spp. on heterotrophic dinoflagellates: an allelochemical defence mechanism independent of PSP-toxin content. Mar. Ecol. Prog. Ser. 230: 47-58.

Torrey, H. B. 1902. An unusual occurrence of dinoflagellata on the California coast. Am. Nat. 36: 187-192.

Toth, G.B., N. Norén, E. Selander and H. Pavia. 2004. Marine dinoflagellates show induced life-history shifts to escape parasite infection in response to water-borne signals. Proc. R. Soc. Lond. B 271: 733-738.

Turpin, D.H., P.E.R. Dobell and F.J.R. Taylor. 1978. Sexuality and cyst formation in Pacific strains of the toxic dinoflagellate *Gonyaulax tamarensis*. J. Phycol. 14: 235-238.

Uchida, T. 1991. Sexual reproduction of *Scrippsiella trochoidea* isolated from Muroran Harbor, Hokkaido. Bull. Jap. Soc. Sci. Fish. 57: 1215.

Uchida, T. 2001. The role of cell contact in the life cycle of some dinoflagellate species. J. Plankton Res. 23: 889-891.

Uchida, T., S. Toda, Y. Matsuyama, M. Yamaguchi, Y. Kotani and T. Honjo. 1999. Interactions between the red tide dinoflagellate *Heterocapsa circularisquama* and *Gymnodinium mikimotoi* in laboratory culture. J. Exp. Mar. Biol. Ecol. 241: 285-299.

Vaulot, D. and S.W. Chisholm. 1987. Flow cytometric analysis of spermatogenesis in the diatoms *Thalassiosira weissflogii* (Bacillariophiceae). J. Phycol. 23: 132-137.

von Stosch, H.A. 1950. Oogamy in a centric diatom. Nature 165: 531-532.

von Stosch, H.A. 1955. Ein morphologischer phasenwechsel by einer Coccolithophoride. Naturwissenschaften 42: 433.

von Stosch, H.A. 1964. Zum Problem der sexuellen Fortpflanzung in der Peridineengattung *Ceratium*. Helgol. wiss. Meeresunters. 10: 140-155.

von Stosch, H.A. 1965a. Manipulierung der Zellgrösse von Diatomeen in Experiment. Phycologia 5: 21-44.

von Stosch, H.A. 1965b. Sexualität bei *Ceratium cornutum* (Dinophyta). Naturwissenschaften 5: 112-113.

von Stosch, H.A. 1973. Observations on vegetative reproduction and sexual life cycles of two freshwater dinoflagellates, *Gymnodinium pseudopalustre* Schiller and *Woloszynskia apiculata* sp. nov. Br. Phycol. J. 8: 105-134.

Wall, D., R.R.L. Guillard and B. Dale. 1967. Marine dinoflagellate cultures from resting spores. Phycologia 6: 83-86.

Wall, D., R.R.L. Guillard, B. Dale, E. Swift and N. Watabe. 1970. Calcitic resting cyst in *Peridinium trochoideum* (Stein) Lemmermann, an autotrophic marine dinoflagellate. Phycologia 9: 151-156.

Watanabe, M.M., M. Watanabe and Y. Fukuyo. 1982. Encystment and excystment of red tide flagellates. I. Induction of encystment of *Scrippsiella trochoidea*. Res. Rep. Nat. Inst. Environ. Stud. 30: 27-42.

Wyatt, T. 2002. How can we combine the population dynamics of life history stages? pp. 112-115. *In* E. Garcés, A. Zingone, M. Montresor, B. Reguera and B. Dale [eds]. LIFEHAB: Life History of Microalgal Species Causing Harmful Blooms. European Commission, Brussels, Belgium. http://www.icm.csic.es/bio/projects/lifehab/.

Wyatt, T. and I.R. Jenkinson. 1997. Notes on *Alexandrium* population dynamics. J. Plankton Res. 19: 551-575.

Xiaoping, G., J.D. Dodge and J. Lewis. 1989. Gamete mating and fusion in the marine dinoflagellate *Scrippsiella* sp. Phycologia 28: 342-351.

Young, J.R. and K. 2003. Henriksen. Biomineralization within vesicles: the calcite of coccoliths. *In* P. M. Dove, J.J. De Yoreo, S. Weiner [eds]. Biomineralisation. Reviews in Mineralogy and Geochemistry 54: 189-215.

Zingone, A., M.-J. Chrétiennot-Dinet, M. Lange and L. Medlin. 1999. Morphological and genetic characterization of *Phaeocystis cordata* and *Phaeocystis jahnii* (Prymnesiophyceae), two new species from the Mediterranean Sea. J. Phycol. 35: 1322-1337.

Zingone, A., P. Licandro, M. Nardella and D. Sarno. 2002. Seasonality and interannual variation in the occurrence of species of the genus *Pseudo-nitzschia* in the Gulf of Naples (Mediterranean Sea). p. 315. *In* Abstract Book of the 10th International Conference on Harmful Algae, October 21-25 2002. St. Pete Beach, Florida, USA.

4

Allelopathic Interactions Among Marine Microalgae

Geneviève Arzul[1] and Patrick Gentien[2]

[1]Ifremer, IFREMER, BE, BP 70, F-29280 Plouzané, France
[2]Ifremer, CREMA, F- 17137 L'Houmeau, France

Abstract

Allelopathy is defined as the interactions mediated by allelochemicals produced by some plant species. Allelochemicals are secondary metabolites actively or passively excreted by living plants, or released by cells on decay, and these substances can alter the metabolism of other organisms. The process is probably involved in the quasi-monospecificity of harmful algal blooms. By limiting the sources of variation (spatial diffusion, temporal decay) phytoplankton cultures constitute a suitable tool for the study of allelopathy. Various bioactive substances from phytoplanktonic origin have been isolated and chemically described. Their biological effects in controlled experiments have been compared in a few cases to observations in the field. Allelopathy expression is an intrinsic property of the species and plays a role at the species, community and biocenose levels. Here, we provide a short review of the allelopathic agents, of the experimental evidence and of the difficulties encountered in extrapolating the results to the field.

INTRODUCTION

In a recent paper, Smetacek (2001) summarized the concept of the 'watery arms race' occurring in the ocean: one of the major processes structuring the planktonic community is thought to be chemical warfare. Allelopathy is the denomination of the chemical interactions between the microalgal species.

Allelopathic interactions between marine microalgae result in growth inhibition, due to substances produced by one or several species (Molisch 1937, Putnam and Tang 1986). The substances are biologically active molecules, and called "allelochemicals" ; they are produced by one organism called the "donor" and the impacted organisms are the "receptors". Some biologically active molecules can be extremely beneficial, but allelochemicals are inhibitors (etymology, allelo: reciprocal, and pathos: disease). Allelopathy plays an important role in the structure, balance and succession of phytoplankton populations (Keating 1977, Vardi et al. 2002, Mulderij et al. 2003, Fistarol et al. 2004a). Allelopathy is often associated with ichthyotoxicity due to the haemolytic activity of seawater induced by allelopathic species (Gentien 1998, Legrand et al 2003, Granéli and Johansson 2003).

Although allelopathy is potentially important in the regulation of phytoplankton communities, its effects have not been extensively studied for two major reasons.

Historically, phytoplankton has been generally studied as a bulk property estimated by the chlorophyll proxy, and stress has been placed on biomass production in trophic web models of an 'average species'. It goes without saying that this 'average species' of phytoplankton does not exist. However, the diversity of species expressed in one year and their interplay is essential to the stability and resilience of ecosystems.

The second reason results from technical problems. Allelopathy expression from a given species is the intrinsic property of this species. Therefore, the results are species-specific and do not allow extrapolation to other species. A study of one single species is long and delicate since it requires the production of a proper blank, testing against different strains and species, isolation and identification of the frequently labile chemicals and the understanding of the physiological conditions leading to the production of active substances. Studies have long been hampered by the lack of a proper experimental blank. As a consequence, there is no general understanding of the role of allelopathy structuring planktonic communities.

Allelopathy success may lead to the temporary dominance of one organism in possibly lowering the growth rate of other species present in the community and in directing selectivity in grazers, thus affecting the whole biocenose. Allelopathy was first observed and described in terrestrial superior plants in 1832 by Augustin de Candolle. Since then, it has become an essential topic in agronomy. Many reviews relating to terrestrial and aquatic environments have been compiled (Molish 1937, Berland et al. 1974, Maestrini and Bonin 1981, Rice 1984, Putnam and Tang 1986, Keating 1999, Legrand et al. 2003). Allelopathy was properly demonstrated in a marine phytoplankton species, by the removal of active substances and the production of proper blanks (Gentien et al. 1991). Recently, more attention has been focussed on some phytoplankton species (harmful ones) and new developments in methods, thus focussing on processes (not only nutritional) leading to the development of quasi-monospecific blooms. Therefore, allelopathy has become a topic of interest when trying to understand phytoplankton bloom formation especially in red tide outbreaks, or the discrimination of grazers against algae. Allelopathy provides a competitive advantage to the donor, which displays specific adaptation to protect itself against the allelochemicals, and may use the nutrient resource (Granéli and Johansson 2003).

Chemical cues are mediated by a large variety of compounds, differing in their chemical structure and their mode of action. These can be pheromones acting at the species level; they can also mediate information between individuals belonging to different species, and even different groups and different realms. This chapter considers allelopathy, as defined above: a repressive action by secondary metabolites produced by plants, micro and macro algae, against other species present in the biocenose. Allelopathy requires at least two prerequisites:

1. extracellular release of biologically active chemicals, the activity of which depends on the balance between the production and the decay rates
2. mediation by secondary metabolites, substances which do not play any obvious role in the energetic activity of the organism (Turner 1971).

In the field, the study of such a process is confronted by spatial (at small scale), temporal and biological variations. Monospecific controlled cultures are the ideal tool for limiting the sources of variation. Thus algal cultures are composed of experimental materials in order to study and understand allelopathy through the identification of the bioactive compounds. Conversely, additional field observations can be explained by the experimental results, and support the *in vitro* models.

We present successively:

- Methods for allelopathic studies
- A short summary of allelochemicals involved in allelopathy
- Allelopathy in natural assemblages and in algal blooms
- Role of allelopathy from species to ecosystem level
- Conclusions

METHODS FOR ALLELOPATHIC STUDIES

Laboratory experiments have been based on experimental cultures that intends to reproduce and interpret the field observations. The effect of allelopathy on growth is difficult to demonstrate, due to the interference of other factors such as: the differences in the ecophysiological requirements of the species (optimal temperature and light, nutrient affinity), the tolerance to biocides of anthropic origin that may be present in the medium (pesticides, antifouling, metals) or the selection by predators (grazing). Some authors have used the complexity of physiological responses in the natural environment to sustain the allelopathy controversy (Moebus 1972a, b, Forsberg et al. 1990), and this fact underlines the importance of experimental studies. These allow the different parts to be broken down: donor action/ receptor response and environmental pressure, clarifying the real action of allelochemicals before understanding the whole process.

The most common method to demonstrate any interaction is cross culturing or plurispecific cultures in batches, or in continuous flow cultures, with additional enrichment inputs to exclude interactions with nutrient competition (Kayser 1979). It should, however, be noted that results on plurispecific cultures are often difficult to interpret. Allelopathic interaction can be estimated by changes in growth lag time, growth rate, motility, or cell lysis and death of the receptor, or grazer deterrence. The study should include several steps,

1. to show that the effect varies with the cell concentration of the donor,
2. to isolate the allelochemical(s) involved,
3. and to verify that the allelochemical(s) produce(s) an effect similar to that observed in the field .

One approach involves using extracts of culture or of cells or culture filtrate, for testing on growth of other species: Pratt (1966) used mixed cultures, and separated media to explain the competition between *Skeletonema costatum* and *Olisthodiscus luteus*. The results seemed to point to the production of ectocrine by *O. luteus*, acting as *S. costatum* growth stimulator at low concentration, and as inhibitor when concentrated. Gentien and Arzul, (1990a) developed a specific technique for the production of blanks. These authors adsorbed the active allelochemical principle onto a Waters™ Florisil® cartridge which allowed them to produce proper blanks (Fig. 4.1). They demonstrated the effect of the *Karenia mikimotoi* filtered culture medium on the growth of diatom, compared to the same medium purified by adsorption (Fig. 4.2). This effect was proportional to the *K. mikimotoi* cell density (Gentien and Arzul 1990a, b). The same method was applied to a bloom of the same species on the Ushant front where the experimental strain had previously been isolated (Arzul et al. 1993). Above a threshold of 10,000 cell L^{-1}, a positive relation between *K. mikimotoi* density and the growth rate

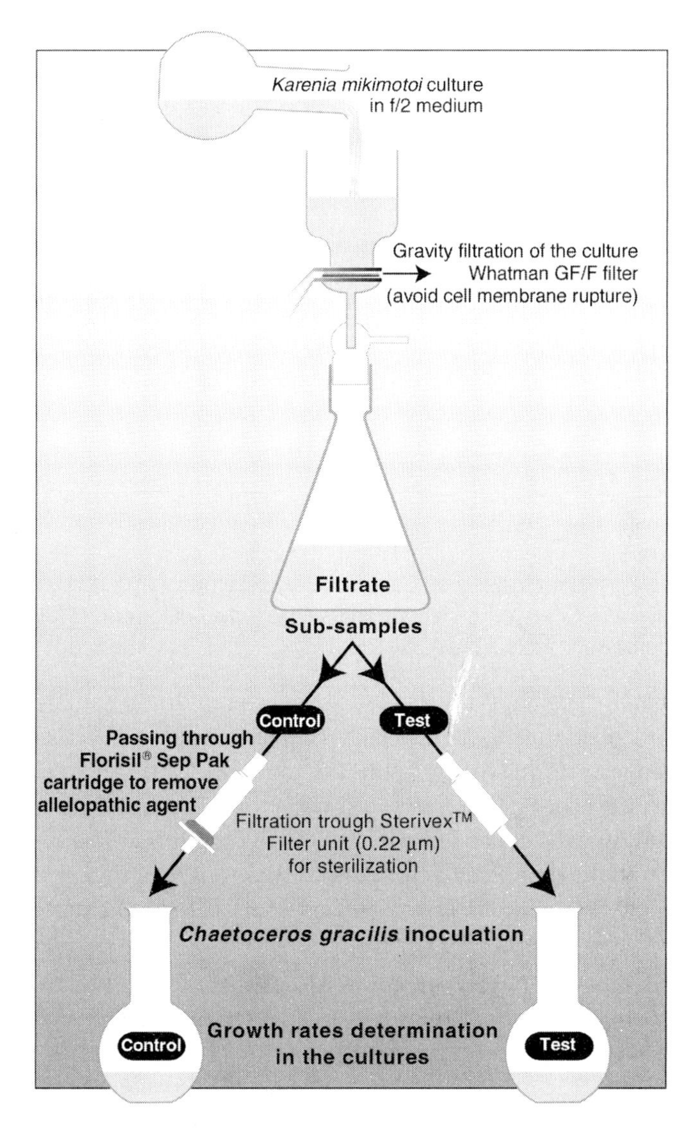

Fig. 4.1 Flow chart of the protocol of experiment by Gentien and Arzul (1990a)

reduction of a diatom was evidenced, demonstrating that, at the onset of a bloom, allelopathy may be an essential process in the temporary establishment of a dominant species.

By using mixed batch cultures, Subba Rao et al. (1995) demonstrated the reciprocal allelopathic activity of *Pseudonitzschia pungens* and *Rhizosolenia alata* cultures. Schmitt and Hansen (2001) studied the allelopathic activity of

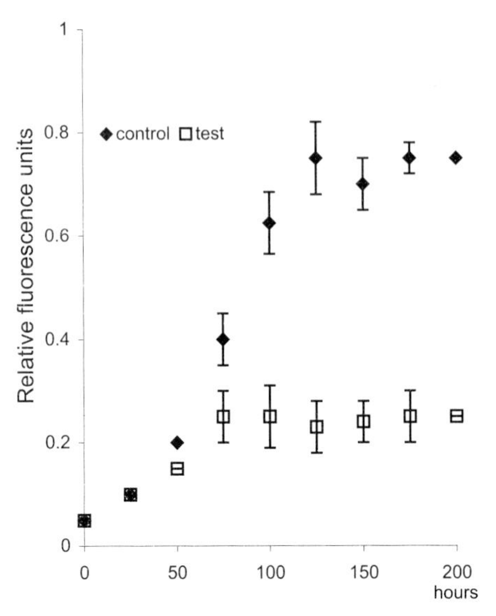

Fig. 4.2 Response growth curve of *Chaetoceros gracilis* grown with crude filtered *Karenia mikimotoi* culture (4.4×10^6 cell L^{-1}) **(test)**, and with detoxified water **(control)** following the protocol of Gentien and Arzul (1990a)

Chrysochromulina polylepis towards seven species of dinoflagellates, testing their motility in relation to the *C. polylepis* density and the pH of the medium. At the exponential phase of growth, the minimal *C. polylepis* cell density producing 40% non-motile *Heterocapsa triquetra* was about 3×10^4cell mL^{-1}, and this activity was enhanced when pH varied up to 9.6. The allelopathic activity tested against algae belonging to different taxa revealed that the dinoflagellate *Prorocentrum minimum* was unaffected by the presence of *C. polylepis*.

The allelopathic property was shown to be specific-dependent by Arzul et al. (1999), in bioassays involving three *Alexandrium* species, *A. minutum*, *A. catenella* and *A. tamarense*. The most potent inhibitor for the growth of the diatom *Chaetoceros gracilis* was *A. tamarense*, and the allelopathic activity varied in parallel to the haemolytic activity in the culture filtrate, which has been found to be related to saxitoxin production. The dinoflagellates *Karenia mikimotoi* (ex *Gymnodinium mikimotoi*) and *Scrippsiella trochoidea* were less affected or totally unaffected by *A. catenella* and *A. tamarense*, *A. minutum* being the least potent. The total repression of the diatom growth observed in the stationary growth phase of *Alexandrium* suggest that the allelopathic activity of *Alexandrium* is due to different active substances produced by the senescent cultures. Pushparaj et al. (1999) studied the allelopathic activity of acetone extract of the cyanobacteria *Nodularia harveyana* towards eubacteria, two chlorophyceae, several cyanobacteria, fungi, rotifers and crustaceans. A high inhibition activity was observed in the cases of *Nannochloris* sp., *Anabaena* genus and *Spirulina platensis*, likewise for *Streptococcus*, *Fusarium*, and the animals.

Another approach uses solid medium (agar): it has been successfully applied for investigating allelopathy among phytoplankton. Plating technique on Petri dish and paper disk method, as used for bacterial bioassays, allows rapid screening tests (Berland et al. 1973, Smith 1994, Tillman and John, 2002). However, some difficulties arise when extrapolating results to their ecological consequences, especially due to possible interferences with organic substrate constituents (Chan et al. 1980).

In order to clearly relate the effect to one compound, it is necessary to identify the chemical composition of this allelochemical and then determine its mode of action. Some examples of allelochemicals chemically identified, produced by marine algae and acting against other marine organisms are given below.

A SHORT SUMMARY OF CHEMICALS INVOLVED IN ALLELOPATHY

For general consideration, it may be assumed that any chemical involved in most cases of allelopathy should have a short half-life in water, either because it is very labile or volatile, in order not to saturate the medium. The *in situ* effect will depend on the production, decay rates and potency. A good example is the case of the exotoxin produced by *K. mikimotoi* (an octadecapentaenoic acid) which has a 20 min half-life in sea water and in the dark (Gentien, unpubl data). This is contrary to what has been observed in the terrestrial environment where allelochemicals can saturate the ground (e.g. the juglone produced by the walnut tree). In aquatic ecological studies, it is therefore necessary to ensure that the active substance is not transformed during isolation.

The characterization of compounds involved in allelopathy has been carried out in some laboratory studies. Various classifications of allelochemicals are presented, according to chemical composition (Whittacker and Feeny 1971, Rice 1984) or mode of action (Legrand et al. 2003). An exhaustive review is outside the scope of this chapter. However, we have grouped allelochemicals and compounds with potential allelopathic properties into three major sets according to their chemical components:

- polyunsaturated fatty acids

- aminoacids
- hydrocarbon

1-Allelochemicals Containing Polyunsaturated Fatty Acids

1-1 APONIN

APparent Oceanic Naturally occurrINg cytolins, (APONINs) are produced by some microorganisms (Table 4.1): e.g. the cyanobacteria *Gomphosphaeria aponina* and the Chlorophyta belonging to *Nannochloris* genus (*N. oculata*, *N. eucaryotum*) (Krienitz et al. 1996, Yamamoto 2001) Deleterious effects of APONINs were observed towards phytoplankton (except for thecate dinoflagellates and *Prymnesium parvum*) and fungi (Moon and Martin 1981). Until now, no effect has been reported on bacteria, crustaceans, molluscs or fish but that remains to be proved, especially for larval stages. Conversely, very low concentrations of APONIN stimulate phytoplankton growth (Perez et al. 2001), and *Nannochloris oculata* is used in the artificial nutrition chain, for example, to raise oyster spat from larvae (Ben-Amotz 1984, Yongmanitchai and Ward 1991).

Table 4.1 Example of APONIN producers involved in allelopathy

Donor	*Receptor*	*References*
Gomphosphaeria aponina	*Gymnodinium breve**	Martin and Martin 1976 McCoy et al. 1979
Gomphosphaeria aponina	*Chattonella subsala*	Halvorson and Martin 1980
Gomphosphaeria aponina	*Ptychodiscus brevis**	Moon and Martin 1981
Nannochloris sp.	Fungi : *Dendryphiella salina*, *Curvularia* sp.	Halvorson et al. 1984
Nannochloris sp.	*Ptychodiscus brevis**	Martin and Martin 1987 Perez et al. 1997
Nannochloris oculata, *N. eucaryotum*	*Gymnodinium breve**	Perez 1999

**Ptychodiscus brevis = Gymnodinium breve = Karenia brevis*

APONIN has been studied by several authors and first described as yellow colored and chloroform extractable from cell-free cultures (Martin and Martin 1976, Martin and Martin 1987, de Majid and Martin 1983). APONIN is made up of more than 30 different fractions including polyunsaturated fatty acids (PUFAs) and their derivatives. APONIN-3 from *Nannochloris oculata* is a mixture of ester palmitate and methyl stearate. These fractions have various biological properties, and act differently: some molecules induce sessile formation activity in *Ptychodiscus brevis*, others are cytolytic (Perez et al. 2001, de Majid and Martin 1983).

Crude APONIN showed no alteration in cytolytic activity towards *K. brevis*, when stored in seawater for 10 d, or submitted to temperatures between 30 and 110 °C, or maintained in acidic condition for 48 h, but APONIN was base labile.

The action mechanisms of the different components have not yet been demonstrated. Cytolytic fractions act as a surfactant towards the cell membrane and the lytic action of APONIN-4 could be explained by the formation of the sterol-APONIN complex (Barltrop and Martin 1984). The active production of APONIN by *Gomphosphaeria aponina* and *Nannochloris* sp. was observed during the exponential growth phase and a linear correlation was obtained for the rates of APONIN and DNA production (Martin and Gonzalez 1978, Derby et al. 2003).

1-2 Allelochemicals with glycolipid derivatives

The major classes of glycolipids found in algae consist of one or two galactose units linked to a glycerol residue containing acyl moieties esterified at the sn-2 and/or sn-3 position of the glyceryl structure. The major classes are monogalactosyldiacylglycerol (MGDG), digalactosyldiacylglycerol (DGDG), and their respective lyso equivalents monogalactosyl monoacylglycerol (MGMG) and digalactosylmonoacylglycerol (DGMG).

Glycolipids display an allelopathic effect when the fatty acids are polyunsaturated (PUFAs). PUFAs are constituents of all vegetal tissues, and play an important role in membrane flexibility and permeability, cell floatability, and the energetic metabolism. PUFAs are also essential for molluscs and fish larval development, genitor fecundity and gonadic maturation (Delauney et al. 1993, St. John et al. 2001). Intracellular PUFA composition in phytoplankton determines its effect as a stimulator or noxious agent for the environment. Some phytoplankton species are considered as excellent forage due to their lipid composition (20:4n6, 20:5n6, 22:5n3) while others are associated with deleterious effects on the accompanying aquatic plants or animals. Lipid content in allelopathic species is characterized by large amounts of unsaturated fatty acids, especially octadecapentaenoic acid (18:5n3), free or associated with sugar (MGDG and DGDG). Glycolipids are sufficiently hydrosoluble to render seawater toxic for surrounding organisms (Yasumoto et al. 1990, Bodennec et al. 1995). Considering the 20 min half-life time and a standard molecular diffusion coefficient, the action radius of the fatty acid octadecapentaenoic acid (18:5n3) is only a few centimeters around the cell. Populations concentrated in thin layers in the pycnocline are commonly observed; therefore, the algae can modify the environment and render it unfit for the growth of phytoplankters sensitive to allelochemicals. Some examples of allelopathic effects between species attributed to PUFAs under glycolipid forms are presented Table 4.2.

Table 4.2 Examples of algal species involved in allelopathy attributed to glycolipids

Donor species	*Receptor species, targets*	*References*
Gyrodinium cf. *aureolum*	*Chaetoceros gracilis*	Gentien and Arzul 1990b, 1991
Chrysochromulina polylepis	12 phytoplankton taxa	Schmidt and Hansen 2001
Phaeodactylum tricornutum	*Cylindrotheca fusiformis*	Chan et al. 1980
Brown alga: *Cladosiphon okamuranus*	*Heterosigma akashiwo*	Kakisawa et al. 1988
Synechococcus sp.	Fish	Mitsui et al. 1989

PUFAs are extracted by chloroform-methanol from the algal cells, or free-cell cultures then purified by HPLC (Shilo 1967, Yasumoto et al. 1990). After partitioning according to polarity, the active fractions were detected in algal bioassays or haemolytic tests (Gentien and Arzul 1990a). Chemical and structural analyses of the compounds by the Chromarod-Iatroscan system, GC-MS and GC-FID revealed that glycolipids were the main hydrosoluble form released by algae in the medium. Glycolipids containing PUFAs are considered to be the bioactive chemicals responsible for allelopathic and ichthyotoxic effects on marine organisms in the case of ichthyotoxic algal species (Yasumoto et al. 1990, Parrish et al. 1994a, b, Bodennec et al. 2000). The antibiotic activity of PUFAs was also demonstrated towards *Vibrio fischeri* and the most potent allelopathic action on the growth of the diatom *Chaetoceros gracilis* was observed with 18:5n3 (Arzul et al. 1995, 2000) (Fig. 4.3).

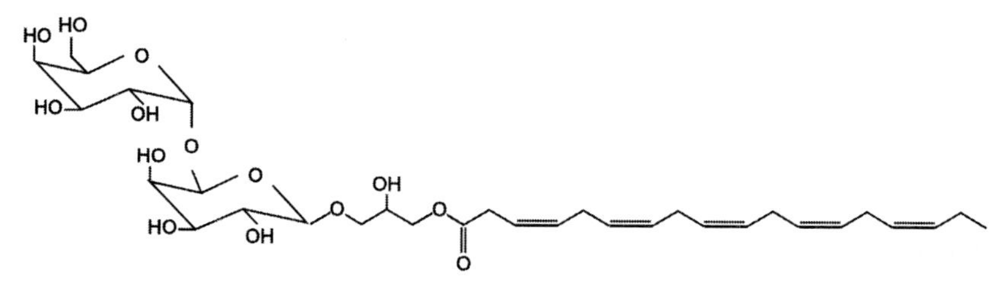

Fig. 4.3 Example of glycolopid involved in allelopathy: general structure of *Karenia mikimotoi* hemolysin; the carbohydrates can differ, but never more than 2 C (according to Yasumoto et al., 1990)

In chains containing the same number of Carbon, the allelopathic potency varies in parallel with the double-bond numbers and depends on their position along the carbon chain (Parrish et al. 1994a, b, Kakisawa et al. 1988, Arzul et al. 1995, 2000).

PUFAs are amphiphilic molecules and can act as a surfactant, producing the saponification of membrane lipids. In addition the reaction of PUFA insaturations in the presence of oxidative substances produces oxygen radicals, known to be highly reactive, and responsible in the destabilization of the cell wall lipid-bilayer (Pacifici et al. 1994). Oxygen appeared to be a co-factor in PUFA haemolytic activity and toxicity toward fish (Arzul et al. 1998). The stability of PUFAs is controversial, and their high reactivity towards oxidative substances involves the production of malonaldehyde, which is not toxic (Ikawa et al. 1997). A pH incease (8.1 to 9.6) in *Chryso-chromulina polylepis* culture filtrate had a dramatic effect on *Heterocapsa triquetra* motility, and previous experiments showed that palmin and acid toxicity were highest in alkalin pH (Schmidt and Hansen 2001). PUFA production depends on the physiological status and environmental conditions of phytoplankton populations, and may explain the differences in allelopathic activity observed *in situ* and in cultures (Gentien and Arzul 1990a). This also explained the difficulties in establishing dose-effect relationships in allelopathic activity through different events. Phosphorus, nitrogen or silicon limitation increased PUFA production in many phytoplankton species (Edvardsen et al. 1990, Roessler 1990). The loss of *C. polylepis* allelopathic activity grown in f/2 medium at pH 8 to 9, was observed at the stationary phase. A lower availability of inorganic carbon could explain the toxicity decrease in strains cultivated in the laboratory for long periods of time (Schmidt and Hansen 2001).

Due to the lability of the 18:5n3, it was impossible to isolate sufficient amounts in order to understand the mode of action. This fatty acid was, therefore, synthesized from 22:6n3 through a γ-iodo lactonisation. The mode of action was then tested on fish gills and tegument (Sola et al. 1999, Fossat et al. 1999). The direct non-specific action of 18:5n3 on membrane ATPases explains not only the allelopathy on other phytoplankton species but also fish kills through action on the chloride cells of the gills.

2-Allelochemicals with Aminoacids

Aminoacids with allelopathic properties which have been isolated mainly in cyanobacteria, include a large number of substances with biological properties (mainly toxic), that offer important perspectives for medical applications. These secondary metabolites are peptidic products with antibiotic and cytotoxic activities, and induce deterrent effects on predators either directly or indirectly, when the excreted substance is transformed by the consumer and becomes either toxic or unpalatable (Table 4.3) (Pennings et al. 1996, 1997). These substances were extracted from cells and cell-free medium in organic solvent or solid support, then purified, separated and identified according to various techniques including HPLC, TLC, HR-FAB mass spectrometry, 2D-NMR spectroscopy and chiral-GC-MS analyses (Burja et al. 2002, Milligan et al. 2000). The isolated metabolites present different chemical structures, thus allowing their characterization.

Table 4.3 Examples of cyanobacteria involved in allelopathy attributed to peptidic allelochemicals

Donor species	*Receptor species, targets*	*References*
Fischerella muscicola	Cyanobacteria, purple bacterium, green alga, higher plants	Gross et al 1991., Hagmann and Jüttner 1996, Srivastava et al., 1998.
Hormothamnion enteromorphoides	fish, urchin and crab	Pennings et al. 1997
Lyngbya majuscula	Fish, shrimp, mollusc, bacteria	Burja et al. 2002, Thacker et al. 1997, Orjala et al. 1995, Milligan et al. 2000).

2-1 Cyclic Peptides

Fischerellin A is the most active allelochemical produced by the cyanobacteria *Fischerella muscicola* (Gross et al 1991, Hagmann and Jüttner 1996). Composed of two heterocycles including amines and a C15 enedyine moiety, pure fischerellin A is a colourless powder resulting from the methanolic extraction of the cells. It inhibits the photosystem II activity in chlorophyllian organisms, and thus presents anti-cyanobacterial, anti-algal and herbicidal activity (Srivastava et al. 1998).

Laxaphycin A is the major component of cyclic peptides mixture produced by a tropical benthic marine cyanobacterium *Hormothamnion enteromorphoides* and was effective in deterring feeding in experiments with parrotfish (Pennings et al. 1997). The marine filamentous cyanobacterium *Lyngbya majuscula* produces a great variety of endocellular and extracellular secondary metabolites made of cyclic or linear lipopeptides, malyngamide A, B and malyngolide, three secondary metabolites different from laxaphycins. These substances have antibiotic, toxic and feeding deterrence properties (Burja et al. 2002, Thacker et al. 1997). Curacin A and lyngbyabellin

$B_{(1)}$, cyclic polypeptides (depsipeptides) produced by *L. majuscula*, are potent brine shrimp toxins (Orjala et al. 1995, Milligan et al. 2000). Nodularin and microcystin produced by *Nodularia spumigena* have a repulsive effect on zooplankton (Stolte et al. 2002) (Fig. 4.4).

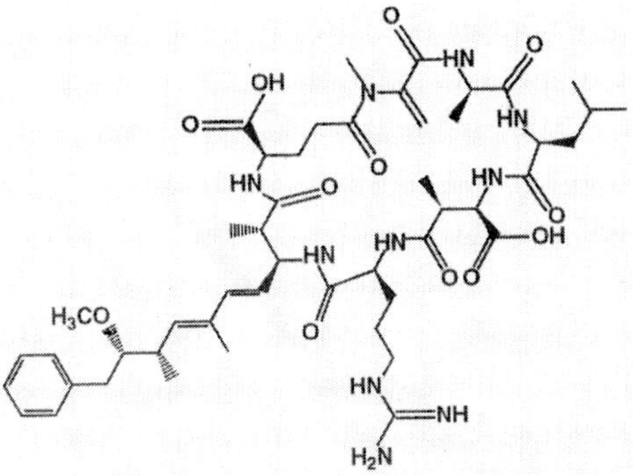

Fig. 4.4 Example of cyclic peptide involved in allelopathy: microcystin, produced by *Nodularia spumigena*

The production of bioactive substances is directly correlated with cyanobacterial growth, and takes place either during the entire culture development, or at the beginning and end of the exponential growth phase (Rossi et al. 1997, Burja et al. 2002,). A possible method for inducing secondary metabolite production by cyanobacteria is stress induction, such as the addition of an antagonist organism (Burja et al. 2002).

2-2 Indirect allelopathic effects associated to aminoacids

Among the various amino acid derivatives produced by phytoplankton, glycoproteins are sometimes released as an accompanying B12 vitamin binder, and render the B12 unavailable for the other phytoplankton competitors (Pintner and Altmeyer 1973, Davies and Leftley 1985).

Domoic acid (Amnesic Shellfish Poison: ASP) produced by *Pseudonitzschia* sp. presents a chemical structure that resembles iron-complexing agents (siderophores) which suggests a possible role in trace metal chelator (Rue and Bruland 2001) (Fig. 4.5).

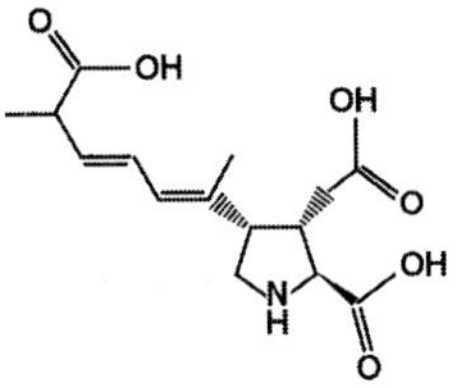

Fig. 4.5 Example of amino acid associated with allelopathy: domoic acid (Amnesic Shellfish Poisoning) produced by *Pseudonitzschia* sp.

The glycocalyx of *Heterosigma akashiwo* (ichthyotoxic) is a complex polysaccharide polypeptide, and presents an inhibiting effect on *Skeletonema costatum* (Honjo 1993). Ichthyotoxicity associated with *H. akashiwo* (alias *H. carterae*) can be mitigated by the use of antioxidant agents (Yang et al. 1995).

3-Allelochemicals Associated with Hydrocarbon

3-1 *Volatile derivates*

Extractible volatile hydrocarbon substances include biogenic alkanes (15 and 17 carbon number), alkenes (17, 19, 21, 27 and 29 carbon number), and terpenoid cyclic polyunsaturated molecules are produced, exuded, or excreted by phytoplankton and several other organisms (seaweeds, zooplankton (*Mesodiunium rubrum*) and bacteria) (Saliot 1975). Some of these substances have a biological effect on the surrounding organisms (Table 4.4).

Table 4.4 Examples of volatile allelochemical producers involved in allelopathy

Donor species	*Receptor species, targets*	*References*
Skeletonema costatum	Bacteria	Bianchi and Varney 1998
Phaeocystis globosa	Bacteria	Noordkamp et al. 1998
Phaeocystis pouchetii	Bacteria	Turner et al. 1996
Mesodinium rubrum	*E. marina*	Bianchi and Varney 1998
Protogoniaulax spp.	Bacteria	Kodama and Ogata 1983
Gymnodinium nagasakiense	Dinoflagellates, raphydophyceae	Kajiwara et al. 1992

A butanedione derivative with antibiotic activities and metal trace chelator is produced by *Prorocentrum minimum* in culture. It is believed that these substances could be degradation products of carotenoids (Andersen et al. 1980).

Cubenol and homologous cadinol sesquiterpene alcohols were identified as characteristic volatile exudate components of *Gymnodinium nagasakiense* (i.e. *Karenia mikimotoi*) cultures (Kajiwara et al. 1992). Cubenol was reported to cause cell destruction of swimming red tide species such as *Heterosigma akashiwo, Chatonella antiqua, C. marina, K. mikimotoi* and *Prorocentrum minimum*, in a similar way to that of polyunsaturated eicosaenoic fatty acids.

DMS is the most abundant volatile sulphur in seawater. With its co-product acrylate, it results in enzymatic conversion by DMSP-lyase of dimethylsulfoniopropionate (DMSP), and ensures the defence of the producer when ingested by predators (Wolfe and Steinke 1996) (Fig. 4.6).

High acrylate concentration in various phytoplankton has been demonstrated to be grazer deterrent (zooplankton, ciliate or other heterotroph phytoplankton) (Wolfe 2000). Moreover the absence of bacteria in *Phaeocystis*

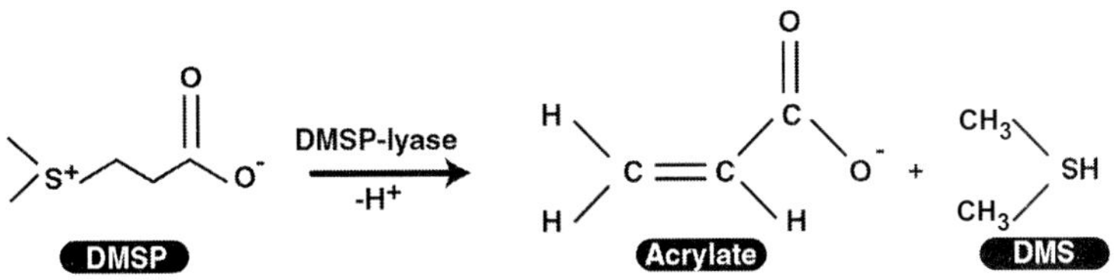

Fig. 4.6 Production of DMS, volatile sulfur, and acrylate from enzymatic conversion of DMSP

cultures in spite of the abundant mucus that embedded the cells, suggested the antibacterial property of their acrylate excretions (Sieburth 1960). Acrylic acid (2-propionic acid $C_3H_4O_2$) antibiotically active against bacteria, was found in scallops contaminated by *Protogonyaulax* spp. (Slezak et al. 1994, Kodama and Ogata 1983). The production of acrylic acid by *Phaeocystis* and its antibacterial action on different bacteria have been studied by several authors (Verity et al. 1988, Noordkamp et al. 2000). In culture at pH 8, acrylic acid at 10 mM reduced bacterial leucine and thymidine incorporation by more than 50%, whereas the reduction was unapparent in uncontrolled pH conditions (Slezak et al. 1994).

3-2 *Halogenated hydrocarbon*

Halogenated hydrocarbons are also synthesized by several marine organisms, and have a deleterious effect on phytoplankton. The red marine algae *Corallina pilulifera* and *Lithophyllum yessoense* produce bromoform, dibromomethane and chlorodibromomethane (Fig. 4.7). Due to their low solubility in seawater, the substances remain at the surface of the algae before transfer to the atmosphere, thus avoiding epiphyte diatom deposition (Ohsawa et al. 2001). The cleaning potentiality of bromoform was confirmed and measured with special equipment giving precise measurements of the elimination rate, necessary to prevent epiphyte microalga fixation on *C. pilulifera* (range 10^{-2} ng min^{-1} cm^{-2}) (Ohsawa et al. 2001).

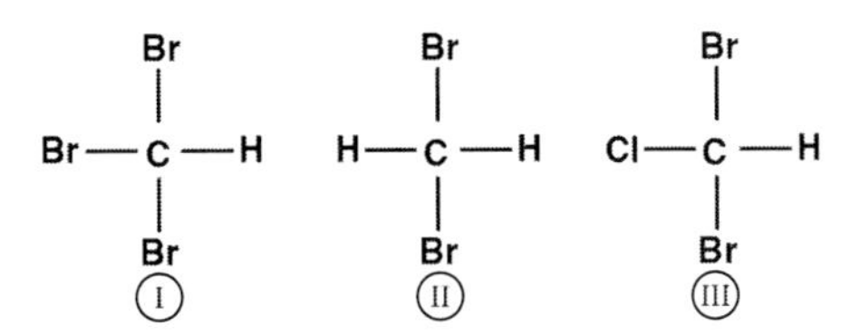

Fig. 4.7 Example of halogenated hydrocarbon presenting allelopathic effect on phytoplankton: bromoform (I), dibromomethan (II) and chlorodibromomethan (III), produced by red marine algae

The production of volatile organohalogens by phytoplankton was followed by isotopic labeling of the inorganic carbon source (Na H $^{13}CO_3$). Labile methyl halides were obtained, chloroform production was checked after extraction and gas chromatography coupled with a mass spectrometry analysis (GC-MS) (Murphy et al. 2000).

3-3 *Polyethers*

Polycyclic ether toxins or closely related compounds from the dinoflagellate *Gambierdiscus toxicus* reduce the growth rate of *Nitzschia longissima* (Bomber et al. 1989). There was a coincidence between the various effects of *Prymnesium parvum* excretions (prymnesin producer) into the growth medium: ichtyotoxicity, haemolysis, cytotoxicity and allelopathy were observed towards various targets, but were not clearly attributed to the same compound (Shilo 1981, Igarashi et al. 1995, Granéli and Johansson 2003, Fistarol et al. 2003, Skovgaard and Hansen 2003). The medium which contained *Prorocentrum lima* (Diarrehic Shellfish Poison producer) inhibited the growth of its potential competitors, the benthic dinoflagellates *Gambierdiscus toxicus, Coolia monotis* and *Ostreopsis lenticularis* but *Amphidinium klebsii* was unaffected. However, the growth-inhibition observed in *P. lima* medium was not entirely attributed to the phosphatase inhibition produced by pure okadaic acid (Sugg and VanDolah 1999) (Fig. 4.8).

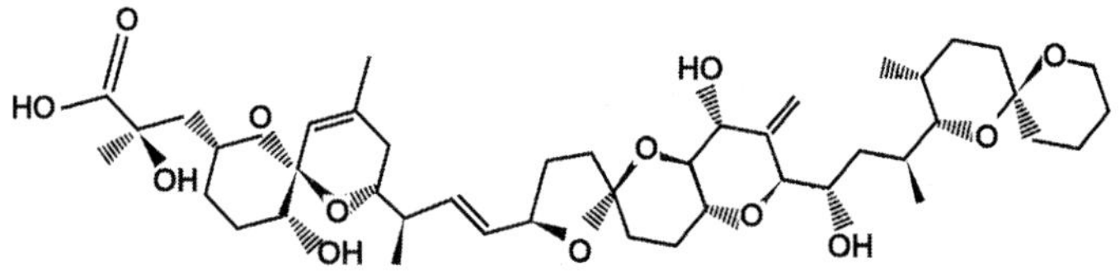

Fig. 4.8 Example of polycyclic ether toxin: Okadaic Acid, from *Prorocentrum lima*, producing phosphatase inhibition on potentially inhibitors

3-4 *Siderophores*

Algae, fungi and bacteria may produce binders for metal sequestration in view of their necessity to cope with low marine iron concentrations. These substances called siderophores, include a complex association of hydroxamates and cyclic molecules such as cathecols, which are iron ligands and favor iron uptake. They are produced mainly by procaryotes (cyanobacteria and bacteria) and the production of hydroxamate siderophore by *Prorocentrum minimum* seems to be an exception (Andersen et al. 1983, Macrellis et al. 2001). Competition for iron complexation by phytoplankton involves a different iron binding system constituted of the porphyrin complex (tetrapyrrolic pigments). Each strategy works differently: iron-siderophores are available to all organisms but preferred by cyanobacteria (specific receptors), while eucaryotes preferentially uptake iron chelated by porphyrins (Kuma et al. 2000; Hutchins et al. 1999).

ALLELOPATHY IN NATURAL ASSEMBLAGES AND IN ALGAL BLOOMS

Experimental results should be validated by observations in the field. However, few descriptions of the biota composition before and after blooms are published, and more often papers demonstrate the effects of allelochemicals during *in vitro* experiments, and explain a posteriori some partial observations in the field. This can be attributed to the fact that biological observations and physico-chemical measurements are not available to describe the situation before the bloom, and often competition for nutrients or light, and concentration by physical entrainment are considered. The following examples of bloom developments suggest that allelopathic interaction between species does play a role in the field.

1-Dinoflagellate Blooms

In 1987, a summer *Karenia mikimotoi* outbreak (alias *Gyrodinium* cf. *aureolum*) occurred in the Bay of Brest (Atlantic coast, France) and extended rapidly, producing massive kills of marine fauna (Gentien and Arzul 1990a). Before the bloom, field observations showed a predominance of diatom *Chaetoceros* sp.. Within two days, *K. mikimotoi* dominated the algal population by 95% with more than 3×10^5 cell L^{-1}. The experimental inoculation of *Chaetoceros gracilis* in bloom seawater filtrate revealed the presence of repressive allelochemicals for diatom growth.

A large-scale bloom of *Gymnodinium* sp. developed around Chiloe Island in Chile during March-April 1999 (Austral autumn), associated with extensive kills in marine fauna (Clément et al. 2001). Initially, brown patches, advected from the ocean entered the coastal areas and inland sea during neap tides during a period of unusual sunny weather. Then the coloured water extended, and displaced following the currents. The algal population was 99.9% *Gymnodinium* sp., with a maximal density of 4 to 8×10^6 cell L^{-1} in the patches. The bloom collapsed during the spring tides period when it mixed with low salinity water mass containing high diatom concentrations. During the bloom event, the biological activity of bloom seawater was studied in the laboratory. The samples presented allelopathic activities towards the diatom *Leptocylindrus minimus*, but no inhibiting effect on the toxic dinoflagellate *Alexandrium catenella* (Tillman and John, 2002). Moreover, strong red blood cell haemolysis was obtained with freshly preserved bloom samples.

2-Diatom Bloom

During the autumn-winter period 1987, a *Pseudonitzschia pungens* bloom developed in Cardigan Bay, Eastern Prince Edward Island (Canada) (Subba Rao et al. 1995). To explain bloom monospecificity, the corresponding seawater filtrate was added to *Rhizosolenia alata*, a species blooming at that time in a nearby area: Hillsborough River estuary. *R. alata* growth was totally inhibited in the *P. pungens* sea water bloom and the allelopathic effect observed against *R. alata* could well explain this monospecificity.

3-Raphydophyte Blooms

In spring 2001, a massive bloom of *Chattonella marina* and *Heterosigma akashiwo* covered the northeast part of Skagerrak, following the annual diatom spring bloom mainly dominated by *Chaetoceros* spp. (Naustvoll et al. 2002). The arrival of water masses with lower salinity resulted in stratification, with a salinity of 22 to 28 psu and temperatures between 1 to 3°C in the 5-10 m upper layer. Maximal cellular concentrations of the raphidophytes reached 9.5×10^6 cell L^{-1} at the end of March, and resulted in fish mortalities. Nitrate and phosphate were low, but silicate was relatively high except near the pycnocline. Inside the area affected by the bloom, few representatives of *Apedinella*, *Pseudopedinella* and *Chrysochromulina* were present, while diatoms stayed in the deeper water mass.

This event suggests that allelopathy was probably involved in the mechanism of raphidophyte local dominance.

4-Prymnesiophyte blooms

During a *Phaeocystis* bloom that occurred during spring 1997 in the Marsdiep tidal inlet, between the North Sea and Dutch Waadden Sea (the Netherlands), acrylate production and antibacterial activity followed (Noordkamp et al. 2000). *Phaeocystis* abundance was around 20 to 45×10^6 cell L^{-1} during the bloom. Acrylate was highly concentrated in the colonies' mucus, and reached 6.5 mM, more than 1000-fold the concentrations measured in unfractionated samples (in culture and in the field) (Noorkamp et al 1998). *Phaeocystis* is known as the most potent acrylate producer. The effect of acrylate on the bacterial populations seems complex: although the antibacterial activity of this compound was observed in culture and in situ (Sieburth 1960, Slezak et al. 1994), bacterial counting during the Marsdiep bloom revealed a five-fold increase in their number. As mentioned previously, the pH value seems to be a determining factor in the bactericidal activity of acrylate, and this could explain the different effects obtained by the authors.

The 1988 *Chrysochromulina polylepis* bloom that developed from the Kattegat and extended to the Skagerrak and along the Norwegian coast produced noxious phenomena of economic importance (Maestrini and Granéli 1991). Diatom spring blooms occurred until April, producing a decrease in silicate concentration. These nutritional conditions associated with the absence of turbulence and stratification, were favourable to non-siliceous species and *C. polylepis* became the dominant primary producer. A possible decrease in grazing pressure due to the repellent effect of the algal toxin content also contributed to the prominence of *C. polylepis* (10^7 to 10^8 cell L^{-1}). The description of the event suggests that the deleterious effects on all components of the food chain most likely contributed to the *C. polylepis* bloom formation, rather than the sole allelochemical repression towards *C. polylepis* competitors.

5-Cyanobacteria bloom

Toxic *Nodularia spumigena* blooms were recorded in Australian estuaries and coastal areas during 1992-1993 period (Blackburn and Jones 1995). The algal population was monospecific in 98% of the total biomass (10^8-10^{10} cell L^{-1}) and lasted during the summer time until February 1993, when the bloom decayed. Diatom concentrations varied inversely compared to the cyanobacteria, and the maximal concentrations in *Nitzschia closterium, Chaetoceros socialis* and *Pleurosigma* spp. were attained when the nodularin content was lower than 10 $\mu g\ g^{-1}$ dry weight in bloom samples.

6-Iron Binders Production

The limitation in primary production due to iron binder allelochemicals was demonstrated in the Southern Ocean (Boye et al. 2001, Hutchins et al. 1999). The chemical speciation of iron (approx. total concentration 0.25 nM) showed that organic complexes are dominant, and organic ligands are in excess of dissolved iron by approx. 0.5 nM. Organic ligands are made up of siderophores excreted mainly by procaryotic cells, and porphyrins produced by phytoplankton. However maximal ligand concentrations do not coincide with bacterial activity and chlorophyll concentration, and the increase in biological activity may result from iron uptake. The uptake of siderophore iron ligands is easier for cyanobacteria than for phytoplankton, whereas porphyrin iron ligands are used preferentially by eucaryotic cells. Siderophore production by bacteria and cyanobacteria could build up a strong complex and render iron less available for eucaryotic phytoplankton, especially in iron-limited environments such as land remote oceanic regions.

7-Volatile Compounds Production

Evidence of phytoplanktonic sources of the volatile compounds alkane and alkenes was obtained in April 1988, during a spring bloom of *Skeletonema costatum* in the Southampton water estuary located on the coastline of Southern England (Bianchi and Varney 1998). In late May and June, a bloom of *Mesodinium rubrum* (autotrophic ciliate) followed the collapse of *Exuviaella marina* (Dinoflagellate). The chemical composition of *M. rubrum* extracts included: n-pentadecane and n-heptadecane. Volatile organosulphide methanethiol (*i.e.* methyl mercaptan), dimethylsulphide (DMS) and dimethyldisulphide (DMDS) concentrations increased 10-fold from spring *S. costatum* blooms to summer blooms with *Scrippsiella* cf. *trochoidea* and *M. rubrum* (Turner et al. 1996, Bianchi and Varney 1998). However in this study, other bioactive molecules could be involved in algal species succession, in addition to volatile compounds.

ROLE OF ALLELOPATHY FROM SPECIES TO THE BIOCENOSE LEVEL

The *in situ* validation of *in vitro* experiments is essential to developing of an understanding of the underlying mechanisms and processes leading not only to monospecific blooms but also to the establishment of typical assemblages. This type of process which directly influences the small scale environment must be integrated at the community level.

1-Species Level

Any monospecific phytoplankton culture exhibits a growth which can be modelled by the logistic equation:

$$\frac{dN}{dt} = rN\left(1 - \frac{N}{K}\right)$$

with N, cell density; r, growth rate and K, the "carrying capacity".

The integration of this equation leads to a typical sigmoid curve with an asymptote which represents the maximal attainable yield in the culture. This equation is the mathematical formulation of the Malthus' population theory. The Malthus' underlying assumptions are that food production can only increase linearly when the population exhibits an exponential growth. It may be the reason why the carrying capacity is in most cases interpreted as a resource limitation. However, the concept is rather fuzzy since any addition of a potentially limiting nutrient and micro-nutrient does not generally lead to any increase in cell density. One could imagine that each cell requires a void volume in its vicinity. The reasons are still unknown, but one factor may well be auto-inhibition.

The production and accumulation of an autoinhibitor by *Skeletonema costatum* have been demonstrated from the exponential growth phase up to the stationary phase (Pratt 1940). The substance (extracted by ethyl acetate) was growth repressive for *Chatonella antiqua* and *Chatonella marina*, but without effect on the dinoflagellates *Prorocentrum minimum* and *Karenia mikimotoi* (Imada et al. 1991).

2-Community Level

The allelopathic potential may be exemplified at the community level in response to environmental conditions: nutritional, physical, biological (Havens et al. 2001).

This interpretation is sustained by the fact that allelopathy increases in nutrient-deprived media, suggesting that allelochemical excretion is stimulated (Havens et al. 2001, Granéli and Johansson 2001). The excretion rate may differ depending on the species considered; it is higher in prymnesiophytes than in prasinophytes and chlorophytes (Reitan et al. 1994). Among the various physiological responses to nutritional shortage, lipid storage is a common process in several phytoplankton species (Reitan et al. 1994, Pernet et al. 2003). Differences in responses could determine competitive advantage, and allelopathic interaction of APONIN has been applied in modelling interactions for *Gymnodinium breve* red tide management (Perez et al. 2000).

The role of allelochemicals in the physiology of the producers is not very well understood. However, the production of fatty acids by *Gymnodinium* cf. *nagasakiense* (now *Karenia mikimotoi*) cultivated under low light conditions and at 18°C corresponded to an increase in floatability and facilitated displacement (Bodennec et al. 1995). To prevent light attenuation due to bacterial and phytoplankton fouling, seaweeds maintain their photosynthetic activity thanks to a clean leaf surface. Diterpene alcohols produced by the brown alga *Dictyota menstrualis* or bromoform produced by the red marine alga *Corallina pilulifera* constitute efficient fouling preventive substances (Schmitt et al. 1995, Ohsawa et al. 2001).

Photosynthesis in the epiphytic diatom *Nitzschia palea* Kützing and natural microalgal communities was diversely inhibited by two sulfuric compounds extracted from the charophytes *Chara* (fresh water macrophyte), dithiolane and trithiane (Mulderij et al. 2003). This was observed in some charophyte culture conditions, depending on the *Chara* strains and their development stage.

Allelopathy has a differential effect on the accompanying species, in the way it eliminates sensitive organisms, while selecting others. In response to *Alexandrium tamarense, Karenia mikimotoi* and *Chrysochromulina polylepis* allelochemical exposure, a temporary cyst formation in the dinoflagellate *Scrippsiella trochoidea* was produced experimentally by Fistarol et al. (2004b). This resistance strategy constitutes an interesting example of coevolution, completing the Lewis' interpretation of allelochemical interactions in microalgae (1986). Sometimes, resistant species can be encountered in species producing similar allelochemical production: *Karenia mikimotoi* is unaffected by *Chrysochromulina polylepis* glycolipids (Schmidt and Hansen 2001) and points to a similar protective mechanism. *Prorocentrum micans* presents particular resistance to several allelochemicals: PUFAs, APONINs, okadaic acid and DTX1 (Grzebyk et al. 1997). The allelopathic effects of plankton excretion can be dose-dependent and complex. Small proportions of *Rhizosolenia alata* culture filtrate stimulated *Pseudonitzschia pungens* divisions, while higher levels were inhibiting (Subba Rao et al. 1995).

3 Biocenose Level

The production of chemicals is used by phytoplankton for protection against predator, competitor and pathogens stress. The consequences for the producer is more safety and less competition. APONIN from *Nannochloris* sp. is fungistatic (Halvorson et al.1984) which could explain fungistasis in seawater.

Allelochemicals play an important role in communities, acting on the structure, balance and succession of populations (Keating 1977, Smith and Doan 1999, Vardi et al. 2002, Fistarol et al. 2003, Mulderij et al. 2003, Fistarol et al. 2004a). The allelopathic effect on grazer predators induces an uptake of non-toxic species and thus the predominance of the allelochemical producer in the medium (Naeem and Li 1998).

Several metabolites produced by microalgae control grazer predation: DMS could be an indicator for seabirds to detect zooplankton and it may constitute an indirect defence compound for phytoplankton (Steinke et al. 2002). Lipophylic substances produced by *Olisthodiscus luteus* and *Dunaliella tertiolecta* are unpalatable for *Mytilus edulis*, likewise *Karenia mikimotoi* ectocrine for the copepods (Gentien 1998). Evidence from *K. mikimotoi* that the same toxins not only affect phytoplankton growth but also kill fish demonstrates that exotoxins may act at different levels of the ecosystem. These effects may even be deferred by several months since fish and shellfish larvae appear to be sensitive to ichtyotoxins.

CONCLUSIONS

As described above, a proper understanding of the role of allelopathy is rather difficult to obtain but, in any case, *in vitro* cultures are essential to the documentation of the processes involved which should then, be validated in the field.

Allelopathy encompasses interactions from different aspects of the environment, and may strongly modify foodweb structures, community composition by reducing biodiversity, and the pathway of biogeochemical cycles (Keating 1977, Vardi et al. 2002, Mulderij et al. 2003, Gross 2003). Allelochemical production should then be considered as a response to environmental signals, and the biological integration of the aggression by the receptors should equally induce their adaptation. The role of receptor sensitivity remains unknown and could correspond to a general ecosystem regulation (Lewis 1986).

Even if seasonal species succession is grossly simulated by models involving nutritional limitations, population biodiversity is relatively dependent on interspecies interactions and allelopathic processes. Moreover, interactions affect several trophic levels and allelochemistry plays a major, but by no means a solitary role in the structuring of the ecosystem's communities (Keating 1999).

ACKNOWLEDGEMENT

We gratefully acknowledge the assistance of Guy Bodennec in offering valuable suggestions in lipid chemistry, Pierre Bodénès for the illustrations and the referee for helpful comments.

REFERENCES

Andersen, R.J., M.J. Le Blanc and F.W. Sum. 1980. 1-(2,6,6-Trimethyl-4-hydroxycyclohexenyl)-1,3-butanedione, an Extracellular Metabolite from the Dinoflagellate *Prorocentrum minimum*. J Org Chem 45: 1169-1170.

Andersen, R.J., C.G. Trick, A. Gillam and P.J. Harrison. 1983. Prorocentrin : An extracellular siderophore produced by the marine dinoflagellate *Prorocentrum minimum*. Science 219 (4582): 306-308.

Armstrong, E., L. Yan, K.G. Boyd, P.C. Wright and J.G. Burgess. 2001. The symbiotic role of marine microbes on living surfaces. Hydrobiologia 461:37-40.

Arzul, G., E. Erard-Le Denn, C.Videau, A.M. Jégou and P. Gentien. 1993. Diatom growth repressing factors during an offshore bloom of *Gyrodinium* cf. *aureolum*. pp 719-724. *In* T.J. Smayda and Y. Shimizu [eds]. Toxic Phytoplankton Blooms in the Sea. Elsevier, Amsterdam, the Netherlands.

Arzul, G., P. Gentien, G. Bodennec, F. Toularastel, A. Youenou and M.-P. Crassous. 1995. Comparison of toxic effects in *Gymnodinium* cf. *nagasakiense* polyunsaturated fatty acids. pp. 394-400. *In* P. Lassus, G. Arzul, E. Erard, P. Gentien and C. Marcaillou [eds]. Harmful Marine Algal Blooms. Technique et documentation-Lavoisier, Intercept Ltd. Paris, France.

Arzul, G., G. Bodennec, P. Gentien, P. Bornens and M.-P. Crassous. 1998. The effect of dissolved oxygen on the haemolytic property of *Gymnodinium* ichthyotoxins. pp. 611-614. *In* B. Reguera, J. Blanco, M.L. Fernandez and T. Wyatt [eds]. Harmful Algae. Xunta de Galicia and Intergovernmental Oceanographic Commission of UNESCO.

Arzul, G., M. Seguel, L. Guzman and E. Erard-Le Denn. 1999. Comparison of allelopathic properties in three toxic *Alexandrium* species. *J Exp Mar Biol Ecol* 232: 285-295

Arzul, G., P. Gentien and G. Bodennec. 2000. Potential toxicity of microalgal polyunsaturated fatty acids (PUFAs). pp. 53-62. *In* G. Baudimant, J. Guezennec, P. Roy and J.-F. Samain [eds]. Marine Lipids. Ifremer, Actes de Colloques, 27. Plouzané, France.

Barltrop, J. and D.F. Martin. 1984. A spectroscopic technique for evaluating sterol-aponin interactions and implications for management of *Ptychodiscus brevis* red tide. Microbios 41 (163): 23-30.

Ben-Amotz, A. 1984. Production of nutritional microphytoplankton for use as food in marine fish hacheries. pp. 195-200. *In* H. Rosenthal and S.Sarig [eds]. Spec. Publ. Eur. Maricult. Soc.. 8

Berland, B.R., D.J. Bonin and S.Y. Maestrini, 1973. Study of bacteria inhibiting marine algae : a method of screening which uses gliding algae. Mem Biol Mar Oceanogr 3: 1-10.

Berland, B.R., D.J. Bonin and S.Y. Maestrini. 1974. The importance of inhibiting substances in the control of algal and bacterial populations of marine plankton. Mem Biol Mar Oceanogr, Messina. 4: 63-97.

Bianchi, A.P. and M.S. Varney. 1998. Volatile organic compounds in the surface waters of a British estuary. Part 1. Occurrence, distribution and variation. Wat Res 2: 352-370.

Blackburn, S. and G.J. Jones. 1995. Toxic *Nodularia spumigena* Mertens blooms in Australia waters – a case study from Orielton Lagoon, Tasmania. pp. 121-126. *In* P. Lassus, G. Arzul, E. Erard, P. Gentien and C. Marcaillou [eds]. Harmful Marine Algal Blooms. Technique et documentation-Lavoisier, Intercept Ltd. Paris. France.

Boye, A., C.M.G. van den Berg, J.T.M. de Jong, H. Leach, P. Croot and H.J.W. de Baar. 2001. Organic complexation of iron in the Southern Ocean. Deep-Sea Research I 48: 1477-1497.

Bodennec, G., P. Gentien, C.C. Parrish, G. Arzul, A. Youenou and M.-P. Crassous. 1995. Production of suspected lipid phycotoxins by *Gymnodinium* cf *nagasakiense* in batch cultures. pp. 407-412. *In* P. Lassus, G. Arzul, E. Erard, P. Gentien and C. Marcaillou [eds]. Harmful Marine Algal Blooms. Technique et documentation-Lavoisier, Intercept Ltd. Paris, France.

Bodennec, G., P. Gentien, C.C. Parrish and M.P. Crassous. 2000. Lipid class and fatty acid compositions of toxic *Gymnodinium* and *Heterosigma* strains : haemolytic and signature compounds. pp. 66-77. *In* G. Baudimant, J. Guezennec, P. Roy and J.-F. Samain [eds]. Marine Lipids. Ifremer, Actes de Colloques, 27. Plouzané. France.

Bomber, J.W., D.R. Tindall and D.R. Norris. 1989. Allelopathy in an epiphytic community. J. Phycol. 25 (summary only).

Burja, A.M., E. Abou-Mansour, B. Banaigs, C. Payri, J.G. Burgess and P.C. Wright. 2002. Culture of the marine cyanobacterium, *Lyngbya majuscula* (Oscillatoriaceae), for bioprocess intensified production of cyclic and linear lipopeptides. J Microbiol Methods 48: 207-219.

Chan, A.T., R.J. Andersen, M.J. Le Blanc and P.J. Harrison. 1980. Algal plating as a tool for investigating allelopathy among marine microalgae. Mar Biol 59: 7-13.

Clément, A., M. Seguel, G. Arzul, L. Guzman and C. Alarcon. 2001. Widespread outbreak of a haemolytic, ichthyotoxic *Gymnodinium* sp. in Southern Chile. pp. 66-69. *In* G.M. Hallegraeff, S.I. Blackburn, C.J. Bolsch and R.J. Lewis [eds]. Harmful Algal Blooms 2000. Intergovernmental of UNESCO.

Davies, A.G. and J.W. Leftley. 1985. Vitamin B12 binding by microalgal ectocrines : Dissociation constant of the vitamin-binder complex determined using an ultrafiltration technique. Mar Ecol Progress Series 21: 267-273.

De Candolle, A., 1832. Physiologie Végétale. 3rd volume in a series of 8, Lausanne, Switzerland.

Delauney, F., Y. Marty, J. Moal and J.-F. Samain. 1993. The effect of monospecific algal diets on growth and fatty acid composition of *Pecten maximus* (L.) larvae. J Exp Mar Biol Ecol 173 : 163-179.

De Majid, L.P. and D.F. Martin. 1983. Induction of sessile stage formation of the red tide organism *Ptychodiscus brevis* by materials elaborated by *Gomphosphaeria aponina*. Microbios Letter 22(86): 59-65.

Derby, M.L., M. Galliano, J.J. Krzanowski, D.F. Martin. 2003. Studies of the effect of - APONIN from *Nannochloris* sp. on the Florida red tide organism *Karenia brevis*, Toxicon 41: 245-249.

Edvardsen, B., F. Moy and E. Paasche. 1990. Haemolytic activity in extracts of *Chrysochromulina polylepis* grown at different levels of selenite and phosphate. pp. 284-289. *In* E. Granéli, B. Sundström, L. Edler and D.M. Anderson. [eds.] 1990. Toxic Marine Phytoplankton. Elsevier, New York, USA.

Fistarol, G.O., C. Legrand, E. Granéli. 2003. Allelopathic effect of *Prymnesium parvum* on a natural plankton community. Mar Ecol Prog Ser 255: 115-125.

Fistarol, G.O., C. Legrand, E. Selander, C. Hummert, W. Stolte and E. Graneli. 2004a. Allelopathy in *Alexandrium* spp.: effect on a natural plankton community and on algal monocultures Aquat Microbial Ecol 35 (1): 45-56.

Fistarol G.O., C. Legrand, K. Rengefors and E. Graneli. 2004b. Temporary cyst formation in phytoplankton : a response to allelopathic competitors? Environ Microbiol 6(8): 791-798.

Forsberg, C., S. Kleiven and T. Willen. 1990. Absence of allelopathic effects of *Chara* on phytoplankton *in situ*. Aquatic Botany 38: 289-294.

Fossat, B., J. Porthe-Nibelle, F. Sola, A. Masoni, P. Gentien and G. Bodennec. 1999. Toxicity of fatty acid 18:5n3 from *Gymnodinium* cf. *mikimotoi* : II. Intracellular pH and K^+ uptake in isolated trout hepatocytes. J Applied Toxicol 19: 275-278.

Gentien, P. and G. Arzul. 1990a. Exotoxin production by *Gyrodinium* cf. *aureolum* (Dinophyceae). J Mar Biol Ass U.K. 70: 571-581.

Gentien, P. and G. Arzul. 1990b. A theoretical case of competition based on the ectocrine production by *Gyrodinium* cf. *aureolum*. pp. 161-164. *In* E. Granéli, B. Sundström, L. Edler and D.M. Anderson. [eds.] 1990. Toxic Marine Phytoplankton. Elsevier, New York, USA.

Gentien, P., G. Arzul and F. Toularastel. 1991. Modes of action of the toxic principle of *Gyrodinium* cf. *aureolum*. pp. 83-86. *In* M. Fremy [ed.]. Proceedings of Symposium on Marine Biotoxins. Paris 30-31 Janvier 1991. CNEVA, Paris, France.

Gentien, P. 1998. Bloom Dynamics and Ecophysiology of the *Gymnodinium mikimotoi* Species Complex. pp. 155-173. *In* D.M. Anderson, A.D. Cembella and G.M. Hallegraeff [eds.]. Physiological Ecology of Harmful Algal Blooms. NATO ASI Series, 41.

Granéli, E. and N. Johansson. 2001. Nitrogen or phosphorus deficiency increases allelopathy in Prymnesium parvum. pp. 358-331. *In* G.M. Hallegraeff, S.I. Blackburn, C.J. Bolch and R.J. Lewis [eds]. Harmful Algal Blooms 2000. Intergov. Oceanografic Commission of UNESCO.

Granéli, E. and N. Johansson. 2003. Increase in the production of allelopathic substances by *Prymnesium parvum* cells grown under N- or P-deficient conditions. *Harmful Algae* 2: 135-145

Gross, E.M., C.P. Wolk, F. Jüttner. 1991. Fischerellin, a new allelochemical from the freshwater cyanobacterium *Fischerella muscicola*. J Phycol 27: 686-692

Gross, E.M. 2003. Allelopathy of aquatic autotrophs. Critical Reviews in Plant Sciences 22(3 & 4): 313-339

Grzebyk, D., A. Denardou, B. Berland and Y. Pouchus. 1997. Evidence of a new toxin in the red-tide dinoflagellate *Prorocentrum minimum*. J Plankton Res 19: 1111-1124.

Hagmann, L. and F. Jüttner. 1996. Fischerellin A, a Novel Photosystem-II-inhibiting Allelochemical of the Cyanobacterium *Fischerella muscicola* with Antifungal and Herbicidal Activity. Tetrahedron Letters 37(36): 6539-6542.

Halvorson, M. and D.F. Martin. 1980. Studies of cytolysis of *Chattonella subsala*. Annual meeting of the academy, Tampa, FL (USA); 23 Mar 1980. Summary only, 43 (suppl. 1), 35.

Halvorson, M.J., D. TeStrate and D.J. Martin. 1984. Effect of aponin, a substance from a green alga *Nannochloris* species, on the spore germination of two fungi. Microbios 41(164): 105-113.

Havens, K.E., J. Hauxwell, A.C. Tyler, S. Thomas, K.J. McGlathery, J. Cebrian, I. Valiela, A.D. Steinman and S.-J. Hwang. 2001. Complex interactions between autotrophs in shallow marine and freshwater ecosystems: implications for community responses to nutrient stress. Environ Poll 113: 95-107.

Honjo, T. 1993. Overview on bloom dynamics and physiological ecology of *Heterosigma akashiwo*. pp. 33-42. *In* T.J. Smayda and Y. Shimizu [eds]. Toxic Phytoplankton Blooms in the Sea. Elsevier, Amsterdam, the Netherlands.

Hutchins, D.A., A.E. Witter, A. Butler and G.W. III Luther. 1999. Competition among marine phytoplankton for different chelated iron species. Nature 400 (6747): 858-861.

Igarashi, T., Y. Oshima, M. Murata and T. Yasumoto. 1995. Chemical studies on prymnesins isolated from Prymnesium parvum. pp. 303-308. *In* P. Lassus, G. Arzul,

E.Erard, P. Gentien and C. Marcaillou [eds]. Harmful Marine Algal Blooms. Technique et documentation-Lavoisier, Intercept Ltd. Paris, France.

Ikawa, M., J.J. Sasner and F.F. Haney. 1997. Inhibition of *Chlorella* growth by degradation and related products of linoleic and linolenic acids and the possible significance of polyunsaturated fatty acids in phytoplankton ecology. Hydrobiologia 356: 143-146.

Imada, N., K. Kobayashi, K. Tahara and Y. Oshima. 1991 Production of an autoinhibitor by *Skeletonema costatum* and its effect on the growth of other phytoplankton. Nippon Suisan Gakkaishi 57: 2285-2290.

Kajiwara, T., S. Ochi, K. Kodama, K. Matsui, A. Hatanaka, T. Fujimura and T. Ikeda. 1992. Cell-destroying sesquiterpenoid from red tide of *Gymnodinium nagasakiense*. Phytochemistry 31(3): 783-785.

Kakisawa, H., F. Asari, T. Kusumi, T. Toma, T. Sakurai, T. Oohusa, Y. Hara and M. Chihara. 1988. An allelopathic fatty acid from th brown alga *Cladosiphon okamuranus*. Phytochemistry 27: 731-735.

Kayser, H. 1979. Growth Interactions Between Marine Dinoflagellates in Multispecies Culture Experiments. Mar Biol 52: 357-369.

Keating K.I. 1977. Allelopathic influence on blue-green bloom sequence in a eutrophic lake. Science 196: 885-886.

Keating, K.I. 1999. Allelochemistry in Plankton Communities. Chapt.11. pp. 165-178. *In* K.M.M. Inderjit, Dakshini and C.L. Foy [eds]. *Principles and Practices in Plant Ecology, Allelochemical Interactions*. CRC Press. Boca Raton, London, New York, Washington, D.C., USA.

Kodama, M. and T. Ogata. 1983. Acrylic acid, as an antibacterial substance in scallop. Bull Jap Soc Sci Fish 49: 1103-1107.

Krienitz, L., V.A.R. Huss and C. Huemmer. 1996. Picoplanktonic *Choricystis* species (Chlorococcales, Chlorophyta) and problems surrounding the morphologically similar "*Nannochloris*-like algae". Phycologia 35: 332-341.

Kuma, K., J. Tanaka, K. Matsunaga and K. Matsunaga. 2000. Effect of hydroxamate ferrisiderophore complex (ferrichrome) on iron uptake and growth of a coastal marine diatom, *Chaetoceros sociale*. Limnol Oceanogr 45: 1235-1244.

Legrand, C., K. Rengefors, G.O. Fistarol, E. Granéli E. 2003. Allelopathy in phytoplankton – biochemical, ecological and evolutionary aspects. Phycologia 42 (4): 406-419.

Lewis, W.M. Jr. 1986. Evolutionary interpretations of allelochemical interactions in phytoplankton algae. The American Naturalist 127: 184-194.

Macrellis, H.M., C.G. Trick, E.L. Rue, G. Smith and K.W. Bruland. 2001. Collection and detection of natural ion-binding ligands from seawater. Mar Chem 76: 175-187.

Maestrini, S.Y. and D. Bonin. 1981. Allelopathic relationships between phytoplankton species. Can Bull Fish 210: 323-338.

Maestrini S.Y. and E. Granéli. 1991. Environmental conditions and ecophysiological machanisms which led to the 1988 *Chrysochromulina polylepis* bloom: an hypothesis. Oceanologica Acta 14: 397-413.

Martin, D.F. and M.H. Gonzalez. 1978. Effects of salinity on synthesis of DNA, acidic polysaccharide, and growth in the blue-green alga *Gomphosphaeria aponina*. Water Res 12: 951-955.

Martin, D.F. and B.B. Martin. 1976. Aponin, a cytolytic factor toward the red tide organism, Gymnodinium breve. Biological assay and preliminary characterization. J. Environ Sci Health, Part A 11(10-11): 613-622.

Martin, B.B. and Martin D.F. 1987. Enhanced activity of a red tide (*Ptychodiscus brevis*) cytolytic fraction (aponin) from the green aga, *Nannochloris* sp.. J Environ Sci Health, Part A. A22(5): 457-462.

McCoy, L.F., D.L. Eng-Wilmot and D.F. Martin 1979. Isolation and partial purification of a red tide (*Gymnodinium breve*) cytolytic factor(s) from cultures of *Gomphosphaeria aponina*. J Agric Food Chem 27: 69-74.

Milligan, K.E., B.L. Marquez, R.T. Williamson and W.H. Gerwick. 2000. Lyngbyabellin B, a toxic and antifungal secondary metabolite from the marine cyanobacterium *Lyngbya majuscula*. J Natural Products 63: 1440-1443.

Mitsui, A., D. Rosner, A. Goodman and G. Reyes-Vasquez. 1989. Hemolytic toxins in marine cyanobacterium *Synechococcus* sp. pp. 367-370. *In* T. Okaichi, D.M. Anderson and T. Nemoto [eds]. Red Tides : Biology, Environmental Science and Toxicology. Elsevier Science Publishing, New York, USA.

Moebus, K. 1972a. Seasonal changes in antibacterial activity of North Sea water. Mar Biol 13: 1-13.

Moebus, K. 1972b. Bactericidal properties of natural and synthetic sea water as influenced by addition of low amounts of organic matter. Mar Biol 15: 81-88.

Molisch, H. 1937. Der Einfluss eine Pflanze auf die andere: Allelopathie. Gustav Fischer, Jena, Germany.

Moon, R.E. and D.F. Martin. 1981. The cytolytic substance Aponin on *Prymnesium parvum* and *Ptychodiscus brevis*, a comparative study. Bot Mar 24: 591-593.

Mulderij, G., E. Van Donk and G.M. Roelofs. 2003. Differential sensitivity of green algae to allelopathic substances from *Chara*. Hydrobiologia 491: 261-271

Murphy, C.D., R.M. Moore and R.L. White. 2000. An isotopic labeling method for determining production of volatile. Limnol Oceanogr 45: 1868-1871.

Naeem, S. and S. Li. 1998. Consumer species richness and autotrophic biomass. Ecology 79: 2603-2615.

Naustvoll, L.-J., E. Dahl and D. Danielssen. 2002. A new bloom of *Chatonella* in Norwegian waters. Harmful Algae News, IOC - UNESCO 23: 3,5.

Noordkamp, D.J.B., M. Schotten, W.W.C. Gieskes, L.J. Forney, J.C. Gottsschal and M. Van Rijssel. 1998. High acrylate concentration in the mucus of *Phaeocystis globosa* colonies. Aquat Microb Ecol 16: 45-52.

Noordkamp, D.J.B., W.W.C. Gieskes, J.C. Gottschal, L.J. Forney and M. van Rijssel. 2000. Acrylate in *Phaeocystis* colonies does not affect the surroundings bacteria. J Sea Res 13: 287-296.

Ohsawa, N., Y. Ogata, N. Okada and N. Itoh. 2001. Physiological function of bromoperoxidase in the red marine alga, *Corallina pilulifera*: production of bromoform as an allelochemical and the simultaneous elimination of hydrogen peroxide. Phytochemistry 58: 683-692.

Orjala, J., D.G. Nagle, V.L. Hsu and W.H. Gerwick. 1995. Antillatoxin : an exceptionnaly ichthyotoxic cyclic lipopeptide from the tropical cyanobacterium *Lyngbya majuscula*. J Am Chem Soc 117: 8281-8282.

Pacifici E.H.K., L.L. McLeod, H. Peterson and A. Sevanian. 1994. Linoleic acid hydroperoxide-induced peroxidation of endothelial cell phospholipids and cytotoxicity. Free Radical Biology, Medicine 17: 285-295.

Parrish, C.C., G. Bodennec, J.L. Sebedio and P. Gentien. 1994a. Intra- and extracellular lipid classes in batch cultures of the toxic dinoflagellate, *Gyrodinium aureolum*. Phytochemistry 32: 291-295.

Parrish, C.C., G. Bodennec and P. Gentien. 1994b. Time courses of intracellular and extracellular lipid classes in batch cultures of the toxic dinoflagellate, *Gymnodinium* cf. *nagasakiense*. Mar Chem 48: 71-82.

Pennings, S.C., A.M. Weiss and V.J. Paul. 1996. Secondary metabolites of the cyanobacterium *Microcoleus lyngbyaceus* and the sea hare *Stylocheilus longicauda* : palatability and toxicity. Mar Biol 126: 735-743.

Pennings, S.C., S.R. Pablo and V.J. Paul. 1997. Chemical defenses of the tropical, benthic marine cyanobacterium *Hormothamnion enteromorphoides* : Diverse consumers ans synergisms. Limnol Oceanogr 42: 911-917.

Perez, E., W.G. Sawyers and D.F. Martin. 1997. Identification of allelopathic substances produced by *Nannochloris oculata* that affect a red tide organism, *Gymnodinium breve*. Biomedical Letters 56(221): 7-14.

Perez, E. 1999. Production of bioactive natural products by the green alga *Nannochloris oculata* and *Nannochloris eucaryotum* that inhibit *Gymnodinium breve* cultures. Thesis, University of South Florida, USA.

Perez, E., F.A. Booth F.A. and D.F. Martin. 2000. Modeling organism-organism interactions as a means of predicting Florida red tide mangement. J Environ Health, Part A. 35: 219-227.

Perez, E., W.G. Sawyers and D.F. Martin. 2001. Lysis of *Gymnodinium breve* by cultures of the green alga *Nannochloris eucaryotum*. Cytobios 104(405): 25-31.

Pernet, F., R. Tremblay, E. Demers and M. Roussy, M. 2003. Variation of lipid class and fatty acid composition of *Chaetoceros muelleri* and *Isochrysis* sp. grown in a semicontinuous system. Aquaculture 221: 393-406.

Pintner, I.J. and V.L. Altmeyer. 1973. Production of vitamin B_{12} binder by marine phytoplankton. J Phycol 9: no. Suppl.

Pratt, R. 1940. Studies on *Chlorella vulgaris*. I. Influence of the size of the inoculumon the growth of *Chlorella vulgaris* in freshly prepared culture medium. Am J Bot 27: 52-56

Pratt, D.M. 1966. Competition between *Skeletonema costatum* and *Olisthodiscus luteus* in Nagarransett Bay and in culture. Limnol Oceanogr 11: 447-455.

Pushparaj, B., E. Pelosi and F. Jüttner. 1999. Toxicological analysis of the marine cyanobacterium *Nodularia harveyana*. J Appl Phycol 10, 527-530.

Putnam, A., Chung-shih Tang, 1986. Allelopathy: The State of the Science. *In* A. Putnam and C. Tang [eds]. The Science of Allelopathy. John Wiley and Sons, New York, USA.

Reitan, K.I., J.R. Rainuzzo and Y. Olsen. 1994. Effect of nutrient limitation on fatty acid and lipid content of marine microalgae. J Phycol 30: 972-979.

Rice, E.L. 1984. Allelopathy, 2nd ed. Academic Press, New York, USA.

Rossi, J.V., M.A. Roberts, H.D. Yoo and W.H. Gerwick. 1997. Pilot scale culture of the marine cyanobacterium *Lyngbya majuscula* for its pharmaceutically-useful natural metabolite curacin A. J Applied Phycol 9(3): 195-204.

Roessler, P.G. 1990. Environmental control of glycerolipid metabolism in microalgae : commercial implication and future research directions. J Phycol 26: 393-399.

Rue, E. and K. Bruland. 2001. Domoic acid binds iron and copper: a possible role for the toxin produced by the marine diatom *Pseudo-nitzschia*. Mar Chem 76: 127-134.

Saliot, A. 1975. Fatty acids, sterols and hydrocarbons in marine environment: inventory, biological and geochemical applications. (Book monograph). Univ. Pierre and Marie Curie, Paris, France.

Schmidt, L.E. and P.J. Hansen. 2001 Allelopathy in the prymnesiophyte *Chrysochromulina polylepis*: Effect of cell concentration, growth phase and pH. Mar Ecol Progress Series 216: 67-81.

Schmitt, T.M., M.E. Hay and N. Lindquist. 1995. Constraints on chemically mediated coevolution : Multiple functions for seaweeds secundary metabolites. Ecology 76: 107-123.

Shilo, M. 1967. Formation and mode of action of algal toxins. Bacteriological reviews 31: 180-193.

Shilo, M. 1981. The toxic principle of *Prymnesium parvum*. pp 37-47. In: Carmichael W. W. (ed) The Water Environment: Algal Toxin and Health. Plenum Press, New York, USA.

Sieburth, J.McN. 1960. Acrylic acid, an "antibiotic" principle in *Phaeocystis* blooms in Antarctic waters. Science 132: 676-677.

Skovgaard, A. and P.J. Hansen. 2003. Food uptake in the harmful *Prymnesium parvum* mediated by excreted toxins. Limnol Oceanogr 48(3): 1161-1166.

Slezak, D.M., S. Puskaric and G.J. Herndl. 1994. Potential role of acrylic acid in bacterioplankton communities in the sea. Mar Ecol Progress Series 105: 191-197.

Smetacek, V. 2001. A watery arm race. Nature 411 (6839): 745.

Smith, B.C. 1994. Investigating allelopathy between the marine microalgae : Some species familliar to aquaculture. J Shellfish Res 13: 319-320.

Smith, G.D. and N.T. Doan. 1999. Cyanobacterial metabolites with bioactivity against photosynthesis in cyanobacteria, algae and higher plants. J Applied Phycol 11: 337-344.

Sola, F., A. Masoni, B. Fossat, J. Porthe-Nibelle, P. Gentien and G. Bodennec. 1999. Toxicity of fatty acid 18:5n3 from *Gymnodinium* cf. *mikimotoi* : I. Morphological and biochemical aspects on *Dicentrarchus labrax* gills and intestine. J Applied Toxicol 19: 179-284.

Srivastava, A., F. Jüttner, R.J. Strasser. 1998. Action of the allelochemical, fischerellin A, on the protosystem II. Biochimica et Biophysica Acta, 1364: 326-336.

Steinke, M., G. Malin and P.S. Liss. 2002. Trophic interactions in the sea : an ecological role for climate relevant volatiles? J Phycol 38: 630-638.

St. John, M.A., C. Clemmensen, T. Lund and T. Koester. 2001. Diatom production in the marine environment : implications for larval fish growth. ICES J Mar Sci 58: 1106-1113.

Stolte, W., C. Karlsson, P. Carlsson and E. Granéli. 2002. Modeling the increase of nodularin content in Baltic Sea *Nodularia spumigena* during stationary phase in phosphorus-limited batch cultures. FEMS Microbiology Ecology 41: 211-220.

Subba Rao, D.V., Y. Pan and S.J. Smith. 1995. Allelopathy between *Rhizosolenia alata* (Brightwell) and the toxinogenic *Pseudonitzschia pungens f. multiseries* (Hasle). pp. 681-686. *In* P. Lassus, G. Arzul, E.Erard, P. Gentien and C. Marcaillou [eds]. Harmful Marine Algal Blooms. Technique et documentation-Lavoisier, Intercept Ltd. Paris, France.

Sugg, L.M. and F.M. VanDolah. 1999. No evidence for an allelopathic role of okadaic acid among ciguatera-associated dinoflagellates. J Phycol 35: 93-103.

Thacker, R.W., D.G. Nagle and V.J. Paul. 1997. Effects of repeated exposures to marine cyanobacterial secondary metabolites on feeding by juvenile rabbitfish and parrotfish. Mar Ecol Progress Series 147: 21-29.

Tillmann U. and U. John. 2002. Toxic effects of *Alexandrium* spp. on heterotrophic dinoflagellates: an allelochemical defence mechanism independent of PSP-toxin content. Mar Ecol Progress Series 230: 47-58.

Turner, W.B. 1971. Fungal metabolites. Academic Press, London, USA.

Turner, S.M., G. Malin, P.D. Nightingale and P.S. Liss. 1996. Seasonal variation of dimethyl sulphide in the North Sea and an assessment of fluxes to the atmosphere. Mar Chem 54: 245-262.

Vardi, A., D. Schatz, K. Beeri, U. Motro, A. Sukenik, A. Levine and A. Kaplan. 2002. Dinoflagellate-cyanobacteria communication may determine the composition of phytoplankton assemblage in a mesotrophic lake. Current Biol 12: 1767-1772.

Verity, P.G., T.A. Villareal and T.J. Smayda. 1988. Ecological investigations of blooms of colonial *Phaeocystis pouchetii*. 2. The role of life-cycle phenomena in bloom termination. J Plankton Res 10: 749-766.

Whittacker, R.H. and P.P. Feeny. 1971. Allelochemics: chemical interactions between species Science. 171(973): 757-70.

Wolfe, G.V. and M. Steinke. 1996. Grazing-activated production of dimethyl sulfide (DMS) by two clones of *Emiliania huxleyi*. Limnol Oceanogr 41: 1151-1160.

Yamamoto, M., 2001. Evolutionary relationships among multiple modes of cell division in the genus *Nannochloris* (Chlorophyta) revealed by genome size, actin gene multiplicity and phylogeny. J Phycol 37: 106-120.

Yang, S.Z., L.J. Albright and A.N. Yousif. 1995. Oxygen-radical-mediated effects of the phytoplankter *Heterosigma carterae* on the juvenile rainbow trout *Oncorhynchus mykiss*. Dis Aquat Org 23: 101-108.

Yasumoto, T., B. Underdahl, T. Aune, V. Hormazabal, O.M. Skulberg and Y. Oshima. 1990. Screening for haemolytic and ichthyotoxic components of *Chrysochromulina polylepis* and *Gymnodinium aureolum* from Norwegian coastal waters. pp. 436-440. *In* E. Granéli, B. Sundström, L.Edler and D.M. Anderson [eds]. Toxic Marine Phytoplankton. Elsevier, New York. USA

Yongmanitchai, W. and O.P. Ward. 1991. Screening of algae potential alternative sources of eicosapentaenoic acid. Phytochemistry 30: 2963-2967.

Wolfe, G.V. 2000. The chemical defence ecology of marine unicellular plankton. Biological Bulletin 198: 225-244.

5

Algal Blooms and Bacterial Interactions

Bopaiah Biddanda[1], *Paulo Abreu*[2] *and Clarisse Odebrecht*[2]

[1]Annis Water Resources Institute and Lake Michigan Center, Grand Valley State University, 740 W Shoreline Drive, Muskegon, MI 49441, USA.
[2]Departmento de Oceanografia, Fundação Universidade Federal do Rio Grande (FURG), Caixa Postal 474, 96201-900 Rio Grande, RS, Brazil.

Abstract

Interaction between phytoplankton and bacteria is the most critical ecological relationship prevalent in pelagic ecosystems. How algae and bacteria interact has major implications for the flow of carbon and nutrients, and the structure and function of the aquatic food web. Frequently, bacterioplankton abundance and activity track those of phytoplankton – even during spring blooms of phytoplankton. Occasionally, however, disconnects (or decoupling) occur between phytoplankton and bacteria – such as those observed during some surf zone diatom blooms and coastal blooms of cyanobacteria. The cause of these disconnects may be bacterial inhibition by low temperatures or bactericidal exudates of the phytoplankton, significant rates bacterivory/viral induced cell lysis, or bacterial dependence primarily on substrates other than those of phytoplankton origin (e.g., terrigenous materials). However, the relationship between bacteria and phytoplankton is not unidirectional, i.e. bacteria too may stimulate or inhibit microalgal growth by the production of growth factors or even algicides. Furthermore, the action of specific bacteria may influence phytoplankton succession, leading to the emergence and control of harmful algal blooms. External factors, like inorganic nutrient availability or UV radiation, can interfere in the balance between phytoplankton and bacteria, with direct influence on the local aquatic food web. Conversely, the phytoplankton-bacteria dynamics during and after algal blooms can affect large-scale phenomena such as dimethyl sulfide (DMS) production and sequestration of carbon dioxide from the atmosphere. By combining field observations and insights gained by using algal cultures as analogs of natural blooms of algae, this chapter attempts to describe the bidirectional interactions between algae and bacteria occurring in freshwater and marine pelagic environments.

INTRODUCTION

The interactions of primary and secondary producers are a central consideration in ecology (Pomeroy 1991). In the vast pelagic environments of both freshwater and marine ecosystems, the principal autotrophs are the phytoplankton, whereas the principal heterotrophs are bacterioplankton (Cotner and Biddanda 2002). Once phytoplankton fix carbon into organic matter in surface waters, heterotrophic processes within the food web dominated by bacterioplankton process carbon and associated elements until it becomes stored in sediments or is exchanged with the atmosphere (Sherr and Sherr 1996, Azam 1998, Falkowski et al. 1998, Cotner and Biddanda 2002). Therefore, understanding the nature of algal-bacterial interactions is central to the study of aquatic ecosystems.

Most of the organic matter in pelagic ecosystems directly or indirectly originates from phytoplankton. Phytoplankton influence the composition of natural waters by the uptake of inorganic carbon and nutrients for synthesis of cellular organic materials (Eppley and Peterson 1979). Furthermore, the occurrence of phytoplankton blooms has been observed to cause physicochemical changes in the water milieu through redistribution of inorganic nutrients (McAllister et al. 1961) and synthesis and release of organic compounds (Mague et al. 1980, Jenkinson and Biddanda 1995, Biddanda and Benner 1997a). As the major sink for dissolved organic matter, heterotrophic bacterioplankton depend on phytoplankton for sustenance (Cole et al. 1988, Van den Meerche et al. 2004).

Measurements routinely show that up to half of pelagic primary production is channeled through bacterioplankton (Cole et al. 1988, Ducklow 2000, Biddanda and Cotner 2002). Consequently, heterotrophic bacteria are now recognized as major consumers of organic matter in natural waters (Williams, 1981, Sherr and Sherr, 1996, Azam 1998, Cole 1999, Karl 1999, Cotner and Biddanda 2002). However, due to their high levels of cellular N and P, bacteria have a higher N and P requirement, and sometimes compete with phytoplankton for dissolved inorganic nutrients (Kirchman 2000, Cotner and Biddanda 2002). Nonetheless, a host of cross-ecosystem observations have demonstrated that microheterotrophic activity is closely associated with that of primary producers (Cole et al. 1988, Cotner and Biddanda 2002). Indeed, numerous studies have also shown that bacterial abundance and activity are enhanced in the vicinity of phytoplankton blooms (Sieburth 1968, Coveny and Wetzel 1995, Simon et al. 1998). But others have demonstrated the opposite influence (viz., decoupling between phytoplankton and bacteria) depending upon phytoplankton community composition (Abreu et al. 2003, Mayali and Azam 2004). For example, Sieburth (1968) has extensively considered the food web consequences of algal antibiosis.

In oligotrophic systems, bacterial production + respiration can approach that of primary production, decreasing in relative significance in eutrophic systems (Williams, 1984, del Giorgio et al. 1997, Cole 1999, Biddanda et al. 2001). Consequently, the relative biomass as well as the magnitude of carbon flux through bacteria is larger in oligotrophic than in eutrophic water bodies (Biddanda et al. 1994, del Giorgio et al. 1997, Biddanda et al. 2001). In oceanographic as well as limnetic literature, there is emerging evidence that the contribution of heterotrophic bacteria to total planktonic biomass and metabolic activity is high in oligotrophic systems and decreases along an increasing productivity gradient (Biddanda et al.1994, 2001, del Giorgio et al. 1997, Cotner and Biddanda 2002). Understanding such trends in the distribution of biomass and activity of autotrophs and heterotrophs across productivity gradients is likely to contribute to our knowledge of phytoplankton-bacterioplankton interactions and biogeochemical cycling occurring within phytoplankton bloom events in nature (Fig. 5.1).

Blooms are, by nature, seasonal. There is emerging evidence in the literature for seasonal cycles of planktonic as well as bacterial metabolic activity (Pomeroy et al. 1991, Pomeroy and Wiebe, 1993, Griffith and Pomeroy1995; Biddanda and Cotner, 2002). Seasonal cycles in planktonic activity have major impacts on the food web structure and carbon balance of natural waters (Pomeroy and Wiebe 1993, Sherr and Sherr 1996). For example, several studies have recorded large buildup of dissolved organic carbon (DOC) in surface waters during the winter-spring period followed by its drawdown during fall-winter (Carlson et al. 1994, Williams 2000, Biddanda and Cotner 2002). Strong positive correlations have been observed between the production of phytoplankton and bacterioplankton over the seasons in intensely monitored Lake Lawrence, Michigan (Coveney and Wetzel 1995), and Lake Constance, Germany (Simon et al. 1998). However, seasonal studies in Newfoundland waters, on the southeastern US continental shelf, and in Lake Michigan have all recorded shifts between net autotrophy during winter-spring and net heterotrophy during summer-fall (Pomeroy et al. 1991, Griffith and Pomeroy, 1995, Biddanda and Cotner, 2002). The formation and fate of blooms have implicit large-scale consequences through its effects on food web dynamics and carbon sequestration in aquatic ecosystems. At present, we are striving to understand these interactions.

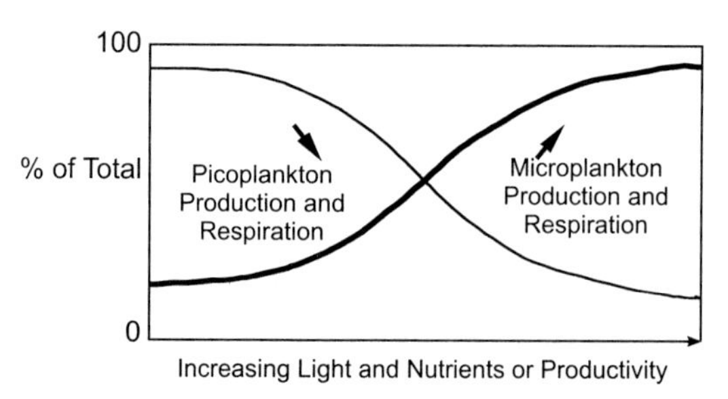

Fig. 5.1 Broadly generalized conceptual diagram of how plankton composition and metabolic rate processes respond to Light, Nutrient and Productivity gradients affecting carbon balance in natural waters. There is a systematic shift in the relative importance of small versus large organisms to planktonic biomass and metabolism across this gradient. High abundance and activity of small picoplankton (< 1.0 μm) prevail under low nutrient-light conditions and low rates of primary production. Autotrophic and heterotrophic processes are in near balance in low-productivity environments (resembling non-bloom conditions) that are dominated by dissolved nutrients and osmotrophic microorganisms. This situation is reversed when the euphotic zone is influenced by high nutrient-light inputs under which conditions, metazoan phagotrophs, and larger microplankton such as large diatoms and colonial cyanobacteria (> 1.0 μm) play a relatively greater role. Autotrophic and heterotrophic processes are not quite in balance in high primary productivity environments (resembling bloom conditions) that are dominated by particulate matter and larger phagotrophic plankton (Figure modified from Biddanda et al. 2001).

A combined approach, one utilizing field observations as well as phytoplankton cultures, can aid in the understanding of the complex interactions taking place between algae and bacteria under bloom conditions. In order to present this case, we have relied on the published literature, and utilized extensively results obtained in our respective laboratories in the USA and Brazil.

METHODS

Standard limnological and oceanographic methods of analysis have been used in the studies that are discussed here. For example, phytoplankton abundance was determined by inverted microscopy (Utermöhl 1958), heterotrophic bacterial abundance by epifluorescence microscopy (Hobbie et al. 1977), Chlorophyll *a* by fluorometry (Parsons et al. 1984), and batch cultures of phytoplankton were grown using standard seawater-based growth media (Guillard and Ryther 1962). Seawater dilution growth cultures were set up by adding 1 part < 1.0 μm filtered water (containing bacteria) to 9 parts < 0.2 μm filtered water (containing few bacteria and viruses, but all of the dissolved nutrients) according to the method of

Ammerman et al. 1984. More recently, molecular biology techniques like fluorescent *in situ* hybridization – FISH, Dot-Blot hybridization, denaturating gel gradient electrophoresis – DGGE, and sequencing of 16sRNA have been applied for the determination of bacterial species composition present in nature and in many diatom and flagellates cultures (Groben et al. 2000, Riemann et al. 2000, Mayali and Doucette, 2002, Shäfer et al. 2002, Green et al. 2004, Hare et al. 2004, Mayali and Azam 2004). Additional methodological details can be found in Biddanda 1988, Biddanda and Pomeroy 1988, Abreu et al. 1992, Odebrecht et al. 1995, Biddanda and Benner 1997a, b, Biddanda et al. 2001, Biddanda and Cotner 2003, and Abreu et al. 2003.

ALGAL BLOOMS IN LAKES AND THE OCEAN

One of the classic seasonal cycles of life observed in temperate and polar seas and lakes is the seasonal increase of phytoplankton in the spring, known as the spring bloom. Although the causal mechanisms of bloom formation in temperate lakes and oceans are not fully understood, they are thought to be a response to increased light in the presence of excess nutrients and increasing water column stability (Sverdrup 1953). Blooms here are usually typified by a temporal disconnect between the autotrophs and heterotrophs, enabling substantial autotrophic biomass to buildup during the initial phases of the bloom. In the subtropics and tropics, blooms are sporadic, forced by physical events (such as upwellings and lateral intrusions of water masses) that inject nutrients that are limiting for plankton growth into the euphotic zone. Blooms here occur under a variety of circumstances, and do not always involve a temporary imbalance between autotrophic phytoplankton and heterotrophic bacteria and zooplankton (Pomeroy 1991). Several studies have monitored the succession of phytoplankton species within bloom events (Barlow et al. 1993, Sieracki et al. 1993), and the immediate (Pomeroy and Deibel 1986) as well as eventual fate of blooms (Smetacek 1985). The formation and fate of blooms in lakes (e.g., *Microcystis, Anabena, Cylindrospermopsis*) and the ocean (e.g., *Alexandrium, Trichodesmium, Phaeocystis, Emiliania*) are of considerable current interest because of the implications for human health and the global carbon cycle.

BACTERIAL INTERACTIONS WITHIN ALGAL BLOOMS

As most of the dissolved organic matter in the pelagic ecosystem originates directly or indirectly from phytoplankton production (Biddanda and Pomeroy 1988), bacterioplankton are dependent on phytoplankton for growth substrates – and typically consume a large fraction of pelagic primary production (Cole et al. 1988, Cotner and Biddanda 2002). In fact, several studies of spring phytoplankton blooms in various locations have demonstrated tight coupling between algal and bacterial biomass and production (Sieburth 1968, Lancelot and Billen 1985, Coveney and Wetzel 1995, Simon et al. 1998, Weisse et al. 2000). However, other studies have noted a characteristic time delay between phytoplankton blooms and increase in bacterial activity (Ducklow 2000; Van den Meersche 2004), as well as a continual offset between bloom-forming algae and heterotrophic bacteria (Abreu et al. 2003). There is also some evidence in the literature that the increase in abundance of algicidal bacteria may coincide with the decline of algal blooms (Mayali and Azam 2004).

Indeed, there may be situations where little or none of the primary production is consumed contemporaneously by bacteria. During the early phases of the spring bloom of phytoplankton in Newfoundland waters, measurements indicated that there was undetectable levels of bacterial activity (presumably inhibited by the low temperatures), whereas phytoplankton continued to grow (Pomeroy and Deibel 1986, Pomeroy et al. 1991) – arguably sustaining the high productivity of spring time fisheries in these waters. However, low temperature inhibition of bacterioplankton can be apparently overcome in the presence of excess dissolved organic substrates (Pomeroy and Wiebe 1993, 2001).

Trichodesmium Bloom – Bacteria Interactions: A Positive Relationship

Trichodesmium is a globally significant colony-forming cyanobacterium that frequently forms blooms in the warm surface waters of tropical and sub-tropical oceans during the summer period – contributing significantly to oceanic biomass, production and N_2 fixation (Capone et al. 1997). In fact, *Trichodesmium* has been implicated in shifting the North Pacific Gyre from a traditionally N-limited system into a P-limited system during El Niño years (Karl et al. 1995) by accounting for up to a third of the new production in this environment through the release of nitrogenous compounds (Letelier and Karl 1996). It is thought that *Trichodesmium* blooms under warm (well-stratified) and nutrient depleted conditions in the N-limited surface ocean, where it has the advantage of being able to fix N_2. Eventually, however, the bloom collapses – probably due to P limitation – and because there are no known consumers of this episodically blooming colonial cyanobacteria, for the most part it sinks or enters the microbial food web where it undergoes decomposition.

We observed massive blooms of *Trichodesmium* during a cruise transecting the Gulf of Mexico during August 1995 (Biddanda 1995, Biddanda and Benner 1997b). Our measurements of *Trichodesmium* bundles and heterotrophic bacterial abundance from surface bucket samples collected during this period suggest there was a significant positive correlation between the colonial cyanobacteria and heterotrophic bacteria (Fig. 6.2). *Trichodesmium* colonies are known to support significant biomass and productivity of heterotrophic bacteria (Nausch 1996), and this may account for the positive relationship we observed between the primary producer and consumer groups in the pelagic ocean during this period (Biddanda 1995). Indeed, high primary production rates sustained by *Trichodesmium* may have fueled the relatively high water column respiration rates that were measured in the Gulf of Mexico during the summer of 1995 (Biddanda and Benner 1997b).

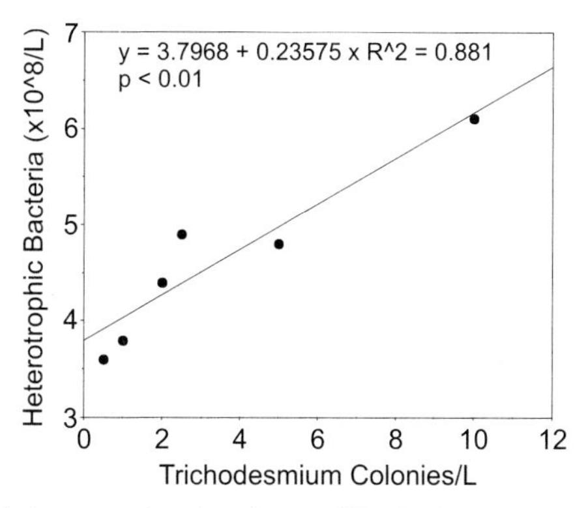

Fig. 5.2 Relationship between the abundance of *Trichodesmium* sp. colonies (1-3 mm long colonies, each composed of 5–10 filaments or trichomes) and heterotrophic bacteria in the surface waters of Gulf of Mexico (transect from Port Aransas, Texas to Key West, Florida) during August, 1995. Data from Biddanda 1995 and Biddanda and Benner 1997b.

Surf-zone Diatom Bloom – Bacteria Interaction: A Negative Relationship

The surf-zone diatom *Asterionellopsis glacialis* (Castracane) Round is the main primary producer in the intermediate/dissipative Cassino Beach in southern Brazil near the entrance to one of the largest coastal lagoons in the world (Abreu et al. 1992, 2003). High diatom abundance ($10^8 - 10^9$ cell l^{-1}) in the surf-zone sustains a huge biomass – up to a maximum of 1647 μg L^{-1} chlorophyll *a* (Odebrecht et al. 1995). This surf-zone diatom biomass is mainly consumed by benthic filter-feeding organisms like the crustacean *Emerita brasiliensis* and the bivalves *Mesodesma mactroides* and *Donax hanleyanus*. These benthic organisms, in turn, serve as food for rich communities of fish and migratory birds (Garcia and Gianuca 1997).

High abundance of *A. glacialis* is not just the result of a bloom process, but it is mainly a consequence of cell accumulation generated by wave and winds that resuspend the diatoms from the sediment and concentrate them in the surf-zone. Due to wind action, most of the cells are deposited on the beach where they are consumed by the benthic macroinvertebrates, but some are transported back to the sediment behind the surf-zone by rip currents (Odebrecht et al. 1995; Rörig and Garcia 2003). Besides the high biomass, *A. glacialis* produce large quantities of dissolved organic carbon, reaching a maximum of 68% of total (particulate +dissolved) diatom carbon production, which may reach 3.44 mg C L^{-1} h^{-1} (Reynaldi 2000). Additionally, riverine input of dissolved organic matter (DOM) from the nearby Patos Lagoon Estuary supplies high amounts of DOM to Cassino Beach waters. During the occurrence of *A. glacialis* patches, DOM can be as high as 7 mg C L^{-1}.

Such enhanced concentrations of dissolved organic carbon should sustain high amount of bacterial biomass. However, this is not what is observed at Cassino Beach, where typical bacterial abundance levels were about an order of magnitude lower than those reported for environments with comparable DOM and Chlorophyll *a* values (Cole et al. 1988). A year round study measuring Chlorophyll *a* and bacterial abundance in the Cassino Beach (August 1992-July 1993, Abreu et al. 2003) showed an inverse relationship between bacteria and *A. glacialis* (Fig. 5.3a). Furthermore, daily observation of events at the end of a bloom from 1–22 July 1997 also confirmed the decoupling between *A. glacialis* and heterotrophic bacteria in this surf-zone environment (Fig. 5.3b). Bacterial abundance did not show any significant increase during the first three days after the decrease in *A. glacialis*. However, following this initial period, bacterial abundance increased steadily thereafter (Abreu et al. 2003).

Previous studies carried out in the dissipative beaches of South Africa, where the surf-zone diatom *Anaulus australis* Drebes et Schulz dominate, concluded that large amounts of dissolved primary production would be consumed by bacteria (McLachlan and Bates 1985; Heymans and McLachlan 1996). However, the results from the Cassino Beach study pointed out a clear and consistent trend of decoupling between bacteria and

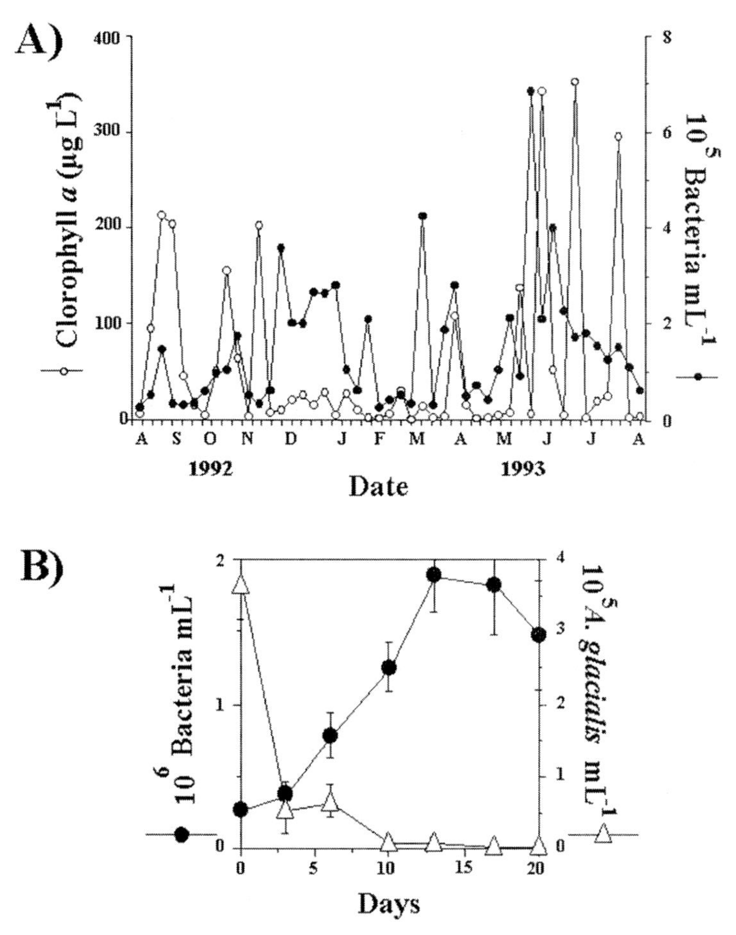

Fig. 5.3 Heterotrophic bacteria abundance variation in Cassino Beach, Brazil during August 1992 to July 1993 (Fig. 5.3a), and immediately following the collapse of the surf-zone diatom *Asterionellopsis glacialis* bloom during July 1-22, 1997 (Fig. 5.3b). Chlorophyll *a* peaks in Fig. 5.3a are caused by the *A. glacialis*. Figures modified from Abreu et al. 2003

the surf-zone diatom *A. glacialis* (Abreu et al. 2003). Laboratory experiments measuring the bacterial growth in *A. glacialis* cultures and in the water from the beach sampled during the occurrence of the surf-zone diatom patches showed a large lag phase in the bacterial growth curve, indicating suppression of bacterial activity (see below).

Abreu et al. 2003 considered five possible reasons for this diatom-bacteria decoupling: 1) viral infection, 2) bacterial grazing, 3) DOC quality, 4) nutrient competition and 5) antibiotic production. Bacterial grazing and nutrient competition were discarded, since measurements of protozoan grazing activity conducted in the Cassino Beach did not justify the low bacterial abundance observed (Hickenbick 2002). Moreover, the dissolved inorganic nutrients measured in this beach were low, but do not characterize an oligotrophic condition, especially after high rainfall, when large amounts of freshwater enter into the coastal region. The possible effect of viral infection (Bratbak et al 1990, Fuhrman 2000) remains to be tested and, though not measured, the antibiotic production by this diatom was considered as the most plausible hypothesis. Indeed, the antibiotic produced by *A. glacialis* was first described by Aubert et al. (1970) as a nucleoside made up of pyrimidinic base and a pentose sugar, probably arabinose. Its structure is analogous to thymidine and it has an antimitotic action.

Despite all evidences that the antibiotic produced by *A. glacialis* could cause the decoupling between this diatom and bacteria, it is very likely that the antibiotic action could be the result of the overall quality of dissolved organic carbon exuded by the surf-zone diatom. For instance, Sander and Purdie (1998) demonstrated that the response of bacteria to blooms of the coccolithophorid *Emiliana huxleyi* and the diatom *Skeletonema costatum* were quite different mainly due to DOC quality produced by both organisms. In the case of *E. huxleyi*, peaks of the coccolithophorid and bacteria were almost synchronous, while bacteria took about one week to start growing after the *S. costatum* attained maximum abundance. These differences were related to the release of simpler monomeric organic compounds by *E. huxleyi* in contrast to the more complex organic matter produced by *S. costatum*. The hypothesis whether DOC quality released by *A. glacialis* could influence the bacterial growth in the Cassino beach was further tested by us in laboratory studies (see below).

Coastal Harmful Algal Blooms – Bacterial Interactions in River-Dominated Systems

The emergence of harmful algal bloom (HAB) species in coastal waters has been an area of increasing interest and study (Anderson 1997, Smayda 1997). Due to toxin production and possible ecological and human health problems, the implications of algae-bacteria relationship in such environments are of concern. Several field studies have pointed out the importance of bacteria in the initiation, development and termination of harmful algal blooms (Doucette et al. 1996). Such possible participation of bacteria in the toxin production or biotransformation has been inferred by the fact that some putative toxigenic bacteria were present in high number when toxic dinoflagellates were blooming (Tobe et al. 2004). Furthermore, field observations have shown that bacteria can participate in the cyst formation of the toxic dinoflagellate *Alexandrium tamarense* (Adachi et al. 1999). On the other hand, this dinoflagellate had no effect on the bacterial community (Fistarol et al. 2004). It is possible that bacteria are also involved in the degradation of toxins in the water that are produced by toxic cyanobacteria (Maruyama et al. 2003). *On the* other hand, by regulating algal community structure, algicidal bacteria may influence the nature of algal bloom dynamics (Mayali and Azam 2004). Recent advances in molecular biology techniques could lead to better understanding of the interactions between bacteria and HAB species interaction.

In river dominated coastal water ecosystems, it is thought that nutrient discharge determines the balance between a predominantly microbial loop driven planktonic community at low nutrient levels, and a highly productive grazer dominated food web composed of larger plankton at high nutrient levels (Legendre and Michaud 1998, Dagg et al. 2004). Bacterial productivity is commonly found to be enhanced at the boundary region where river water mixes with coastal water, wherein bacteria appear to utilize both riverine organic matter as well as organic matter produced by the local phytoplankton (Chin-Leo and Benner 1992, Pakulski et al. 2000). Furthermore, studies suggest that terrestrial organic matter undergoes rapid transformations during its transit to coastal waters (Hedges et al. 1997), and that bacteria and sunlight interactions likely play a critical role in this process (Miller and Moran 1997, Biddanda and Cotner 2003).

We observed a weak positive relationship between the distribution of Chlorophyll *a* and heterotrophic bacteria along the Maumee River – Lake Erie transect in August 2003 (Biddanda and Tester, unpublished data, Fig. 5.4). Maumee River is a major tributary to Western Lake Erie (Wilhelm et al. 2003), where large blooms of the cyanobacteria *Microcystis* were occurring during the study period. Extremely high phytoplankton biomass values (> 100 μg Chlorophyll a l^{-1}) and heterotrophic bacterial production rates (> 300 μg C l^{-1} d^{-1}) were measured in parts of western Lake Erie. However, the apparent lack of a strong relationship between algal biomass and bacterial abundance that prevailed here (Fig. 5.4) may reflect the possibility that bacteria are sustained in part by terrigenous inputs via the tributary, and that they could also be negatively influenced by toxins that are known to be produced by *Microcystis* (Christoffersen 1996). Any production of algicides by bacteria would likely further weaken the prevailing relationship between bacteria and algae (Mayali and Azam 2004).

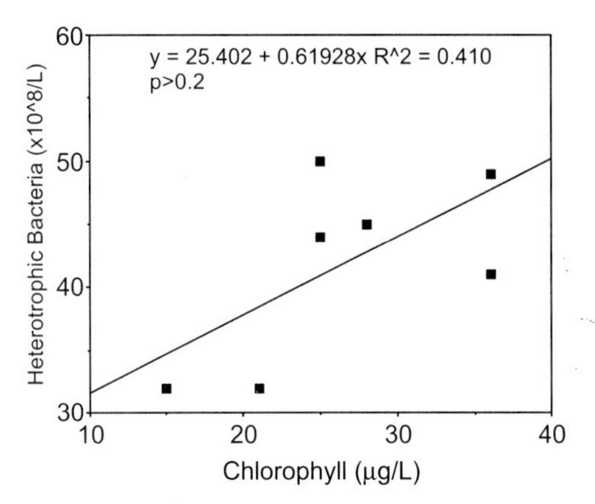

Fig. 5.4 Relationship between Chlorophyll *a* and heterotrophic bacterial abundance in Western Lake Erie in the vicinity of the Maumee River during August, 2003. Unpublished data of Biddanda and Tester.

CULTURES AS ANALOGS OF BLOOMS – INSIGHTS INTO COMMUNITY AND CARBON DYNAMICS

Cultures of algae can serve as useful analogs of blooms in nature by providing a controllable environment where algal growth and decomposition may be studied under well-defined conditions. Studies relating to succession of organisms as well as nutrient dynamics can be performed under controlled conditions using algal cultures in defined growth media (Biddanda 1988, Biddanda and Pomeroy 1988) or dilution cultures of bacteria in natural filtered water (Reynaldi 2000, Abreu et al. 2003).

Succession of Microorganisms and Fate of Phytoplankton Bloom in Cultures

Several studies have examined the succession of microorganisms and nutrient dynamics during the growth and death of phytoplankton using batch cultures of select algae (Fukami et al. 1985). In their studies, Biddanda and Pomeroy (1988) and Biddanda (1988) examined the microbial succession and carbon dynamics associated with detritus derived from three phytoplankton belonging to diverse taxanomic groups (*Synechococcus, Dunaliella and Cylindrotheca*) in natural seawater. The authors found a remarkably similar pattern of microbial succession involving heterotrophic bacteria and bacterivorous protozoa occurring in all three cases with resultant aggregation and disaggregation of the algal detritus – the same kind of scenario that is frequently observed following the death of algal blooms in natural waters.

Tracing the fate of algal derived carbon within the same experiment, Biddanda (1988) found that nearly half of the carbon in both the dissolved and particulate fractions was consumed by the bacteria with a higher proportion of the carbon being assimilated in the earlier stages of algal degradation than at the later stages. This suggests that dying blooms in nature initially fuel heterotrophic production, but will eventually fuel heterotrophic respiration. Additionally, the early aggregation phases of the algae may serve as food for metazoan consumers or as vehicles for vertical transport of surface derived carbon, whereas the later disaggregated phases may serve as substrates for microbial mineralization of limiting nutrients within the water-column.

Carbon, Nitrogen and Carbohydrate Fluxes During the Growth of Phytoplankton

Although the principal source of carbon in pelagic waters is phytoplankton, experimental data on carbon and nitrogen mass balance during their growth cycle is seriously lacking. The first study to address the carbon mass balance of phytoplankton during their growth cycle was that of Eppley and Sloan (1965). Biddanda and Benner (1997a) grew batch cultures of phytoplankton from widely different taxanomic groups (*Synechococcus, Phaeocystis, Emiliania,* and *Skeletonema*) in defined synthetic seawater media, and monitored changes in particulate and dissolved carbon, nitrogen and carbohydrates during the entire growth cycle. Over 14 days of the study, there was a close molar balance between dissolved inorganic carbon (DIC) uptake and total organic carbon (TOC = dissolved + particulate organic carbon) production in all the phytoplankton except *Emiliania*, which synthesizes carbonate-containing coccoliths. Dissolved organic carbon production by the phytoplankton was dominated by the carbohydrate fraction (primarily polysaccharides), constituted a substantial 10-30% of the TOC production, and was maximal during the senescent phases of the cultures. Furthermore, the phytoplankton produced dissolved organic matter consisted of ~35% high molecular weight compounds having high C:N ratios (~21) and ~65% low molecular weight compounds having low C:N ratios (~6). Excess carbohydrate production towards the later phases of the bloom, and the variable nitrogenous composition of the high and low molecular weight DOM fractions could have important consequences for the microbial food web and the eventual fate of carbon produced by algal blooms (Kepkay et al. 1993, Amon and Benner 1996, Biddanda and Benner 1997a).

Use of Cultures to Demonstrate Decoupling Between Surf-Zone Diatoms and Bacterioplankton

Most of the insights about the interaction between the surf-zone diatom *Asterionellopsis glacialis* and bacteria come from laboratory experiments. Regarding the influence of bacteria on the diatom growth, it was observed that a strain named *Pseudomonas* sp. 022 added to *A. glacialis* culture was able to stimulate the growth of this diatom, while the opposite occurred with the addition of bacteria classified as *Vibrio* sp. 05 (Riquelme et al. 1987). It was found that the *Pseudomonas* sp.022 produced a glycoprotein that serves as a growth factor to the algae. On the other hand, *A. glacialis* exudate inhibited the growth of *Staphylococus aureus* (Aubert et al. 1970) and *Vibrio* spp., while it stimulated the growth of a bacteria belonging to the *Pseudomonas* genera (Riquelme et al. 1989).

Such results demonstrate that bacteria can *potentially* influence the development and decline of algal blooms (Fukami et al. 1997, Mayali and Azam 2004). Furthermore, the different responses of bacteria species to *A. glacialis* could indicate that the antibiotic produced by this diatom, as described by Aubert et al. (1970), is very specific (i.e., not of a broad spectrum), or that different bacteria species could use the exudates produced by the algae in distinctly different ways. To test the hypothesis whether the decoupling between *A. glacialis* and bacteria in the Cassino Beach was caused by the DOC quality, Reynaldi (2000) conducted a series of seawater dilution culture experiments where bacterial growth was measured in filtered water collected inside and outside the diatom patches. The results were compared to those growing in dilution cultures with filtered water samples from the diatom patch that were enriched with Glucose (~100 µM). Bacteria in the glucose enriched water showed a faster growth with maximum abundance occurring about 48 h following bacteria inoculation. Subsequently, the next highest abundance of bacteria occurred in the water collected outside the diatom patch, followed by the water sampled directly in the region of maximum diatom abundance (inside the patch). Like the observations made by Sanders and Purdie (1998) with blooms of phytoplankton in nutrient-enriched microcosms, the exudate produced by *A. glacialis* is also of complex composition – such that it would restrict its immediate utilization by the bacteria, leading to temporary accumulation of dissolved organic carbon in the water.

Abreu et al. 2003 reexamined the issue of the inhibition of bacterial growth by the surf-zone diatom in two separate culture studies. In the first study, they utilized seawater dilution cultures that were established using filtered water from Cassino Beach. In dilution cultures set up during the period of maximum *A. glacialis* abundance (Day 0, in Fig. 5.3b), it was shown that the bacterial lag- phase extended up to 4 days. On the other hand, in dilution cultures set up during the period of minimum *A. glacialis* abundance following the collapse of the bloom at Cassino Beach (Day 14, in Fig. 5.3b), the initial bacterial lag-phase was reduced to just one day.

In the second study, Abreu et al. 2003 isolated *A. glacialis* from Cassino Beach and grew it in batch culture using F2 medium. Dilution cultures were set up using filtrate from the cultures (9 parts < 0.2 μm culture water) and an inoculum of bacteria from Cassino Beach Water (1 part < 1 μm water from Cassino Beach). In dilution cultures set up during the growth phase of *A. glacialis* cultures (Day 7), the bacterial lag-phase extended for a minimum of one full day. On the other hand, in dilution cultures set up during the senescent phase of *A. glacialis* cultures (Day 14), there was no discernible bacterial lag-phase at all. The above two experimental studies clearly demonstrate the inhibitory effect of *A. glacialis* on heterotrophic bacterial growth – the effect being significant during the growth phase of the surf-zone diatom when it is most abundant, and insignificant during the senescent phases of the surf-zone diatom life cycle. The authors hypothesized that such decoupling between the surf-zone diatom and bacteria may be a mechanism developed by the diatom to surpass bacteria in the direct competition for dissolved inorganic nutrients. If true, this is probably of vital importance to *A. glacialis*, considering the small window of time (few days) available for its growth between resuspension from the sediment and concentration of the algae in the surf-zone. One implication of these field observations and laboratory studies is that, due to bacterial growth suppression, large amount of diatom production is directly transferred to metazoan consumers, making the Cassino Beach and adjacent near shore habitat one of the most productive systems along the 8,500 km Brazilian coast (Garcia and Gianuca 1997, Seeliger et al. 1997).

Study of Bacteria – Harmful Algal Blooms (HAB) Interaction in Algal Cultures

The interaction between phytoplankton and bacteria can be positive as well as negative: algae and bacteria can influence the growth, activity, and even succession of each other, characterizing a very complex relationship, mostly developed by chemical signals of elements exuded into the water (Fukami et al. 1997, Gross 2003). It is estimated that the impacts of harmful algal blooms (HAB) have increased worldwide in the last years, especially due to anthropogenic inputs of nutrients in most coastal regions (Hallegraeff 1993, Anderson 1997, Smayda 1997). According to Green et al. (2004), bacteria play an important role in the formation and development of HAB not only by participating in the process of toxin production and biotransformation, but also influencing the algae population dynamics due to positive and negative allelopathic interactions.

Although the possible influence of bacteria on toxin production is still a matter of controversy, some laboratory studies have analyzed the importance of these microorganisms in, for example, the Paralytic Shellfish Toxin – PST production after germination of *Gymnodinium catenatum* cysts (Green et al. 2004), Domoic Acid (DA) production in *Pseudo-nitzschia multiseries* cultures (Bates et al. 2004), although in the latter case, no effect of bacteria on toxin production could be detected. More information about laboratory studies of bacterial phycotoxins production can be found in Doucette et al. (1996).

Other studies have examined the bacterial production of algicidal compounds that affect several HAB species (Doucette et al. 1999, Mayali and Doucette, 2002, Hare et al. 2004). According to Hare et al. (2004), algicidal bacteria could be used as a 'short-term' solution for HAB problems in coastal waters. Algicidal bacteria may show different degrees of specificity to HAB species, varying from a broad range of effects to specific action against one or more closely related species (Doucette et al. 1999). There are also differences regarding the way bacteria can act - some need direct contact with algal cells, while others excrete active compounds into the water (Doucette et al. 1999, Bates et al. 2004).

Most studies on bacterial algicidal effect begin with the isolation of bacteria present in the environment or in the water where microalgae is growing. Afterwards, a large amount of cultivated bacteria are added to the algae. Positive (stimulatory) or negative (inhibitory) action of bacteria is determined by the variation of microalgae abundance over time. It is important to stress that such high abundance of a specific group of bacteria rarely occurs in nature. Mayali and Doucette (2002) observed that algicidal action of the bacteria 41-DBG2 on the dinoflagellate *Karenia brevis* was only effective when specific bacterial abundance reached 10^6 cells ml^{-1}. The authors considered the possibility of a bacterial 'threshold concentration'. In this scenario, bacteria must reach high threshold abundance in order to trigger the algicidal response. In their survey of the literature, Mayali and

Azam (2004) conclude that evidence for bacterial algicidy causing the decline of algal blooms does exist, but it is circumstantial.

Another important aspect that has emerged in the recent years from laboratory studies is the phylogenetic identification of bacteria in microalgae cultures, especially HAB cells, using molecular biology techniques. Shäfer et al. (2002) studied the diversity of accompanying bacterial communities, the so-called 'satellite' bacteria, in six algal cultures. In their study most bacterial communities had representative species of the ∝-Proteobacteria and Cytophaga-Flavobacterium-Bacteroides (CFB) groups, i.e., typical marine phylotype representatives. However, the bacterial community composition was unique for each phytoplankton culture and the bacterial diversity was always less in culture than in nature. The loss of bacterial species diversity in cultures probably reflects the lack of a complex suite of algal exudates within the cultures as opposed to that produced by a relatively diverse community of producers in natural waters.

According to Hare et al. 2004, all marine algicidal bacteria belong to Proteobacteria and CFB phylogenetic groups and similarities observed for bacterial flora among dinoflagellates cultures could be explained by selective mechanisms running in laboratories. The other possibility is that specific groups of bacteria may be favored, since they could be important to the growth and physiology of dinoflagellates cells (Green et al. 2004).

CONCEPTUAL MODEL OF ALGAE-BACTERIAL INTERACTIONS WITHIN BLOOMS

According to conceptual food web models developed by Legendre and LeFevre (1995), Legendre and Rassoulzadegan (1995), Biddanda et al. (2001), Cotner and Biddanda (2002), the non-bloom low productivity environments are likely to be characterized by microbial food webs composed of small cyanobacteria and phytoplankton, heterotrophic bacteria and protozoans – a system sustained by regenerated production. On the other hand, high productivity bloom environments are likely to be characterized by herbivorous food webs composed of larger phytoplankton and zooplankton – a system sustained by new production. Microbial food web dominated low productivity environments may produce little or no export of carbon to the lake/ocean floor, whereas herbivorous food web dominated bloom environments may result in significant vertical export flux of carbon (Peinert et al. 1989, Biddanda et al. 2001, Fig. 5.1).

From the many examples cited above, it is clear that the phytoplankton-bacteria interaction is bidirectional, and understanding the balance between both groups of microorganisms is essential if we are to comprehend the functioning of microbial food web and biogeochemical cycles in aquatic ecosystems. For example, as early as 1968 (Sieburth1968), investigators have considered the food web consequences of algal antibiosis. More recently, there has been considerable progress regarding the ecological and evolutionary role of algicidy by bacteria (Mayali and Azam 2004). In the revised conceptual model presented here, we emphasize the bidirectional characteristic of phytoplankton-bacteria interaction, and point to some external factors that can influence both microorganisms (Fig. 5.5). Conversely, due to their high biomass and ubiquitous presence in all aquatic systems, the results of phytoplankton-bacteria interaction have worldwide implications, influencing large-scale phenomena like global climate.

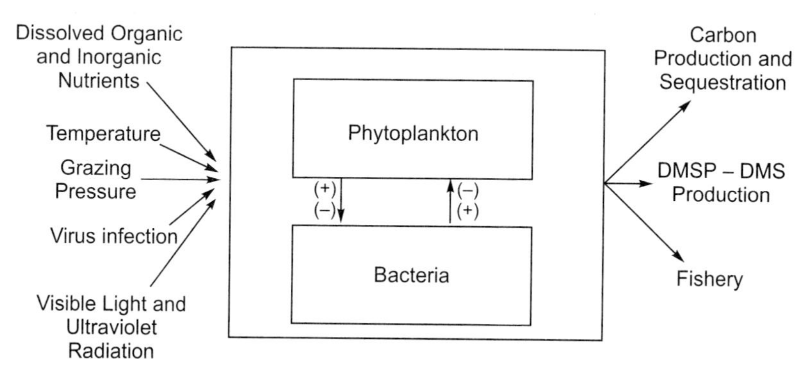

Fig. 5.5 Conceptual diagram emphasizing the bidirectional interactions taking place between phytoplankton and bacteria in natural waters. Phytoplankton and bacteria are both being influenced by biotic (competition for resources, mutualism/symbiosis, grazing pressure, virus infection, bactericidy, algicidy) and abiotic (temperature, dissolved organic and inorganic nutrients, light and ultraviolet radiation) factors. The outcome of these interactions, influence large scale events like global climate due to carbon production and sequestration, as well as, planetary cooling due to cloud formation nucleated by dimethyl sulfide (DMS), originating from dimethyl sulfoniopropionate (DMSP) produced by haptophytes and dinoflagellates. Similarly, the balance between phytoplankton and bacteria may have strong influence on the world fisheries. Positive and negative signals beside the arrows indicate the resulting positive and negative influences.

Although bacteria depend on phytoplankton as the primary source of organic matter for their nutrition, bacteria can significantly influence phytoplankton growth, succession, and decomposition. The way both organisms respond to each other, and the main consequences of this interaction, is highly influenced by external biotic and abiotic factors. For instance, bacteria and phytoplankton can respond in different ways to water temperature (Pomeroy and Wiebe 1993), dissolved inorganic and organic nutrients supply (Kirchman 2000, Cotner and Biddanda 2002), and the amount of visible light and ultraviolet radiation (Carrillo et al. 2002). Similarly, consumer grazing pressure (Strom 2000) and viral infection (Bratbak et al. 1990, Fuhrman 2000) can impact phytoplankton and bacteria differently.

Different outcomes of the phytoplankton-bacteria interaction may have considerable influence on large-scale events, such as global warming, due to carbon dioxide production and uptake (Hoppe et al. 2002, Biddanda and Cotner 2002), or due to the albedo generated by clouds formed by dimethyl sulfide (DMS). Major DMS source comes from dimethyl sulfoniopropionate (DMSP) found in high concentration in some haptophytes and dinoflagellates. The conversion of DMSP to DMS depends of the action of the enzyme DMSPlyase present in high amount in bacteria and in some microalgae (Burkill et al. 2002). Similarly, depending on the kind of interaction between microalgae and bacteria, microbial food webs may enhance or reduce, fisheries production (Pomeroy and Deibel 1986, Azam 1998, Azam and Worden 2004, Osidele and Beck 2004). The challenge for the future is to elucidate the complex ecological interactions amongst microbial plankton because they constitute the very bases of biogeochemical cycling of elements in natural waters.

CONCLUSIONS

Autotrophic phytoplankton and heterotrophic bacteria constitute two of the most fundamental and complementary functional units in pelagic ecosystems. Consequently, how algae and bacteria interact has important consequences to carbon and nutrient flow and food webs in aquatic ecosystems. In the pelagic environment, bacteria primarily depend on phytoplankton for organic matter. Bacterial mineralization of organic matter supplies limiting nutrients to the algae. Consequently both bacterial abundance and activity are usually closely linked to phytoplankton abundance and production. Exceptions to this rule do occur when algal exudates inhibit bacterial growth, resulting in reduced bacterial growth and accumulation of dissolved organic matter. On the other hand, bacteria may compete with phytoplankton for limiting nutrients, and the presence and activity of specific bacteria may influence species succession in the phytoplankton communities – potentially affecting the outcome of bloom composition including the emergence and control of harmful algal blooms.

Combining field and laboratory culture studies can be expected to provide increased understanding of the multi-faceted and vitally important interactions between algae and bacteria in nature. Recent advances in new molecular biology techniques has helped to refine the study of microalgae-bacteria interaction, enhancing our capacity to understand possible relationships at the species levels of both algae and bacteria. Nevertheless, laboratory studies allow us to conduct well-controlled experiments in order to the test hypotheses generated by observations made in nature.

ACKNOWLEDGEMENTS

We are thankful to an anonymous reviewer (Editorial Board), Pat Tester (NOAA), Ying Hong (University of Michigan), Scott Kendall and Sarah Barnhard (Grand Valley State University), for comments that helped improve the first draft of the manuscript. Authors are thankful to Gary Fahnenstiel (NOAA) for ship time on board the R/V Laurentian, and Pat Tester for the use of unpublished Chlorophyll *a* data for western Lake Erie. BB's work was supported through grants from NASA and NOAA. PA and CO were supported by the CNPq (Brazilian Science Foundation).

REFERENCES

Abreu, P.C., B.A. Biddanda and C. Odebrecht. 1992. Bacterial dynamics in the Patos Lagoon Estuary, southern Brazil: Relationship with phytoplankton production and suspended material. Est. Coast. Shelf. Sci. 35: 621-635.

Abreu, P.C., L.R. Rörig, V. Garcia and C. Odebrecht. 2003: Decoupling between bacteria and the surf-zone diatom *Asterionellospis glacialis* at Casssino Beach, Brazil. Aquat. Microbial Ecol. 32: 219-228.

Adachi M., T. Kanno, T. Matsubara, T. Nishijima, S. Itakura and M. Yamaguchi. 1999: Promotion of cyst formation in the toxic dinoflagellate *Alexandrium* (Dinophyceae) by natural bacterial assemblages from Hiroshima Bay, Japan. Mar. Ecol. Prog. Ser. 191: 175-185.

Ammerman, J.W., J. Fuhrman, A. Hagstrom and F. Azam. 1984. Bacterioplankton growth in seawater: 1. Growth kinetics and cellular characteristics in seawater cultures. Mar. Ecol. Prog. Ser. 18: 31-39.

Amon, R.M.W. and R. Benner. 1996: Bacterial utilization of different size classes of dissolved organic matter. Limnol. Oceanogr. 41: 41-51.

Anderson, D.M. 1997. Bloom dynamics of toxic *Alexandrium* species in the northeastern U.S. Limnol. Ocenogr. 42: 1009-1022.

Aubert, M., D. Pesando and M. Gauthier. 1970: Phenomenes d'antibiosi d'origine phytoplanctonique en milieu marin. Rev. Int. Océanogr Méd. 18-19: 69-76.

Azam, F. 1998: Microbial control of oceanic carbon flux: the plot thickens. Science 280: 694-696.

Azam, F. and A.Z. Worden. 2004. Microbes, molecules and marine ecosystems. Science 303: 1622-1624.

Barlow, R.G., R. Mantoura, M. Gough, and W. Fileman. 1993. Pigment signatures of phytoplankton composition in the northeastern Atlantic during the 1990 spring bloom. Deep-Sea Res. II. 40: 459-477.

Bates, S.S., J. Gaudet, I. Kaczmarska and J.M. Ehrman. 2004. Interaction between bacteria and the domoic-acid-producing diatom *Pseudo-nitzschia multiseries* (Hasle) Hasle; can bacteria produce domoic acid autonomously? Harmful Algae 3: 11-20.

Biddanda, B.A. 1988. Microbial aggregation and degradation of phytoplankton-derived detritus in seawater. II. Microbial metabolism. Mar. Ecol. Prog. Ser. 42: 89-95.

Biddanda, B.A. 1995. *Trichodesmium* bloom in Gulf of Mexico, summer 1995. Harmful Algal News (IOC-UNESCO Newsletter on toxic algae and algal blooms), 12-13: 2.

Biddanda, B.A. and L.R. Pomeroy. 1988. Microbial aggregation and degradation of phytoplankton-derived detritus in seawater. I. Microbial succession. Mar. Ecol. Prog. Ser. P42: 79-88.

Biddanda, B.A., S. Opsahl and R. Benner. 1994. Plankton respiration and carbon flux through bacterioplankton on the Louisiana shelf. Limnol. Oceanogr. 39: 1259-1275.

Biddanda, B.A., and R. Benner. 1997a. Carbon, nitrogen and carbohydrate fluxes during the production of particulate and dissolved organic matter by marine phytoplankton. Limnol. Oceanogr. 42: 506-518.

Biddanda, B.A., and R. Benner. 1997b. Major contribution from mesopelagic plankton to heterotrophic metabolism in the upper ocean. Deep-Sea Res I, 44: 2069-2085.

Biddanda, B.A., M. Ogdahl and J.B. Cotner. 2001. Dominance of bacterial metabolism in oligotrophic relative to eutrophic waters. Limnol. Oceanogr. 46: 730-739.

Biddanda, B.A., and J.B. Cotner. 2002. Love handles in aquatic ecosystems: Role of dissolved organic carbon drawdown, resuspended sediments and terrigenous inputs in the carbon balance of Lake Michigan. Ecosystems 5: 431-445.

Biddanda, B.A., and J.B. Cotner. 2003. Enhancement of dissolved organic matter bioavailability by sunlight and its role in the carbon cycle of Lakes Superior and Michigan. J. Great Lakes Res. 29: 228-241.

Bratbak, G., M. Hedal, S. Norland and T. Thingstad. 1990. Viruses as partners in spring bloom microbial trophodynamics. Appl. Environ. Microbiol. 56: 1400-1450.

Burkill, P.H., S.D. Archer, C. Robinson, P.D. Nightingale, S.B. Groom, G.A. Tarran and M.V. Zubkov. 2002. Dimethyl sulphide biogeochemistry within a coccolithophore bloom (DISCO): an overview. Deep-Sea Research II 49: 2863-2885.

Capone, D.G., J. Zehr, H. Paerl, B. Bergman and E. Carpenter. 1997. *Trichodesmium,* a globally significant marine cyanobacterium. Science 276: 1221-1229.

Carlson, C.A., H. Ducklow and A.F. Michaels. 1994. Annual flux of dissolved organic carbon from the euphotic zone in the northwestern Sargasso Sea. Nature 371: 405-408.

Carrillo, P., J. M. Medina-Sánchez and M. Villar-Argaiz. 2002: The interaction of phytoplankton and bacteria in a high mountain lake: Importance of the spectral composition of solar radiation. Limnol. Oceanogr. 47: 1294-1306.

Chin-Leo, G., and R. Benner. 1992. Enhanced bacterioplankton production and respiration at intermediate salinities in the Mississippi River plume. Mar. Ecol. Prog. Ser. 87: 87-103.

Christoffersen, K. 1996. Ecological implications of cyanobacterial toxins in aquatic food webs. Phycologia 35: 42-50.

Cole, J.J. 1999. Aquatic microbiology for ecosystem scientists: New and recycled paradigms in ecological microbiology. Ecosystems 2: 215-225.

Cole, J.J., S. Finlay and M.L. Pace. 1988. Bacterial production in fresh and saltwater ecosystems: a cross-system overview. Mar. Ecol. Prog. Ser. 43: 1-10.

Cotner, J.B., and B.A. Biddanda. 2002. Small players, large role: Microbial influence on biogeochemical processes in pelagic aquatic ecosystems. Ecosystems 5: 105-121.

Coveney, M.F., and R.G. Wetzel. 1995. Biomass, production and specific growth rate of bacterioplankton and coupling to phytoplankton in an oligotrophic lake. Limnol. Oceanogr. 40: 1187-1200.

Dagg, M., R. Benner, S. Lohrenz and D. Lawrence. 2004. Transformation of dissolved and particulate materials on continental shelves influenced by large rivers; plume processes. Cont. Shelf. Res. (in press).

del Giorgio, P.A., J.J. Cole and A. Cimbleris. 1997. Respiration rates in bacteria exceed phytoplankton production in unproductive aquatic ecosystems. Nature. 385: 148-151.

Doucette, G.J., M. Kodama, S. Franca and S. Gallacher. 1996. Bacterial interaction with harmful algal bloom species: bloom ecology, toxigenesis, and cytology. pp. 619-647. *In* D.M. Anderson, A.D. Cembella and G.M. Hallergraeff (eds.) Physiological Ecology of Harmful Blooms. NATO ASI Series, Springer, Berlin, Germany.

Doucette, G.J., E.R. McGovern and J.A. Babinchak. 1999. Algicidal bacteria active against *Gymnodinium breve* (Dinophyceae). I. Bacterial isolation and characterization of killing activity. J. Phycol. 35: 1447–1454.

Ducklow, H. 2000. Bacterial production and biomass in the oceans. pp. 85-120. *In* D. L. Kirchman [ed] Microbial ecology of the oceans. Wiley-Liss, New York, USA.

Eppley, R.W. and P.R. Sloan. 1965. Carbon balance experiments with marine phytoplankton. J. Fish. Res. Bd. Can. 22: 1083-1097.

Eppley, R.W. and B. J. Peterson. 1979. Particulate organic matter flux and planktonic new production. Nature 282: 677-680.

Falkowski, P.G., R.T. Barber and V. Smetacek. 1998. Biogeochemical controls and feedbacks on ocean primary production. Science 281: 200-206.

Fistarol, G.O., C. Legrand, E. Selander, C. Hummert, W. Stolte and Graneli, E. 2004. Allelopathy in *Alexandrium* spp.: effect on a natural plankton community and on algal monocultures. Aquat. Microbial Ecol. 35: 45-56.

Fuhrman, J. 2000. Impact of viruses on bacterial processes. pp. 327-350. *In*, D.L. Kirchman (ed). Microbial Ecology of the Oceans, Wiley, New York, USA.

Fukami, K., U. Shimidu and N. Taga. 1985. Microbial decomposition of phyto- and zooplankton in sea water. Mar. Ecol. Prog. Ser. 21: 1-13.

Fukami, K., T. Nishijima and Y. Ishida. 1997. Stimulative and inhibitory effects of bacteria on the growth of microalgae. Hydrobiologia 358: 185-191.

Garcia, V.M.T. and N.M. Gianuca. 1997. The beach and surf zone. pp. 168-170. *In*: Subtropical Convergence U. Seeliger, C. Odebrecht and J.P. Castello. (eds). Environments: the Coast and Sea in the Southwestern Atlantic. Springer Verlag, Berlin, Germany.

Green, D.H., L.E. Llewellyn, A.P. Negri, S.I. Blackburn and C.J.S. Bolch. 2004. Phylogenetic and functional diversity of the cultivable bacterial community associated with the paralytic shellfish poisoning dinoflagellate *Gymnodinium catenatum*. FEMS Microbial Ecol. 39: 186-196.

Griffith, P.C., and L.R. Pomeroy. 1995. Seasonal and spatial variations in pelagic community respiration on the southeastern U.S. continental shelf. Deep-sea Res. 15: 815-825.

Groben, R., G.J. Doucette, M. Kopp, M. Kodama, R. Amann and L.K. Medlin. 2000. 16S rRNA Targeted probes for the identification of bacterial strains isolated from cultures of the toxic dinoflagellate *Alexandrium tamarense*. Microbial Ecol. 39: 186-196.

Gross, E.M. 2003. Allelopathy of aquatic autotrophs. Critical Reviews in Plant Sciences, 22(3&4): 313–339.

Guillard, N.M. and J.H. Ryther. 1962. Studies of marine diatoms. 1. *Cyuclotella nana* Husted and *Detonula confervacea* (Cleve) Gran. Can. J. Microbiol. 8: 229-239.

Hallegraeff, G.M. 1993. A review of harmful algal blooms and their apparent global increase. Phycologia 32: 79-99.

Hare, C.E., E. Demir, K.J. Coyne, S.C. Cary, D.L. Kirchman and D.A. Hutchins. 2004. A bacterium that inhibits the growth of *Pfiesteria piscicida* and other dinoflagellates. Harmful Algae 4: 221-234.

Hedges, J.I., R.G. Keil and R. Benner. 1997. What happens to terrestrial organic matter in the ocean? Org. Geochem. 27: 195-212.

Heymans, J. and McLachlan, A. 1996: Carbon budget and network analysis of a high energy beach/surf-zone ecosystem. Estuar. Coast Shelf Sci. 43: 485-505.

Hickenbick, G.R. 2002. Predação de bactérias pelo protozooplâncton em um gradiente de salinidade e seu efeito no balanço de massa bacteriano. MSc thesis, Federal University of Rio Grande, Brazil.

Hobbie, J.E., R. Daley and S. Jasper. 1977. Use of nuclepore filters for counting bacteria by epifluorescence microscopy. Appl. Environ. Microbiol. 22: 1225-1228.

Hoppe, H.G., K. Gocke, R. Koppe and C. Begler. 2002. Bacterial growth and primary production along a north-south transect of the Atlantic Ocean. Nature 416: 168-171.

Jenkinson, I.R. and B. Biddanda. 1995. Bulk-phase viscoelastic properties of seawater: relationship with plankton components. J. Plankton. Res. 17: 2251-2274.

Karl, D.M. 1999. A sea of change: biogeochemical variability in the North Pacific subtropical gyre. Ecosystems 2: 181-214.

Karl, D.M., R. Letelier, D. Hebel, L. Tupas, J. Dore, J. Christian and C. Winn. 1995. Ecosystem changes in the North Pacific subtropical gyre attributed to the 1990-92 El Niño. Nature 373; 230-234.

Kepkay, P.E., S. Niven and T. Milligan. 1993. Low molecular weight and colloidal organic carbon production during a phytoplankton bloom. Mar. Ecol. Prog. Ser. 100: 233-244.

Kirchman, D.L. 2000. Uptake and regeneration of inorganic nutrients by marine heterotrophic bacteria. pp. 261-288. *In*, D.L. Kirchman (ed). Microbial Ecology of the Oceans, Wiley, New York, USA.

Lancelot, C. and G. Billen. 1985. Carbon-nitrogen relationships in nutrient metabolism of coastal marine ecosystems. Adv. Aquat. Microbiol. 3: 263-321.

Legendre, L. and J. Le Fevre. 1995. Microbial food webs and the export of biogenic carbon in oceans. Aquat. Microbial. Ecol. 9: 69-77.

Legendre, L. and F. Rassoulzadegan. 1995. Plankton and nutrient dynamics in marine waters. Ophelia 14: 153-172.

Legendre, L. and J. Michaud. 1998. Flux of biogenic carbon in oceans: size dependent regulation by pelagic food webs. Mar. Ecol. Prog. Ser. 164: 1-11.

Letelier, R.M. and D.M. Karl. 1996. Role of *Trichodesmium* spp. In the productivity of the subtropical North Pacific Ocean. Mar. Ecol. Prog. Ser. 133: 263-273.

Mague, T.H., E. Friberg, D. Hughes and I. Morris. 1980. Extracellular release of carbon by marine phytoplankton: a physiological approach. Limnol. Oceanogr. 25: 262-279.

Maruyama T., K. Kato, A. Yokoyama, T. Tanaka, A. Hiraishi and H.D. Park. 2003. Dynamics of microcystin-degrading bacteria in mucilage of *Microcystis*. Microbial Ecol. 46 (2): 279-288.

Mayali, X. and G.J. Doucette. 2002. Microbial community interactions and population dynamics of an algicidal bacterium active against *Karenia brevis* (Dinophyceae). Harmful Algae 1: 277–293.

Mayali, X. and F. Azam. 2004. Algicidal bacteria in the sea and their impact on algal blooms. J. Eukaryotic Microbiology. 51: 139-144.

McAllister, C.D., T. Parsons, K. Stephen and J. Strickland. 1961. Measurements of primary production in coastal seawater using a large-volume plastic sphere. Limnol. Oceoanogr. 6: 237-258.

McLachlan, A. and G. Bates. 1985. Carbon budget for a high-energy surf zone. Vie Milieu 34: 67-77.

Miller, M.L. and M.A. Moran. 1997. Interaction of photochemical and microbial processes in the degradation of refractory dissolved organic matter from a coastal marine environment. Limnol. Oceanogr. 42: 1317-1324.

Nausch, M. 1996. Microbial activities on *Trichodesmium* colonies. Mar. Ecol. Prog. Ser. 141: 173-181.

Odebrecht, C., A.Z. Segato and C.A. Freitas. 1995: Surf-zone chlorophyll *a* variability at Cassino Beach, southern Brazil. Est. Coast. Shelf Sci. 41: 81-90.

Osidele, O.O. and M.B. Beck. 2004. Food web modelling for investigating ecosystem behaviour in large reservoirs of the south-eastern United States: Lessons from Lake Lanier, Georgia. Ecological Modelling 173: 129-158.

Pakulski, J.D., R. Benner, T. Whitledge, R. Amon, B. Eadie, L. Cifuentes, J. Ammerman and D. Stockwell. 2000. Microbial metabolism and nutrient cycling in the Mississippi and Atchafalya River plumes. Est. Coast. Shelf. Sci. 50: 173-184.

Parsons, T.R., Y. Maita and C. Lalli. 1984. A Manual of chemical and biological methods for sewater analysis. Pergamon Press, Oxford, UK.

Peinert, R., B. von Bodungen and V. Smetacek. 1989. Food web structure and loss rate. pp. 35-48. *In*: W.H. Berger, V., Smetacek, G. Wefer, (eds). Productivity of the Ocean: Present and Past. Wiley, New York, USA.

Pomeroy, L.R. 1991. Relationships of primary and secondary production in lakes and marine ecosystems. pp. 97-119. *In*: J. Cole, G. Lovett, S. Findlay (eds) Comparative analyses of Ecosystems. Springer-Verlag, New York, USA.

Pomeroy, L.R. and D. Deibel. 1986. Temperature regulation of bacterial activity during the spring bloom in Newfoundland coastal waters. Science 233: 359-361.

Pomeroy, L.R., and W.J. Wiebe. 1993. Seasonal uncoupling of the microbial loop and its potential significance for the global carbon cycle. pp. 407-409. *In*: R. Guerrero and C. Pedros-Alio (eds), Trends in Microbial Ecology. Spanish Society for Microbiology. Madrid, Spain

Pomeroy, L.R., and W.J. Wiebe. 2001. Temperature and substrates as interactive limiting factors for marine heterotrophic bacteria. Aquat. Microb. Ecol. 23: 187-204.

Pomeroy, L.R., W. Wiebe, D. Deibel, R. Thompson, G. Rowe and J. Pakulski. 1991. Bacterial responses to temperature and substrate concentration during Newfoundland spring bloom. Mar. Ecol. Prog. Ser. 75: 143-159.

Reynaldi, S. 2000. Efeito da diatomácea de zona de arrebentação *Asterionelopsis glacialis* (Castracane) Round, sobre o crescimento bacteriano na Praia do Cassino, RS, Brasil. MSc thesis, Federal University of Rio Grande, Brazil.

Riemann, L., G.F. Steward and F. Azam. 2000. Dynamics of bacterial community composition and activity during a mesocosm diatom bloom. Applied and Environmental Microbiology 66(2): 578-587.

Riquelme, C.E., K. Fukami and Y. Ishida, 1987. Annual fluctuations of phytoplankton and bacterial communities in Maizuru Bay and their interrelationship. Bull. Japan. Soc. Microb. Ecol. 2: 29-37.

Riquelme, C.E., K. Fukami and Y. Ishida. 1989. Growth response of bacteria to extracellular products of bloom algae. Nippon Suisan Gakkaishi 55: 349-355.

Rörig, L.R. and V.M.T. Garcia. 2003. Accumulations of the surf-zone diatom *Asterionellopsis glacialis* in Cassino Beach, southern Brazil, and relationship with environmental factors. J. Coastal Res. 35: 167-177.

Sanders, R. and D. A. Purdie. 1998: Bacterial response to blooms dominated by diatoms and *Emiliania huxleyi* in nutrient-enriched mesocosms. Est. Coast. Shelf Sci. 46 (Supplement A): 35-48.

Seeliger, U., C. Odebrecht and J.P. Castello (eds.). 1997. Subtropical Convergence Environments: the Coast and Sea in the Southwestern Atlantic. Springer Verlag, Berlin, Germany.

Shäfer, H., B. Abbas, H. Witte and G. Muyzer. 2002. Genetic diversity of 'satellite' bacteria present in cultures of marine diatoms. FEMS Microbiology Ecology 42: 25-35.

Sherr, E.B. and B.F. Sherr. 1996. Temporal offset in oceanic production and respiration processes implied by seasonal changes in atmospheric oxygen: the role of heterotrophic microbes. Aquat. Microbial. Ecol. 11: 91-100.

Sieburth, J. McN. 1968. The influence of algal antibiosis on the ecology of marine microorganisms. pp. 63-94. *In*: M. R. Droop (ed), Advances in Microbiology, Vol 1. Academic Press, New York, USA.

Sieracki, M.E., P.G. Verity and D. Stoecker. 1993. Plankton community response to sequential silicate and nitrate depletion during the 1989 North Atlantic spring bloom. Deep Sea Res. II 40: 213-225.

Simon, M., M. Tilzer and H. Muller. 1998. Bacterioplankton dynamics in a large mesotrophic lake 1. abundance, production and growth control. Arch. Hydrobiol. 143-407.

Smayda, T.J. 1997. Harmful algal blooms: their physiology and general relevance to phytoplankton blooms in the sea. Limnol. Oceanogr. 42: 1137-1153.

Smetacek, V. 1985. Role of sinking in diatom life-history cycles; ecological, evolutionary and geological significance. Mar. boil. 84: 239-251.

Strom, S.L. 2000. Bacterivory: interactions between bacterial and their grazers. pp. 351-386. *In*, D. L. Kirchman (ed), Microbial Ecology of the Oceans, Wiley, New York, USA.

Sverdrup, H.U. 1953. On conditions for the vernal blooming of phytoplankton. J. Cons. Int. Explor. Mer. 18: 287-295.

Tobe K, C. Ferguson, M. Kelly, S. Gallacher and L.K. Medlin. 2004. Seasonal occurrence at a Scottish PSP monitoring site of purportedly toxic bacteria originally isolated from the toxic dinoflagellate genus *Alexandrium*. European Journal of Phycology 36: 243-256.

Utermohl, H. 1958. Zur Vervollkommnung der quantitativen. Phytoplanton Methodik. Mit. Int. Ver. Theor. Angew. Limnol. 9: 1-38.

Van den Meersche, K., J. Middlebeurg, K. Soetaert, P. van Rijswijk, H. Boschker and C. Heip. 2004. Carbon-nitrogen coupling and algal-bacterial interactions during an experimental bloom: modeling a ^{13}C tracer experiment. Limnol. Oceanogr. 49: 862-878.

Weisse, T., H. Muller, R. Pinto-Coelho, A. Schweizer, D. Springmann and G. Baldringer. 2000. Response of microbial loop to the phytoplankton spring bloom in a large prealpine lake. Limnol. Oceanogr. 35: 781-794.

Wilhelm, S.W., J. DeBryun, O. Gillor, M. Twiss, K. Livingston, R. Bourbonaiere, L. Pickell, C. Trick, A. Dean and A. McKay. 2003. Effect of phosphorus amendments on present day plankton communities in pelagic Lake Erie. Aquat. Microb. Ecol. 32: 275-285.

Williams, P.J. leB. 1981. Microbial contribution to overall plankton metabolism: direct measurements of respiration. Oceanol. Acta. 4: 359-364.

Williams, P.J. leB. 1984. A review of measurements of respiration rates of marine plankton populations. pp. 357-389. *In* J.E. Hobbie and P.J. leB. Williams (eds), Heterotrophic Activity in the Sea. Plenum, Press, New York, USA.

Williams, P.J. leB. 2000. Heterotrophic bacteria and the dynamics of dissolved organic material. pp. 153-200. *In*, D. L. Kirchman (ed), Microbial Ecology of the Oceans. Wiley, New York, USA.

6

Viral Infection in Marine Eucaryotic Microalgae

Keizo Nagasaki

Harmful Algae Control Section, Harmful Algal Bloom Division, National Research Institute of Fisheries and Environment of Inland Sea, 2-17-5 Maruishi, Ohno, Saeki, Hiroshima 739-0452, JAPAN

Abstract

Viruses are the most abundant biological agents in aquatic environments, and their ecological significance has been highlighted recently. A wide variety of phytoplankton including Chlorophyceae, Chrysophyceae, Prasinophyceae, Prymnesiophyceae, Dinophyceae, Raphidophyceae, and Bacillariophyceae have been shown to be exposed to viral attacks. So far, more than 14 viruses infecting marine eucaryotic microalgae have been isolated and successfully cultured in the laboratory. They are all polyhedral in shape, but they are highly diverse both in size (25–220 nm in diameter), genome type (dsDNA, dsRNA, or ssRNA), and genome size (4.4 kb – 560 kbp). Some of these viruses have been examined from the viewpoint of their ecological roles. Evidence has accumulated showing that viral infection may affect not only the quantity (biomass) of algal blooms to cause interspecies-succession, but also the quality (clonal composition) to cause intraspecies-succession. In other words, even within a bloom where a single species is dominant, it is composed of clones that are diverse in terms of virus sensitivity, and the virus population attacking the bloom is highly diverse in terms of intraspecies host specificity; thus, the ecological interaction between algal hosts and viruses in natural waters is considered to be highly complex. In this chapter, characteristics of the marine microalgal viruses isolated up until now, phylogenetic studies of microalgal viruses, methods to make microalgal viruses into culture, and the possibilities of their practical use are summarized.

INTRODUCTION

With the recent realization that viruses and virus-like particles (VLPs) are highly abundant in various aquatic environments, interests in aquatic viruses have been increasing. They should include viruses infectious to bacteria, phytoplankton, zooplankton, seaweeds, crustacean, fishes, and any other larger organisms; thus, several studies have highlighted their ecological significance (Bergh et al. 1989, Proctor and Fuhrman 1990, Thingstad et al. 1993, Suttle 2000, Wommack and Colwell 2000, Brussaard 2004a). Among them, VLPs or viruses have been found in more than 50 species in 12 of the 14 recognized classes of eucaryotic algae (Van Etten et al. 1991, Zingone 1995, Castberg 2001, Brussaard 2004a). The successful isolation of these microalgal viruses has accelerated the progress of 'algal virology'. Viruses in culture that are infectious to marine eucaryotic algae are listed in Table 6.1; they are diverse in size (25–220 nm in diameter), genome type (dsDNA, dsRNA or ssRNA), genome structure (single molecule or fragmentary), and genome size (4.4 kb – 560 kbp). Although there have been no reports of algal viruses harboring a single-stranded DNA genome, considering the high diversity of these viral isolates, their future findings will not be surprising at all. In the sections below, studies on microalgal viruses isolated so far are summarized.

Viruses Infecting Marine Raphidophytes

Heterosigma akashiwo (Hada) Hada (Raphidophyceae) is a typical HAB (harmful algal bloom) -causing microalga. It occurs in coastal waters of subarctic and temperate areas of both northern and southern hemispheres, and often causes mortality of cultured fish such as salmon, yellowtail, red sea bream, and greater amberjack (Honjo 1993, Smayda 1998). As *H. akashiwo* has a wide distribution, causes annual dense blooms, and is easy to culture, a great number of studies have been made with regard to this species. Three distinct types of virus infecting *H. akashiwo* have ever been reported; *H. akashiwo* virus (HaV), *H. akashiwo* nuclear inclusion virus (HaNIV), and *H. akashiwo* RNA virus (HaRNAV).

HaV

HaV is icosahedral, large (202 nm in diameter), and replicates in the cytoplasm of *H. akashiwo* (Fig. 6.1; Nagasaki and Yamaguchi 1997). By means of one-step growth experiments, its latent period and burst size of HaV were estimated at 30–33 h and ca. 770 infectious units cell^{-1}, respectively (Nagasaki et al. 1999). Pulse-field gel electrophoresis revealed

Table 6.1 Viruses isolated so far which are infections to maribe eucaryotic algae

Virus	*Algal host*	*Size*	*Latent period*	*Burst size*	*Genome*	*References*
BtV[a]	*Aureococcus anophagefferens*	140 nm	–[b]	500	dsDNA	Gastrich et al. (1998)
CbV	*Chrysochromulina brevifilum*	145–170 nm	–	> 320	dsDNA	Suttle and Chan (1995)
CeV	*Chrysochromulina ericina*	160 nm	14–19 h	1800–4100	dsDNA, 510 kbp	Sandaa et al. (2001)
EhV	*Emiliania huxleyi*	170–200 nm	12–14 h	400–1000	dsDNA, 410–415 kbp	Wilson et al. (2002), Castberg et al. (2002)
HaV	*Heterosigma akashiwo*	202 nm	30–33 h	770	dsDNA, 294 kbp[c]	Nagasaki and Yamaguchi (1997)
HaNIV	*Heterosigma akashiwo*	30 nm	> 42 h	ca. 10^5	–	Lawrence et al. (2001)
HaRNAV	*Heterosigma akashiwo*	25 nm	< 12 d	–	ssRNA, 9.1 kb	Tai et al. (2003)
HcRNAV	*Heterocapsa circularisquama*	30 nm	33–48 h	7200–43000	ssRNA, 4.4 kb	Tomaru et al. (2004a)
HcV	*Heterocapsa circularisquama*	197 nm	40–56 h	1300–2440	dsDNA, 356 kbp[c]	Tarutani et al. (2001), Nagasaki et al. (2003)
MpV	*Micromonas pusilla*	115 nm	7–14 h	72	dsDNA, 200 kbp	Cottrell and Suttle (1991)
MpRNAV	*Micromonas pusilla*	65–80 nm	36 h	460–520	dsRNA, 25.5 kbp[d]	Brussaard et al. (2004b)
PoV	*Pyramimonas orientalis*	180–220 nm	14–19 h	800–1000	dsDNA, 560 kbp	Sandaa et al. (2001)
PpV	*Phaeocystis pouchetii*	130–160 nm	12–18 h	350–600	dsDNA, 485 kbp	Jacobsen et al. (1996)
PgV	*Phaeocystis globosa*	100–170 nm	10–16	–	dsDNA	Brussaard et al. (2004a)
RsRNAV	*Rhizosolenia setigera*	32 nm	48 h	1010–3100	ssRNA, 11.2 kb[e]	Nagasaki et al. (2004b)

[a] Two types of viruses infecting *Aureococcus anophagefferens* have been ever reported, AaV (Milligan and Cosper 1994) and BtV (Gastrich et al. 1998). However, the presence of AaV is now under discussion (Suttle 2000), therefore we have referred to only BtV in this table, [b] No data, [c] Nagasaki et al. (2005a), [d] total length of eleven dsRNA segments, [e] smaller RNA molecules (0.6–1.5 kb) are occasionally observed.

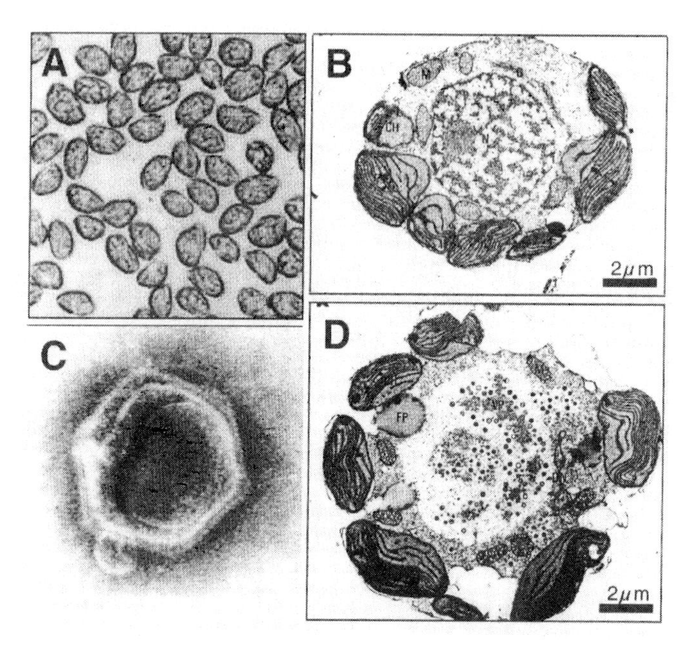

Fig. 6.1 *Heterosigma akashiwo* and its virus, HaV. (A) An optical micrograph of vegetative cells of *H. akashiwo* (ca. 10–15 μm in length), (B) a transmission electron micrograph of an intact *H. akashiwo* cell, (C) a negatively stained HaV particle (ca. 0.2 μm in diameter), and (D) a HaV-infected *H. akashiwo* cell. (B and D: reproduced from Nagasaki et al. 1994a with publisher's permission).

that HaV has a dsDNA genome about 294 kbp in length (Nagasaki et al. 2005a). Fragmental data of the putative ATPase found in HaV-DNA has demonstrated a close relationship between HaV and a typical algal virus PBCV-1 (*Paramecium bursaria Chlorella virus 1*) that infects a *Chlorella*-like alga hosting *P. bursaria* (Nagasaki et al. 2001). Recently, the DNA polymerase gene of HaV was identified and its high similarity to PBCV-1 and the other dsDNA viruses (see below) infecting marine microalgae was found (Nagasaki et al. 2005a). These data suggest that HaV very likely belongs to the new family *Phycodnaviridae* (Van Etten 2000). Since several distinct viruses, including HaV, that are infectious to *H. akashiwo* have recently been isolated, its nomenclature should be reconsidered (e.g. HaDNAV = *Heterosigma akashiwo* DNA virus).

HaV was first found through a transmission electron microscopic (TEM) study of field material. Nagasaki et al. (1994a) found that ca. 5 % of natural cells in a *H. akashiwo* population were visibly infected by HaV-like large particles. Furthermore, a specific increase of virus-harboring cells within natural *H. akashiwo* population at the final stage of its bloom was soon deduced by Nagasaki et al. (1994b). Based on these observations, the possible interrelationship between viral infection and the *H. akashiwo* bloom termination was predicted. Following studies revealed that the relationship between viruses and *H. akashiwo* was much more complex than had been expected. Although HaV is specifically infectious to *H. akashiwo*, all HaV strains do not always infect all *H. akashiwo* strains, i.e., the viral infectivity is clearly strain-specific (Nagasaki and Yamaguchi 1998, Nagasaki et al. 2000). Tarutani et al. (2000) and Tomaru et al. (2004a) revealed the coexistence of multiple types of host and HaV (at least three types of HaV and six types of *H. akashiwo*) in natural *H. akashiwo* blooms. To date, however, the mechanism determining the intra-species host specificity has not yet been clarified.

Ecological relationship between *H. akashiwo* and HaV can be summarized as follows: i) HaV infection influences both the total abundance and the clonal composition of *H. akashiwo* blooms, ii) *H. akashiwo* is diverse among clones in terms of sensitivity to HaV infection, and distinct types of *H. akashiwo* clone coexist in natural environments (even within a bloom), iii) HaV is diverse among clones in terms of infectivity to *H. akashiwo*, and distinct types of HaV clone coexist in natural environments (even within a bloom), iv) HaV infection is one of the significant factors causing the *H. akashiwo* bloom termination. It is not as simple as the relationship between predator and prey because of the diversity in the viruses' infection spectra and the hosts' sensitivity spectra. Based on these observations, Tarutani et al. (2000) predicted that viral infection contributes to the maintenance of the genetic and physiological diversity within *H. akashiwo* populations, and it may allow populations to thrive over a broad range of environmental conditions; hence, viral infection might play a role in enhancing the ecological fitness of the host species.

HaNIV

HaNIV (*H. akashiwo* nuclear inclusion virus) is ca. 30 nm in diameter, and it replicates in the nucleus of *H. akashiwo*, often forming paracrystalline arrays (Lawrence et al. 2001). Its infectivity is also strain-specific. Although the genome type of HaNIV has not been clarified yet, its morphology and replication site demonstrate that HaNIV is clearly different from HaV and the other viruses in the family *Phycodnaviridae* (or probably belong to the family *Phycodnaviridae*). Its latent period was 24 h, and the number of HaNIV particles within an infected cell was estimated at 3×10^5 based on TEM images. In HaNIV-infected cells, an apoptotic feature (margination of heterochromatin within the nucleoplasm) was detected, but its association with the viral-induced apoptosis or programmed cell death has not been elucidated. This host-virus system is expected to be interesting material to study the mechanism of algal cell lysis.

HaRNAV

HaRNAV, the first RNA virus which infects microalgae that was made into culture, assembles within the cytoplasm in infected *H. akashiwo* cells (Tai et al. 2003). Its infection is also strain-specific. In HaRNAV-infected cells, apoptotic features such as swelling of the endoplasmic reticulum, vacuolation of the cytoplasm, and appearance of numerous clusters of fibrils are observed. This host-virus system is also expected to be interesting material to study the mechanism of algal cell lysis. This small virus (ca. 25 nm in diameter) has a ssRNA (positive strand) genome that is 8587 nucleotides (nts) long, plus poly(A) tail (length unknown). The genome contains one open reading frame (ORF) 7743-nts in length encoding a polyprotein that presumably codes for RNA-dependent RNA polymerase, helicase, and conserved picorna-like protein(s). The result of genome analysis did not support that HaRNAV belongs within any currently defined virus family. In the viral lysate, both noninfectious and infectious HaRNAV particles are included, and they can be separated by a sucrose gradient centrifugation. By comparing the structural proteins of the noninfectious and infectious HaRNAV particles, Lang et al. (2004) predicted the maturation process (i.e., protein processing events to achieve infectivity) of the virus particles. Based on the specific features of HaRNAV, a new virus family *Marnaviridae* was proposed by Lang et al. (2004) in which HaRNAV should be the first member.

Finding of HaRNAV not only emphasized the diversity of *H. akashiwo* pathogens but the algal virus pathogens, and also indicated the complexity of virus-host interactions in natural waters. By using RT-PCR methods specific to picorna-like viruses targeting RNA-dependent RNA polymerase fragments, amplicons highly similar to the HaRNAV sequence were repeatedly obtained from the same geographic location (Straight of Georgia, British Columbia) over a four-year period, which suggest persistent HaRNAV infection in the area (Culley et al. 2003). The RT-PCR technique is expected to be a sensitive tool to monitor the dynamics of RNA viruses in marine environments in future studies.

Viruses Infecting Marine Dinophytes (Dinoflagellates)

Heterocapsa circularisquama Horiguchi sp. nov. is a small thecate dinoflagellate (Fig. 6.2 A.B: 20–29 mm in length, 14–20 mm in width), which was recently discovered from Ago Bay, central Japan (Horiguchi 1995). Since this dinoflagellate was recorded for the first time in Uranouchi Bay located in the western part of Japan in 1988, the distribution area has ex-

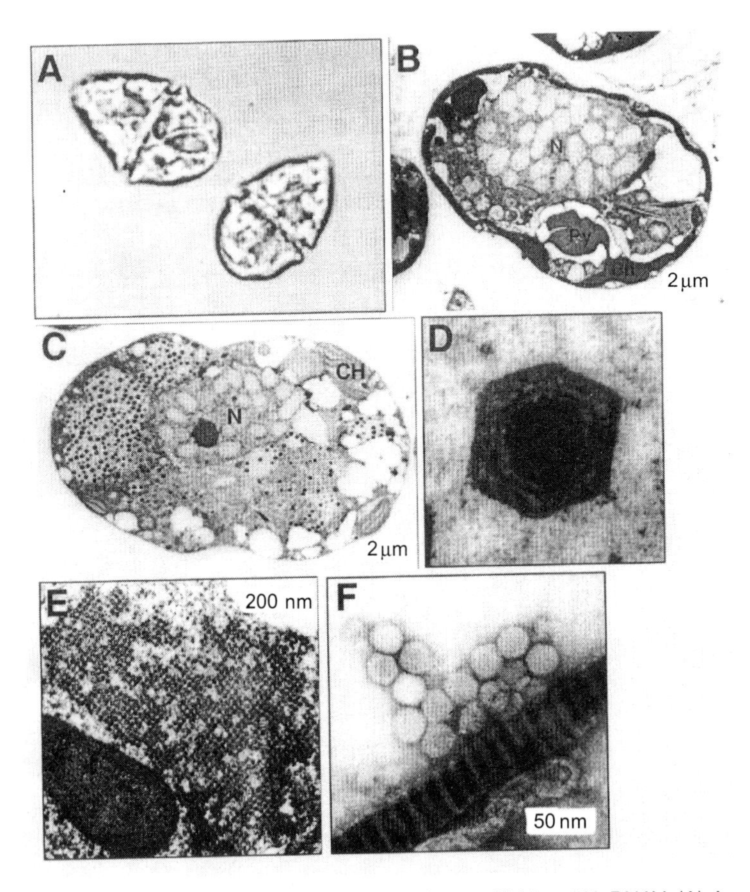

Fig. 6.2 *Heterocapsa circularisquama* and its viruses, HcV and HcRNAV. (A) An optical micrograph of vegetative cells of *H. circularisquama* (ca. 20 μm in length), (B) a transmission electron micrograph of an intact *H. circularisquama* cell, (C) a HcV-infected *H. circularisquama* cell, (D) a thin section of HcV (ca. 0.2 μm in diameter), (E) a crystalline array of HcRNAV within a *H. circularisquama* cytoplasm, and (F) negatively stained HcRNAV particles. (B: reproduced from Tarutani et al. (2001) with permission of the publisher, E and F: reproduced from Tomaru et al. (2004a) with publisher's permission).

panded rapidly into embayments throughout central and western Japan (Matsuyama 1999). With this distribution expansion, this species has often formed large-scale red tides and caused mass mortality of bivalves such as pearl oysters *Pinctada fucata*, oysters *Crassostrea gigas*, and short-necked clams *Tapes philippinarum* (Matsuyama et al. 1996, Nagai et al. 1996, Matsuyama 1999). The most notable case occurred in 1998 when a large bloom resulted in severe financial losses of approximately 4 billion yen to the oyster culture industry in Hiroshima Bay (Matsuyama 1999). There are two viruses infecting *H. circularisquama* that have been reported so far; *H. circularisquama* virus (HcV) and *H. circularisquama* RNA virus (HcRNAV). Both viruses can affect natural *H. circularisquama* blooms (Tomaru and Nagasaki 2004).

HcV

HcV was first isolated from the western coast of Japan. It is icosahedral, lacking tail, ca. 197 nm in diameter, and contains an electron dense core of dsDNA (Fig. 6.2 C, D) (Tarutani et al. 2001). Nagasaki et al. (2003) reported that HcV was widely infectious to *H. circularisquama* strains isolated in the western part of Japan, i.e., algal lysis occurred in 94.7 % of host-virus combination between *H. circularisquama* clones and HcV clones (n = 530). Considering the morphological features, replication site, genome size (Nagasaki et al. 2005a), and host range, HcV is also a member of the family *Phycodnaviridae*. However, determination of any phylogenetic relationships of HcV requires further genomic analysis of its DNA. As HcV isolation was followed by HcRNAV infecting the same host alga (see below), its nomenclature should be reconsidered (e.g. HcDNAV = *H. circularisquama* DNA virus).

HcRNAV

HcRNAV was isolated from coastal waters of Japan (Tomaru et al. 2004a). Its infection is highly strain-specific; i.e., HcRNAV strains can be divided into two types having complementary host ranges. Typical strains of each type, HcRNAV34 and HcRNAV109, were icosahedral, ca. 30 nm in diameter, (Fig. 6.2 E)and harbored a single molecule of ssRNA approximately 4.4 kb in size. Thus, in morphology, nucleic acid type, and genome size, HcRNAV is apparently distinct from HcV. Virus particles appeared in the cytoplasm of the host cells within 24 h post infection, and crystalline arrays (Fig. 6.2 F) or unordered aggregations of virus particles were observed. The burst size and latent period were estimated at 3.4×10^3 to 2.1×10^4 infectious particles cell^{-1}and

24–48 h, respectively. Based on the analysis on nucleotide sequence of genomic RNA, HcRNAV34 and HcRNAV109 were highly similar to each other (similarity = ca. 97 %), and in addition, had some similarities to plant viruses belonging to genera Sobemovirus and Luteovirus. The genome sequence predicts two open reading frames (ORFs): ORF-1 encoding a polyprotein that contains putative RNA-dependent RNA polymerase region, and ORF-2 coding for a single major structural protein of HcRNAV ca. 38 kDa in molecular weight. Comparative studies of HcRNAV sequence predict that a highly distinct region in amino acid sequence within ORF-2 may be responsible for determining their complementary strain-specificities (Nagasaki et al. unpublished data). This host-virus system would be promising material to study the mechanism of specificity in virus infection.

Nagasaki et al. (2004a) examined the possible relationship between *H. circularisquama* and HcRNAV through a field survey in Ago Bay, Japan, in 2001, when a *H. circularisquama* bloom occurred in July. The bloom peaked in mid July and disintegrated during a few days at the end of July; then, the abundance of viruses infectious to *H. circularisquama* was high from the peak of the bloom and throughout in the post-bloom period, but became negligible by the end of August. At the peak of the bloom, 88% of the *H. circularisquama* cells in the population harbored small virus-like particles (VLPs). Based on transmission electron microscopic observations, a morphological resemblance between these VLPs and HcRNAV isolated from the bloom was noticeable. The fluctuation patterns of the viruses indicated that there coexisted at least two distinct types of virus having different host specificity spectra, which agreed with the results of laboratory experiments on HcRNAV types mentioned above (Tomaru et al. 2004a). Specific increase in viral abundance in the sediments was also observed in the middle of the bloom, and they were likely able to maintain their infectivity for more than three months. These results support the idea that viruses have a significant impact on the biomass and clonal composition of algal populations in the natural environment given by Lawrence et al. (2002).

Viruses Infecting Marine Prymnesiophytes

Emiliania huxleyi virus (EhV)

Emiliania huxleyi (Lohmann) Hay and Mohler (Prymnesiophyceae) is a bloom-forming coccolithophore, which is well known as a key phytoplankton species for the current studies on global biogeochemical cycles (Westbroek et al. 1994). Bratbak et al. (1993) first found VLPs in the cytoplasm of *E. huxleyi,* and pointed out the possibility that viral infection is an important factor in eliminating *E. huxleyi* blooms because morphologically distinct large VLPs specifically increased following the peak of *E. huxleyi* blooms. Later, Brussaard et al. (1996) also found the possible infection of two different types of virus in *E. huxleyi* cells based on electron microscopic observations of natural field samples.

Recently, EhV was isolated and characterized by European research groups (Castberg et al. 2002, Wilson et al. 2002). Castberg et al. (2002) reported that EhV is a dsDNA virus with a genome size of ca. 415 kbp. Its latent period and burst size are 12–14 h and 400–1000 viral particles per cell. They also demonstrated EhV should be assigned to the *Phycodnaviridae* virus family based on its DNA polymerase amino acid sequences. Schroeder et al. (2002) also succeeded in isolating EhV, and demonstrated the diversity among EhV clones by host range analysis and sequence analysis of a gene fragment encoding part of their putative major capsid protein. By using a well-designed denaturing gradient gel electrophoresis (DGGE) of the putative major capsid protein gene fragments amplified by PCR from water samples, Schroeder et al. (2003) demonstrated the succession of EhV clones during an *E. huxleyi* bloom in a mesocosm experiment, and they found that only a few genotypes of EhV might be responsible for the decay of the *E. huxleyi* bloom. Combination of virus-specific PCR and DGGE is a promising tool to monitor the change in clonal composition of algal viruses in marine systems.

Viruses infecting the genus *Phaeocystis*

Phaeocystis Pouchetii (Hariot) Lagerheim (Prymnesiophyceae) is a colony-forming microalga with virtually a world-wide distribution often forming massive mucilaginous blooms (Lancelot et al. 1994). *P. globosa* is also one of the typical causative agents of dense bloom occurrences in temperate coastal waters of the North Sea. They have different life stages of which the single flagellated and colonial nonflagellated cells are the most prominent. There are two viruses infecting the genus *Phaeocystis* that have been reported so far; *P. pouchetii* virus (PpV) and *P. globosa* virus (PgV).

PpV, a dsDNA virus infecting *P. pouchetii* was isolated from Norwegian coastal waters (Jacobsen et al. 1996). It is icosahedral, 130–160 nm in diameter, and its latent period and burst size were estimated at 12–18 h and 350–600 viral particles per lysed cells, respectively (Jacobsen et al. 1996). It has a large dsDNA genome ca. 485 kbp in length, and its partial sequence suggested similarity to *Chlorella* virus PBCV–1 (Castberg 2001). Intensive studies on PpV's replication were conducted: Bratback et al. (1998)

examined the relationship between viral replication and physiological condition of host cultures, and demonstrated that (i) *P. pouchetii* was susceptible to viral infection in all stages of growth, (ii) condition of the host cells growth did not affect the infectivity of the progeny viruses and the length of the lytic cycle, and (iii) it had a significant impact on the burst size, i.e., high and low burst sizes were found in exponentially growing cultures and stationary phase cultures, respectively. Later, Thyrhaug et al. (2002) revealed that the production and the latent period of PpV were independent of the host's cell cycle.

On the other hand, a novel virus infecting *P. globosa* (PgV) propagating within its cytoplasm was isolated from Dutch coastal waters and characterized (Brussaard et al. 2004a). Phylogenetic analysis of DNA polymerase gene fragments of PgV clones revealed that they formed a closely related monophyletic group within the family *Phycodnaviridae*, which grouped most closely with those of viruses infecting *Chrysochromulina brevifilum* and *C. strobilus*. Based on these data, Brussaard et al. (2004a) suggested that *Chrysochromulina* and *Phaeocystis* viruses share a common ancestor.

Viruses infecting the genus Chrysochromulina

Chrysochromulina is a cosmopolitan phytoplankter which can comprise >50% of the photosynthetic nanoplanktonic cells in the ocean, and some species of *Chrysochromulina* can cause toxic blooms (Estep and MacIntyre 1989). There are two viruses infecting the genus *Chrysochromulina* that have been reported so far; *C. brevifilum* virus (CbV) and *C. ericina* virus (CeV).

Suttle and Chan (1995) isolated a virus infecting *C. brevifilum*. It is large (145–170 nm in diameter), polyhedral, and proliferates itself within the host's cytoplasm. Considering that this virus is also lytic to *C. strobilus*, this may be an interesting system to study the mechanism determining its host specificity. In order to examine genetic relatedness among the viruses infecting eucaryotic algae, Chen and Suttle (1995, 1996) compared DNA polymerase gene fragments that were amplified by specific primers, and revealed the close relatedness among *Chlorella* viruses, *C. brevifilum* virus, and *Micromonas pusilla* virus.

On the other hand, Sandaa et al. (2001) isolated a virus (CeV) that is infectious to *C. ericina*, which has a large dsDNA genome (ca. 510 kbp). By means of the one-step growth experiments, they estimated the latent period and burst size of CeV at 14–19 h and 1800–4100 infectious units cell^{-1}, respectively.

Viruses infecting marine prasinophytes

Micromonas pusilla (Butcher) Manton et Parke is a small (1.5–3.0 μm) photosynthetic flagellate belonging to Prasinophyceae. It occurs widely in both coastal and oceanic waters ordinarily in low abundances. There are two viruses infecting the cosmopolitan phytoplankter M. pusilla that have been reported so far; M. pusilla virus (MpV) and M. pusilla RNA virus (MpRNAV). Other than these, a virus infecting Pyramimonas orientalis was also isolated and studied (Sandaa et al. 2001).

MpV

MpV, a dsDNA virus infecting *M. pusilla* was the first marine microalgal virus isolated (Mayer and Taylor 1979). It is icosahedral, ca. 115 nm in diameter, lacking tails, and the genome size is ca. 200 kbp (Cottrell and Suttle 1991, Suttle pers. comm.). Considerable variation among MpV clones was demonstrated by restriction fragment analysis, SDS polyacrylamide gel electrophoresis (Cottrell and Suttle 1991), DNA hybridization (Cottrell and Suttle 1995), phylogenetic analysis of DNA polymerase gene fragments (Chen and Suttle 1995), and cross-reactivity tests (Sahlsten 1998). The ecological implication of the diversity among virus clones is of great interest. Waters and Chan (1982) reported that *M. pusilla* cells can mutate to virus resistance at the cell surface and that MpV is also changeable exhibiting variable infectivity in different strains of *M. pusilla.* Although its mechanism has not been experimentally clarified yet, mutations can be one of the causes increasing the diversity of MpV and its host. This host virus system is expected as promising material to study the host range mutation events.

MpRNAV

A novel dsRNA virus MpRNAV infecting *M. pusilla* was isolated by Brussaard et al. (2004b), which proliferates itself in the cytoplasm. It is medium sized (65–80 nm in diameter) among algal viruses isolated so far. The dsRNA genome was composed of 11 segments ranging between 0.8 and 5.8 kb, with a total size of approximately 25.5 kb, suggesting some similarity to dsRNA genera Rotavirus and Aquareovirus. MpRNAV is the only algal virus harboring dsRNA genome that has been isolated so far, and the finding of microalgal viruses harboring dsRNA also emphasizes the diversity of the algal-virus pathogens.

PoV

P. orientalis Butcher is a potential bloom-causer whose cell is covered with numerous scales and widely distributed in the world. Sandaa et al. (2001) isolated a virus (PoV) that is infectious to *P. orientalis*, which has the largest dsDNA genome among algal viruses (ca. 560 kbp). By means of the one-step growth experiments, they estimated the latent period and burst size of PoV at 14–19 h and 800–1000 infectious units cell^{-1}, respectively.

Viruses Infecting Marine Bacillariophytes (Diatoms)

A bloom-forming diatom *Rhizosolenia setigera* belongs to the order Centrales and occurs widely throughout the world: in the North Atlantic Ocean, North Sea, Baltic Sea, English Channel, Mediterranean Sea and Pacific Ocean. In 2002, a ssRNA virus specifically infecting the bloom-forming diatom *R. setigera* (*R. setigera* RNA virus: RsRNAV) was isolated from the Ariake Sea, Japan (Nagasaki et al. 2004b). Viral replication occurred within the cytoplasm, and the virus particle was icosahedral, lacking a tail, 32 nm in diameter on the average. The major nucleic acid extracted from the RsRNAV particles was a single-stranded RNA (ssRNA) molecule 11.2 kb in length, although smaller RNA molecules (ranging in size of 0.6, 1.2, and 1.5 kb) were occasionally observed. The latent period of RsRNAV was 2 d, and the burst size was 3,100 and 1,010 viruses per host cell when viruses had been inoculated to the host culture at the exponentially growing phase and stationary phase, respectively. More recently, Nagasaki et al. (2005b) isolated the second diatom-infecting virus, CsNIV (*Chaetoceros salsugineum* nuclear inclusion virus). CsNIV genome consists of a single molecule of covalently closed circular single-stranded DNA (ssDNA, 6005 nt) as well as a segment of linear ssDNA (997 nt). As for the detailed feature of this virus, refer to Nagasaki et al. (2005b).

Viruses Infecting Marine Chrysophyte

Aureococcus anophagefferens gen et sp. nov. is a picoplankter ca. 2.5 μm in diameter which causes brown tide (Cosper et al. 1987). BtV is a large (140–160 nm in diameter) dsDNA virus infecting *A. anophagefferens*. It proliferates itself within the host's cytoplasm, and host cells are lysed within 24 h. Considering these features, BtV most likely belongs to the family *Phycodnaviridae* (Milligan and Cosper 1994, Gastrich et al. 1998). Gastrich et al. (2004) reported that as high as 37.5% of VLP-infected *A. anophagefferens* cells occurred at the termination of the brown tide bloom in New Jersey in 2002, indicating that viruses may be a major source of mortality for brown tide blooms.

PHYLOGENY OF MICROALGAL VIRUSES

In the last decade of the 20th century, more than 10 marine viruses infecting microalgae were isolated. They were all large (> 100 nm in diameter) and harbored dsDNA genome. Before the 1990's, algal viruses infecting symbiont *Chlorella*-like alga had been isolated and intensively studied (Van Etten et al. 1991). Van Etten and Ghabrial (1991) established the new virus family *Phycodnaviridae* and positioned the *Chlorella* viruses in it. Studies on the other large algal viruses demonstrated that several had properties in common with the *Chlorella* viruses in terms of the morphology, host range, genome size, and some other characteristics. Chen and Suttle (1995, 1996) examined the genetic relatedness among microalgal viruses by comparing sequences of the DNA polymerase gene fragments amplified by a refined nested-PCR technique. They designed degenerate primers targeting three conserved regions found within the DNA polymerase gene of some microalgal viruses. This technique was applicable to *Chlorella* viruses, MpV, and CbV, and also able to amplify possible DNA polymerase gene fragments of unknown viruses from natural seawater samples (Chen and Suttle 1995); therefore, the technique has been frequently used as a powerful tool for fingerprinting natural virus communities combined with denaturing gradient gel electrophoresis (Short and Suttle 2000, 2002, 2003). Schroeder et al. (2002) also succeeded in amplifying the DNA polymerase gene fragment of EhV by a similar technique, and examined its phylogenetic relationship with other algal viruses. Now, many algal virologists regard DNA polymerase gene as a key region to discuss the phylogeny of algal viruses, and now four genera (Chlorovirus, Prasinovirus, Prymnesiovirus, and Phaeovirus) are categorized within the family *Phycodnaviridae* (Van Etten 2000, Van Etten et al. 2002). However, in some algal viruses, amplification of DNA polymerase gene fragments by the degenerate primers was unsuccessful (Chen and Suttle 1995, Castberg 2001, Nagasaki et al. 2005a). In the case of *Ectocarpus siliculosus* virus (a virus infecting multicellular alga) and HaV, a shotgun cloning technique was used to determine the nucleotide sequence of its genomic DNA, and phylogenetic comparison of DNA polymerase gene was conducted (Delaroque et al. 2001, Nagasaki et al. 2005a). Phylogenic analysis for RNA viruses infecting microalgae will be a future problem.

FUNDAMENTAL TECHNIQUES TO STUDY MICROALGAL VIRUSES

Thanks to algal virologists' efforts to date, evidence showing that viruses play a key role affecting the dynamics of microalgae in marine systems has been accumulated. To investigate microalgal viruses, it is necessary to make target viruses into laboratory cultures. Thus, isolation, cultivation, and maintenance of microalgal viruses in a suitable condition are important procedures that sustain the progress of algal virology. The flow of necessary procedures is simple: screening, cloning, and maintenance, and the techniques used in each procedure are described in this section. Also refer to some papers describing methods for microalgal virus isolation (Suttle et al. 1990, Suttle 1993, Nagasaki 2001).

Preparations of Algal Host Cultures

Needless to say, preparation of host microalgal cultures is essential to isolate microalgal viruses. For isolation experiments, it is preferable to use unialgal and axenic cultures as hosts for making estimation of viral effect easy and clear.

To establish axenic microalgal cultures, a micropipetting method or the other washing methods are commonly employed, often using some antibiotics (Connell and Cattolico 1996). Also refer methods given by Lee (1993). A simple technique proposed by Imai and Yamaguchi (1994) is recommended for washing phytoflagellates.

It is important to use multiple clones of host alga for virus isolation, because virus sensitivity spectra can differ among host clones (Nagasaki et al. 2000, Tomaru et al. 2004a). It is also preferable to keep host algae under optimal conditions for their growth. Generally, exponentially growing cultures tend to be more sensitive to viral attacks than stationary phase cultures (Nagasaki et al. 2003). It is presumably because viruses utilize the biosynthetic function of hosts such as DNA synthesis and protein synthesis, and host cells in the vigorously growing phase have a higher biosynthesis activity.

Screening of Microalgal Viruses

Microalgal viruses are included not only in seawaters but also in marine sediments. Recent studies suggest sediments are also a reservoir of viruses (Lawrence et al. 2002, Nagasaki et al. 2004a). By shaking sediments with a suitable medium (e.g. SWM3, Chen et al. 1969), viruses can be easily suspended, and low-speed centrifugation separates the virus suspension from the sediment debris. Prior to inoculation of the virus suspension to host cultures, aggregation of bacteria and larger organisms (zooplankton, phytoplankton, nanoflagellates, fungi, etc.) should be removed by filtration. Considering the size of microalgal viruses (25–220 nm in diameter), 0.2 μm, 0.22 μm or 0.45 μm nominal pore-size polycarbonate membrane filter (Nuclepore) is recommended. Then, the resultant filtrate (viral fraction) is added to vigorously growing potential host cultures. Growth of hosts is traced and compared to that of control cultures without virus inoculation.

To isolate microalgal viruses from natural seawater, two interesting methods have been proposed: the concentration method and ultraviolet (UV)-treatment. In the former method, the virus-size fraction is concentrated using a specialized ultrafiltration column (Suttle et al. 1990). Some viruses have been successfully isolated by this ultrafiltration technique: viruses infecting *M. pusilla* (Cottrell and Suttle 1991), *A. anophagefferens* (Milligan and Cosper 1994), *C. brevifilum* (Suttle and Chan 1995), *H. akashiwo* (Nagasaki and Yamaguchi 1998), HaNIV (Lawrence et al. 2001), and HaRNAV (Tai et al. 2003). A detailed protocol of the ultrafiltration technique is given by Suttle (1993).

The UV-treatment technique is intended to induce possibly latent viruses in natural algal cells, but its actual mechanism has not been verified yet. Briefly, natural seawater containing the target phytoplankton species is collected and centrifuged to obtain a cell concentrate which is then poured into a petri dish, and exposed to UV-irradiation for an appropriate time (Jacobsen et al. 1996). Next, the algal cell suspension is incubated under appropriate condition (Jacobsen et al. 1996), and the resultant cell debris is removed by centrifugation before inoculation into a fresh host culture. By using this method, EhV (Bratbak et al. 1996) and PpV (Jacobsen et al. 1996) were successfully isolated.

On the other hand, Brussaard et al. (2004a) reported that incubation of natural seawater for a week at *in situ* temperatures promoted the isolation of viruses infecting *P. globosa*. This procedure may enhance the maturation of viruses in the natural seawater samples.

Cloning and Maintenance of Microalgal Viruses

Following the procedures given above, when a decay of the tested host algal cultures is detected, cloning of the algicidal factor should be tried as soon as possible. In many cases, the extinction dilution method is used for cloning microalgal viruses (Suttle 1993). In the author's group, two cycles of the extinction dilution procedure are often used for cloning viral pathogens; briefly, the culture lysate is diluted with modified SWM3 medium in a series of 10-fold dilution steps. Aliquots (100 μl) of each dilution are added to 8 wells in cell-culture plates with 96 round bottom wells, mixed with 150 μl of exponentially growing host culture, and incubated under the conditions suitable for the host's growth. Lysed cultures are removed from the most diluted wells in which lysis occurred and the above entire procedure is repeated. The lysate in the most diluted wells of the second assay is sterilized by filtration through 0.1 μm or 0.2 μm pore size polycarbonate membrane filters and transferred into an exponentially growing host culture. After removing cell debris from the resulting lysate by low-speed centrifugation, the supernatant is used as the clonal pathogen suspension.

In many cases, microalgal viruses are maintained by repetitious transfer of the viral lysate to fresh host cultures. However, it should be noted that it might cause a loss of infectivity, which is presumably caused by the increase of defective interfering particles (Bratbak et al. 1996).

Microalgal viruses are diverse in terms of stability. For example, the titer of PpV, HcV, and HaV gradually decreases even during storage at 4°C in the dark. Thus, the establishment of a suitable method to keep the infectivity of each virus is necessary. For example, conditions for cryopreserving HaV are the addition of 10–20% dimethyl sulfoxide and storage at –196°C (Nagasaki and Yamaguchi 1999), and those for PpV are the addition of 10–20% sucrose and storage at –70°C (Nagasaki 2001).

Enumeration of Microalgal Viruses

To examine the basic characteristics of microalgal viruses in terms of their replication ability, techniques for counting viruses are essential. Currently, several methods are used for enumerating viruses: plaque assays, most probable number (MPN) assays, transmission electron microscopy (TEM), epifluorescence microscopy (FM), and flow cytometry (FCM). According to the experimental objectives, the most suitable method should be selected and employed, because the results of each method can be interpreted differently. The abundance of viruses that are infectious and lytic to the algal strain used in the measurement can be quantified by means of plaque assays and MPN assays. Plaque assays are available only when the potential host alga grows on solid phase medium to form a lawn (e.g. *M. pusilla*, *E. huxleyi*). However, given the counting results of the plaque assay method for MpV, it appears less efficient than the other methods employing liquid media (Cottrell and Suttle 1995). Moreover, cultivation of marine algal flagellates on solid phase medium is generally difficult (Nagasaki and Imai 1994). In contrast, TEM gives the abundance of 'particles that are virus-like in appearance' and FM gives that of 'particles that harbor nucleic acids stainable with the dye (DAPI, SYBR GREEN I, etc.)' used in the measurement. Refer to the detailed protocols given by Suttle (1993) and Wilhelm and Poorvin (2001). The recent rise of accessibility to FCM has eased the labor requirement for counting fluorescence-labeled large DNA virus particles, and has significantly improved the speed of analysis and accuracy of counting (Marie et al. 1999, Brussaard et al. 2000). Refer to the conditions reported by Brussaard (2004b) who conducted a detailed optimization of procedures for counting viruses by FCM. FCM is also applicable to study phytoplankton viability following viral infection by specifically staining using a membrane impermeant nucleic acid dye such as SYTOX-Green (Brussaard et al. 2001).

Estimation of the latent period and the burst size by means of the one-step growth experiment is an essential process in the characterization of microalgal viruses. Then, the choice of enumeration techniques for viruses is an important point determining the implication of resultant data. When a plaque assay or an MPN assay is used for viral enumeration, the resultant burst size indicates the number of infective centers (an infectious virus particle and its aggregation) that are lytic to the host strain per host cell. In contrast, when TEM, FM, or FCM are used for viral enumeration, the resultant burst size indicates the total number of virus particles harboring its genomic nucleic acids; thus, it may also include defective particles (virus particles that lack infectivity). Currently, both methods are employed to enumerate burst size of microalgal viruses; hence, the results are shown as either "infectious units per infected cell" or "particles per infected cell". In the case of ssRNA microalgal virus HcRNAV, as it was difficult to detect and count by FM even if they are stained with an appropriate dye (SYBR GREEN II or SYBR GOLD), the MPN assay was used for its enumeration (Tomaru et al. 2004a).

FUTURE STUDIES FOR MICROALGAL VIRUSES

So far, more than 14 microalgal viruses have been isolated and characterized, and at present, researchers' likely interest centers on their ecological roles, molecular aspects, genomes, etc. Indeed this field of study is still young and lots of insights into the marine ecosystem have been revealed year by year, it is necessary to consider in which industrial field and how we can apply and utilize the data in future. In the last section, a few examples of studies on possible use of microalgal viruses are introduced.

Possible Use of Viruses to Control HABs

As mentioned above, recent studies revealed that viral infection is one of the most significant factors controlling the dynamics of algal blooms and especially causing their termination. These observations led to an idea for the possible use of viruses to control HABs. It is a concept of a biotic pesticide to control a harmful organism using its natural enemy based on the relationship between the host and the infecting microorganism. In the field of agriculture, some biological agents have already been used practically, and utilization of natural organisms to control pests has been widely studied. By promoting the studies on microalgal viruses, the possibility of using a natural organism to control a pest in the marine field can be assessed. The possibility of their use in preventing outbreaks of HABs and problems involved in the challenge are summarized as follows; 1) Basically, microalgal viruses are regarded as the most specifically infectious agents to the HAB-causing microalgae. Their infection to organisms other than their hosts has not been reported yet (except for the case of CbV that is infectious to both *C. brevifilum* and *C. strobilus*), 2) In many cases, they originated from the same place as their host algae, i.e., host-virus systems existing in nature. They were derived from natural seawater and have not been subjected to any genetic manipulation, 3) They have considerably high replication abilities, and are able to replicate as long as their hosts exist in the environment, 4) Production of microalgal viruses does not require any special equipment or expensive reagents.

Since microalgal viruses do exist in natural waters and are involved in the natural ecosystems, it appears reasonable to utilize these 'natural anti-HAB factors' to reduce damage to fisheries. In the field of studies of viruses infecting cyanophages, several attempts have over the years been made to control cyanobacterial blooms, but with rather limited success (Martin and Benson 1988). The predictable problem is that, in many cases, viral infectivity is normally 'strain-specific' rather than 'species-specific'; in other words, a single clone of microalgal virus cannot eliminate the whole host algal population which is composed of different types of clones in terms of virus sensitivity spectra. However, in the case of studies on HaV, it seems most types of virus clones can be collected by using multiple types of host clones (Tomaru et al. 2004b). As we all know, this applied biological science is in its infant stage, and it is necessary to accumulate fundamental information on various host-virus systems to assess the possibility of this new technique. It is also essential to give enough explanation on viruses' safety for obtaining the public acceptance.

Source of Novel Enzymes

Algal viruses are also expected as a source of various novel enzymes (Yamada et al. 1999). Recently, the 331 kbp dsDNA genome of *Chlorella* virus (PBCV-1) and the 336 kb dsDNA genome of *E. siliculosus* virus (EsV-1) were sequenced, and 231 and 375 possible protein-encoding genes were found, respectively (Van Etten et al. 2002). They contain a variety of enzymes involved in the metabolism of polysaccharides, amino acids, lipids, nucleotides and nucleosides. Some of the viral enzymes have already been successfully utilized in biotechnological fields; e.g., *Chlorella* virus endonucleases and methylases having specific restriction sites. Yamada et al. (1999) suggested the possible applications of viral chitinase and chitosanase in the recycling of natural resources. Further studies on the genome of microalgal viruses would lead to more interest in their novel enzymes.

ACKNOWLEDGMENTS

Works on HaV, HcV, HcRNAV, CsNIV, and RsRNAV were supported by funding from the Fisheries Agency of Japan, the Ministry of Agriculture, Forestry and Fisheries of Japan, the Japan Science and Technology Corporation, the Industrial Technology Research Grant Program in 2000–2004 from the New Energy and Industrial Technology Development Organization (NEDO) of Japan, and the Society for Techno-Innovation of Agriculture, Forestry and Fisheries (STAFF). I am grateful to Dr. K. Tarutani for his critical reading of the manuscript and useful suggestions.

REFERENCES

Bergh, Ø., K.Y. Børsheim, G. Bratbak and M. Heldal. 1989. High abundance of viruses found in aquatic environments. Nature 340: 467-468.

Bratbak, G., J.K. Egge and M. Heldal. 1993. Viral mortality of the marine alga *Emiliania huxleyi* (Haptophyceae) and termination of algal blooms. Mar. Ecol. Prog. Ser. 93: 39-48.

Bratbak, G., W. Wilson and M. Heldal. 1996. Viral control of *Emiliania huxleyi* blooms? J. Mar. Syst. 9: 75-81.

Bratbak, G., A. Jacobsen, M. Heldal, K. Nagasaki and F. Thingstad. 1998. Virus production in *Phaeocystis pouchetii* and its relation to host cell growth and nutrition. Aquat. Microb. Ecol. 16: 1-9.

Brussaard, C.P.D., R.S. Kempers, A.J. Kop, R. Riegman and M. Heldal. 1996. Virus-like particles in a summer bloom of *Emiliania huxleyi* in the North Sea. Aquat. Microb. Ecol. 10: 105-113.

Brussaard, C.P.D., D. Marie and G. Bratbak. 2000. Flow cytometric detection of viruses. J. Virol. Meth. 85:175-182.

Brussaard, C.P.D., D. Marie, R. Thyrhaug and G. Bratbak. 2001. Flow cytometric analysis of phytoplankton viability following viral infection. Aquat. Microb. Ecol. 26: 157-166.

Brussaard, C.P.D. 2004a. Viral control of phytoplankton populations – a review. J. Eukaryot. Microbiol. 51: 125-138.

Brussaard, C.P.D. 2004b. Optimization of procedures for counting viruses by flow cytometry. Appl. Environ. Microbiol. 70(3): 1506-1513.

Brussaard, C.P.D., S.M. Short, C.F. Frederickson and C.A. Suttle. 2004a. Isolation and phylogenetic analysis of novel viruses infecting the phytoplankton *Phaeocystis globosa* (Prymnesiophyceae). Appl. Environ. Microbiol. 70: 3700-3705.

Brussaard, C.P.D., A.A.M. Noordeloos, R.A. Sandaa, M. Heldal and G. Bratbak. 2004b. Discovery of a dsRNA virus infecting the marine photosynthetic protist *Micromonas pusilla*. Virology, 319: 280-291.

Castberg, T. 2001. Algal viruses – characteristics and ecological effects. D.S. Thesis, University of Bergen, Bergen, Norway.

Castberg, T., R. Thyrhaug, A. Larsen, R.-A. Sandaa, M. Heldal, J.L. Van Etten and G. Bratbak. 2002. Isolation and characterization of a virus that infects *Emiliania huxleyi* (Haptophyta). J. Phycol. 38: 767-774.

Chen, L.C.M., T. Edelstein and J. McLachlan, 1969. *Bonnemaisonia hamifera* Hariot in nature and in culture. J. Phycol. 5: 211-220.

Chen, F. and C.A. Suttle. 1995. Amplification of DNA polymerase gene fragments from viruses infecting microalgae. Appl. Environ. Microbiol. 61: 1274-1278.

Chen, F. and C.A. Suttle. 1996. Evolutionary relationships among large double-stranded DNA viruses that infect microalgae and other organisms as inferred from DNA polymerase genes. Virology 219: 170-178.

Connell, L. and R.A. Cattolico. 1996. Fragile algae: axenic culture of field-collected samples of *Heterosigma carterae*. Mar. Biol. 125: 421-426.

Cosper, E.M., W.C. Dennison, E.J. Carpenter, V.M. Bricelj, J.G. Mitchell and S.H. Kuenstner. 1987. Recurrent and persistent brown tide blooms perturb coastal marine ecosystem. Estuaries 10: 284-290.

Cottrell, M.T. and C.A. Suttle. 1991. Wide-spread occurrence and clonal variation in viruses which cause lysis of a cosmopolitan, eukaryotic marine phytoplankter *Micromonas pusilla*. Mar. Ecol. Prog. Ser. 78: 1-9.

Cottrell, M.T. and C.A. Suttle. 1995. Genetic diversities of algal viruses which lyse the photosynthetic picoflagellate *Micromonas pusilla*. Appl. Environ. Microbiol. 61: 3088-3091.

Culley, A.I., A.S. Lang and C.A. Suttle. 2003. High diversity of unknown picorna-like viruses in the sea. Nature, 424: 1054-1057.

Delaroque, N., D.G. Müller, G. Bothe, T. Pohl, R. Knippers and W. Boland. 2001. The complete DNA sequence of the *Ectocarpus siliculosus* Virus EsV-1 genome. Virology 287: 112-132.

Estep, K.W. and F. MacIntyre 1989. Taxonomy, life cycle, distribution and dasmotrophy of *Chrysochromulina*: A theory accounting for scales, haptonema, muciferous bodies and toxicity. Mar. Ecol. Prog. Ser. 57: 11-21.

Gastrich, M.D., O.R. Anderson, S.S. Benmayor and E.M. Cosper. 1998. Fine structure analysis of viral infection in the harmful brown alga *Aureococcus anophagefferens*.

pp. 419-421. *In* B. Reguera, J. Blanco, M. L. Fernandez, and T. Wyatt [eds.]. Harmful Algae. Xunta de Galicia and IOC of UNESCO, Spain.

Gastrich, M.D., J.A. Leigh-Bell, C.J. Gobler, O.R. Anderson, S.W. Wilhelm and M. Bryan. 2004. Viruses as potential regulators of regional brown tide blooms caused by the alga, *Aureococcus anophagefferens*. Estuaries, 27(1): 112-119.

Honjo, T. 1993. Overview on bloom dynamics and physiological ecology of *Heterosigma akashiwo*. pp. 33-41. *In* T.J. Smayda and Y. Shimizu [eds.] Toxic Phytoplankton Blooms in the Sea. Elsevier, Amsterdam.

Horiguchi, T. 1995. *Heterocapsa circularisquama* sp. nov. (Peridinales, Dinophyceae): A new marine dinoflagellate causing mass mortality of bivalves in Japan. Phycol. Res. 43: 129-136.

Imai, I. and M. Yamaguchi. 1994. A simple technique for establishing axenic cultures of phytoflagellates. Bull. Jap. Soc. Microb. Ecol. 9: 15-17.

Jacobsen, A., G. Bratbak and M. Heldal. 1996. Isolation and characterization of a virus infecting *Phaeocystis pouchetii* (Prymnesiophyceae). J. Phycol. 32: 923-927.

Lancelot, C., P. Wassmann and H. Burth. 1994. Ecology of *Phaeocystis*-dominated ecosystems. J. Mar. Syst. 5: 1-4.

Lang, A.S., A.I. Culley and C.A. Suttle. 2004. Genome sequence and characterization of a virus (HaRNAV) related to picorna-like viruses that infects the marine toxic bloom-forming alga *Heterosigma akashiwo*. Virology, 320: 206-217.

Lawrence, J.E. A.M. Chan and C.A. Suttle. 2001. A novel virus (HaNIV) causes lysis of the toxic bloom-forming alga *Heterosigma akashiwo* (Raphidophyceae). J. Phycol. 37: 216-222.

Lawrence, J.E., A.M. Chan and C.A. Suttle. 2002. Viruses causing lysis of the toxic bloom-forming alga *Heterosigma akashiwo* (Raphidophyceae) are widespread in coastal sediments of British Columbia, Canada. Limnol. Oceanogr. 47: 545-550.

Lee, J.J. 1993. General techniques for the isolation and culture of marine protists from estuarine, littoral, psammolittoral, and sublittoral waters. pp. 41-50. *In* P.F. Kemp, B. Sherr, E. Sherr, and J.J. Cole [eds.]. Handbook of Methods in Aquatic Microbial Ecology. Lewis Publishers, Boca Raton, USA.

Marie, D., C.P.D. Brussaard, R. Thyrhaug, G. Bratbak and D. Vaulot. 1999. Enumeration of marine viruses in culture and natural samples by flow cytometry. Appl. Environ. Microbiol. 65: 45-52.

Martin, E. and R. Benson. 1988. Phages of cyanobacteria. pp. 607-645. *In* R. Calendar [ed.]. The Bacteriophages. Volume 2. Plenum Press, New York, USA.

Matsuyama, Y., T. Uchida, K. Nagai, M. Ishimura, A. Nishimura, M. Yamaguchi and T. Honjo. 1996. Biological and environmental aspects of noxious dinoflagellate red tides by *Heterocapsa circularisquama* in west Japan. pp. 247-250. *In* T. Yasumoto, Y. Oshima, and Y. Fukuyo [eds.]. Harmful and Toxic Algal Blooms. IOC of UNESCO, Paris, France.

Matsuyama, Y. 1999. Harmful effect of dinoflagellate *Heterocapsa circularisquama* on shellfish aquaculture in Japan. Jpn. Agr. Res. Q. 33: 283-293.

Mayer, J.A. and F.J.R. Taylor. 1979. A virus which lyses the marine nanoflagellate *Micromonas pusilla*. Nature 281: 299-301.

Milligan, K.L.D. and E.M. Cosper. 1994. Isolation of virus capable of lysing the brown tide microalga, *Aureococcus anophagefferens*. Science 266: 805-807.

Nagai, K., Y. Matsuyama, T. Uchida, M. Yamaguchi, M. Ishimura, A. Nishimura, S. Akamatsu and T. Honjo. 1996. Toxicity and LD_{50} levels of the red tide dinoflagellate *Heterocapsa circularisquama* on juvenile pearl oysters. Aquaculture 144: 149-154.

Nagasaki, K., M. Ando, I. Imai, S. Itakura and Y. Ishida. 1994a. Virus-like particles in *Heterosigma akashiwo* (Raphidophyceae): a possible red tide disintegration mechanism. Mar. Biol. 119: 307-312.

Nagasaki, K., M. Ando, S. Itakura, I. Imai and Y. Ishida. 1994b. Viral mortality in the final stage of *Heterosigma akashiwo* (Raphidophyceae) red tide. J. Plankton Res. 16: 1595-1599.

Nagasaki, K. and I. Imai. 1994. Solid-phase culture of marine phytoflagellates. Bull. Jap. Soc. Microb. Ecol. 9: 37-43 (in Japanese with English abstract).

Nagasaki, K. and M. Yamaguchi. 1997. Isolation of a virus infectious to the harmful bloom causing microalga *Heterosigma akashiwo* (Raphidophyceae). Aquat. Microb. Ecol. 13: 135-140.

Nagasaki, K. and M. Yamaguchi. 1998. Intra-species host specificity of HaV (*Heterosigma akashiwo* virus) clones. Aquat. Microb. Ecol. 14: 109-112.

Nagasaki, K., K. Tarutani and M. Yamaguchi. 1999. Growth characteristics *of Heterosigma akashiwo* virus and its possible use as a microbiological agent for red tide control. Appl. Environ. Microbiol. 65: 898-902.

Nagasaki. K. and M. Yamaguchi. 1999. Cryopreservation of a virus (HaV) infecting a harmful bloom causing microalga, *Heterosigma akashiwo* (Raphidophyceae). Fisheries Sci. 65: 319-320.

Nagasaki. K., K. Tarutani and M. Yamaguchi. 2000. Cluster analysis on algicidal activity of HaV clones and virus sensitivity of *Heterosigma akashiwo* (Raphidophyceae). J. Plankton Res. 21: 2219-2226.

Nagasaki, K. 2001. Domestication of eucaryotic microalgal viruses from the marine environments. Microbes Environ. 16: 3-8.

Nagasaki, K., K. Tarutani, M. Hamaguchi and M. Yamaguchi. 2001. Preliminary analysis on a *Heterosigma akashiwo* virus DNA. Microbes Environ. 16: 147-154.

Nagasaki, K., Y. Tomaru, K. Tarutani, N. Katanozaka, S. Yamanaka, H. Tanabe and M. Yamaguchi. 2003. Growth characteristics and intra-species host specificity of a large virus infecting the dinoflagellate *Heterocapsa circularisquama*. Appl. Environ. Microbiol. 69: 2580-2586.

Nagasaki, K., Y. Tomaru, K. Nakanishi, N. Hata, N. Katanozaka and M. Yamaguchi. 2004a. Dynamics of *Heterocapsa circularisquama* (Dinophyceae) and its viruses in Ago Bay, Japan. *Aquat. Microb. Ecol.* 34: 219-226.

Nagasaki, K., Y. Tomaru, N. Katanozaka, Y. Shirai, K. Nishida, S. Itakura and M. Yamaguchi. 2004b. Isolation and characterization of a novel single-stranded RNA virus infecting the bloom-forming diatom *Rhizosolenia setigera*. Appl. Environ. Microbiol. 70: 704-711.

Nagasaki, K., Y. Shirai, Y. Tomaru, K. Nishida, S. Pietrovski. 2005a. Algal viruses with distinct intraspecies host specificities include identical intein elements. Appl. Environ. Microbial. 71: In Press.

Nagasaki, K., Y. Tomaru, Y. Takao, K. Nishida, Y. Shirai, H. Suzuki, T. Nagumo. 2005b. Previously unknown virus infects marine diatom. Appl. Environ. Microbial. 71: In Press.

Proctor, L.M. and J.A. Fuhrman. 1990. A viral mortality of marine bacteria and cyanobacteria. Nature 343: 60-62.

Sahlsten, E. 1998. Seasonal abundance in Skagerrak-Kattegat coastal waters and host specificity of viruses infecting the marine photosynthetic flagellate *Micromonas pusilla*. Aquat. Microb. Ecol. 16: 103-108.

Sandaa, R.-A., M. Heldal, T. Castberg, R. Thyrhaug and G. Bratbak. 2001. Isolation and characterization of two viruses with large genome size infecting *Chrysochromulina ericina* (Prymnesiophyceae) and *Pyramimonas orientalis* (Prasinophyceae). Virology 290: 272-280.

Schroeder, D.C., J. Oke, G. Malin and W.H. Wilson. 2002. Coccolithovirus (Phycodnaviridae): characterisation of a new large dsDNA algal virus that infects *Emiliania huxleyi*. Arch. Virol. 147: 1685-1698.

Schroeder, D.C., J. Oke, M. Hall, G. Malin and W.H. Wilson. 2003. Virus succession observed during an *Emiliania huxleyi* bloom. Appl. Environ. Microbiol. 69: 2484-2490.

Short, S.M. and C.A. Suttle. 2000. Denaturing gradient gel electrophoresis resolves virus sequences amplified with degenerate primers. BioTechniques 28: 20-26.

Short, S.M. and C.A. Suttle. 2002. Sequence analysis of marine virus communities reveals that group of related algal viruses are widely distributed in nature. Appl. Environ. Microbiol. 68: 1290-1296.

Short, S.M. and C.A. Suttle. 2003. Temporal dynamics of natural communities of marine algal viruses and eukaryotes. Aquat. Microb. Ecol. 32: 107-119.

Smayda, T.J. 1998. Ecophysiology and bloom dynamics of *Heterosigma akashiwo* (Raphidophyceae). pp. 113-131. *In* D.M. Anderson, A.D. Cembella, and G.M. Hallegraeff [eds.] Physiological Ecology of Harmful Algal Blooms. Springer-Verlag, Berlin, Germany.

Suttle, C.A., A.M. Chan and M.T. Cottrell. 1990. Infection of phytoplankton by viruses and reduction of primary productivity. Nature 347: 467-469.

Suttle, C.A. 1993. Enumeration and isolation of viruses. pp. 121-137. *In* P.F. Kemp, B. Sherr, E. Sherr, and J.J. Cole [eds.] Handbook of Methods in Aquatic Microbial Ecology. Lewis Publishers, Boca Raton, USA.

Suttle, C.A. and A.M. Chan. 1995. Viruses infecting the marine Prymnesiophyte *Chrysochromulina* spp.: isolation, preliminary characterization and natural abundance. Mar. Ecol. Prog. Ser. 118: 275-282.

Suttle, C.A. 2000. The ecological, evolutionary and geochemical consequences of viral infection of cyanobacteria and eukaryotic algae. pp. 247-296. *In* C.J. Hurst [ed.] Viral Ecology. Academic Press, London, UK.

Tai, V., J.E. Lawrence, A.S. Lang, A.M. Chan, A.I. Culley and C.A. Suttle. 2003. Characterization of HaRNAV, a single-stranded RNA virus causing lysis of *Heterosigma akashiwo* (Raphidophyceae). J. Phycol. 39: 343-352.

Tarutani K., K. Nagasaki and M. Yamaguchi. 2000. Viral impacts on total abundance and clonal composition of the harmful bloom-forming phytoplankton *Heterosigma akashiwo*. Appl. Environ. Microbiol. 66: 4916-4920.

Tarutani, K., K. Nagasaki, S. Itakura and M. Yamaguchi. 2001. Isolation of a virus infecting the novel shellfish-killing dinoflagellate *Heterocapsa circularisquama*. Aquat. Microb. Ecol. 23: 103-111.

Thingstad, T.F., M. Heldal, G. Bratbak and I. Dundas. 1993. Are viruses important partners in pelagic food webs? Trends Ecol. Evol. 8: 209-213.

Thyrhaug, R., A. Larsen, C.P.D. Brussaard and G. Bratbak. 2002. Cell cycle dependent virus production in marine phytoplankton. J. Phycol. 38: 338-343.

Tomaru, Y., N., Katanozaka, K., Nishida, Y., Shirai, K., Tarutani, M., Yamaguchi and K. Nagasaki. 2004a. Isolation and characterization of two distinct types of HcRNAV, a single-stranded RNA virus infecting the bivalve-killing microalga *Heterocapsa circularisquama*. Aquat. Microb. Ecol. 34(3): 207-218.

Tomaru, Y., K. Tarutani, M. Yamaguchi and K. Nagasaki. 2004b. Quantitative and qualitative impacts of viral infection on a *Heterosigma akashiwo* (Raphidophyceae) bloom in Hiroshima Bay, Japan. Aquat. Microb. Ecol. 34: 227-238.

Tomaru, Y. and K. Nagasaki. 2004c. Widespread occurrence of viruses lytic to the bivalve-killing dinoflagellate *Heterocapsa circularisquama* in the western coast of Japan. Plankton Biol. Ecol. 51, 1-6.

Van Etten, J.L. and S.A. Ghabrial. 1991. Phycodnaviridae. pp. 137-139. *In* R.I.B. Francki, C.M. Fauguet, D.L. Knudson, and F. Brown [eds.] Classification and Nomenclature of Viruses. Arch. Virol. Supple. 2, Springer-Verlag, Vienna, Austria.

Van Etten, J.L., L.C. Lane and R.H. Meints. 1991. Viruses and viruslike particles of eukaryotic algae. Microb. Rev. 55: 586-620.

Van Etten, J.L. 2000. Phycodnaviridae. pp. 183-193. *In* M.H.V. Van Regenmortel, C.M. Fauquet, D.H.L. Bishop, E.B. Carsten, M.K. Estes, S.M. Lemon, J. Maniloff, M.A. Mayo, D.J. McGeoch, C.R. Pringle, and R.B. Wickner [eds.] Virus Taxonomy, Classification and Nomenclature of Viruses, Seventh Report. Academic Press, San Diego, USA.

Van Etten, J.L., M.V. Graves, D.G. Müller, W. Boland and N. Delaroque. 2002. Phycodnaviridae - large DNA algal viruses. Arch. Virol. 147: 1479-1516.

Waters, R.E. and A.T. Chan. 1982. *Micromonas pusilla* virus: the virus growth cycle and associated physiological events within the host cells; host range mutation. J. Gen. Virol. 63: 199-206.

Westbroek, P., J.E. van Hinte, G.J. Brummer, M. Veldhuis, C. Brownlee, J.C. Green, R. Harris and B.R. Heimdal. 1994. *Emiliania huxleyi* as a key to biosphere-geosphere interactions. pp. 321-334. *In* J.C. Green and B.S.C. Leaderbeater [eds.] The Haptophyte Algae, vol 51. Clarendon Press, Oxford, USA.

Wilhelm, S.W. and L. Poorvin. 2001. Quantification of algal viruses in marine samples. Met. Microbiol. 30: 53-65.

Wilson, W.H., G.A. Tarran, D. Schroeder, M. Cox, J. Oke and G. Malin. 2002. Isolation of viruses responsible for the demise of an *Emiliania huxleyi* bloom in the English Channel. J. Mar. Biol. Assoc. UK. 82: 369-377

Wommack, K.E. and R.R. Colwell. 2000. Virioplankton: Viruses in aquatic ecosystems. Microbiol. Mol. Biol. Rev. 64: 69-114.

Yamada, T., N. Chuchird, T. Kawasaki, K. Nishida and S. Hiramatsu. 1999. *Chlorella* viruses as a source of novel enzymes. J. Biosci. Bioeng. 88: 353-361.

Zingone, A. 1995. The role of viruses in the dynamics of phytoplankton blooms. Giorn. Bot. It. 129: 415-423.

7

Autecology of Bloom-Forming Microalgae: Extrapolation of Laboratory Results to Field Populations and the Redfield-Braarud Debate Revisited

Theodore J. Smayda[1]

[1]Graduate School of Oceanography, University of Rhode Island, Kingston, RI 02881, USA

Abstract

The relevance and limits of autecological experiments extrapolated to *in situ* phytoplankton behavior are evaluated from the perspective of the unresolved Redfield-Braarud debate. Redfield discouraged autecological experiments, being skeptical of their ecological relevance. He argued that field-descriptions of natural populations are incomplete and compromise experimental design. Braarud dismissed this view, arguing that autecological experiments were not premature, but essential to quantify *in situ* phytoplankton behavior. In revisiting their debate, frequent reference is made to harmful algal bloom dynamics, a phenomenon that is highly species-specific, and the current focus of much experimentation. Some major issues dealt with include: the influence of early experimentation and new discovery in molding the opposing views of Redfield and Braarud, and the advances made in these areas since their debate; the conceptual differences between autecological and synecological approaches to phytoplankton ecology, and their respective experimental merits and limitations to quantify natural behavior; the three modes of phytoplankton growth (cellular, population, community growth); the two types of nutrient limitation (yield and physiological limitation); the various nutrient limitation models, and experiments examining the nutrient affinity, resource and ratio theories of phytoplankton species selection and competition. The nutrient issues are discussed as part of a general evaluation whether *in situ* nutrient limitation occurs and, if so, whether it is single nutrient limitation, as posited by Liebig's Law of the Minimum and Droop's model. Experiments with clones, on the suite of responses to nutrients, and on growth rates are used as tests of fidelity of *in situ* extrapolation of autecological experimental results.

It is suggested that the classical definition of the phytoplankton needs to be modified for conceptual and experimental design purposes. Significant differences in ecophysiology distinguish flagellates (swim strategists) from diatoms (sink strategists). The ecological behavior of diatoms is in greater accord with the classical definition of the phytoplankton, but is an inadequate template when applied to flagellate ecology and experimental design. Conflation of the definitions of primary production and phytoplankton ecology also presents significant conceptual and methodological problems affecting experimentation. Methodological and ecophysiological impediments to ecologically relevant experiments include: the occurrence of physiological strains (clones); difficulties in simulating *in situ* nutrient and grazing conditions; experimental enclosure dampening dispersion and vertical microstructure; substitution of static, batch culture conditions for the more dynamic *in situ* nutrient recycling mode captured in continuous culture experiments. Incubation artifacts always compromise ecological extrapolation of experimental results. Laboratory-based experiments intrinsically are inadequate to explain *in situ* behavior: some major ecological behavior and processes cannot be addressed adequately, either in 'bottle experiments' or in large enclosures such as mesocosms. *In situ* experimentation also presents formidable practical and analytical difficulties, has problems of ecological fidelity, and its efficacy is unconvincing.

Considerable evidence, including mass balance calculations and extrapolation, supports Braarud's view that autecological experiments on species ecophysiology provide an ecological 'orientation' that is not available from field measurements, but are essential to quantify *in situ* phytoplankton behavior and oceanic biogeochemical issues. Multiple evidence supports Redfield's countering argument that laboratory-based and *in situ* experimentation are inadequate to explain *in situ* ecological behavior. Despite this apparent contradiction, it is argued that Redfield and Braarud's opposing views are complementary rather than exclusive, and reflect their focus on different aspects and experimental needs of phytoplankton ecology. This does not negate that ecologically relevant experiments for some major features of *in situ* behavior may be beyond reach, whether based on laboratory experiments or *in situ* manipulations, and whether examined as 'top-down' or 'bottom-up' effects. Phytoplankton experimentation can avoid some of Redfield's concerns over the adequacy of ecological experiments and take fuller value from Braarud's advocated approach by applying a functional group, or phylogenetic, experimental strategy to *in situ* behavior. Functional group and phylogenetic experimentation should combine autecological and synecological concepts, field and laboratory approaches, and use advanced technology for measurements of *in situ* molecular and biochemical properties diagnostic of physiological status, and for description of habitat growth conditions and micro-structure. Other guidelines in following this new experimental approach to phytoplankton ecology are also given.

INTRODUCTION

The Redfield-Braarud Debate

Four decades ago, Alfred Redfield, of Redfield Ratio fame (Redfield 1934, 1958), and Trygve Braarud, a phytoplankton ecologist recognized (Mills 1989) "as one of the pioneers of quantitative biological oceanography", debated the relevance of experimentation to biological oceanography (Redfield 1960, Braarud 1961). Redfield's perspective (p. 22 in Redfield 1960) was that marine biology (his term) is "primarily ecological [with the] objective to develop effective models, and ultimately a single, consistent model, of life in the sea". He stated that these models should be based on a combination of field observations, experiments and inference. The first task in this effort, Redfield advised, was to acquire representative descriptions of the *in situ* behavior of natural populations because building 'effective models' requires substantial knowledge of its prototype. Knowledge of *in situ* behavior is needed also to design ecologically relevant experiments. And since new discovery of *in situ* behavior and habitat features can greatly alter general concepts, this also influences experimental design. Redfield concluded that current (i.e. then) knowledge was very incomplete – "a tremendous range of phenomena still awaits simple observation" – presenting a major stumbling block to the design of ecologically relevant experiments and model building. This diagnosis led to Redfield's (1960) opinion highlighted in the title of his essay "the inadequacy of experiment in marine biology".

Redfield was not averse to experimentation; he believed "marine biology ... can profit from the experimental approach in defining its problems and in elaborating the details of its solutions". But, he was skeptical that ecologically relevant experimentation was within reach. He cautioned that experiments usually are unnatural because they often are designed to give yes - no answers, or seek to quantify behavior in response to only one or two of the variables relevant to the process being examined. The other variables supposedly are then kept under experimental control or assigned neutral effects. Redfield (p. 18 in 1960) expressed a skeptical view of experiments: they are "essentially observations made under conditions rigged so as to answer particular questions". This intrinsic bias, Redfield added, is aggravated by the incomplete field-based descriptions of *in situ* behavior which obstruct the proper design of ecologically relevant experiments. Redfield then advanced his startling conclusion: ecological extrapolations of laboratory results to field populations were inadequate, could even mislead, and should be discontinued. He suggested (p. 21 in 1960) "the greater need at the moment is more knowledge of the phenomenon to be explained".

Braarud (1961), in response to Redfield, countered that autecological (= species-specific) experiments were neither premature nor invalid, but essential to quantify *in situ* phytoplankton dynamics. Braarud (1962) asserted that the inadequate understanding of phytoplankton distribution patterns was "mainly due to the deplorably small advances which have been made in [their] autecology". Braarud based his opposing view on his extensive field experience and single-species laboratory experiments that he carried out to assess environmental influences on *in situ* behavior. He even discussed problems in phytoplankton ecology that required experimentation for their resolution (Braarud 1961). Notwithstanding his advocacy of experimentation, Braarud acknowledged the ecological limitations of single factor experiments, particularly the problems of extrapolating laboratory results with monospecific cultures to field behavior. He puzzled over the discrepancies found between the *in situ* behavior of species in response to temperature and that expected from their laboratory growth (Karentz and Smayda 1984), and he emphasized the need for factor-interaction experiments. Although Braarud shared some of Redfield's concerns over the inadequacy of experiments, he touted their value in providing ecological perspectives (= '*orientation*') not revealed by field investigation. Braarud concluded (p. 294 in 1961), at least with regard to the phytoplankton, that "Dr. Redfield's warning against too one-sided an experimental approach does not apply at all".

Redfield did not focus on the phytoplankton in challenging the adequacy of ecological experiments to explain *in situ* behavior; his approach was more general. But the thrust of his arguments and the experimental methods and results then available suggest that he would not have been dissuaded by Braarud's opposing view, to which he apparently did not respond. Since Redfield and Braarud were experimentalists well versed in oceanographic principles and plankton ecology, their divergent views warrant serious consideration. Revisiting their discourse is further desirable for several other reasons. Their conflicting views were not reconciled then, nor have been rigorously assessed since; there has been a huge increase in laboratory and field-based experimentation based on more sophisticated methods and ecological insights; major discoveries of overlooked trophic communities and processes have been made; and possibly an antiquated trophic definition of the phytoplankton currently is being applied which compromises experimental approaches to *in situ* phytoplankton bloom and other behavior.

This chapter evaluates the issue of the reliability and the limits of laboratory-based experiments, where extrapolated to *in situ* phytoplankton behavior, from the perspective of the Redfield-Braarud discourse. There will be frequent reference to harmful algal bloom dynamics, a phenomenon that is highly species-specific and currently the focus of considerable laboratory-based autecological experimentation.

SOME ELEMENTS OF THE REDFIELD-BRAARUD DISCOURSE

New Discovery

Redfield's anticipated discovery of unknown biotic groups and trophic behavior that would lead to more representative ecological experiments was very prescient. Major discoveries made since include: the abundant and ubiquitous picophytoplankton (prochlorophyte/synechococcoid) community (Johnson and Sieburth 1979, Waterbury et al. 1979); discovery of the microbial loop (Azam et al. 1983, Laybourn-Parry and Parry 2000) and thin-layer communities (Rines et al. 2002); and recognition of Fe-limited phytoplankton growth in regions previously thought to be N-limited or free of nutrient limitation (Martin et al. 1991, Kolber et al. 1994, Behrenfeld et al. 1996, Boyd et al. 2000, Tsuda et al. 2003). These revelations, neither predicted, nor anticipated from models and experimental data, have bolstered Redfield's argument that incomplete qualitative descriptions of *in situ* population behavior compromise experimental design, data extrapolation, and concept development. Redfield's call to anticipate unrecognized phenomena has been reiterated (Parsons 1985, Longhurst 1998). Parsons gives special importance to this, contending (p. 110 in Parsons 1985) that "it is the anomaly that has been the true driving force for new science and not hypothesis testing and rigorous statistics". The unexpected discovery that Fe may be limiting in erstwhile N-limited regions (Martin et al. 1991) has stimulated intense field experimentation (Kolber et al. 1994, Behrenfeld et al. 1996, Boyd et al. 2000, Tsuda et al. 2003), consistent with Parsons' view.

A current phenomenon falling into the category of 'new discovery' is the global spreading of harmful algal blooms (HABs), which include blooms of cryptic, novel and newly discovered species (Smayda 1990). This phenomenon no longer can be considered an outbreak of rogue blooms having minor trophic significance. Their increased frequency, global behavior, and adverse trophic consequences are an emergent biotic event driven by unknown stimuli. I share Redfield's belief that a prerequisite to incorporating new *in situ* discovery into experimental approaches is first to establish, through comparative ecological investigation, the spatial and temporal distribution and abundance of the biotic component(s) newly revealed. Since field descriptions of HAB behavior are still in this 'information-gathering' phase, continuing, comparative field study of this phenomenon is needed to design appropriate ecological experiments. The important feature of unexpected biotic discoveries relevant to the Redfield-Braarud discourse is not taxonomic discovery, but awareness of the presence, the functional group significance, and trophic role of the new species being discovered.

Experimental Techniques

Culture media

Redfield and Braarud obviously could not consider experimental results obtained subsequently using better phytoplankton culturing media and improved methodology, including chemostats and mesocosms. Inadequate culturing media, techniques, and experimental dependence on 'weed' species were a major, early experimental stumbling block that precluded more exacting experiments on ecologically relevant species and processes. Early investigators distinguished between 'good' vs. 'bad' sea water after establishing that sea water varies seasonally and regionally in its capacity to support phytoplankton growth, even when emended with the essential macro-nutrients nitrogen and phosphorus (Provasoli et al. 1957). Given that natural sea water is an unreliable basal medium, many investigators relied on chemically undefined, soil extract (Erdschreiber) enrichments of sea water to rear cultures (Schreiber 1927, Sweeney 1951), but continued the long-term quest begun already in the early 1900s (Allen and Nelson 1910, Allen 1914) to develop artificial media for more reliable cultivation of representative species. Redfield, himself, contributed to this development (Ketchum and Redfield 1938). Just prior to the Redfield-Braarud debate, Provasoli et al. (1957) made a major advance with their artifical media formulations which contained essential vitamins and chelating agents to control micro-nutrient availability, such as Fe and Mn. This breakthrough, Guillard's equally famous 'medium f/2' (Guillard and Ryther 1962) and other media (Morel et al. 1979, Berges et al. 2001) have greatly stimulated ecological experimentation on representative phytoplankton species and cellular processes. Nonetheless, and despite numerous attempts, some species still resist cultivation, such as autotrophic species in the dinoflagellate genus *Dinophysis* which cause diarrhetic shellfish poisoning (Maestrini 1998). A *Dinophysis* species was last brought successfully into sustained culture (though briefly) 70 years ago (Barker 1935). And, only relatively recently the fastidious N-fixing cyanobacterial bloom species

Trichodesmium erythraeum and *T. thiebautii* have been established in culture (Ohki et al. 1986).

Static vs. Pulsed Experiments as in situ *Analogues*

Despite the advances in culture media formulations, unresolved experimental impediments and issues relevant to the Redfield-Braarud debate remain. Laboratory experiments have the methodological problem of simulating the natural *in situ* pulsing of cellular and population growth factors that occurs. The half-life of cellular exposure to a fixed level of irradiance, nutrients and grazer abundance, for example, is very brief. Exposure and access to a given nutrient or irradiance level can last only seconds-to-minutes, while grazing intensity varies hourly or less. In contrast, the short-term fluctuations in growth factors such as temperature and salinity experienced by cells are relatively insignificant. The *in situ* flux of nutrients from physical and biological (remineralization) processes, essential to phytoplankton growth, is particularly difficult to reproduce experimentally. The standard practice of using 'batch' culture techniques in which a fixed amount of nutrient is added, and then decreases with growth, precludes this dynamic. Another common, but unnatural experimental procedure is to expose cultures to a constant irradiance level and photoperiod. This deprives the cells of a natural (= sinusoidal) diel irradiance cycle and, within this, exposure to *in situ* variations in irradiance induced by water column mixing (Denman and Gargett 1983). Brzezinski and Nelson (1988b) have shown that different rates of nutrient pulsing in combination with photoperiod variations affect the outcome in diatom competition experiments. Batch culture experiments at the time of Redfield and Braarud, and since, have been the primary source of ecological data available for *in situ* extrapolation. A major shortcoming of the batch culture approach is that it presents a static, progressively deteriorating experimental environment for growth. It can be difficult to determine whether the observed, averaged behavior is primarily a response to stress or is ecologically valid.

Chemostats, i.e. continuous culture, in contrast, allow growth to be assessed in response to nutrients fluxed at fixed supply rates that can be varied from experiment-to-experiment to obtain responses to a suite of flux rates. Chemostats have provided very useful data on nutrient uptake kinetics, experimentally the best quantified of the processes that regulate *in situ* phytoplankton growth (Droop 1974, Conway and Harrison 1977, Harrison et al. 1976, Pan et al. 1996b, among many others). Semi-continuous culturing techniques, a hybrid of the 'batch' and 'chemostat' approaches, to mimic *in situ* dynamics (Paasche 1975, Tilman and Kilham 1976, Takahashi and Fukazawa 1982, among others), and *in situ* incubations of phytoplankton enclosed within permeable dialysis sacs (Sakshaug and Jensen 1978) have been used less frequently. A problem that experimentalists confront is whether a 'batch' vs. 'chemostat' type experiment is more suitable to examine the behavior of interest. Nutrient uptake rates, the effects of nutrient ratios on species selection, and species competition experiments are better approached, i.e. have greater ecological value, using continuous culture than batch culture methodology. Batch culture technique is probably more suitable for cellular-level studies of a species' life history; to establish its essential nutrient requirements and ecological minima, maxima and optima for irradiance, temperature and salinity, i.e. the ecological limits of its tolerance to those parameters; and its suitability as prey. However, since *in situ* population dynamics are driven by multiple, interactive growth factors within a macro- and micro-habitat structure operating in pulsed rather than static mode, the extrapolation of batch culture behavior to natural populations is compromised, particularly if the species' autecological profile is based primarily on monospecific experimentation. Some investigators have employed both experimental approaches to confirm a result, or to examine the process from different perspectives (Pan et al. 1996a, b).

Standardized techniques for ecological experimentation on phytoplankton do not exist, nor should be expected. The complexities and regulation of the phytoplankton life mode, and the diverse adaptations of species in response to pelagic survival in basically a nutritionally dilute medium and regionally variable ocean (Longhurst 1998) preclude standardized approaches. The remarkable buoyancy-regulated, nutrient gathering migrations of large diatoms in oligotrophic seas is one expression of the complex ecology and adaptive capacity of the phytoplankton (Villareal et al. 1996). Other nutrient-inspired behavior includes the nutrient-gathering migrations of dinoflagellates (Smayda 1997a) and diatom sinking behavior to overcome nutrient limitation (Brzezinski and Nelson 1988a). Standardization of methods is less a problem than the need for investigators to design and carry out ecologically relevant experiments, a reliance often unfulfilled. Two notable features of contemporary phytoplankton ecology and experimentation are the traditional and imbalanced experimental focus on whole community (synecological) issues over organismal behavior, and the limited efforts at hypothesis testing and experimental proof of concept. Available experimental results largely validate Braarud's (1961) view that the primary value of experiments still is in providing 'ecological orientations', rather than meeting Redfield's (Redfield 1960) more exacting criterion of facilitating representative model building.

EARLY EXPERIMENTATION

The Redfield-Braarud discourse did not take place in an era when experiments and quantitative approaches were ignored. It was predated by a long tradition of experimental research on phytoplankton cultures and natural populations to quantify *in situ* behavior. The Plymouth School and other groups in the UK, particularly, pioneered monospecific culture experimentation to determine bloom species' growth rates; depth of the euphotic zone; photosynthesis-irradiance relationships; nutrient requirements; suitability as prey for copepods, and extrapolated the results to natural populations (Harvey et al. 1935, Harvey 1933, 1942, Jenkin 1937, Marshall and Orr 1928, and Mills 1989). Decades earlier, from 1916 to 1922, the Norwegian phytoplankton ecologist H.H. Gran enriched spring bloom and autumn samples with NH_4, NO_3 and PO_4 to determine species-level, nutrient-growth relationships (Gaarder and Gran 1927, Gran 1927). Gran also pioneered measurement of primary production, having developed the light and dark bottle O_2 method, and carrying out such measurements already in 1916. This effort ultimately led to Steemann Nielsen's introduction in 1952 of the now ubiquitously used ^{14}C method (Barber and Hilting 2002).

Redfield undoubtedly was aware of the major results of plankton experiments in recommending their abeyance until more representative descriptions of *in situ* behavior were obtained. His recommendation is probably not attributable to the instructions that attendees at the 1956 Symposium on Perspectives in Marine Biology received: to focus on problems "of marine life ready for experimental attack [and not to] survey past accomplishments" (Buzzati-Traverso 1960). Note the operative instruction: problems "ready for experimental attack". Redfield and Braarud both were aware that poor methods impeded early investigators from establishing key species into culture, and that their experiments were compromised by limited insights into key ecological processes such as nutrient-regulated behavior, bloom dynamics and grazing. Although early experimental results often were of limited ecological value, and their *in situ* extrapolations were tenuous, important concepts were being developed from this crude data bank. Gran (p. 5 in 1929), for example, stated "we may consider it proved that the limiting factor in the production of plankton, both in marine and fresh waters, in the majority of cases is the available amount of nutritive salts in solution". Conversely, Gran characterized nutrient enrichment experiments to evaluate in situ species responses as 'a statistical investigation' making it necessary to "renounce any possible critical account of the limitations of the species" (Gaarder and Gran 1927). Redfield expressed a similar view regarding the role of correlation in efforts to understand the distribution and abundance of life in the sea. He stated (p. 22 in Redfield 1960) the approach to be followed "must be primarily statistical [based on] significant relationships between large quantities of observations on biological and physical events... in widely scattered places". Phytoplankton field ecologists traditionally have followed this statistically-based, descriptive approach by measurement of pre-selected environmental and biotic variables, and using the correlations obtained as evidence for, or against the relationship being assessed.

The insights gained from field investigations have greatly influenced experimental design and paradigm formulation. Phytoplankton experimentation, historically, has evolved primarily from field-based observations, and rarely from organismal perspectives. There is a great difference between these two approaches in their expectations and use of experimental data. Gran's contradiction in accepting experimental data as explaining *in situ* behavior is a case in point (Gran 1929, Gaarder and Gran 1927). In evaluating his nutrient enrichment experiments to explain species behavior, Gran seemingly agreed with Braarud (1961) that the primary value of autecological experiments is in providing an ecological 'orientation'. Yet, he accepted the equally limited (semi-quantitative) experimental and field-based data then available as proof for the nutrient limitation hypothesis (Gaarder and Gran 1927). Ambivalent acceptance and application of experimental data persists in contemporary phytoplankton ecology. This is particularly evident in the use of experimental data and field evidence to evaluate *in situ* nutrient limitation, and examined in a later section.

CONCEPTS, PARADIGMS AND EXPERIMENTS

Current plankton concepts derive largely from those in formulation at the time of the Redfield-Braarud discourse. This raises the question of whether these concepts and derivative paradigms have a firm experimental basis. Neither Redfield nor Braarud considered the role of experimentation in hypothesis testing and concept development, a surprising neglect given Redfield's skepticism of ecological experiments. Redfield stressed that field observations had yet to provide representative descriptions and statistically based correlations of ocean-wide *in situ* behavior, insights needed to design experiments to explain this behavior. An intense, subsequent regional oceanographic effort has reduced this deficiency (Longhurst 1998). Field efforts continue to focus on mapping and measuring the distributions, abundance and productivity of plankton, and underlying habitat conditions.

To a large extent, this is a techno-driven effort. The emergence of satellite oceanography has energized this effort. In contrast, there is remarkably little effort at ecological hypothesis testing, either during field investigations or derivative from such efforts. A notable exception is the experimentation directed to the Fe-limitation hypothesis (Kolber et al. 1994, Behrenfeld et al. 1996, Boyd et al. 2000, Tsuda et al. 2003). The major thrust of experimentation is, seemingly, directed towards establishing the rates of processes viewed as isolated events rather than in interaction with the combined physical, chemical and biological milieu that underlies and regulates plankton behavior.

Recognition of flawed and biased experimentation, whether old or more recent, is relatively easy. It is more difficult to detect unwarranted, broad macroecological extrapolations of experimental data because the ecological connections being made often appear reasonable, but mask actual relationships. Concepts usually result from reductionist searches for common behavior and mechanisms and, when formulated (= presumed), such erstwhile uniformity is applied collectively to functional groups (diatoms, dinoflagellates, etc.), to processes (blooms, primary production, etc.), and mechanistically (critical depth, grazer control, etc.). This reductionism tends to focus on whole community (synecology) behavior over autecological investigation of individual species, from which synecological extrapolations are made. Both approaches carry serious risk of unwarranted extrapolation of data and concepts. For example, while there is some support (Pratt 1966) for the long-standing 'ectocrine theory' (Lucas 1947, 1955), which seeks to explain species succession, grazing and other species-level behavior, the frequent general application of such allelochemical and allelopathic regulation lacks experimental confirmation, and is unwarranted (Smayda 1980). A global expansion in harmful algal blooms, including blooms of unusual species and novel events, is in progress (Smayda 1990). This perplexing phenomenon is often attributed to ballast water seedings of starter populations, but supporting field and experimental evidence, as for the 'ectocrine theory', is very limited (Smayda 2002b).

Phytoplankton ecologists tend to use unverified and attractive unifying concepts such as the 'ectocrine' and 'ballast water' theories (both experimentally difficult to validate) to account for events not explained applying existing dogma. Such efforts are not without value. The weakly silicified, intertidal benthic diatom *Phaeodactylum tricornutum* (= *Nitzschia closterium f. minutissima*), a rock pool species established into culture nearly a century ago (Allen 1910), was a major, early experimental 'work horse' to help resolve *in situ* phytoplankton behavior. Early experimental results obtained with this 'weed' species (and later with other non-fastidious species) were extrapolated to more representative species applying the premise of shared ecological requirements and physiological behavior. Subsequent experiments validated this extrapolation from such 'unexacting species' (Harvey 1955). The widely applied premise of a 'common autecology' is consistent with Redfield's view that inference is important in model building; the inferred behavior (response) is then subject to experimental validation. The notion that biochemical and ecological uniformity characterizes the phytoplankton is discussed in a later section.

There is a more basic conceptual and experimental issue relevant to Redfield's (1960) view that the ultimate objective of biological oceanography is to build a single, consistent model of life in the sea. The traditional definition of the phytoplankton is based on their taxonomy and ecological role as primary producers. I suggest that this should be modified, both conceptually and for experimental design. Redefinition particularly is needed to deal with harmful algal species' blooms, which primarily are blooms of phylogenetically diverse flagellates. Two basic strategies characterize the phytoplanktonic life mode: the sink strategy of diatoms and the swim strategy of flagellate species (Smayda 1997a). Diatoms and flagellates differ also in nitrate metabolism (Lomas and Glibert 2000), and many (if not most) flagellate species, unlike diatoms, are not obligate autotrophs but capable of both photosynthesis and mixotrophy (Granéli and Carlson 1998). That is, many are both primary producers and grazers (often bactivorous). Moreover, many non-toxic dinoflagellates assumed to be photosynthetic are heterotrophic (Hansen 1991, Larsen and Sournia 1991). Yet, laboratory experiments with flagellate species rarely accommodate both nutritional modes, and experimental procedures usually repress ecologically significant behavior associated with motility, i.e. diel and nutrient gathering migrations, thin-layer accumulations, etc. This frequent experimental mismatch with flagellate life-form requirements is further aggravated by application of the 'diatom template' bias – the assumption that the experimental approaches commonly applied to diatom ecological issues are valid for the phytoplankton generally. Diatoms, the most recently evolved, major phytoplankton group, unquestionably are important in trophic and bloom processes which justifies their intense experimentation. However, while diatoms most closely adhere to the classical definition of the phytoplankton, flagellates diverge from this conformity in many important ecophysiological features and *in situ* behavioral patterns (Smayda 1997a, Smayda and Reynolds 2003). Dinoflagellates appear to have incorporated into their ecology and adaptive strategies some of the swarming behavioral features of insects (Smayda 2002a). In dealing with the capacity of autecological experiments to help quantify harmful algal bloom dynamics, functional group distinctions must be recognized.

SYNECOLOGY, AUTECOLOGY AND PHYTOPLANKTON GROWTH MODES

Redfield and Braarud's opposing views reflected their different approaches to plankton processes. Redfield emphasized 'whole community' processes, i.e. a synecological perspective. Braarud pursued an organismal (autecological) approach to community dynamics. The focus, inferences and experimental methodology of these two approaches differ. Current knowledge and paradigms of *in situ* phytoplankton processes are based primarily on field measurements that have applied mass balance, reductionist techniques and synecological concepts. While this effort has been partially guided by experimental results on organismal ecophysiology, the autecological (organismal) approach has developed virtually as a separate and secondary line of enquiry within biological oceanography. This divergence has become increasingly evident since the Redfield-Braarud debate, but it is not a direct consequence. The separation of community and organismal based ecology in field and experimental studies contrasts with the historical inclusion of species assemblages, their blooms and successions in early field investigations and primary productivity assessments (Mills 1989). Gaarder and Gran (1927) partitioned among individual species the primary production measured for the total community. And, it is generally overlooked that Hensen (1887) in inspiring the reductionist, mass balance approach reached his conclusions based on detailed analyses of the planktonic flora and fauna in applying his perspectives as a physiologist interested in fishery yields. There is reason to believe (Barber and Hilting 2002) that the introduction, in 1952, of Steemann Nielsen's ^{14}C method to measure primary production hastened the natural tendency of the autecological and synecological approaches to diverge. The ability of the powerful ^{14}C method to measure primary productivity with precision, accuracy and ease, combined with more quantitative phytoplankton and nutrient analytical techniques (Strickland, 1960, Strickland and Parsons 1972), elevated synoptic field investigations to a quantitative level not yet matched by routine floristic assessments. Emphasis on whole community dynamics over individual species dynamics has been further propelled by interests in oceanic biogeochemical processes; the need to quantify food web dynamics and trophic transfer (Cushing 1975) and, more recently, concern over potentially adverse impacts of global warming, hypernutrification of coastal waters, and over-fishing. These concerns are best approached from a community, rather than a species-specific perspective.

This skewed emphasis, however, has begun to change. The global expansion of harmful algal blooms (Anderson 1989, Smayda 1990, Hallegraeff 1993), a highly species-specific phenomenon, has refocused attention on organismal biology because mass balance, synecological approaches are inadequate to quantify individual species' bloom selection, behavior and toxicity. An interesting dichotomy has resulted: harmful algal bloom (HAB) studies have become intensely autecological, whereas biological oceanography continues to develop through, and to emphasize synecological approaches. The species-based knowledge to be gained from field and experimental study is essential to development of predictive models, mitigation techniques, and early warning criteria for aquacultural and human health protection. The concerns expressed in the Redfield - Braarud discourse are relevant to these needs, and to the issues of the reliability and limits of experimentation applied to harmful bloom dynamics. To evaluate this, the methodological and conceptual divergence between autecological and synecological approaches must be recognized, since in reality both ecologies combine to structure and regulate marine community dynamics.

Synecological Field and Experimental Approaches

The synecological approach – the 'whole community' approach – is based on a 'Notion of Equivalence'. Conceptually, the abundance and processes of the phytoplankton community are viewed in aggregate. Operationally, total community abundance is measured in bulk, usually as chlorophyll (= biomass), and primary production is the usual surrogate measurement for the collective physiological processes. A key assumption made is that the assembled species have similar kinetics and food web value, and respond synchronously. Species-specific processes are ignored, and the untested assumption applied that the measured whole community responses reflect the behavior of the dominant species. The contributions of minor species, usually so categorized because of their low cell abundance, are considered to be insignificant. This assumption, where applied to harmful blooms, can be risky (Smayda 1997b). *Dinophysis* species at abundance levels of < 600 cells L^{-1} can have toxic effects (Maestrini 1998). Another synecological bias is the traditional focus on the spring diatom bloom as the major annual bloom event driving pelagic processes. Seasonal and aperiodic blooms of other taxa, and blooms that are ephemeral, unpredictable or of low magnitude are considered trophodynamically insignificant irrespective of their frequency or ecological impact. Field and experimental synecological approaches basically have been developed using a diatom-based template of dynamics and environmental regulation.

In essence, synecology is a mass balance and biogeochemical approach in which biomass and primary production are the major expressions of phytoplankton dynamics, and nutrient regulation is assumed to be the dominant control mechanism of these two properties. This conceptual and practical approach to phytoplankton ecology has been in force for nearly 125 years, ever since Hensen's (1887) brilliant efforts to quantify phytoplankton production, and reinforced by the classical Redfield Ratio which codifies the mass balance approach (Redfield 1934, 1958). Conceptually, the mass balance approach applies the photosynthesis - respiration reaction (Eqs. 1, 2) in combination with the stoichiometric relationships (Redfield Ratio) that occur between nutrients (N, P) and the photosynthesis and respiration of organic matter. In this relationship, nitrogen (N) and phosphorus (P) are bound and released during the synthesis and respiration of organic carbon (C) which, in turn, results either in the release or utilization of oxygen (O). The quantities of the four elements processed in this biochemistry can be expressed in terms of atoms, or by weight:

$$\text{O:C:N:P} = 212:106:16:1 \text{ (by atoms)} \qquad \text{(Eq. 1)}$$

$$\text{O:C:N:P} = 109:41:17.2:1 \text{ (by weight)} \qquad \text{(Eq. 2)}$$

Thus, during photosynthesis (primary production), for every atom of P assimilated 16 atoms of N will be assimilated and 106 atoms of C fixed into organic matter (from CO_2), liberating 212 atoms of O. This assimilation of N and P leads to phytoplankton growth (i.e. biomass = carbon, chlorophyll, etc.) and oxygenates the watermass.

The biogeochemical thrust of synecology has been fundamental to the development of biological oceanographic principles. An unfortunate consequence has been conflation of the definitions of primary production and phytoplankton ecology. Primary production and phytoplankton ecology are regularly used synonymously, which influences concepts and methodology, including experimental approaches. This extravagance is not surprising, accepting that determination of primary production "has been a major, if not the major goal of biological oceanography [since] the mid nineteenth century" (p. 16 in Barber and Hilting 2002). This synonymic confusion, I believe, has promoted the divergence of autecological and synecological approaches, rather than their needed fusion.

The ecology of a phytoplankton species and its assembly into communities is a triad of its cellular requirements and population (both autecological) and community-based (synecological) behavior and regulation. Primary productivity is only one aspect of phytoplankton ecology, albeit an important one, but it is only the outcome and not the defining basis of this ecology. It is the behavior of the species comprising the marine phytoplankton *in toto*- their taxonomy, life cycles, distribution, seasonalities, physiology, photosynthesis, trophic impacts, etc. – that constitutes their ecology, and not a particular subset of behavior such as primary production or bloom patterns. Fundamentally, phytoplankton ecology is the composite of species-level processes that occur as genetically programmed cellular behavior in response to habitat conditions and disturbances. This requires fusion of the autecological and synecological approaches, both in field studies and experimentally, to quantify phytoplankton processes such as primary production and blooms. Neither Redfield nor Braarud in their discourse commented on this dual approach; Redfield seemingly advocated a whole community approach and Braarud an organismal approach.

The Autecological Approach

While mass balance, synecological experiments are primarily field oriented, autecology is primarily laboratory-based. Autecology is the study of an individual organism or species and its interactions with its abiotic and biotic environment. The autecology of phytoplankton species is a composite of their phylogenetic, genetic and biophysical traits. Phylogeny influences their behavioral and physiological adaptations, including pigment complexes, motility, and nutritional idiosyncrasies. Genetics influences their growth capacity, nutrient uptake rates and clonal (strain) variability, etc. Biophysical constraints are imposed on tolerance to turbulence, motility and migratory behavior; metabolic rates are influenced by cell size and shape. The 'Notion of Biodiversity' is fundamental in the autecological approach, i.e. recognition that phytoplankton communities are polymixtures of species of overlapping, yet differing cell size, life-form and ecological tolerances. The taxonomic diversity of the assembled species leads to ecophysiological diversity and species-specific differences, rather than to uniform biochemical and ecological behavior, unlike the assumption of ecological equivalence that underlies the synecological approach. The autecological approach assumes that the species making up the community are in different growth stages, i.e. that succession is taking place; that different factors can regulate the assembled species, and that a given species may be under multifactorial rather than single factor regulation. The autecological approach focuses on the *in situ* interrelationships among species, which vary in their abundance and seasonality, at four different degrees of complexity: organism-environment; organism-organism; cellular vs. population dynamics; and population - population interactions. This differs from the synecological, mass balance focus on group dynamics (i.e. whole community) in response to a given habitat parameter. The autecological approach seeks to integrate the dynamics of individual species, including their successions, blooms, growth requirements and food web value, with whole community behavior. In this, while it is assumed that the overall properties of the communities assembled during succession are a composite of the species then selected for growth and assembly, the dynamics of a given species cannot be deduced from knowledge of overall community behavior, or from the behavior of a competing species.

Major assumptions underly autecological experimentation and *in situ* extrapolations: the behavior of a species in laboratory culture is assumed to be manifested *in situ* with regard to essential nutrient requirements, uptake kinetics and cellular quotas; the optima and range of tolerance for temperature and salinity; its growth rates, and photosynthesis-irradiance relationships. Thus, if a particular vitamin must be provided for growth in culture, it is assumed to be required *in situ*. And, a species highly efficient in culture at taking up NH_4 at low concentrations, i.e. it has a low K_s coefficient, is expected to have a similar capacity *in situ*. A species exhibiting a very high growth rate in culture, for example $\mu = 2.5\ d^{-1}$, is assumed capable of similar *in situ* growth at equivalent environmental conditions. It is also assumed that species do not mutate while in culture, and that the responses of the cultured strain are representative of that species generally. These assumptions are examined in a later section.

Autecological *in situ* behavior poses great experimental difficulties, as Braarud (1961) acknowledged. Its species-specific focus requires that the natural community with its populations of the individual species assembled be isolated for *in situ* experimental manipulation. Since incubation of the entire community without loss of key, regulatory habitat features is virtually impossible, this limitation has favored the reductionist (synecological) approach. The usual alternative strategy is to isolate into culture the species of experimental interest. However, this greatly limits the types of ecologically relevant experiments that can then be carried out. Monospecific laboratory experiments, whether in batch or in continuous culture mode, and factor interaction studies, usually do not provide data useful to synecological assessments. A compromise is to enclose natural communities into mesocosms for experimental manipulation to observe the induced behavior of the process or species of interest (Grice and Reeve 1982, van der Wal et al. 1994, Verity et al. 1988). Discussion of mesocosm experiments falls outside the scope of this chapter. This omission does not alter the conclusions to be reached.

Before evaluating the evidence for whether meaningful autecological and synecological experiments can be achieved, the two different types of phytoplankton nutrient-limitation and the three modes of phytoplankton growth are considered.

Modes of Phytoplankton Growth and Limitation

Growth rate and biomass accumulation are the traditional measures of the *in situ* success of phytoplankton species and communities, with primary production rate also used for the latter. When growth rates and biomass levels are low, the concept of 'limited growth' is applied, and efforts are then made to identify the specific limiting factor. A problem with this approach, and primarily a synecological shortcoming, is that investigators generally do not distinguish between the two types of growth limitation that occur: biomass (= population) limitation and physiological (= cellular) limitation (Falkowski et al. 1992). *In situ* population growth follows the well known growth minus loss relationship:

$$N_t = N_o e(K + K_i - K_a - K_m - K_s - k_g)t \qquad \text{(Eq. 3)}$$

where the population growth rate (N_t) is the difference between the sum of the cellular growth (K) and immigrant cell recruitment (K_i) rates, and the combined population losses due to washout or flushing (K_a), mortality (K_m), sinking (K_s), and grazing (K_g). Ignoring the loss terms in Equation 3, cellular growth rate is influenced by the combined effects of the physical (turbulence, temperature, irradiance), chemical (nutrients) and biological (allelochemical effects) factors making up the environment. The presence and success of a species can be controlled either by qualitative and quantitative deficiences in these growth factors, or if they exceed the ecological limits of tolerance for that species. In the case of nutrients, the rate of supply determines the amount of biomass (cells) produced in a dose-yield relationship, i.e. the standing stock (biomass) increases in direct, linear response to the amount of nutrient supplied. Thus, nutrients regulate both a species' growth rate (an autecological effect) and determine the environmental carrying capacity for its biomass (a synecological effect), i.e. nutrients influence both cellular and population behavior.

Phytoplankton growth in regions and at times of low nutrient supply is often considered to be nutrient-limited, if not stressed. Numerous exceptions to this generalization occur, as Goldman et al. (1979) showed and who stated (p. 4 in 1979) "at low, steady state nutrient levels it is possible to have simultaneously low or undectable residual nutrient levels and high growth rates regardless of the biomass concentrations". That is, the *in situ* standing stock of a population (or a species in culture) can be limited by the amount and rate of supply of required nutrients, but the population (species) is not then necessarily impaired physiologically. Physiological limitation sets in when cellular nutrient requirements, i.e. cell quota (Droop 1974), exceed nutrient supply rates. Physiologically limited cells not only decline in abundance, they are outcompeted and replaced by other species in the local succession.

Physiological limitation is a cellular response, while biomass (bloom) accumulation is the residual population response reflective of overall community dynamics, i.e. it is a synecologically influenced response. The convergence of these two ecologies compromises *in situ* experiments seeking to relate bloom and successional behavior to specific environmental conditions. Classically, statistical analyses have been used to relate changes in phytoplankton abundance and bloom events to nutrient levels. Experimental mass balance approaches (Eqs. 1, 2) have provided a firmer foundation for nutrient regulated growth (Dugdale 1967), but the problems of distinguishing between biomass and physiological limitation, or identification of the specific, limiting nutrient within the macro- and micro-nutrient pool remain. Tracers, i.e. an isotope of the nutrient element of interest, used to measure the pathways of uptake, transfer and release have helped to establish nutrient limitation effects (Dugdale and Wilkerson 1992, Nelson and Brzezinski 1990). Falkowski et al. (1992) have suggested that empirical identification of the symptoms of physiologically limited growth in natural populations and the specific limiting factor (i.e. nutrients, temperature, irradiance, etc.) should be possible using cellular biochemical markers symptomatic of the specific deficiency and impaired process. An advantage of this assay would be the analytical detection of physiological limitation without the need for experimental incubation.

The *in situ* dynamics of phytoplankton species and communities are shaped by combinations of fixed and flexible autecological parameters, modified by variable synecological processes. Eventually, growth of all species organizing the community becomes limited, both in biomass and physiologically. Given the occurrence of succession, it seems unlikely that communities per se become physiologically limited, although nutrient availability might restrict the gross biomass level attainable, i.e. prior to the loss terms in Equation 3. The consequences of the two different impacts of nutrient limitation vary with the mode of phytoplankton growth and the stage within these modes. Three different, concurrent phytoplankton growth modes occur and must be distinguished experimentally: cellular, population and community growth (Smayda 1997a).

Celluler, Population and Community Growth

Cellular growth is the active, basic growth unit; the outcome of coupled physiological processes under genetic and multifactorial control. The specific growth requirements, adaptive responses and the rates of response to growth factors such as nutrients, irradiance, and tolerance to factors that influence this physiology, such as temperature, are also under genetic control. Cellular growth is represented by the term, K, in the population growth equation (Eq. 3).

Population growth (N_t) is the environmentally modified outcome of cellular growth; it is the recruitment term and therefore the bloom unit. While it is dependent on cellular growth (K), the factors regulating cellular and population growth rates are not identical. Zooplankton grazing and advective washout, for example K_g and K_a, respectively in Equation 3, influence population growth rate, but not cellular growth. A given factor may influence cellular and population growth rates in different ways. Temperature for example, affects cellular growth rate via photosynthesis and nutrient uptake, but influences population growth through an affect on grazing rates. Nutrients also influence both cellular and population responses. At the cellular (autecological) level, nutrients regulate life cycle changes (Garcés et al. 2002) and growth rate, while the amount and supply rate of nutrients determine the population abundance (= carrying capacity).

Community growth rate can be expressed in various ways: from changes in pro-rated carbon assimilation rate (i.e. the amount of carbon fixed per unit chlorophyll and time [mgC $mgChl^{-1}$ h^{-1}]); community biomass, or as the succession rates of the species that organize the community (Smayda 1980). Cellular and population growth rates are the usual experimental focus. Phytoplankton communities are transient assemblages of multiple species, each in different bloom cycle stages and regulated (potentially) by different combinations of growth factors. Community organization and growth obviously are the outcome of the cellular and population growth rates of the species organizing the community, with interspecific competition also an important structuring element.

While community dynamics are a direct outcome of the three types of growth rates recognized, there are unpredictable stochastic contributions. Smayda and Reynolds (2001), based on a detailed analysis, concluded that the highly unpredictable blooms of harmful species often appear to be the result of the species *"being in the right place at the right time"*. Stochastic bloom events can result from variable excystment of resting stages deposited onto seed banks, or during unpredictable innoculations of cells from pelagic seed banks during watermass intrusions, K_i in Equation 3 (Smayda 2002a). Unusual local meteorological and ephemeral habitat disturbances can selectively stimulate or repress species, and unpredictably alter community dynamics. Smayda (2002b), based on niche considerations, has hypothesized that the selection of bloom species is heirarchical, which poses experimental problems. He has proposed that harmful bloom-species selection follows a taxonomic heirarchical pathway progressing from phylogenetic to generic to species selection, in that sequence and with considerable unpredictability as to which phylogenetic group(s) will bloom; the bloom genus (genera) selected from within that group, and the selection of the bloom-species from within the genus.

Autecological Extrapolations: Fidelity to *in situ* Behavior

Phytoplankton ecologists have had to simplify the complexities of *in situ* phytoplankton growth for methodological and other practical reasons en route to elucidation of first-order processes. Field observations have focused on specific events (e.g. the spring bloom), processes (e.g. primary production; grazing), and selected habitat conditions perceived to be important regulators of those events and processes (e.g. nutrient regulation). The conceptual and methodological advances applying this approach, and the large body of ecophysiological data acquired, no longer support the traditional ecological approach to treat the biota, their events and processes in isolation. Species' populations are heterogeneous in life cycles, physiological states and genetic makeup, and they vary in space and time as a result of interactions with other, equally heterogeneous and dynamic populations and physical conditions. We also know that the great reliance on mass balance, synecological approaches has overlooked relevant biotic contributions. This complexity poses great problems to the design of ecologically relevant experiments to explain *in situ* behavior, and raises questions as to the reliability, not only of field observations – Redfield's concern – but also extrapolation of laboratory-based autecological data to *in situ* behavior, which Braarud encouraged.

The following sections examine the fidelity of experimental results extrapolated to *in situ* behavior for three major ecological issues: are the results obtained with a clone (strain) applicable to the species at large? Are laboratory responses to nutrients, including type, amount and ratios, and also experimental growth rates representative of *in situ* behavior? Other autecological assumptions could also be tested, but evaluation of these three basic and major issues provides a reliable insight into the overall efficacy of autecological experiments.

Clones and Autecological Extrapolation

Braarud (1961) addressed a fundamental, dual question: are experimental results obtained with single clones (strains) of microalgae representative of the species at large, and of the population from which they were isolated? Embedded within this is the question: do strains remain constant in their responses to environmental factors? Braarud's experiments on the salinity and temperature responses of dinoflagellate strains isolated into culture from different geographic regions and from within the same locality led him to conclude (p. 278 in Braarud 1961): "even with the use of a clone from only one locality, the results may be applied for populations of a large area". Braarud dismissed as experimental artifact the different temperature optima for growth that he found for geographically isolated strains. Since his cultures were not bacteria-free, he thought this might explain the variability found. The corollary of Braarud's conclusion – that different physiological strains of phytoplankton do not seem to occur – had a major impact. This notion that the responses of strains are generally representative of the species energized autecological experimentation and encouraged ecologists to extrapolate the results for a given strain to that species generally. This autecological reductionism was, and continues to be applied despite subsequent contradictory evidence. Jensen et al. (1974) for example, found that strains of the diatom *Skeletonema costatum* varied in their Zn tolerance based on the proximity of the source population to waste discharge from a smeltering factory. Zinc tolerance was greatest in strains isolated from Zn-enriched waters. This finding challenged the notion that strains are conservative, and raised basic questions concerning the mechanisms and rates of adaptation among phytoplankton species.

Gallagher's (1980, 1982) classical study on strains of *S. costatum* isolated from Narragansett Bay demonstrated that significant interclonal differences in physiological behavior indeed occur, at least among diatoms. She concluded (p. 473 in Gallagher 1980) "no single clone is ever representative of the entire species *Skeletonema costatum*, or even, necessarily, of the majority of the population from which it was isolated". The responses of strains can be influenced by culture media, but Gallagher 1982) found that genetic differences, not experimental manipulation, explained her results. The physiological variability among the *S. costatum* strains potentially was of great ecological significance: strain growth rates ranged 50-fold; cellular chlorophyll levels varied 8-fold, and their hourly carbon uptake rate per unit cellular chlorophyll varied 7-fold (Gallagher 1982). Significant genetic, ecophysiological and behavioral differences have since been found among populations and strains of all phytoplankton species. The diatom genus *Pseudo-nitzschia*, which includes species that produce the toxin domoic acid (DA), exhibits high genetic variation and forms complex genetic groups (Parsons et al. 1999, Stehr et al. 2002). Multiple strains of toxic and non-toxic *Pseudo-nitzschia pseudodelicatissima* coexist in Louisiana coastal waters (Parsons et al. 1999). A bloom of this species in Danish waters was composed of at least six different isozyme subgroups, and another bloom in that region was also polyclonal (Skov et al. 1997). Wood and Leatham (1992) list 13 physiological and biochemical traits that vary among phytoplankton strains, including temperature-dependent growth rate, nitrogen and silicon metabolism, vitamin requirements, and toxicity.

Clearly, available data do not support Braarud's still generally applied view that phytoplankton strain differences are relatively unimportant ecologically. The more valid conclusion has been expressed by Wood and Leatham (p. 723 in 1992): "for nearly every physiological character examined, significant interclonal variability is found essentially every time that strains from the same putative taxon are compared". This trait presents major impediments to confident autecological extrapolation of experimental results to species-specific *in situ* behavior, particularly environmental regulation of bloom patterns and successions. There is also an experimental design problem. The magnitude of within-species variation must be determined when seeking to identify the significant physiological differences among competing species in efforts to explain their bloom selection and regulation. As Wood and Leatham (1992) pointed out, without such 'in-group' estimates it is impossible to determine whether single-strain differences found among competing species are any greater than might be observed in random sampling of the individual populations of the competing species. Modelers are also affected given their reliance on experimentally derived rates of cellular-based growth, photosynthesis, nutrient uptake, etc.

It is generally assumed that species brought into culture do not lose or modify their ecophysiological traits. This may not apply to the synthesis of secondary substances that have allelochemical effects. Gentien and Arzul (1990) report that the dinoflagellate *Karenia mikimotoi* lost most of its diatom inhibiting capacity after six to nine months in culture. Another strain of this species demonstrated a loss in toxicity against juvenile and adult scallop during nine months in culture (Erard-Le Denn et al. 1990). It may be that where a trait is sensitive to environmental conditions, it can vary in its expression depending on the suitability of the culturing conditions to elicit and support that trait.

Clones: Environmental Regulation vs. Stochastic Occurrences

Do strains occur stochastically, or are they environmentally regulated in a manner analogous to the succession of taxonomic species? Definitive evidence requires determination of the temporal (seasonal) and spatial scales at which differences in strain behavior occur. This requires clonal isolations along time-space gradients. Available evidence suggests that strains are environmentally regulated, rather than occur as stochastic 'pop-ups'. Seasonal patterns in the genetic composition of a *S. costatum* population correlated with changes in environmental conditions (Gallagher 1982). The geographic origin of dinoflagellate strains influences their genetic features (Scholin 1998), and considerable genetic diversity can occur among regional sub-populations of a species. Maranda et al. (1985) reported a north-to-south latitudinal decline in the level of paralytic shellfish poison (PSP) toxicity in *Alexandrium* populations. Clones from northern populations contained carbamate, the most potent saxitoxin derivative, and southern populations contained the less potent (sulfamate) saxitoxins (Anderson 1997). The UK populations of *Alexandrium tamarense*, common in European waters, exhibit a north-south gradient in PSP toxicity (Higman et al. 2001), with two distinct lineages represented and based on molecular evidence (Higman et al. 2001, Medlin et al. 1998). Strains isolated from northern (Scottish) populations were toxic in culture and assigned to the toxic 'North American lineage' because of shared toxicity and molecular features. Southern populations (England, Ireland) were not toxic, but molecularly related to the non-toxic 'Western European Lineage'. Such distributional patterns and gradients in physiological strains provide further evidence that their emergence and occurrence are not primarily stochastic events, but represent adaptive behavior.

Clones: Physiological Variants or Taxonomic Species?

Ecologists and taxonomists face a basic problem: are strains physiological variants of a given species or are they different species? The genetic variations that accompany diversification of a putative species into strains (Gallagher 1982, Medlin et al. 1991, Scholin 1998, Wood and Leatham 1992) suggest that they are evolved features rather than transient (opportunistic) adaptive behavior. Coexistence of genetically distinct sub-populations is ecologically beneficial. It increases the chances that species will grow and survive (Eq. 3) in temporally and spatially variable environments. On the other hand, the putative strains may, in fact, represent different species difficult to distinguish applying standard taxonomic techniques. The problem this poses to *in situ* extrapolations of experimental results is obvious. The assignment of toxic dinoflagellates to various 'species complexes' because of morphologic and genetic overlap, rather than to a distinct species, is an example of this unresolved issue (Bravo et al. 1995). Some strains of *Alexandrium tamarense, A. catenella* and *A. fundyense* isolated from North American coastal waters are genetically very similar despite morphological differences (Scholin 1998). Conversely, some strains of *A. tamarense* isolated from western Europe and North America are morphologically indistiguishable, but genetically distinct. Uncertainty over the taxonomic status of genetic strains vs. "morphospecies" can compromise autecological extrapolation of experimental results. There is need to know whether a putative strain is, in fact, an infraspecific unit or a different species, and this should be based on robust taxonomic and ecophysiological criteria (Gallagher 1998). Coefficients used in numerical models are usually taken at random from laboratory experiments. The choice of the experimental clone used, and whether it is a bonafide strain of that species, a cryptic member of some other species, or is a member of a 'species complex' could significantly affect the modeled outcome and extrapolations. Differences in geographical bloom behavior of a species despite a common presence may reflect the absence of the strain (sub-population) adapted to bloom under those conditions. Physiological differences occur among geographically segregated populations (meta-populations) of ichthyotoxic *Heterosigma akashiwo* (Smayda 1998a).

The taxonomic confusion generated by variable morphological and genetic overlap among strains and species of harmful dinoflagellates (Scholin 1998) is not unique to flagellates. Three clones of the nano-diatom *Thalassiosira pseudonana* (Ø = ca. 1.5 - 14 µm; Hasle 1983), considered initially to be taxonomically identical because of minor differences in morphology, differed significantly in their temperature, salinity and vitamin requirements, and nutrient uptake kinetics (Guillard and Ryther 1962, Murphy and Guillard 1976). The physiological results obtained using these clones were commonly extrapolated to *in situ* behavior, and underpin many current concepts. The clones were isolated originally from distinct locations along an onshore-offshore nutrient gradient extending from southern New England to the Sargasso Sea: from a nutrient enriched bay (clone 3H), continental slope water (clone 7-15) and oligotrophic Sargasso Sea (clone 13-1). The physiological divergence among these clones, despite their morphological similarity, initially was interpreted as the adaptive capacity of phytoplankton species to establish physiological races in response to different habitat conditions. In electrophoretic tests, Clone 7-15 exhibited morphs intermediate between the other two clones. This was interpreted as evidence for genetic exchange between the neritic and oceanic clonal populations (Murphy and Guillard 1976), and that *T. pseudonana* consisted of neritic and oceanic "races" (Brand et al. 1981). Electron microscope analyses of cellular morphology later revealed that the clonal differences in ecophysiology presumed to characterize a single species, *T. pseudonana*, are, in fact, the behavior of three distinct species, *T. pseudonana* (3H), *T. guillardi* (7-15), and *T. oceanica* (13-1) (Hasle 1983). Notwithstanding the potential confusion of taxonomic species with infraspecific cloning, ecophysiological clones within a given species are an important feature of the phytoplanktonic life mode (Wood and Leatham 1992). No single clone of a taxonomic species is truly representative of that species.

Autecological Experimentation: Nutrients

Nutrient limitation models

An enormous number and variety of field studies and experiments have examined a central issue in phytoplankton ecology: nutrient regulation of the distribution, abundance and growth of phytoplankton. Two basic notions anchor this effort: that nutrient limitation occurs in the sea and this is usually by a single nutrient. The assumption of single nutrient limitation is based on Liebig's Law of the Minimum formulated for agricultural systems (de Baar 1994) and applied to the sea a century ago (Brandt 1899). Since nutrients are consumable, they vary in their concentrations and ratios (Eqs. 1, 2), with their specific instantaneous, seasonal and regional variations dependent on uptake and resupply rates. Nutrient regulation has a family of effects – on the life cycle, distribution, growth, and biomass – and functions in multiple ways: by element, concentration, and ratios. All this, together with the axiom that nutrient limitation cannot occur in the absence of competition for nutrients. Nutrient uptake, limitation and competition, and growth rate are linked processes; the competition among species for nutrients within this linkage determines community structure and dynamics. Tilman (1977) has termed the interspecific competition for nutrients "resource competition".

It has been taken for granted that nutrient limitation occurs in the sea, but definitive proof of this hypothesis is lacking. Models of nutrient limitation are based on laboratory experiments using batch or continuous culture techniques and, as discussed earlier, differ in the type of nutrient impairment induced (Cullen et al. 1992). Growth in batch culture experiments is unbalanced and leads to nutrient starvation, while growth in continuous culture under photocycle control is balanced and the nutrient supply rate sets the degree of nutrient limitation. Four types of nutrient limitation models have been developed (Droop 1983; Sommer 1989). The 'dose-yield' model, referred to earlier, is a log-log regression of the amount of biomass produced in reponse to the concentration of limiting nutrient. This model applies the mass balance approach of Eqs. 1 and 2. The three other limitation models focus on the linked processes of nutrient uptake and growth. They are saturation-based, i.e. nutrient uptake and growth rate exhibit a rectangular hyperbolic response to increasing nutrient concentrations up to an upper, asymptotic maximal response level. The Dugdale model (Dugdale 1967) describes the dependence of the uptake rate to the external concentration of the limiting nutrient:

$$v = V_{max} S/(S + K_s) \qquad \text{(Eq. 4)}$$

where v is the specific uptake rate per cell or unit of biomass; V_{max} is the maximum velocity of uptake; S is the nutrient concentration, and K_s is the half-saturation coefficient at which the velocity of uptake is one half V_{max}, i.e. $V_{max}/2$. The Dugdale model derives from the Monod model, which describes the dependence of the *growth rate* (μ) on the *external* concentration of the limiting nutrient (S):

$$\mu = \mu_{max} S/(S + K_s) \qquad \text{(Eq. 5)}$$

where μ_{max} is the maximum growth rate. Note that both models emphasize *external* nutrients as the source, and treat nutrient uptake and growth rate as kinetically similar.

Continuous culture (chemostat) experiments have demonstrated that the uptake of ammonia, nitrate, phosphorus, iron and vitamin B_{12} is a hyperbolic function of their concentration (Droop 1974, 1983; Davies 1988). The growth rate (μ) of a monospecific phytoplankton population is also a function of the internal concentration (= cell quota) of the rate limiting nutrient. These general findings led to Droop's (1974) internal stores, or cell quota model, the third type of saturation model:

$$\mu = \mu_{max}(1 - q_o/q) \quad \text{(Eq. 6)}$$

where growth rate (μ) is dependent on the intracellular concentration of the limiting nutrient ("cell quota", *q*), and *qo* is the minimal (or subsistence) cell quota, below which the cell is physiologically impaired and mortality exceeds growth. The relationship $1 - q_o/q$ can be defined as the intensity of nutrient limitation, which is inversely proportional to the cell quota (Sommer 1989). Droop stressed that the relation between cell quota (q) and growth rate (μ) has a much firmer experimental basis (from continuous culture studies) than that between external nutrient concentrations and growth rate. Determination of which nutrient is rate limiting should be based on their relative concentrations within internal cellular pools, and not their external (extracellular) concentrations. This is a major conceptual and methological revision of traditional field approaches and a divergence from the Dugdale and Monod models which use external nutrient concentrations as the measure of their potential limitation. Droop also pointed out that in seeking to identify the effects of nutrient limitation, a cell's nutrient uptake history ideally should be known since this prehistory influences instantaneous uptake behavior. As discussed earlier, *in situ* nutrient supply rates occur as a series of transient episodes of varying inputs, to which phytoplankton are physiologically adapted through their evolved capacity for surge uptake (= above steady state rates) during pulsed nutrient delivery (McCarthy and Goldman 1979, French and Smayda 1995). While nutrient uptake can be continuous and the velocity of uptake varies with concentration (Eq. 4), the biochemical products of this metabolism will accumulate until reaching the biophysical threshold triggering cell division (μ).

Droop's cell quota model is a reasonable representation of nutrient-limited growth in continuous culture experiments, and is widely used (Hecky and Kilham 1988). However, while cell quotas are easily measured in laboratory experiments, the utility of Droop's model to *in situ* behavior (Jones et al. 1978) is impeded. Not only is there the problem of measuring the cell quota (q) for individual species in natural populations, these measurements are contaminated by the presence of detritus, micro-heterotrophs, and other seston components. Measurements of the oscillating nutrient supply at appropriate ecological scales are also usually beyond analytical detection or sampling procedures. Contamination of *in situ* population cell quota measurements and failure to measure nutrient supply rates and particulate (i.e. biomass) nutrient can compromise ecological conclusions regarding nutrient limitation. The interdependence of specific growth rate and internal nutrient concentration led Droop (p. 852 in 1974) to state: the "most important implication of this is that the potential of a body of water for supporting further growth may depend as much on the nutrient already inside the cells, as that yet to be taken up".

Cellular chemical composition is frequently used (Eqs. 1, 2) to assess phytoplankton nutritional status, and for biomass and nutrient element conversions. Brzezinski (1985), for example, evaluated the Si:C:N ratio for 27 diatoms reared at various experimental conditions to assess the effects of nutrient limitation on diatom elemental composition. His objective was to convert field estimates of biogenic Si into C and N units and to estimate silicate production from primary production (^{14}C). Sakshaug and Holm-Hansen (1977) evaluated the influence of nitrate, phosphate and iron limitation on the chemical composition of experimental cultures for use as indicators of nutrient deficiency. Goldman (Goldman et al. 1979, Goldman 1980) was among the first to apply deviations from the Redfield Ratio for N:P of 16:1, by atoms, as an indicator of the degree of nutrient limitation and stress on oceanic phytoplankton growth rates. Sakshaug et al. (1983) used a similar stoichiometric approach to assess nutrient limitation in Norwegian coastal waters. Such efforts to establish physiological status and degree of nutrient stress based on cellular stoichiometric ratios are relatively insensitive and subject to considerable variability. A more sensitive and preferable alternative approach would be to use cellular biochemical and physiological diagnostic techniques to establish the nutrient status of natural populations (Falkowski et al. 1992, Cullen et al. 1992, La Roche et al. 1995).

Nutrient Limitation—Field and Experimental Evidence

Two major, venerable hypotheses of the *in situ* regulation of the growth and temporal and spatial distribution of phytoplankton have been advanced: the irradiance-nutrient hypothesis, and the grazer hypothesis (Yentsch 1980). In reality, these different control mechanisms interact and blur detection of the active factor or combinations of factors that control the type of nutrient-biomass relationship characterizing a given watermass or bloom event. Of these, nutrient limitation has been the principal focus and hypothesis, prompting the question of does nutrient limitation occur *in situ*; has this been confirmed experimentally and, if so, does it follow Liebig's Law of the Minimum? The origins of the nutrient-limitation hypothesis lie in field measurements of the temporal and spatial variations in nutrient concentrations and accompanying phytoplankton abundance. Three basic nutrient-biomass conditions and regions have been used to infer nutrient-limited behavior: high nutrient levels and low phytoplankton biomass (HNLB); low nutrients and high biomass (LNHB), and low nutrients and low biomass (LNLB). [Chlorophyll is usually the biomass estimate.] The progressive decline in nutrients during spring blooms, for example, that accompanies increasing chlorophyll biomass, and whose continued build-up eventually depletes nutrient concentrations and the bloom collapses, results in a progression from HNLB to LNHB. The low phytoplankton abundance typical of oligotrophic seas (LNLB) has also been considered symptomatic of nutrient- limitation, either through dose-yield control or physiologically (Goldman et al. 1979). Where a condition of HNLB persists, some other factor is assumed to be limiting, such as irradiance, in delaying inception of the spring bloom or, a more recent revelation, low iron concentrations that prevent complete assimilation of available macro-nutrients (de Baar 1994). Grazing can confound the nutrient-biomass relationship via cropping of the biomass and lead either to a condition of HNLB or LNLB. Ever since Brandt's (1899) application of Liebig's Law of the Minimum a century ago, this notion has had, and continues to have a major conceptual influence on nutrient-limitation research and to encourage quasi-statistical, descriptive field approaches in search of validation. This simplistic view is based primarily on relating phytoplankton growth to the inorganic macronutrients, N, P and Si.

Liebig's Law and the nutrient limitation models (Eqs. 4, 5, 6) apply an important assumption: when (if) nutrient limitation occurs, it is because of a single nutrient (Davies 1988). Continuous culture experiments with nutrient pairs – vitamin B_{12} + PO_4 (Droop 1974) and NO_3 + PO_4 (Rhee 1978) – led to the conclusion that only one of the nutrient pairs was limiting, and did not interact with the other. This interpretation is in accord with Liebig's Law of the Minimum, i.e. the specific growth rate is related to the cell quota (q_o) of a single nutrient rather than is an additive or multiplicative function of two or more nutrient cell quotas (Droop 1974, Davies 1988). Droop (p. 825 in 1974) expressed this, as follows: "non-limiting nutrients exert no control at all over the patterns of growth — control follows a threshold rather than multiplicative pattern — the limiting nutrient is the one that shows the smallest cell quota: subsistence ratio". This conclusion may hold for monospecific laboratory experiments, but extrapolation of the single, uniform nutrient limitation hypothesis to *in situ* behavior is suspect for several reasons. Natural communities are assemblies of species representing different phylogenies and, hence, have differing nutrient requirements.

When nitrogen-fixing cyanobacteria and diatoms occur simultaneously different nutrients can become limiting contemporaneously. For the cyanobacteria, phosphorus may become the limiting nutrient, while silicon may become limiting to the diatoms. When diatoms co-occur with dinoflagellates, the latter may become either N- or P-limited, and the diatoms N-, P- or Si-limited. When oceanic populations of coccolithophores and diatoms co-occur, the latter may be Fe-limited and the coccolithophores P-limited. Thus, communities composed of phylogenetically diverse species, having different nutritional requirements potentially can become limited simultaneously by two or more different nutrients. Based on ambient field measurements, intracellular nutrient concentrations and K_s considerations (Eq. 4), (Levasseur et al. 1990) concluded that nitrogen and silicate deficiency simultaneously limited the phytoplankton community in a coastal jet-front. In Antarctic ecosystems, experiments suggest that Fe is one of a suite of limiting factors, rather than the sole factor controlling productivity (de Baar 1994).

Nutrient enrichment experiments provide evidence that nutrient co-limitation occurs (Davies 1988). The joint addition of phosphate with nitrate or ammonia had a synergistic, stimulatory effect on photosynthesis rates and chlorophyll synthesis, an interaction consistent with the stoichiometric relationship found between the essential nutrients N and P (Eqs. 1, 2). Iron-limited cells of certain *Pseudo-nitzschia* species lose their ability to assimilate nitrate (Maldonado et al. 2002). Nitrogen-fixing cyanobacteria require large amounts of iron which may stimulate oceanic blooms of *Trichodesmium* which then become vulnerable to P-limitation (Walsh and Steidinger 2001). Nutrient uptake can be co-limited by non-nutritional factors, such as irradiance. Cells exposed to low irradiance require much more iron for growth (Raven 1990). The bloom dinoflagellate *Karenia mikimotoi* cannot take up nitrate in the dark when N-sufficient, unlike the bloom species *Prorocentrum minimum* (Paasche et al. 1984). Experiments on nutrient co-limitation in natural communities and on celluar processes of individual species are very limited, as are factor interaction experiments, even though both inter- and intraspecific co-limitation are expected to occur in natural populations. This expectation compromises application of Liebig's Law of the Minimum, and use of the nutrient limitation models (Eqs. 4-6) and monospecific experimentation for *in situ* extrapolation. Droop (p. 894 in 1974) concluded that "mathematical models have great relevance in the laboratory, but their predictive value in the field is severely limited... being concerned with single species or even single clones". He added "their value would be greatly advanced if... they could be applied to mixed populations and even to adapting species and species successions". Three decades have passed since Droop's statement without development of autecological and synecological experimental techniques to overcome these persistent limitations. This compromises ecologically relevant experiments and satisfaction of Redfield's (1960) criterion that the objective of biological oceanography "is to develop effective models, and ultimately a single, consistent model, of life in the sea". Since nutrient limitation is a major paradigm, what is the evidence that it occurs *in situ*, and is it amenable to experimental verification? The extensive laboratory evidence leaves no doubt that cellular nutrient limitation can be induced experimentally, but does it occur in natural communities?

Does Nutrient Limitation Occur *in situ*?

The specific nutrient hypothesized to be limiting *in situ* largely has been a technique-driven conclusion. The presumptive limiting nutrient has changed with analytical improvements and improved field descriptions of *in situ* nutrient distributions and behavior. Experiments largely have sought to confirm field-based conclusions, rather than evaluate other potentially limiting nutrients, or other growth-limiting factors. At one time or another, ammonia, nitrate, phosphate, silicate, iron and vitamin B_{12}, each, was projected to be, either directly or potentially, the nutrient most limiting to phytoplankton growth and biomass. Ryther and Dunstan's (1971) nutrient enrichment bioassays in a polluted coastal embayment have led to the general acceptance that nitrogen is the primary limiting nutrient in marine waters rather than phosphorus, and unlike in freshwater systems. (Hecky and Kilham 1988). This prompted Howarth and Cole's (1985) unsuccessful experimental effort to demonstrate that Mo insufficiency underlay this presumptive nitrogen-limitation. Hecky and Kilham (1988) have vigorously challenged the field and experimental evidence that marine waters are nitrogen-limited. Numerous nutrient enrichment experiments of different spatial and temporal coverage, mostly single species bioassays, consistently show that when enriched with nitrogen, abundance increases relative to phosphorus enrichment and the unenriched control (Johnston 1963, Smayda 1971, 1974, Maestrini et al. 1997, Bonin et al. 1986). However, Hecky and Kilham contend that difficulties in extrapolating these results to natural systems and the lack of experimental rigour compromise the nitrogen-limitation hypothesis. They argue that the experimental evidence for phosphorus limitation of freshwater systems is much stronger, and concluded (p. 796 in Hecky and Kilham 1988) "the extent and severity of nitrogen limitation in the marine environment remain an open question".

Similar reservations have been expressed regarding the Fe-limitation hypothesis, which Cullen et al. (1992) rejected.

The *a priori* invocation of uniform nitrogen limitation in marine waters continues to be challenged. Peeters and Peperzak (2002) in their review of the literature cite evidence for nitrogen, phosphorus and silicon limitation, and within-region variations in the type of limitation. For example, western Mediterranean waters are believed to be phosphate-limited, but eastern (oligotrophic) Mediterranean waters nitrogen-limited. The *Phaeocystis* spring bloom in Belgian coastal waters appears to be nitrogen-limited, but phosphate-limited off the Dutch coast. In defense of nutrient enrichment bioassays, Peeters and Peperzak (2002) argue that it is the only method that allows determination of which nutrient potentially is the first to become limiting, and also allows ranking of the other nutrients in order of likelihood to become limiting. However, the results and interpretation of nutrient enrichment experiments are influenced by the choice of bioassay species, sampling locations and depths, experimental protocol, and unrecognized changes and responses that occur in the chemical milieu of modified sea water media (Smayda 1970). In nutrient-limited systems, the better competitors survive and the most limited species become rare, if not en route to successional replacement (Cullen et al. 1992).

In 'species-blind' experiments with nutrient-enriched natural populations, i.e. biomass or primary production is the measured response. The enhanced growth obtained may reflect the disproportionate responses of physiologically stressed, rarer cells to the enrichment, and not the dominant species less likely to show nutrient-limited behavior. The many permutations of this prospect, further influenced by the bloom-species and successional stages, complicate the design and interpretation of enrichment experiments with natural populations. This is very evident in the *in situ* Fe-enrichment experiment carried out by Kolber et al. (1994). Increases in chlorophyll fluorescence of all phytoplankton size classes in response to iron enrichment relative to *unenriched* control waters led them to conclude that control waters led them to conclude that the community was Fe-limited. Despite this cellular response to added iron, Fe-enrichment did not lead to *'large'* increases in biomass or a corresponding reduction in macronutrient concentrations. This led to Kolber et al.'s dubious conclusion that perhaps "grazing maintained a low standing stock of phytoplankton". This, in turn, led to their broader ecological extrapolation, unwarranted either by the experimental results or design (p. 148 in Kolber et al. 1994): "Hence, the growth of phytoplankton in the equatorial Pacific is physiologically limited by iron rather than by extrinsic factors such as grazing".

Autecological laboratory experimentation also suffers the potential of enthusiastic ecologists to over-extrapolate experimental results. To be sure, this is encouraged by the evidence that, in general, microalgae from the diverse phylogenies making up natural communities have remarkably similar physiological and cellular chemical compositional responses to nutrient limitation (Hecky and Kilham 1988). This notion of physiological and biochemical uniformity can lull experimentalists and ecologists into over-extrapolation of experimental results to *in situ* responses, a danger recognized by Redfield (1960). The inorganic nitrogen uptake by several bloom-forming dinoflagellates exhibited significant interspecific differences that prompted Paasche et al. (1984), who sought to extrapolate their results to *in situ* behavior, to point out the difficulty (and danger) of associating mass occurrences of harmful and benign dinoflagellates in nature with any particular nutritional mode. There is also need for caution because of discrepancies that can occur between laboratory vs. field behavior, such as the demonstration that natural populations of *Skeletonema costatum* took up urea 30-times faster than a laboratory culture (Wheeler et al. 1974). This may be related both to clonal issues and differing growth conditions. These diverse observations lead me to concur with Hecky and Kilham (1988) that nutrient enrichment experiments and the limitation models (Eqs. 4-6) intrinsically are unsuited to establish ecologically relevant nutrient limitation of *in situ* phytoplankton communities.

The premise of nutrient experimentation is that limitation occurs. However, there is growing evidence that this may not be so or, at least, this has not been confirmed experimentally. It might be expected that oligotrophic regions would provide a very strong signal of nutrient limitation, but this is not the case. Massive blooms of the ichthyotoxic dinoflagellate *Karenia brevis* occur regularly in the oligotrophic Gulf of Mexico (Steidinger et al. 1998). Goldman (1980) showed that near maximum growth rates of phytoplankton can be supported when ambient nutrient concentrations are below detectable concentrations and biomass levels are exceedingly low, i.e. a condition of LNLB. Goldman concluded (p. 190 in 1980) that "descriptive water quality data on gross biomass and nutrients provide absolutely no insight as to the magnitude of [phytoplankton] growth rates". That is, residual nutrient levels cannot be used as evidence for nutrient limitation. Cullen et al. (p. 69 in 1992) reviewed the literature on nutrient regulation of phytoplankton photosynthesis and concluded: "When it comes to nutrient limitation of marine photosynthesis, a good paradigm is hard to find". The photosynthesis assimilation number (P^B) can remain high despite nutrient limitation. Just as phytoplankton have an enormous capacity to exploit nutrient micropatches (Goldman 1980), they also adapt to limiting nutrient levels by regulating their cellular chlorophyll levels (Cullen et al. 1992). These capacities help to overcome potential physiological impairment because of reduced nutrient concentrations. Sommer, who approached this issue from the absence of steady state conditions *in situ*, concluded, based partly on experimental evidence, that "nutrient limitation [may occur] more as a bottleneck for only a few generations rather than as a constant steady state " (p. 66 in Sommer 1989). Harris (p. 156 in 1986) concluded "the response to nutrient depletion in surface waters... is a community response and the only time when severe nutrient depletion and a reduction in growth should be observed is when a single species bloom occurs". In this instance, the subsistence cell quota (Eq. 6) exceeds the nutrient supply from the nutrient stocks depleted by the bloom. The population loss rate then exceeds the growth rate (Eq. 3). Monospecific blooms are characteristic harmful algal species behavior (Smayda 1997a).

The cumulative evidence suggests that the nutrient limitation hypothesis remains to be validated experimentally, and that extant field and experimental results and approaches are inadequate to this task. The alternative hypothesis that grazing, rather than nutrients, is the primary regulator of the phytoplankton dynamics (Yentsch 1980) presents the same experimental autecological and synecological impediments as those for nutrient regulation. [Discussion of experimental grazing studies lies beyond the scope of this chapter.] Some additional complications of applying experimental approaches to resolve *in situ* phytoplankton behavior will be briefly considered prior to assessment of the current merits and relevance of the Redfied-Braarud debate.

Nutrient Competition among Species

Competition among species for nutrient resources influences their distributional and abundance patterns. When several species compete for the same resource, the species with the lowest requirement for that nutrient should become dominant at equilibrium (Tilman et al. 1981). Its growth rate becomes superior to that of competing species and leads to its dominance. Two experimental and theoretical approaches to interspecific competition have been applied: the efficiency of nutrient uptake based on K_s coefficients (Eqs. 4, 5; Eppley et al. 1969) and resource-ratio kinetics (Tillman 1977, Tilman et al. 1981, Grover 1989, Smith 1993). Dinoflagellate occurrences and abundance are generally associated with low nutrient concentrations and highly stratified watermasses which dampen upward pulsing of nutrients from deeper layers (Smayda 2002a). Yet, blooms often develop and persist in oligotrophic waters such as the enormous blooms of ichthyotoxic *Karenia brevis* in the Gulf of Mexico (Steidinger et al. 1998). This behavior has led to the assumption that dinoflagellates have efficient nutrient uptake mechanisms that allow their survival and growth in nutrient-poor watermasses. In contrast, the spring and upwelling blooms of diatoms are associated with high nutrient concentrations. This has led to the view that diatoms are inefficient nutrient-gatherers and require high nutrient concentrations to bloom.

Eppley et al. (1969) proposed that the efficiency of a species' nutrient uptake capacity could be gauged from its half-saturation coefficient for uptake, K_S, as formalized in Equation 4. K_S is considered an index of a species' affinity for nutrient uptake and its potential competitive ability at low nutrient concentrations. Species with high nutrient affinity have low K_S coefficients. Their selection is hypothesized to be favored in seasons and regions of chronically low nutrient supply rates, when they are expected to out-compete species less able to take up nutrients at low nutrient concentrations, i.e. they have a high K_S coefficient (Smayda 1997a). Thus, extrapolating Monod type kinetics (Eqs. 4, 5) to field distributions and bloom behavior suggests that dinoflagellates should have low K_S coefficients and diatoms high coefficients. Available experimental data (Smayda 1997a) reveal the opposite: collectively, harmful bloom flagellates lack the expected high affinity for nutrient uptake, while diatoms have a considerably greater affinity for nutrients. This is seemingly a needless adaptation since diatoms thrive in enriched seas. The ecological basis of this paradox is obscure, but in the case of flagellates their motility-based behavior and facultative mixotrophy may be adaptive (compensating) corrections to such physiological limitation. The point being made is the dual difficulty encountered: making accurate physiological extrapolations based on field observations (this also influences experimental design), and accurate ecological extrapolations to field behavior based on experiments.

Nutrient Ratios

Cultural eutrophication of coastal waters is one of four primary causation theories advanced to explain the global expansion in harmful algal blooms (Smayda 2002b). Two aspects of nutrient enrichment have been focused on: the increase in nutrient concentration, and changes in ratios of the macronutrients N, P and Si. A review of long-term, regional relationships between blooms and anthropogenic enrichment of N and P led Smayda (1990) to hypothesize that the accompanying decline in ratios of Si:N and

Si:P promoted flagellate blooms. This hypothesis has stimulated considerable interest and research (Anderson et al. 2002, Hodgkiss and Lu 2004). The proposed selection for flagellate blooms extended (Officer and Ryther's 1980) suggestion that coastal eutrophication could lead to seasonal Si depletion and eliminate diatoms from the communities. In Smayda's view, the remaining N and P would then become available to promote blooms of undesirable flagellate species no longer in competition with diatoms. A considerable experimental literature has documented the sensitivity of diatoms to Si availability, depletion of which or very low N:Si ratios (< 1:1) select against diatoms (Riegman et al. 1992). Smayda did not propose that all harmful blooms are nutrient-regulated events, only that the potential for such Si:N and Si:P nutrient ratio effects is greatest in waters anthropogenically enriched. This hypothesis is a variant of Tilman's (1977) 'resource-ratio theory'. Confirmation of the theory and its application require experimentation.

Resource-ratio Theory

Tilman's theory posits that the changes in resource-supply ratios, e.g. N:P, N:Si, P:Si, should change the principal limiting nutrient and, in so doing, regulate phytoplankton community structure (Smith 1993). This mechanistic theory of competition is based on the Monod model (Eq. 5). Tilman's (1977) steady state experiments showed that diatom species have an optimal Si:P ratio, above which the species becomes limited by one of the paired nutrients, and below which by the other nutrient. His mixed-diatom cultures experiments confirmed that two species will either co-exist, or one or the other will become competitively dominant dependent on the varying proportions of the Si:P ratios supplied. The general features of these competitive relationships were reliably modeled by the Monod (Eq. 5) and Droop (Eq. 6) models. Thus, laboratory experiments tend to validate the nutrient-ratio theory or, in the words of Sommer (1989) "*make it plausible*". Competition experiments apply steady state dynamics to mixtures of competing species free from grazers and advective losses, e.g. some of the terms in Eq. 3. None of these experimental conditions is a realistic simulation of *in situ* conditions. Steady state does not occur *in situ*; nutrient concentrations vary greatly, and the frequency and magnitude of their pulsed supply influences uptake and growth kinetics; nutrient gathering migrations of flagellates (Smayda 1997a) and certain diatoms occur (Villareal et al. 1996); and there is a plethora of grazer types ranging from copepods to nanoplanktonic, heterotrophic flagellates. Moreover, species replacement rates during successions within mixed-species blooms are much more rapid than when competing in steady state chemostats, suggestive of more important regulatory factors (Smayda and Krawiec in prep.). Grover (1989), however, found neither nutrient pulsing, nor grazing on daily to weekly scales altered the taxonomic responses of natural fresh water communities exposed to variations in Si:P supply ratios in semi-continuous culture. He concluded that the observed responses were generally consistent with expectations based on resource competition theory.

Although the potential for resource-ratio regulation of *in situ* species selection is present, it probably is rarely realized in marine communities. Numerous confounding factors influence the outcome of species competition, experimental simulation of which is difficult. Sommer (1989) is more optimistic. His mesocosm experiments to assess the effects of altered nutrient ratios led him to suggest that nutrient ratios can successfully predict the gross taxonomic composition of marine phytoplankton communities "*even during periods of elevated grazing*" (Sommer et al. 2004). The lack of grazer influences on the outcome of resource ratio perturbations on competing species selections that Sommer et al. (2004) and Grover (1989) claim to have observed is puzzling.

Experimental Growth Rates

The most important rate measurement in efforts to model phytoplankton behavior is their rate of growth (Eq. 3). The scientific literature abounds with growth rate measurements based on cell abundance, biomass (chlorophyll; cell volume), paired cell frequency, assimilation numbers from photosynthesis experiments, etc. These rates are often by-products of experiments designed to assess other physiological feature; they have been made for all stages of the population growth curve; are usually averaged over various periods of growth, and less often are based on experiments specifically designed to establish the growth rate of species in response to a given factor, or the full range of ecological factors influencing growth. Growth rates have been reviewed by Furnas (1990); related to temperature (Eppley 1972, Goldman and Carpenter 1974) and compared allometrically (Tang 1995). The results consistently show that diatom growth rates are higher than those for flagellates, which suggests a significant phylogenetic difference. Tang (1996) has addressed the issue of why dinoflagellate growth rates are so low. The impressive feature of the dinoflagellate growth literature is the surprisingly high number of estimates suggestive of growth rates equivalent to a cell division every three days. In contrast, many diatoms can divide two and even three times daily. The apparently sluggish growth rate of dinoflagellates poses a problem in dealing with harmful blooms. The question is whether their often precipitous and ephemeral blooms result from the physical accumulation of slow growing, poorly grazed species, or whether they reflect aperiodically high growth rates that lead to blooms and discolored water. Alternatively, are they largely an experimental artifact associated with finicky growth requirements difficult to provide in laboratory culture?

In an effort to resolve this, Smayda (1996) determined the daily growth rates of 14 taxa of bloom dinoflagellates incubated in monospecific culture outdoors in flow-through incubators at five different incident irradiance levels over the population lag, exponential and stationary stages in media heavily enriched to avoid nutrient limitation. Maximal growth of seven of the species exceeded $\mu = 2.0\ d^{-1}$, and only one species failed to grow at $\mu = \geq 1.0\ d^{-1}$. While this does not resolve the issue of slow vs. rapid growth rate as a factor in harmful bloom formation, the results of this outdoor incubation suggest that the majority of growth rates on dinoflagellates obtained in culture may be primarily stress-related responses because of inadequate growth conditions, particularly in irradiance quality and intensity, and failure to provide experimental conditions suitable to motility-based behavior. Caution should be exercised in their extrapolation to *in situ* harmful bloom behavior. This conclusion certainly needs to be confirmed. The point is that it is ecologically risky to use cellular growth rate data for a species selected at random from autecological experiments, particularly when estimates of μ_{max} are needed for modeling purposes (Eqs. 3, 5).

THE REDFIELD-BRAARUD DEBATE REVISITED: CONCLUSIONS

Ecological Limitations of Autecological Experiments

The major advances in phytoplankton ecology made since Redfield and Braarud debated the relevance of autecological experiments to *in situ* behavior prompt the question: is their discourse now primarily of historical interest or do their opposing views need to be reconciled? Evaluation of this builds upon the material presented in earlier sections of this chapter i.e. the distinctions between autecological and synecological approaches to phytoplankton ecology; the merits and limitations of their experimental capacity to quantify natural behavior; the three types of *in situ* growth; the two types of nutrient limitation; the sink (diatom) vs. swim (flagellate) strategies; experimental results with clones; on the suite of responses to nutrients, and growth rates as indicators of the overall efficacy of autecological experimentation extrapolated to *in situ* behavior.

The autecological objective to generate experimental data to explain *in situ* phytoplankton behavior has several rigid requirements. Failure to meet these criteria compromises the ecological value and extrapolation of experimental results. The *in situ* behavior of interest should have quasi-predictable occurrence patterns, be a general feature, and amenable to laboratory and field experimentation. The spring diatom bloom largely meets these criteria, unlike harmful algal blooms. Harmful flagellate blooms, often monospecific, are highly unpredicable as to when, where and which species will bloom (Smayda and Reynolds 2001, 2003, Smayda 2002b). At least 11 distinct types of variability characterize the bloom and occurrence patterns of individual species (Smayda 1998b). The irregular and unpredictable bloom behavior of HAB species poses significant experimental problems in efforts to quantify their ecology. Insights from field-based behavior are essential to experimental design (Redfield 1960), yet how does one design experiments to explain bloom behavior when the behavior itself is unpredictable, erratic and even stochastic? Experiments on HAB ecology are additionally compromised by the still incomplete descriptions and regional comparisons of bloom behavior of the various types of toxic species found, i.e. those responsible for paralytic, diarrhetic and amnesic shellfish poisoning, and ichthyotoxicity. New species, toxins, toxicity types and bloom events are continuously being reported.

The autecological approach, itself, is an impediment, and there are ecophysiological complications. As discussed earlier, experimental problems include the occurrence of clones; difficulties in simulating *in situ* nutrient and grazing conditions; experimental enclosure which dampens dispersion and vertical microstructure; and substitution of static, batch culture conditions for the more dynamic *in situ* nutrient recycling mode. Incubation artifacts always compromise ecological extrapolation of experimental results; laboratory-based experiments intrinsically are inadequate to explain *in situ* behavior. Major ecological behavior and processes cannot be addressed adequately by 'bottle experiments', or even in large enclosures such as mesocosms. These include blooms, species competition, succession and *in situ* grazing control. In essence, ecological problems that are beyond experimental approach are insolvable. To increase the 'naturalness' of experimental systems decreases the experimenter's control (Hecky and Kilham 1988). Experimental control requires simplification of experimental design, but this compromises ecological relevance and extrapolation. There is also vulnerability to Redfield's (p. 18 in 1960) concern that experiments can become "essentially observations made under conditions rigged so as to answer particular questions". The great difficulty in carrying out relevant and reliable ecological experiments on population and community dynamics, historically, has led to an emphasis on cellular ecophysiology based on "bottle experiments". This may have led to an unrecognized, derivative problem: the bias of applying cellular experimental approaches and techniques to *in situ* population and community dynamics where new conceptual and different experimental approaches are required.

In situ experimentation – the alternative to 'bottle experiments' – also presents problems of ecological fidelity. Efforts to quantify bloom regulation, bloom-species selection and species succession, for example, require knowledge of the population losses from grazing, mortality and advective washout (Eq. 3). Evaluation of *in situ* nutrient-regulation requires experimental provision of the diverse, key biological and physical sources and pulsed nutrient resupply. This, in turn, requires provision of the full suite of relevant physical, chemical and biotic factors in synecological combination, and at suitable spatial and temporal scales. This general protocol for the design of *in situ* experiments with natural populations (e.g. Eq. 3) is partly evident in Martin's (1992) recognition that tests of the Fe-limitation hypothesis are not amenable to 'bottle experiments', but require large scale *in situ* enrichment experiments (he recommended a 100 km^2 patch study). The use of tracers, such as sulfur hexafluoride (SF6), to track patch dispersion and mixing is essential in experimental deployments *in situ* (Ledwell et al. 1993, Rees et al. 2001), and have been incorporated into Fe-enrichment experiments (Kolber et al. 1994, Tsuda et al. 2003). The practical and analytical difficulties encountered in *in situ* experimentation are formidable. There are problems associated with suitable controls, replication, experimental patch dispersion, mixing of the experimental habitat with surrounding waters, variable population structure, with the results obtained further compromised by differing experimental responses of the entrained species in competition with each other and exposure to selective grazing (Kolber et al. 1994, Tsuda et al. 2003). The efficacy of *in situ* experimentation remains unconvincing. This impacts a large class of ecological problems whose resolution requires *in situ* manipulation of natural populations; the required methodology is unavailable, or the process itself is not amenable to such experimentation. How does one conduct a relevant *in situ* grazing experiment, or one on regulation of species succession?

There is also indirect evidence that the autecological approach is inadequate. The progress in phytoplankton ecology since the Redfield-

Braarud debate suggests that experiments have not greatly altered insights into the mechanisms influencing the temporal and spatial regulation of phytoplankton distribution, their abundance, blooms and production. The persistent failure of ecophysiological knowledge and models to predict field-based discovery of major, new biotic groups and trophic behavior is indicative of this marginal, experimental contribution. Field-based new discovery has been the primary driver of advances in phytoplankton ecology, and not autecological experimentation. This continues the trend noted by Redfield (1960) and Parsons (1985). These advances largely have been facilitated by an increasingly sophisticated technology. The growing reliance on technology has focused on descriptive measurements of *in situ* biotic behavior and habitat features; in contrast, experiments designed to test hypotheses and to develop theory have been relatively neglected. In many respects, phytoplankton ecology is an immature science (particularly HAB ecology) still in the information gathering stage, rather than hypothesis driven. Longhurst (1998) has discussed the role of remote sensing, a particularly important new technology, in expanding our descriptive knowledge base, stating (p. 17 in 1998): "it is difficult now to remember how ignorant we were of the extent, variability, and seasonal algal blooms prior to their global visualization by the CZCS (= Coastal Zone Color Scanner)".

Evidence for, and Against Braarud and Redfield's Views

The diverse body of evidence, collectively, favors Redfield's view that autecological experiments have limited *in situ* application, and counter to Braarud's view. Notwithstanding, important aspects of their seeminly contradictory views remain relevant. Braarud (1961, 1962) emphasized that autecological experiments on cellular behavior provide an ecological 'orientation' not available from field measurements that is needed to understand *in situ* species behavior. This perspective is well founded. Subsequent experimental advances in understanding cellular processes and life histories, and application of this knowledge have strengthened Braarud's view. The cumulative value of species-specific autecological profiles, already great, will increase as the data bank expands and facilitate insights into harmful algae, whose blooms and harmful affects are highly species-specific. It is beginning to allow grouping of phytoplankton species into life-form types (Smayda and Reynolds 2001, Smayda 2002b) that better reflect *in situ* ecophysiological behavior than that based on the autecology of a very limited number of key species. This facilitates identification of functional groups, experimental design, concept development, and improves modelling. It also provides insights into neglected issues, such as the rules of assembly of phytoplankton communities (Smayda 2002b, Smayda and Reynolds 2003). The value of these physiological data is also evident in their generalized use as rate functions (photosynthesis, growth, etc.) in global ocean, mass balance and other oceanic and planetary biogeochemical issues [see also below].

Redfield's argument that laboratory-based approaches (and here I include *in situ* experiments) are inadequate to explain individual species *in situ* ecological behavior is supported by the evidence. Laboratory-based results on cellular physiology and life history behavior only marginally have advanced knowledge of *in situ* behavior, such as the selection of bloom-species, their bloom dynamics, nutrient regulated behavior, successions, distributions, assembly into communities, and partitioning among grazers exhibiting prey-preferences. The difficulties of *in situ* experimentation exposed by the Fe-enrichment experiments have been pointed out. At best, some very general ecological extrapolations can be obtained from such experiments, but site-specific events, whole community behavior and anomalous behavior are not tractable for the most part. The difficulties in such experimentation are intrinsic: they persist independent of potential new discovery, increased knowledge of cellular-based kinetics, physiology, life histories, and *in situ* phytoplankton behavior.

Redfield's disparagement of autecological experimentation does not diminish Braarud's advocacy that experiments provide valuable ecological "orientation", nor negate the evidence since Braarud that physiological rates generated in autecological experiments are important to mass balance and modelling efforts. Rather than being contradictory, their differences, which reflect their focus on different aspects of phytoplankton ecology, are complementary rather than exclusive. Hence, I agree both with the views of Braarud and Redfield. With Braarud (1961), that autecological experiments provide valuable ecological orientations and information on species-based cellular behavior such as nutrient requirements and uptake rates; photosynthesis-irradiance relationships; growth rates and diel division patterns; ecological tolerance profiles to minimum, maximum and optimal temperature, salinity, trace metal concentrations; clonal variations; life cycle; prey value to a given grazer, etc. With Redfield (1960), that experimental autecological information alone is inadequate to explain the *in situ* population and community behavior of the phytoplankton. Braarud's focus was on species behavior; Redfield's focused on population and community behavior, with the ultimate objective of building representative models. Redfield, in advocating discontinuance of experimentation until representative descriptions of natural population behavior was achieved, reached his conclusion as a corollary rather than from analysis of the experimental literature. His position was not that experiments per se were invalid, but pre-mature. He (p. 26 in Redfield 1960) believed that biological oceanography (phytoplankton ecology) "can develop fully only if it employs all means of discovery at its disposal, observation, comparison, correlation and experiments". He reasoned that proper experimental design requires representative knowledge of natural population behavior, but since this knowledge is incomplete, ecological experiments are therefore inadequate and should be discontinued. One might argue, following this reasoning, that given the inadequacy of field measurements they, too, should be discontinued because of the faulty impressions gained. However, no one would seriously recommend cessation of field study until all the needed methodology became available.

It is important to distinguish between the need to understand organisms in the ocean and the ocean per se. With regard to the ocean per se, Redfield's view is too rigid in calling for a halt to experimentation until descriptions of *in* situ behavior are complete. Since new discovery depends on field-based study, how does one know when the essential details of *in situ* behavior have been acquired? And, there is the need to start somewhere with experiments. The mass balance approaches and use of extant experimental data to quantify oceanic processes relevant to planetary ecology are both proper and tractable using current techniques and insights. An added impetus is that anthropogenic modifications, realized and potential, associated with climate change, hypernutrification, aquaculture and over-fishing have increased the need to apply experimentally-based solutions to minimize ecosystem disruption. This 'whole system' approach, to be sure, reflects organismal biology and ecosystem function. However, the smoothing of species-based physiological behavior to generalized rates is expected to have a lower error factor than application of autecological data to explain *in situ* blooms, bloom-species selection, species succession and grazing control of biomass.

In agreeing with Redfield on the ecological inadequacies of autecological experiments, I suggest that ecologically relevant experiments leading to definitive explanation of some major *in situ* behavior may be beyond reach, including species' bloom dynamics, co-limitation effects, species succession, and *in situ* grazer regulation, whether based on laboratory experiments or *in situ* manipulations, and whether examined as 'top-down' or 'bottom-up' effects. This incapacity is not because knowledge gaps in natural population behavior compromise experimental design, or that the experimental approach is invalid. Rather, because this class of first-order behavior has a particularly complex ecology regulated by multiple oceanographic features, neither the behavior, nor the habitat drivers are amenable to experimental manipulation. In these cases, combining autecological and synecological field and experimental approaches with better *in situ* molecular and biochemical diagnostic, and habitat descriptor technology, at best, can help to establish candidate cause and effect relationships, but not definitive proof. Even then, the complex regional oceanography (Longhurst 1998) limits general ecological extrapolation.

Functional Group Experiments: Coalescing Redfield and Braarud's Views

Critics of this endorsement of Redfield's view might point to the highly predictable spring diatom bloom, and its well-established initial regulation by critical depth conditions (Sverdrup 1953), followed by a decline in nutrients and increases in grazing pressure which contribute to bloom termination. The mass balance approach applying Sverdrup's critical depth concept and the nutrient-grazer role has been of exceptional value – but its value is not in accounting for individual species' blooms. The predictive power is of another type – a functional group prediction, i.e. that the spring bloom will be a *diatom* bloom. While certain species typically make up the bloom community, the predictive power is not of their occurrence, but that they occur because they are diatoms. It just so happens that certain diatom species (*Skeletonema costatum*, for example) usually bloom during spring, an appearance that we then come to expect. There is greater certainty that the spring diatom bloom will be a diatom bloom, and not a dinoflagellate or coccolithophore bloom, than which species will bloom.

Phytoplankton experimentation can minimize some of Redfield's concerns and take fuller value from Braarud's recommended approach by applying a functional group, or phylogenetic experimental approach to *in situ* behavior. This is more tractable and ecologically relevant than current experimental manipulations of species for *in situ* extrapolation. The nitrogen-fixing cyanobacteria, the calcium (coccolithophorids) and silicon (diatoms) mineralizers, the phylogenetically diverse flagellates, and the non-motile vs. motile strategies are key examples of the types of functional group and phylogenetic behavior that have major ecophysiological, biochemical and biophysical differences amenable both to field experimentation and laboratory study. There is already considerable baseline data on *in situ* distributions and dynamics of these ubiquitous groups to minimize problems associated with new discovery. Conceptually and methodologically, experimental designs and results on functional (phylogenetic) groups are more tractable ecologically, and have greater first-order relevance and validation prospects than species-based experiments or those based on cell size fractionation. The functional group approach allows experimental focus on the sites, bloom periods, dominance and dynamics of diatom vs. dinoflagellate vs. coccolithophore, etc. without regard to which species will bloom, which carries significant 'biological noise' and stochastic behavior. Among the experimental approaches that could be followed are resource-ratio and competive exclusion type experiments refined to address phylogenetically behavior driven by nutritional differences (Tilman et al. 1986, Grover 1989, Sommer 1989, Sommer et al. 2004). The influence of nutrient ratios on the competitive outcome between two or more vying phylogenies (e.g. diatoms vs. flagellates vs. N-fixing cyanobacteria) may be more tractable and ecologically relevant, unlike experiments on the competitive outcome between two, or more species from within the same genus, or between genera. The experimental matrix could be modified to examine co-limitation effects such as between nutrients, or nutrients + grazing, and other growth-factor combinations. The heirarchical selection of bloom species proposed for harmful blooms could serve as a template for functional group experiments (Smayda 2002b, Smayda and Reynolds 2003). The very successful combination of field and experimental studies on haptophyte nuisance species of *Phaeocystis* is an encouraging example of the value of the functional group approach (Lancelot et al. 1991, Peperzak 2002). *Phaeocystis* and its blooms have become persistent in the Dutch Wadden Sea since the mid-1970s, out-competing diatoms as the dominant spring bloom floral element as it established itself within the community. *Phaeocystis globosa* is probably the best understood and modelled species based on its behavior in those waters. This has been the result of an effective, combined use of field observations, time series behavior, experimental data from "bottle experiments" and mesocosms, and ecophysiological contrasts with diatoms and their local behavior. In advocating a functional group approach, it is not recommended that laboratory-based autecological experimentation on species be discontinued. Such experimentation is very important for the reasons already discussed, and very relevant to the proposed functional group and phylogenetic approach.

Whatever the ecological problem of interest, it is essential that the experimental approach followed be designed in harmony with the oceanographic processes that control *in situ* phytoplankton behavior, also bearing in mind Harvey's guideline (p. 80 in Harvey 1955): in "considering how each factor may affect a single plant, it is helpful to consider how the tangle of variables affects the whole population of marine plants in nature, a population which waxes and wanes, changing its constitution through the seasons and differs both in constitution and in quantity from one sea area to another".

ACKNOWLEDGEMENTS

This research was funded by the EPA's Science to Achieve Results (STAR) Program, supported by EPA Grant No. R82-9368-010 awarded to Dr. Smayda. STAR is managed by the EPA's Office of Research and Development (ORD), National Center for Environmental Research and Quality Assurance (NCERQA). STAR research supports the Agency's mission to safequard human health and the environment. I thank Dr. Subba Rao for his kind invitation to prepare this chapter.

REFERENCES

Allen, E.J. and E.W. Nelson. 1910. On the artificial culture of marine plankton organisms. J. Mar. Biol. Assoc. U.K. 8: 421-474.

Allen, E.J. 1914. On the culture of the plankton diatom *Thalassiosira gravida* Cleve in artificial sea-water. J. Mar. Biol. Assoc. U.K. 10: 417-439.

Anderson, D.M. 1989. Toxic algal blooms: a global perspective. pp. 11-16. *In* T. Okaichi, D. Anderson and T. Nemoto. [eds.] Red Tides: Biology, Environmental Science and Toxicology. Elsevier, New York, USA.

Anderson, D.M. 1997. Bloom dynamics of toxic *Alexandrium* species in the northeastern U.S. Limnol. Oceanogr. 42: 1009-1022.

Anderson, D.M., P.M. Glibert and J.M. Burkholder. 2002. Harmful algal blooms and eutrophication: Nutrient sources, composition and consequences. Estuaries 25: 704-726.

Azam, F., T. Fenchel, J.G. Field, J.S. Gray, R.A. Meyer-Reil and F. Thingstad. 1983. The ecological role of water column microbes in the sea. Mar. Ecol. Prog. Ser. 10: 257-263.

Barber, R.T. and A.K. Hilting. 2002. History of the study of plankton productivity. pp. 16-43. *In* P.J. le B. Williams, D.N. Thomas, and C.S. Reynolds [eds.] 2002. Phytoplankton Productivity. Blackwell Science, Oxford, UK.

Barker, H.A. 1935. The culture and physiology of marine dinoflagellates. Arch. Mikrobiol. 6: 157-181.

Behrenfeld, M.J., A.J. Bale, Z.S. Kolber, J. Aiken and P.G. Falkowski. 1996. Confirmation of iron limitation of phytoplankton photosynthesis in the equatorial Pacific Ocean. Nature 383: 508-511.

Berges, J.A., D.J. Franklin and P.J. Harrison. 2001. Evolution of an artificial seawater medium: Improvements in enriched seawater, artificial water over the last two decades. J. Phycology 37: 1138-1145.

Bonin, D.J., M.R. Droop, S.Y. Maestrini and M.C. Bonin. 1986. Physiological features of six micro-algae to be used as indicators of seawater quality. Cryptogamie, Algologie 7: 23-83.

Boyd, P.W. et al. 2000. A mesoscale phytoplankton bloom in the polar Southern Ocean stimulated by iron fertilization. Nature 407: 695-702.

Braarud, T. 1961. Cultivation of marine organisms as a means of understanding environmental influences on populations. pp. 271-298. *In* M. Sears [ed.] 1961. Oceanography. American Association Advancement Science, Washington, DC, USA.

Braarud, T. 1962. Species distribution in marine phytoplankton. pp. 628-649. *In* J. Oceanogr. Soc. Japan, 20th Anniversary Volume, 1962.

Brandt, K. 1899. Ueber den Stoffwechsel im Meere. Wiss. Meeresuntersuch., Abt. Kiel, N.F. 4: 215-230.

Brand, L.E., L.S. Murphy, R.R.L. Guillard, and H.-t. Lee. 1981. Genetic variability and differentiation in the temperature niche component of the diatom *Thalassiosira pseudonana*. Mar. Biol. 62: 103-110.

Bravo, I., B. Reguera, and S. Fraga. 1995. Description of different morphotypes of *Dinophysis acuminata* complex in the Galician Rias Baixas in 1991. pp. 21-26. *In* P. Lassus, G. Arzul, E. Erard, P. Gentien and C. Marcaillou [eds.] Harmful Algal Blooms. Lavoisier Intercept Ltd., Paris, France.

Brzezinski, M. 1985. The Si:C:N ratio of marine diatoms: interspecific variability and the effect of some environmental variables. J. Phycol. 21: 347-357.

Brzezinski, M. and D.M. Nelson. 1988a. Differential cell sinking as a factor influencing diatom species competition for limiting nutrients. J. Exp. Mar. Biol. Ecol. 119: 179-200.

Brzezinski, M. and D.M. Nelson. 1988b. Interactions between pulsed nutrient supplies and a photocycle affect phytoplankton competiton for limiting nutrients in long-term culture. J. Phycol. 24: 346-356.

Buzzati-Traverso, A.A. [ed.] 1960. Perspectives in Marine Biology. Univ. California Press, Berkeley, CA, USA.

Conway, H.L. and P.J. Harrison. 1977. Marine diatoms grown in chemostats under silicate or ammonium limitiation. IV.Transient response of *Skeletonema costatum* to a single addition of the limiting nutrient. Mar. Biol. 43: 33-43.

Cullen, J.J., X. Yang and H.L. MacIntyre. 1992. Nutrient limitation of marine photosynthesis. pp. 69-88. *In* P.G. Falkowski and A.D. Woodhead [eds.] 1992. Primary Productivity and Biogeochemical Cycles, Plenum Press, New York, USA.

Cushing, D.H. 1975. Marine Ecology and Fisheries. Cambridge Univ. Press, Cambridge, UK.

Davies, A.G. 1988. Nutrient interactions in the marine environment. pp. 241-256. *In* L.J. Rogers and J.R. Gallon [eds.] 1988. Biochemistry of the Algae and Cyanobacteria. Clarendon Press, Oxford, UK.

Denman, K.L. and A.E. Gargett. 1983. Time and space scales of vertical mixing and advection of phytoplankton in the upper ocean. Limnol. Oceanogr. 28: 801-815.

de Baar, H.J.W. 1994. Von Liebig's Law of the Minimum and Plankton Ecology (1899-1991). Progress in Oceanography 33: 347-386.

Droop, M.R. 1974. The nutrient status of algal cells in continuo[illegible]ture. J. Mar. Biol. Assoc. U.K. 54: 825-855.

Droop, M.R. 1983. Twenty-five years of algal growth kinetics. Botanica Marina 26: 99-112.

Dugdale, R.C. 1967. Nutrient limitation in the sea: dynamics, identification and significance. Limnol. Oceanogr. 12: 685-695.

Dugdale, R.C. and F. Wilkerson. 1992. Nutrient limitation of new production in the sea. pp. 107-122. *In* P.G. Falkowski and A.D. Woodhead [eds.] 1992. Primary Productivity and Biogeochemical Cycles. Plenum Press, New York, USA.

Eppley, R.W. 1972. Temperature and phytoplankton growth in the sea. Fish. Bull. U.S. 70: 1063-1085.

Eppley, R.W., J.N. Rogers and J.J. McCarthy. 1969. Half-saturation constants for uptake of nitrate and ammonium by marine phytoplankton. Limnol. Oceanogr. 14: 912-920.

Erard-Le Denn, E., M. Morlaix and J.C. Dao. 1990. Effects of *Gyrodinium* cf. *aureolum* on *Pecten maximus* (post larvae, juveniles and adults), pp. 132-136. *In* E. Granéli, B. Sundström, L. Edler and D.M. Anderson [eds.] 1990. Toxic Marine Phytoplankton. Elsevier, New York, USA.

Falkowski, P.G., R.M. Greene and R.J. Geider. 1992. Physiological limitations on phytoplankton productivity in the ocean. Oceanology 5: 84-91.

French, D. and T.J. Smayda 1995. Temperature regulated responses of nitrogen limited *Heterosigma akashiwo,* with relevance to its blooms. pp. 585-590. *In* P. Lassus, G. Arzul, E.Erard, P. Gentien and C. Marcaillou [eds.] 1995. Harmful Algal Blooms. Lavoisier Intercept Ltd., Paris, France.

Furnas, M.J. 1990. *In situ* growth rates of marine phytoplankton: approaches to measurement, community and species growth rates. J. Plankton Res. 12: 117-1151.

Gaarder, T. and H.H. Gran. 1927. Investigations of the production of plankton in the Oslo Fjord. Cons. Perm. Int. L'Explor. Mer Rapp. Proc. Verb. d. Reun. 42: 1-48.

Gallagher, J.C. 1980. Population genetics of *Skeletonema costatum* (Bacillariophyceae) in Narragansett Bay. J. Phycol. 16: 464-474.

Gallagher, J.C. 1982. Physiological variation and electrophoretic banding patterns of genetically different seasonal populations of *Skeletonema costatum* (Bacillariophyceae). J. Phycol. 18: 148-162.

Gallagher, J.C. 1998. Genetic variation in harmful algal bloom species: An evolutionary ecology approach. pp. 225-242. *In* D.M. Anderson, A.D. Cembella and G.M. Hallegraeff [eds.] 1998. Physiological Ecology of Harmful Algal Blooms. NATO ASI Series, Springer-Verlag, Berlin, Germany.

Garcés, E., A. Zingone, M. Montresor, B. Reguera, and B. Dale [eds.]. 2002. Report of the Workshop on: LIFEHAB: Life histories of microalgal species causing harmful blooms. Research In Enclosed Seas series 12, European Commission, EUR 20361: 188 pp.

Gentien, P. and G. Arzul. 1990. Exotoxin production by *Gyrodinium* cf. *aureolum* (Dinophyceae). J. Mar. Biol. Assoc. U.K. 70: 571-581.

Gieskes, W.W. and G.W. Kraay. 1986. Analysis of phytoplankton pigments by HPLC before, during and after mass occurrence of the microflagellate *Corymbellus aureus* during the spring bloom in the open North Sea in 1983. Mar. Biol. 92: 45-52.

Goldman, J.C. 1980. Physiological processes, nutrient availability, and the concept of relative growth rate in marine phytoplankton ecology, pp. 179-194. *In* P.G. Falkowski [ed.] 1980. Primary Productivity in the Sea. Plenum Press, New York, USA.

Goldman, J.C. and E.J. Carpenter. 1974. A kinetic approach to the effect of temperature on algal growth. Limnol. Oceanogr. 19: 756-766.

Goldman, J.C., J.J. McCarthy and D.G. Peavey. 1979. Growth rate influence on the chemical composition of phytoplankton in oceanic waters. Nature 279: 210-215.

Gran, H.H. 1927. The production of plankton in the coastal waters off Bergen March - April 1922. Rept. Norw. Fish. Mar. Invest. 3: 1-74.

Gran, H.H. 1929. Investigation of the production of plankton outside the Romsdalsfjord 1926-1927. Cons. Perm. Int. L'Explor. Mer Rapp. Proc. Verb. d. Reun. 56: 1-112.

Granéli, E. and P. Carlsson. 1998. The ecological significance of phagotrophy in photosynthetic flagellates. pp. 539-557. *In* D.M. Anderson, A.D. Cembella and G.M. Hallegraeff [eds.] 1998. Physiological Ecology of Harmful Algal Blooms. NATO ASI Series, Springer-Verlag, Berlin, Germany.

Grice, G.D. and M.R. Reeve [eds.]. 1982. Marine Mesocosms. Spring-Verlag, Berlin, Germany.

Grover, J.P. 1989. Effects of Si:P supply ratio, supply variability, and selective grazing in the plankton: An experiment with a natural algal and prostistan assemblage. Limnol. Oceanogr. 34: 349-367.

Guillard, R.R.L. and J.H. Ryther. 1962. Studies of marine planktonic diatoms. I. *Cyclotella nana* Hustedt and *Detonula confervacea* (Cleve) Gran. Can. J. Microbiol. 8: 229-239.

Hallegraeff, G.M. 1993. A review of harmful algal blooms and their apparent global increase. Phycologia 32: 79-99.

Hansen, P.J. 1991. Qualitative importance and trophic role of heterotrophic dinoflagellates in a coastal pelagial food web. Mar. Ecol. Prog. Ser. 73: 253-271.

Hansen, P.J. 1998. Phagotrophic mechanisms and prey selection in mixotrophic phytoflagellates. pp. 525-537. *In* D.M. Anderson, A.D. Cembella and G.M. Hallegraeff [eds.] 1998. Physiological Ecology of Harmful Algal Blooms. NATO ASI Series, Springer-Verlag, Berlin.

Harris, G.P. 1986. Phytoplankton Ecology. Chapman and Hall, New York, USA.

Harrison, P.J., H.L. Conway and R.C. Dugdale. 1976. Marine diatoms grown in chemostats under silicate or ammonium limitation. I. Cellular chemical composition and steady state growth kinetics of *Skeletonema costatum*. Mar. Biol. 35: 177-186.

Harvey, H.J. 1933. On the rate of diatom growth. J. Mar. Biol. Assoc. U.K. 19: 253-275.

Harvey, H.J. 1942. Production of life in the sea. Biol. Rev. 17: 221-246.

Harvey, H.J. 1955. The Chemistry and Fertility of Sea Waters. Cambridge University Press, London, UK.

Harvey, H.J., L.H.N Cooper, M.V. Lebour and F.S. Russell 1935. Plankton production and its control. J. Mar. Biol. Assoc. UK 20: 407-441.

Hasle G.R. 1983. The marine, planktonic diatoms *Thalassiosira oceanica* sp. nov. and *T. partheneia*. J. Phycol. 19: 220-229.

Hecky, R.E. and P. Kilham 1988. Nutrient limitation of phytoplankton in freshwater and marine environments: A review of recent evidence on the effects of enrichment. Limnol. Oceanogr. 33: 796-822.

Hensen, V. 1887. Ueber die Bestimmung des Plankton's oder des im Meere treibenden Materials an Pflanzen und Thieren. Fünfter Ber. Komm. wissenschaft. Untersuch. der deutschen Meere, in Kiel für die Jahre 1882 bis 1886, Jahrgang XII - XVI, 1-147.

Higman, W.A., D.M. Stone, and J.M. Lewis. 2001. Sequence comparisons of toxic and non-toxic *Alexandrium tamarense* (Dinophyceae) isolates from UK waters. Phycologia 40: 256-262.

Hodgkiss, I.J. and S. Lu. 2002. The effects of nutrients and their ratios on phytoplankton abundance in Junk Bay. Hong Kong. Hydrobiologia 512: 215-229.

Howarth, R.W. and J.J. Cole. 1985. Molybdenum availability, nitrogen limitation, and phytoplankton growth in natural waters. Science 229: 653-655.

Jenkin, P. 1937. Oxygen production by the diatom *Coscinodiscus excentricus* Ehr. in relation to submarine illumination in the English Channel. J. Mar. Biol. Assoc. U.K. 22: 301-343.

Jensen, A., B. Rystad and S. Melsom. 1974. Heavy metal tolerance of marine phytoplankton. I. The tolerance of three algal species to zinc in coastal seawater. J. Exp. Mar. Biol. Ecol. 15: 145-157.

Johnston, R. 1963. Sea water, the natural medium of phytoplankton. I. General features. J. Mar. Biol. Assoc. UK 43: 427-456.

Johnson, P.W. and J. McN. Sieburth. 1979. Chroococcoid cyanobacteria in the sea: a ubiquitous and diverse phototrophic biomass. Limnol. Oceanogr. 24: 928-935.

Jones, K.J., P. Tett, A.C. Wallis and B.J.B. Wood. 1978. Investigation of a nutrient-growth model using a continuous culture of natural phytoplankton. J. Mar. Biol. Assoc. UK. 58: 923-941.

Karentz, D. and T.J. Smayda. 1984. Temperature and the seasonal occurrence pattern of 30 dominant phytoplankton species in Narragansett Bay over a 22-year period (1959-1980). Mar. Ecol. Prog. Ser. 18: 277-293.

Ketchum, B.H. and A.C. Redfield. 1938. A method for maintaining a continuous supply of marine diatoms by culture. Biol. Bull. 75: 165-169.

Kolber, Z.B., R.T. Barber, K.H. Coale, S.E. Fitzwater, R.M. Greene, K.S. Johnson, S. Lindley and P.G. Falkowski. 1994. Iron limitation of phytoplankton photosynthesis in the equatorial Pacific Ocean. Nature. 371: 145-149.

Lancelot, C., G. Billen, and B. Barth. 1991. The dynamics of *Phaeocystis* blooms in nutrient enriched coastal zones. Commission of European Communities, Water Pollution Research Report. 23: 1-106.

LaRoche, J., H. Murray, M. Oreliana and J. Newton. 1995. Flavodoxin expresssion as an indicator of iron limitation in marine diatoms. J. Phycol. 31: 520-530.

Larsen, J. and A. Sournia. 1991. The diversity of heterotrophic flagellates. pp. 313-332. *In* D.J. Patterson and J. Larsen [eds.] 1991. The Biology of Free-living Heterotrophic Flagellates. Clarendon Press, Oxford, UK.

Laybourn-Parry, J. and J. Parry. 2000. Flagellates and the microbial loop. pp. 216-239. *In* B.S.C. Leadbeater and J.C. Green [eds.] 2000. The Flagellates: Unity, Diversity and Evolution. Taylor and Francis, London, UK.

Ledwell, J.R., A.J. Watson and C.S. Law. 1993. Evidence for slow mixing across the pycnocline from open-ocean tracer-release experiments. Nature 364: 701-703.

Levasseur, M.E., P.J. Harrison, B.R. Heimdal and J.-C. Therriault. 1990. Simultaneous nitrogen and silicate deficiency of a phytoplankton community in a coastal jet-front. Mar. Biol. 104: 329-338.

Lomas, M.W. and P.M. Glibert. 2000. Comparisons of nitrate uptake, storage, and reduction in marine diatoms and flagellates. J. Phycol. 36: 903-913.

Longhurst, A. 1998. Ecological Geography of the Sea. Academic Press, London, UK.

Lucas, C.E. 1947. Ecological effects of external metabolites. Biol. Rev. 22: 270-295.

Lucas, C.E. 1955. External metabolites in the sea. Pap. Mar. Biol. Oceanogr. Deep Sea Res. Suppl. 3: 139-148.

Maestrini, S.Y. 1998. Bloom dynamics and ecophysiology of *Dinophysis* spp. pp. 243-266. *In* D.M. Anderson, A.D. Cembella and G.M. Hallegraeff [eds.] 1998. Physiological Ecology of Harmful Algal Blooms. NATO ASI Series, Springer-Verlag, Berlin, Germany.

Maestrini, S.Y., B.R. Berland, M. Bréret, C. Béchemin, R. Poletti and A. Rinaldi. 1997. Nutrient limiting the algal growth potential (AGP) in the Po River plume and an adjacent area, Northwest Adriatic Sea: Enrichment bioassays with the test algae *Nitzschia closterium* and *Thalassiosira pseudonana*. Estuaries 20: 416-429.

Maldonado, M.T., M.P. Hughes, E.L. Rue and M.L. Wells. 2002. The effects of Fe and Cu on growth and domoic acid production by *Pseudo-nitzschia multiseries* and *Pseudo-nitzschia australis*. Limnol. Oceanogr. 47: 515-526.

Marshall, S.M. and A.P. Orr. 1928. The photosynthesis of diatom cultures in the sea. J. Mar. Biol. Assoc. UK 15: 321-360.

Maranda, L., D.M. Anderson and Y. Shimizu. 1985. Comparison of toxicity between populations of *Gonyaulax tamarensis* of eastern North American waters. Estuarine Coastal Shelf Sci. 21: 401-410.

Martin, J.H. 1992. Iron as a limiting factor in oceanic productivity. pp. 123-137. *In* P.G. Falkowski and A.D. Woodhead [eds.] 1992. Primary Productivity and Biogeochemical Cycles, Plenum Press, New York, USA.

Martin, J.H., R.M Gordon and S.E. Fitzwater. 1991. The case for iron. Limnol. Oceanogr. 36: 1793-1802.

McCarthy, J.J. and J.C. Goldman. 1979. Nitrogenous nutrition of marine phytoplankton in nutrient-depleted waters. Science 203: 670-672.

Medlin, L.K., H.J. Elwood, S. Stickel and M.L Sogin. 1991. Morphological and Genetic variation within the diatom *Skeletonema costatum* (Bacillariophceae): evidence for a new species of *Skeletonema pseudocostatum*. J. Phycol. 27: 514-24.

Medlin, L.K., M. Lange, U. Wellbrock, G. Donner, M. Elbrächter, M. Hummert and B. Luckas. 1998. Sequence comparisons link toxic European isolates of *A. tamarense* from the Orkney Islands to toxic North American stocks. Eur. J. Protistology 34: 329-335.

Mills, E.L. 1989. Biological Oceanography An Early History, 1970-1960. Cornell University Press, Ithaca, USA.

Morel, F.M.M., J.G. Rueter, D.M. Anderson and R.R.L. Guillard. 1979. Aquil: a chemically defined phytoplankton culture medium for trace metal studies. J. Phycol. 15: 135-141.

Murphy, L. and R.R.L. Guillard. 1976. Biochemical taxonomy of marine phytoplankton by electrophoresis of enzymes. 1. The centric diatoms *Thalassiosira pseudonana* and *T. fluviatilis*. J. Phycol. 12: 9-13.

Nelson, D.M. and M. Brzezinski. 1990. Kinetics of silicic acid uptake by natural diatom assemblages in two Gulf Stream warm-core rings. Mar. Ecol. Prog. Ser. 62: 283-292.

Officer, C.B. and J.H. Ryther. 1980. The possible importance of silicon in marine eutrophication. Mar. Ecol. Prog. Ser. 3: 83-91.

Ohki, K., J.G. Rueter, and Y. Fujita. 1986. Culture of the pelagic cyanophytes *Trichodesmium erythraeum* and *T. thiebautii* in synthetic medium. Mar. Biol. 91: 9-13.

Paasche, E. 1975. Growth of the plankton diatom *Thalassiosira nordenskioeldii* Cleve at low silicate concentrations. J. Exp. Mar. Biol. Ecol. 18: 173-183.

Paasche, E., I. Bryceson and K. Tangen. 1984. Interspecific variation in dark nitrogen uptake by dinoflagellates. J. Phycol. 20: 394-401.

Pan, Y., D.V. Subba Rao, K.H. Mann, R.G. Brown and R. Pocklington. 1996a. Effects of silicate limitation on production of domoic acid, a neurotoxin, by the diatom *Pseudo-nitzschia multiseries*. I. Batch culture studies. Mar. Ecol. Prog. Ser. 131: 225-233.

Pan, Y., D.V. Subba Rao, K.H. Mann, W.K. Li and W.G. Harrison. 1996b. Effects of silicate limitation on production of domoic acid, a neurotoxin, by the diatom *Pseudo-nitzschia multiseries*. II. Continuous culture studies. Mar. Ecol. Prog. Ser. 131: 235-243.

Parsons, M.L., C.A. Scholin, P.E. Miller, G.J. Doucette, C.L. Powell, G.A. Fryxell, Q. Dortch and T.M. Soniat. 1999. *Pseudo-nitzschia* species (Bacillariophyceae) in

Louisiana coastal waters: molecular probe field trials, genetic variabiilty, and domoic acid analyses. J. Phycol. 35: 1368-1378.

Parsons, T.R. 1985. Hypothesis testing and rigorous statistics as criteria for marine research proposals. La mer, 23: 109-110.

Peeters, J.C.H. and L. Peperzak. 2002. Nutrient limitation in the North Sea: A bioassay approach. Neth. J. Sea Res. 26: 61-73.

Peperzak, L. 2002. The wax and wane of *Phaeocystis globosa* blooms. Ph. D. Thesis, Groningen University, the Netherlands.

Pratt, D.M. 1966. Competition between *Skeletonema costatum* and *Olisthodiscus luteus* in Narragansett Bay and in culture. Limnol. Oceanogr. 11: 447-455.

Provasoli, L., J.J.A. McLaughlin, and M.R. Droop. 1957. The development of artificial media for marine algae. Arch. Mikrobiol. 25: 392-428.

Raven, J.A. 1990. Predictions of Mn and Fe use efficiencies of phototrophic growth as a function of light availability for growth. New Phytology 116: 1-18.

Redfield, A.C. 1934. On the proportions of organic derivatives in sea water and their relation to the composition of plankton. pp. 176-192. *In* James Johnston Memorial Volume. Liverpool Press, UK.

Redfield, A.C. 1958. The biological control of chemical factors in the environment. Amer. Scientist 46: 205-221.

Redfield, A.C. 1960 The inadequacy of experiment in marine biology. pp. 17-26. *In* A.A. Buzzati-Traverso [ed.] 1960. Perspectives in Marine Biology. Univ. California Press, Berkeley, CA, USA.

Rees, A.P., I. Joint, E.M.S. Woodward and K.M. Donald. 2001. Carbon, nitrogen budgets within a mesoscale eddy: comparison with in vitro determinations. Deep-Sea Res. II 48: 859-872.

Rhee, G-Y. 1978. Effects of N:P atomic ratios and nitrate limitation on algal growth, cell composition, and nitrate uptake. Limnol. Oceanogr. 23: 10-25.

Riegman, R., A.A.M. Noordeloos and G.C. Cadée. 1992. *Phaeocystis* blooms and eutrophication of the continental zones of the North Sea. Mar. Biol. 112: 479-484.

Rines, J.E.B., P.L. Donaghay, M.M. Dekshenicks, J.M. Sullivan and M.S. Twardowski. 2002. Thin layers and camouflage: hidden *Pseudo-nitzschia* spp. (Bacillariophyceae) populations in a fjord in the San Juan Islands, Washington, USA. Mar. Ecol. Prog. Ser. 225: 123-137.

Ryther, J.H. and W.M. Dunstan. 1971. Nitrogen, phosphorus, and eutrophication in the coastal marine environment. Science 171: 1008-1013.

Sakshaug, E., K. Andresen, S. Myklestad and Y. Olsen. 1983. Nutrient status of phytoplankton communities in Norwegian waters (marine, brackish, and fresh) as revealed by their chemical composition. J. Plankton Res. 5: 175-196.

Sakshaug, E. and O. Holm-Hansen. 1977. Chemical composition of *Skeletonema costatum* and *Pavlova (Monochrysis) lutheri* as a function of nitrate-, phosphate- and iron-limited growth. J. Exp. Mar. Biol. Ecol. 29:1-34.

Sakshaug, E. and A. Jensen. 1978. The use of dialysis cultures in studies of marine phytoplankton. Oceanogr. Mar. Biol. Ann. Rev. 16: 81-106.

Scholin, C.A. 1998. Morphological, genetic, and biogeographic relationships of the toxic dinoflagellates *Alexandrium tamarense, A. catenella, and A. fundyense*. pp. 13-27. *In* D.M. Anderson, A.D. Cembella, and G.M. Hallegraeff [eds.] 1998. Physiological Ecology of Harmful Algal Blooms. Springer-Verlag, Berlin, Germany:

Schreiber, E. 1927. Die Reinkultur von marinen Phytoplankton und deren Bedeutung für die Erforschung der Produktionsfähigkeit des Meerwassers. Wiss. Meeresuntersuch., N.F. 16: 1-34.

Skov, J., N. Lundholm, R. Pocklington, S. Rosendahl and Ø. Moestrup. 1997. Studies on the marine planktonic diatom *Pseudo-nitzschia*. 1. Isozyme variation among isolates of *P. pseudodelicatissima* during a bloom in Danish coastal waters. Phycologia 36: 374-380.

Smayda, T.J. 1970. Growth potential bioassay of water masses using diatom cultures: Phosphorescent Bay (Puerto Rico) and Caribbean waters. Helgoländer wissenschaft. Meeresuntersuch. 20: 172-194.

Smayda, T.J. 1971. Further enrichment experiments using the centric diatom *Cyclotella nana* (clone 13-1) as an assay organism. pp. 493-511. *In* J. Costlow [ed.]. Fertility of the Sea. Gordon and Breach, Ltd., London, UK.

Smayda, T.J. 1974. Bioassay of the growth potential of the surface waters of lower Narragansett Bay over an annual cycle using the diatom *Thalassiosira pseudonana* (oceanic clone 13-1). Limnol. Oceanogr. 19:889-901.

Smayda, T.J. 1980. Species succession. pp. 493-570. *In* I. Morris [ed.] 1980. The Physiological Ecology of Phytoplankton. Univ. Calif. Press, Berkeley, CA, USA.

Smayda, T.J. 1990. Novel and nuisance phytoplankton blooms in the sea: Evidence for a global epidemic. pp. 29-40. *In* E. Granéli, B. Sundström, L. Edler and D.M. Anderson [eds.] 1990, Toxic Marine Phytoplankton. Elsevier. New York, USA.

Smayda, T.J. 1996. Dinoflagellate bloom cycles: What is the role of cellular growth rate and bacteria? pp. 331-334. *In* T. Yasumoto, Y. Oshima and Y. Fukuyo [eds.] 1996. Harmful and Toxic Algal Blooms. Intergovernmental Commission on Oceanography of UNESCO, Paris, France.

Smayda, T.J. 1997a. Harmful algal blooms: their ecophysiology and general relevance to phytoplankton blooms in the sea. Limn. Oceanogr. 42: 1137-1153.

Smayda, T.J. 1997b. What is a bloom? A commentary. Limnol. Oceanogr. 42: 1132-1136.

Smayda, T.J. 1998a. Ecophysiology and bloom dynamics of *Heterosigma akashiwo* (Raphidophyceae). pp. 113-131. *In* D.M. Anderson, A.D. Cembella and G.M. Hallegraeff [eds.] 1998. Physiological Ecology of Harmful Algal Blooms. NATO ASI Series, Springer-Verlag, Berlin, Germany.

Smayda, T.J. 1998b. Patterns of variability characterizing marine phytoplankton, with examples from Narragansett Bay. ICES J. Mar. Sci. 55: 562-573.

Smayda, T.J. 2002a. Turbulence, watermass stratification and harmful algal blooms: An alternative view and frontal zones as "pelagic seed banks". Harmful Algae 1: 95-112.

Smayda, T.J. 2002b. Adaptive ecology, growth strategies, and the global bloom expansion of dinoflagellates. J. Ocean. 58: 281-294.

Smayda, T.J. and C.S. Reynolds. 2001. Community assembly in marine phytoplankton: Application of recent models to harmful dinoflagellate blooms. J. Plankton Res. 23: 447-461.

Smayda, T.J. and C.S. Reynolds. 2003. Strategies of marine dinoflagellate survival and some rules of assembly. J. Sea Res. 49: 95-106.

Smith, V. H. 1993. Applicability of resource-ratio theory to microbial ecology. Limnol. Oceanogr. 38: 239-249.

Sommer, U. 1989. The role of competition for resources in phytoplankton succession. pp. 57-106. *In* Sommer, U. [ed.] 1989. Plankton Ecology. Succession in Plankton Communities. Springer Verlag, Berlin, Germany.

Sommer, U., T. Hansen, H. Stibord and O. Vadstein. 2004. Persistence of phytoplankton responses to different Si: N ratios under mesozooplankton grazing pressure: a mesocosm study with northeast Atlantic plankton. Mar. Ecol. Prog. Ser.278: 67-75.

Stehr, C.M., L. Connell, K.A., Baugh, B.D. Bill, N.G. Adams and V.I. Trainer. 2002. Morphological, toxicological, and genetic differences among *Pseudo-nitzschia* (Bacillariophyceae) species in inland embayments and outer coastal waters of Washington State, USA. J. Phycol. 38: 55-65.

Steidinger, K.A., G.A. Vargo, P.A. Tester and C.R. Tomas. 1998. Bloom dynamics and physiology of *Gymnodinium breve* with emphasis on the Gulf of Mexico. pp. 133-153. *In* D.M. Anderson, A.D. Cembella and G.M. Hallegraeff [eds.] 1998. Physiological Ecology of Harmful Algal Blooms. NATO ASI Series, Springer-Verlag, Berlin, Germany.

Strickland, J.D.H. 1960. Measuring the production of marine phytoplankton. Fish. Res. Bd. Canada Bull. 122: 1-172.

Strickland, J.D.H. and T.R. Parsons 1972. A practical handbook of seawater analysis. Fish. Res. Bd. Canada Bull. 167: 1-311.

Sweeney, B. 1951. Culture of the dinoflagellate *Gymnodinium* with soil extract. Amer. J. Bot. 38: 669-677.

Sverdrup, H.U. 1953. On conditions for vernal blooming of phytoplankton. J. Cons. Explor. Mer 18: 287-295.

Takahashi, M. and N. Fukazawa. 1982. A mechanism of "red-tide" formation. II. Effect of selective nutrient stimulation on the growth of different phytoplankton species in natural water. Mar. Biol. 70: 267-273.

Tang, E.P.Y. 1995. The allometry of algal growth rates. J. Plankton Res. 17: 1325-1335.

Tang, E.P.Y. 1996. Why do dinoflagellates have lower growth rates? J. Phycol. 32: 80-84.

Tilman, D. 1977. Resource competition between planktonic algae: an experimental and theoretical approach. Ecology 58: 338-348.

Tilman, D., R. Kiesling, R. Sterner, S.S. Kilham and F.A. Johnson. 1986. Green, blue-green and diatom algae: Taxonomic differences in competitive ability for phosphorus, silicon, and nitrogen. Arch. Hydrobiol. 106: 473-485.

Tilman, D. and S.S. Kilham. 1976. Phosphate and silicate growth and uptake kinetics of the diatoms *Asterionella formosa* and *Cyclotella meneghiniana* in batch and semicontinuous culture. J. Phycol. 12: 375-383.

Tilman, D., M. Mattson, and S. Langer. 1981. Competition and nutrient kinetics along a temperature gradient: An experimental test of a mechanistic approach to niche theory. Limnol. Oceanogr. 26: 1020-1033.

Tsuda, A. et al. 2003. A mesoscale iron enrichment in the western Subarctic Pacific induces a large centric diatom bloom. Science 300: 958-961.

van der Wal, P., G.W. Kraay and J. van der Meer. 1994. Community structure and nutritional state of phytoplankton growing in mesocosms with different N:P ratios studied with high performance liquid chromatography. Sarsia 79: 409-416.

Verity, P.G., T.A. Villareal and T.J. Smayda. 1988. Ecological investigations of blooms of colonial *Phaeocystis pouchetii* - 1. Abundance, biochemical composition, and metabolic rate. J. Plankton Res. 10: 219-248.

Villareal, T.A., S. Woods, J.K. Moore and K. Culver-Rymsza. 1996. Vertical migration of *Rhizosolenia* mats and their significance to NO_3- fluxes in the central North Pacific gyre. J. Plankton Res. 18: 1103-1121.

Waterbury, J.B., S.W. Watson, R.R.L. Guillard and L.E. Brand. 1979. Widespread occurrence of a unicellular, marine, planktonic, cyanobacterium. Nature. 277: 293-294.

Walsh, J.J. and K.A. Steidinger. 2001. Saharan dust and Florida red tides: The cyanophyte connection. J. Geophys. Res. 106(C6): 11,597-11,611.

Wheeler, P.A., B.B. North and G.C. Stephens. 1974. Amino acid uptake by marine phytoplantkers. Limnol. Oceanogr. 19: 249-259.

Wood, A.M. and T. Leatham. 1992. The species concept in phytoplankton ecology. J. Phycol. 28: 723-729.

Yentsch, C.S. 1980. Phytoplankton growth in the sea: A coalescence of disciplines. pp. 17-32. *In* P.G. Falkowski [ed.] 1980. Primary Productivity in the Sea. Plenum Press, NY, USA.

8

The Trace Metal Composition of Marine Microalgae in Cultures and Natural Assemblages

Tung-Yuan Ho[1]

[1]Department of Earth and Environmental Sciences, National Chung Cheng University, Ming-Hsiung, 621, Chia-Yi, Taiwan

Abstract

Marine microalgae deplete some essential trace metals in surface oceans and thus influence their distribution and cycling in the ocean; relatively, due to the scarce concentrations in the surface waters, the trace metals can be important in controlling algal production and regulating their community structure. However, there is limited information available for the average trace metal composition either from culture or field studies when compared to our understanding on macronutrient composition in marine microalgae. This chapter reviews the reliable field and culture studies to assess the trace metal composition (Fe, Mn, Zn, Cu, Ni, Co, and Cd) and its variability in marine microalgae. Remarkable similarities of the average trace metal composition among different culture and field studies are observed. The average composition estimated from the selected field studies can be expressed as P_{1000}, $Fe_{5.1\pm1.6}$, $Mn_{0.68\pm0.54}$, $Zn_{2.1\pm0.88}$, $Cu_{0.41\pm0.16}$, $Ni_{0.70\pm0.54}$, $Co_{0.15\pm0.06}$, $Cd_{0.42\pm0.20}$. The similar metal/P composition ratios between deep ocean water and phytoplankton for Zn and Cd indicate that the two metals are regulated by microalgae like macronutrients. The metal/P ratios for Fe and Mn are about one order of magnitude smaller in deep oceanic water than in microalgae, suggesting the relative importance of scavenging processes in controlling the cycling of Fe and Mn in the ocean. Overall, these evidences support that the constant stoichiometric concept for algal elemental composition may be usefully broadened from the major macronutrients to the trace metals. The stoichiometric ratios then can provide a quantitative basis to understand how marine microalgae regulate or influence the cycling and distribution of the metals in the ocean.

INTRODUCTION

Elemental Composition of Marine Microalgae

Marine microalgae (phytoplankton), with their large biomass and high growth turnover rates in the ocean, are responsible for almost half of global primary production (Field et al. 1998). Considering their critical role on regulating elemental cycling in the ocean, knowledge of elemental composition of marine microalgae is essential to quantitatively understand how marine microalgae influence elemental cycling in the ocean. As early as 1930, based on some field measurements on macronutrient composition of the bulk plankton assemblages and the concentrations of nitrate and phosphate in the seawater samples, Redfield et al. (1963) concluded that the macronutrient composition of microalgae assemblages follow the constant ratios –C:N:P=106:16:1– and the N/P ratio are close to the corresponding nutrient ratios in seawater (Redfield 1934; Redfield et al. 1963; Flemming 1940). Over the past 70 years, this constant stoichiometry concept has been widely applied in numerous oceanography and marine biogeochemistry studies as reviewed in the studies (Broecker and Peng 1982; Hecky et al. 1993; Falkowski 2000; Geider and La Roche 2002; Li and Peng 2002). The ratios are especially useful for modeling plankton processes, nutrient cycling, and estimating primary productivity from the supply of a known limiting nutrient. Likewise, the information of micronutrient (essential trace metals) composition of marine phytoplankton may also provide a basis for examining how microalgae influence the relative distribution and the vertical transport of the trace metals in the ocean. It would thus be very useful if the traditional Redfield-type elemental composition ratios can be extended from the major nutrients to the trace metals–Fe, Zn, Mn, Cu, Ni, Co, and Cd.

Roles of the Trace Metals in Regulating Oceanic Productivity

It is now well known that the availability of some certain trace metals in seawater are tightly linked with the growth and community structure of marine phytoplankton. Trace metals, including Fe, Zn, Mn, Cu, Ni, Co, Mo, are cofactors of many essential metalloenzymes and proteins that carry out various metabolic processes such as photosynthesis, respiration and major nutrient assimilation in phytoplankton. However, the dissolved concentrations of many of the trace metals are extremely low in oceanic surface waters, ranging from low nM to sub-nM levels in general (Bruland 1980; Martin and Gordon 1988). Moreover, most of the metals are chelated by organic ligands with strong chelating ability (Bruland 1989; Rue and Bruland 1995; Moffett 1995; Saito and Moffett 2001), which are not bioavailable to most of the marine microalgae. The bioavailable concentrations of some of the essential metals are so low in the surface waters that the growth of marine algae may be limited or colimited by the metals under certain circumstance (Martin and Gordon 1988, Morel et al. 1994; Saito et al. 2002). For example, iron, the most well studied trace element in the ocean, is known to be a major limiting nutrient for algal growth in the high nutrient low chlorophyll (HNLC) regions, which at least include the Southern Ocean, the Northeast Pacific Ocean, and the equatorial Pacific Ocean (Martin and Fitzwater 1988; Martin and Gordon 1988; Martin et al. 1994; Coale et al. 1996; Boyd et al. 2000). In spite of the increasing understanding for the importance of trace metals on influencing algal productivity and regulating community structure in the ocean, compared to the abundant studies on major nutrient composition in marine phytoplankton, there is relatively little information on marine cellular metal composition from field studies.

Previous Culture and Field Studies

While discovering the important roles of the trace metals in the ocean, a variety of culture studies have been carried out to study marine phytoplankton-trace metal interactions during the past two decades (Brand et al. 1983; Harrison and Morel 1986; Price and Morel 1990, Lee and Morel 1995, Sunda and Huntsman 1983, 1995a, b, c; 1996; 1998 a, b, c; 2000). These algal culture experiments were mainly focused on the following three studies: the interaction between medium concentrations and growth rates, interaction between growth conditions and algal metal requirement, and metal-metal interaction. These culture studies were generally limited to a few model species and to one or two essential trace metals. Among the studies, Sunda and Huntsman have published systematic detailed trace metal-algae culture studies using the model algal species, particularly the coccolithophore *Emiliania huxleyi* and the diatoms *Thalassiosira pseudonana* (neritic) and *Thalassiosira oceanica* (oceanic) in addition to *Thalassiosira weissflogii* (Sunda and Huntsman 1995a, b, c; 1996; 1997; 1998a, b; 2000). These studies have greatly enhanced our understanding on microalgae-trace metal interaction and also reported some trace metal composition for the model species under certain growth conditions.

However, it appears to be inadequate to use the trace metal composition obtained from the model species in the culture studies mentioned above to represent the composition of the whole phytoplankton assemblages in the ocean. The culture studies to determine the average metal composition simultaneously in various marine microalgae were scarce, which is mainly due to the difficulty in preparing an ideal algal culture medium to simulate the metal concentrations in natural oceanic surface waters, the difficulty in precisely determining the various extremely low trace metal composition in the cells, and the difficulty in choosing representative algal species to represent natural algal assemblage.

The ideal approach is to directly determine the metal composition in natural phytoplankton assemblage. Nevertheless, given the sampling and contamination problems on collecting marine microalgae assemblages in oceanic surface water, there have been only a few reliable field studies reported so far. Bruland et al. (1991) compiled the results of three field studies (Martin and Knauer 1973; Martin et al. 1976; Collier and Edmond 1984) from 15 plankton samples with low Al contents and gave a rough approximate for the metal composition, representing as $P_{1000}Fe_5Zn_2(Cu, Mn, Ni, Cd)_{0.4}$. Kuss and Kremling (1999), using large volume pumping sampling technique, also proposed a Redfield-type metal composition from 9 biogenic samples collected in the North Atlantic Ocean, expressed as $P_{1000}Fe_5(Zn, Mn)_2Ni_1Cd_{0.5}Cu_{0.4}Co_{0.2}$. It should be noted that the trace metal composition obtained from field studies are not only easily biased by the interferences of lithogenic and other biogenic particles but also influenced by different sampling methods. Intercomparison of the trace metal data among the field studies using different sampling methods needs to consider the influence of their sampling methods on the data. Alternatively, comparison of the trace metal data between culture and field studies may be an appropriate approach to establish a representative trace metal composition in marine phytoplankton. A recent algal-trace metal culture study (Ho et al. 2003; 2004) was designed to test the feasibility of this idea. The study examined trace metal composition in a variety of marine microalgae species –15 marine eukaryotic phytoplankton representing five major marine phyla – and grew the species under an identical culture medium designed to mimic natural seawater condition. The major findings of the study would be briefed in the following section.

In 1985, Morel and Hudson has already initiated the concept of extending the Redfield ratio to the trace metals, expressed as P_{1000} (Fe, Zn, Mn)$_{10}$ (Cu, Cd, Ni)$_1$, based on scant culture studies on a couple of algal species. After two decades, limited field data sets and the possibility of the field data biased by the contamination of abiogenic particles still hinder our confidence to conclude whether it is appropriate and useful to establish an average trace metal composition for marine microalgae. By comparing the trace metal composition and its variability obtained from reliable culture and field studies, and also comparing the average algal composition with deep water metal/P ratios, whether the Redfield ratio may also be usefully extended to the trace metals is evaluated.

METHODS

Preparation of Algae-trace Metal Culture Medium

Due to the extraordinarily low bioavailable concentrations in natural oceanic surface waters, the trace metal concentrations rarely act as toxicants in the ocean, though many of them can be at elevated levels (Sunda and Guillard 1976; Brand et al. 1986). The studies at the high and toxic metal concentration levels would not be the interest of this chapter. To mimic natural surface seawater conditions, bioavailable trace metal concentrations in trace metal culture medium, compared to traditional algal culture media like f/2 (Guillard 1975), have to be maintained at relatively low and constant concentration levels. Trace metal ion buffer system has been proved to be an effective and easy approach to sustain low and constant supply for the bioavailable metal ions in the culture medium (Sunda and Guillard 1976). The bioavailable (inorganic) concentrations in culture medium, denoted as M′, can be buffered by organic chelators like EDTA (ethylenediamine-tetraacetic acid), which have strong chelation ability on the transition metals to keep relatively low bioavailable metal concentrations in batch culture medium. By using organic chelators to regulate M′ in algal culture medium, it was revealed that both algal growth rates and metal uptake rates of the model algal species used are directly related to M′ but are independent of the total metal concentrations (Sunda and Guillard 1976; Anderson et al. 1978). Given the concentrations of the total trace metals and organic chelators, the bioavailable metal concentrations and aquated free metal ions concentrations can be calculated with the complexation constants. The bioavailable and aquated free concentrations for all of the trace metals used may be precisely calculated by using chemical equilibrium computer models like MINEQL (Westall et al. 1986). Aquil, an EDTA-trace metal ion buffered medium designed for algae-trace metal studies, has been widely used in marine microalgae-trace metal studies (Morel et al. 1979; Price et al. 1988/1989; Price and Morel 1990; Yee and Morel 1996; Morel et al. 1994). The procedures for preparing the Aquil medium are briefly described here. Details for the preparation of the medium and trace metal ion buffer system may be found in Price et al. (1989/1990) and Sunda et al. (2004).

The seawater used for trace metal-algae growth studies should contain relatively low trace metal impurities, which can be obtained either directly from natural oceanic surface waters with low metal concentrations or from trace metal free synthetic ocean water (SOW). To obtain the natural seawater, trace metal clean techniques are required during the collection and storage procedures to avoid contamination. To prepare the trace metal free SOW, the trace metal impurities in raw SOW can be removed by passing SOW through trace metal chelating resins like Chelex-100 (Price et al. 1989/1990). The low trace metal seawater is then first enriched with sterile and metal free major nutrients and vitamins. In Aquil medium, the final concentrations of the major nutrients and vitamins are 150 µM $NaNO_3$, 10 µM Na_2HPO_4, and 40 µM Na_2SiO_3, plus 0.1 µM vitamin B_{12}, 0.1 µM biotin, 20 µM thiamin. After adding EDTA stock solution to reach a final 100 µM final concentration, sterile trace metal stock standards are then added into the culture medium. The medium should stay for at least 3h to reach complete complexation equilibrium before inoculating algae (Price et al. 1988/1989). In Aquil medium, total trace metal concentrations are prepared to the following concentrations: Fe_T = 8600 nM, Mn_T = 120 nM, Zn_T = 80 nM, Cu_T = 20 nM, and Co_T = 50 nM, calculated in the presence of 100 µM EDTA to yield unchelated constant concentrations of Fe' = 20 nM, Mn' = 10 nM, Zn' = 20 pM, Cu' = 0.2 pM, and Co' = 20 pM (Westall et al. 1986, Price et al. 1988/1989). Due to its broad application in marine biogeochemistry, cadmium is usually added with Cd_T = 15 nM to yield unchelated concentrations of Cd' = 20 pM (Westall et al. 1986). It should be noted there are uncertainties regarding the exact unchelated concentrations in the medium because the complexation constants for metal-EDTA in seawater are not all precisely known. Comparison among different data sets thus requires careful matching of total metal and ligand concentrations (Ho et al. 2003).

Choice of Fe' in the Culture Medium

With the exception of Fe, the choices of M' in Aquil medium are fairly low and generally near what are thought to be the bioavailable concentrations in natural surface seawater: in the range of 1-20 pM for Cd' and Zn' (Bruland 1989, 1992), 0.4-5 pM for Cu' (Moffett 1995), and 0.1-100 fM for Co' (Saito and Moffett 2001). Previous studies have shown that the metal concentration levels in Aquil medium are high enough to allow maximum growth rates for the model microalgae (Brand et al. 1983, 1986, Brand 1991, Sunda and Huntsman 1992; 1997). However, to avoid limiting the growth of coastal species, the Mn' is regulated at 10 nM in Aquil medium, a value that is typical of coastal waters but about ten times higher than in the open ocean (Landing and Bruland 1980). Likewise, the Fe concentration maintained in

Aquil culture medium is set to be extremely high to avoid limiting the growth of some coastal algal species (Morel et al. 1979; Price et al. 1989/1990). The bioavailable Fe concentration (Fe′) is set at 20 nM in Aquil medium, which is much higher than the bioavailable concentration in the open ocean – ca 0.05-1 pM (Rue and Bruland 1995). Precipitation of Fe oxides on algal cell surfaces is unavoidable under the high concentration, and the extracellular wash for the precipitate is thus required (Hudson and Morel 1989).

To obtain meaningful trace metal concentration of microalgae to mimic natural algal assemblage, it is critical to choose an appropriate Fe concentration in algal culture medium. On one hand, low unchelated Fe medium concentrations usually lead to low growth rates for coastal algal species (Brand et al. 1983); on the other hand, high Fe medium concentrations unavoidably result in precipitation of hydrous ferric oxide onto cell surfaces (Hudson and Morel 1989). Since ferric oxide may adsorb other trace metals, without removing the surface Fe oxide, the precipitation can lead to a significant overestimation for the adsorbed trace metals in addition to Fe (Morel and Hering 1993). To obtain high growth rates and minimize the problem of extracellular Fe precipitation, Ho et al. (2003) examined the change of intracellular and extracellular Fe concentrations and growth rates for the model diatom *Thalassiosira weissflogii* by varying Fe′ in culture medium. The results showed that more than 98% of Fe measured under the Aquil recipe (a total Fe concentration of 8200 nM with 100 μM EDTA) was extracellular Fe and the percentage decreased sharply with the decreasing Fe′. At the Fe concentration of a total Fe concentration of 82 nM with 100 μM EDTA, corresponding to an unchelated Fe concentration, Fe′= 0.2 nM at 250 μE light intensity, *Thalassiosira weissflogii* grew at 85% of its maximum growth rate and the extracellular Fe were less than 30% of the total Fe measured. This choice of Fe′ (0.2 nM) thus allows one to obtain reasonably accurate value for algal cellular Fe concentrations without the tedious procedures to wash the cells for removing Fe oxides precipitate (Ho et al. 2003).

Culturing and Sampling

To avoid contamination during culturing and pretreatment, all apparatus used for the culture medium preparation, algal culturing and sampling, and elemental analysis are prepared according to rigorous acid cleaning procedures (Cullen and Sherrell 1999; Ho et al. 2003). The algae are usually grown in polycarbonate bottles at 20°C using a 12h:12h light dark cycle with a Cool White light source between 125 and 250 μmol quanta $m^{-2}s^{-1}$ light intensity. Replicate culture samples (bottles) are usually maintained in exponential growth through a minimum of 6 generations before harvesting. Cell density and size are determined daily, usually with a Coulter particle counter to calculate growth rates and cell volumes. Cells are usually harvested with a plastic filtration apparatus consisting of polypropylene filter holder and filters, which are also pre-washed with 10% trace metal grade HCl to avoid trace metal contamination. Right after harvesting the cells, the filters with the cells should be thoroughly rinsed with trace metal free seawater or NaCl solution to remove the residues of the culture medium. The filters with the cells are then digested with trace metal grade concentrated HNO_3 in pre-cleaned Teflon vials for metal analysis (Ho et al. 2003).

Trace Metal Analysis

The trace metal composition in the digested algal culture samples may be determined by any analytical instruments with high sensitivity for determination of trace metals. Sector field high resolution inductively coupled plasma mass spectrometry (HR-ICPMS) and graphite furnace atomic absorption spectrometry (GFAAS) are the most common analytical methods for the purpose. Especially, HR-ICPMS, providing an efficient means for simultaneous multi-elemental analysis including some major elements like P and S, is an ideal analytical tool for the study.

Sampling Methods for Phytoplankton Assemblages in the Ocean

The trace metal composition in marine microalgae may be obtained directly by determining the composition in phytoplankton assemblages collected in the field. However, due to the extremely low trace metal contents in marine microalgae and extremely high trace metal concentrations in lithogenic particles in natural seawater, it is unavoidable encountering the interference from abiogenic particles in plankton assemblage samples (Fig. 8.1). The problems were clearly noted by Martin and Knauer in 1973 in their classic study for elemental composition of plankton, "the reasons for the scarcity of information concerning the elemental (metal) composition of the primary producers and primary consumers of the sea are not difficult to ascertain...", which include the contamination from terrestrial particles, difficulty to separate phytoplankton from zooplankton, and contamination from vessels and sampling apparatus. After three decades, the sampling problems proposed by Martin and Knauer for the shortage of the field data in

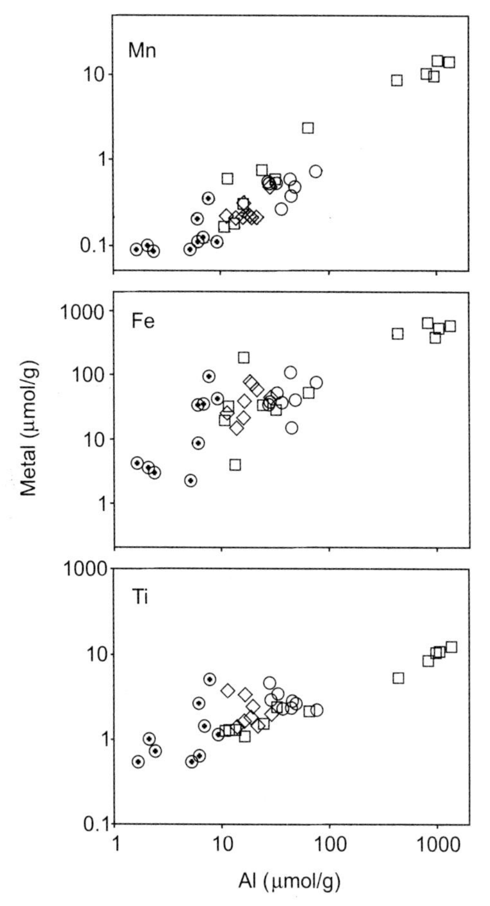

Fig. 8.1 Comparison of the Al, Mn, Fe, and Ti concentration normalized to dry weight of the suspended particles collected in the surface waters of the South China Sea by size fractionated plankton nets with 10, 60, and 150 μm net apertures (Ho et al. in prep.). The samples were collected from 5 stations across the coastal region, through continental shelf, to deep water regions. The symbols of open circle, diamond, and cross circle represent the data obtained from 10, 60, and 150 μm nets, respectively. Square symbols stand for the data of the samples collected by 0.4 μm polycarbonate filters.

phytoplankton do not lessen much. Although trace metal clean methodology may effectively decrease the contamination from vessels and sampling procedures, it is still challenging to remove the interferences of abiogenic materials and larger particles from the collected suspended particles. While applying the metal data of suspended particles obtained in the field to represent the composition of plankton assemblages, it is essential to exclude the influence of abiogenic composition from the samples.

Moreover, while comparing trace metal composition of suspended particles among different data sets, it should be noticed that different sampling methods represent different size particles collected. Algal assemblages and suspended particles in the oceans are usually collected by planktonic tows (Martin and Knauer 1973; Martin et al. 1976; Collier and Edmond 1984), pumping and centrifuge (Kuss and Kremling 1999), or pumping with membrane filtration (Sherrell and Boyle 1992; Cullen and Sherrell 1999; Cullen et al. 2003). Given the relatively large net aperture (several tenths of μm in usual), plankton nets are prone to retain larger particles like zooplankton; instead, membrane filtration methods (with 0.45 μm pore size in general) would collect almost all particulate particles, and centrifuge method may collect even smaller particles. Compared to the true phytoplankton elemental composition, the composition in the samples collected by plankton nets are biased with larger particles like zooplankton; samples collected by centrifuge with membrane filtration methods tend to be biased by smaller abiogenic particles as well as larger zooplankton. As witnessed by Al concentration, Figure 8.1 clearly shows that the influence of abiogenic particles on the essential trace metal composition in the suspended particles is severe and the influence is strongly correlated to the particle size, the smaller the particles; the higher percentage of the abiogenic particles in the samples (Fig. 8.1).

The contamination problem from abiogenic particles in natural algal assemblages is commonly observed both in coastal and open oceans studies (Martin and Knauer 1973; Martin et al. 1976; Sherrell and Boyle 1992; Kuss and Kremling 1999). By 76 μm plankton nets, Martin and Knauer (1973), collected the phytoplankton samples in Monterey Bay, California, on 28 different dates. With relative constant concentrations on major cations (e.g., Na, K, Mg), there were more than 80% of the 'phytoplankton' samples in the study containing extremely high amount of Al, Fe, and/or Ti. As Al or Ti are indicators for lithogenic particle concentration, the samples with high Al or Ti are highly likely to be severely biased by the composition of lithogenic particles due to the high concentration of the essential trace metals in lithogenic particles. It was found in the data set of Martin and Knauer (1973) that the essential trace metals of plankton, including Fe, Mn, Cu, Ni, Zn, were all elevated in the samples containing high Al and/or Ti, indicating that the trace metal data set from the planktonic samples with high Al and/ or Ti concentration are contaminated by lithogenic particles. By using Al or Ti as an indicator for lithogenic particles, it is clearly shown that most of the particulate materials collected in the field studies, both in neritic and open ocean regions, contained considerable amount of lithogenic particles

(Martin and Knauer 1973; Sherrell and Boyle 1992; Kuss and Kremling 1999).

Since natural planktonic assemblage samples are likely to be hampered by the contamination of lithogenic particles, reliable data for natural algal metal composition must be cautiously assessed from the studies that had sufficient information of the indicative elements like P, Al, and Fe. Given the known and precise metal composition of the crust dust in sampling regions, the influence of lithogenic particles on samples might be corrected. In this chapter, the samples that had a low fraction of lithogenic material and high algal biomass as witnessed by low Al concentrations and high P concentrations are chosen. Following the constraint of Bruland et al. (1991), a cutoff of Al < 100 μg/g dry weight, which corresponds to ca. 1 μmol Fe/g dry weight in crustal rock, equal to the lowest algal cellular Fe concentration (Ho et al. 2003), would be used to exclude the samples that are likely biased by lithogenic particles.

TRACE METAL COMPOSITION AND ITS VARIABILITY IN MARINE MICROALGAL CULTURE

Control of the Trace Metal Composition: Internal and External Factors

Before examining the metal composition and its variability in marine microalgal culture, the major factors affecting the composition and the variability are briefly discussed here. Trace metal composition in marine microalgae reflects an interactive balance between external environmental conditions and internal cellular biochemical responses. The variability may come from the change of the external factors such as the nutrient concentrations, light intensity, and temperature; or due to the internal factors like the difference of algal physiological demand on the micronutrients, cellular adaptation to the habitat, or evolutionary heritage.

Bioavailable metal concentrations (M′)

Bioavailable metal concentration, M′, is one of the major external factors controlling the cellular metal quotas. The essential trace metals are transported into cells by membrane proteins, thus algal metal uptake mechanism generally follows the Michaelis-Menten kinetics (Hudson and Morel 1993). Under steady state exponential growth condition, algal metal uptake rates are equal to specific growth rate (μ) multiplied by cellular metal quotas (Q). As the bioavailable metal concentration (M′) in surface seawater are far smaller than the Michaelis-Menten half saturation constant, the Michaelis-Menten equation can thus be simplified and metal uptake rate would be proportional to M′. Laboratory culture studies also found that the metal uptake rates and cellular metal quotas were both proportionally related to external labile inorganic concentrations (M′) when the cellular growth rates are limited by the metals (Sunda and Huntsman 1998c). Under metal limitation condition, as specific growth rates are also regulated by M′ and are proportional to M′, the metal quota increasing extent is lessened due to the simultaneous increase of cellular growth rates.

Metal-Metal Interaction

Due to the non-specific property of the metal transport sites on algal cellular membrane, trace metals with similar size/charge ratios or chemical nature would compete for the same transport sites as well as intracellular metal binding sites (Hudson and Morel 1993; Bruland et al 1991). The relative concentrations of different metals in culture medium would thus affect cellular uptake for the 'similar' metals. Depending on the relative concentrations of the metals, the interactive relationship among trace metals can be either synergistic or antagonistic (Bruland et al. 1991). For example, the cellular concentrations of Zn, Co and Cd in marine microalgae are highly interdependent because the metals have similar size/charge ratios (Zn and Co) or chemical nature (Zn and Cd) and thus can substitute with each other to carry out same biochemical functions (Price and Morel 1990; Lee et al. 1995; Yee and Morel 1996). In the case of *Thalassiosira weissflogii*, Zn, Co and Cd are known to substitute for each other in carbonic anhydrases which are involved in inorganic carbon acquisition (Lane and Morel 2000; Cullen et al. 1999). In contrast, when Zn′ is depleted and Co is available, coccolithophores is able to utilize Co to reach maxiumum growth rate (Sunda and Huntsman 1995b; Yee and Morel 1996). However, preference exists if three of the metals are all available in culture medium. The Co quota of *Emiliania huxleyi* decreased by a factor of 2 as Zn′ increased from 1 to 25 pM (Sunda and Huntsman 1995b) and the Cd quota of *Thalassiosira weissflogii* decreased five fold when Zn′ increased from 10 to 100 pM (Sunda and Huntsman 2000), suggesting these two species prefer to take up Zn if available. Due to the metal substitutions, we should expect elevated cellular concentrations of Co and Cd in open ocean phytoplankton growing at very low Zn′ concentrations. The variability of Zn, Co, and Cd quotas observed between diatoms and coccolithophores reflects in part the difference of the cellular response to metal-metal substitution (Price and Morel 1990, Morel et al. 1994, Sunda and

Huntsman 1996). Not only for Zn, Co, and Cd, different essential metals groups express various metal-metal interaction patterns, which have been studied in details especially by Sunda and Huntsman (1992; 1995b; 1996; 1998a; 2000) and by Morel and his colleagues. These metal-metal interaction patterns were well discussed and reviewed in the studies (Bruland et al. 1991; Sunda and Huntsman 1998c; Whitfield 2001).

Major Growth Conditions (Light, Temperature, Macronutrient Availability et al.)

Since cellular metal quotas are related to cellular growth rates, the major growth factors like light, macronutrients, temperature which influence growth rates can directly or indirectly influence cellular metal quotas. For example, the Fe and Mn quotas in marine algae are highly dependent on the intensity of the light regime. Light intensity affects the cellular concentration of Fe as it is an integral part of a host of electron carriers involved in photosynthesis. A change in light intensity from 50 μmol quanta m^{-2} s^{-1} to 500 μmol quanta m^{-2} s^{-1} resulted in a decrease by a factor of two in the Fe quota of the dinoflagellate *Prorocentrum minimum* (Sunda and Huntsman 1997). As many of the essential metals are involved in the uptake of the major nutrients (C, N, and P), the availability of the major nutrients is also an important parameter affecting cellular metal concentrations. For example, the Zn, Co and Cd quotas in diatoms also depend on the pCO_2 in the growth medium due to the demand on synthesizing carbonic anhydrases. Under low pCO_2 condition in culture medium, Zn, Co and Cd quotas in diatoms tend to increase, and vice versa. The concentration and chemical form of nitrogen in the medium also influence Fe requirements and quotas algal. For example, *Thalassiosira weissflogii* cultures growing nitrate were found to have a 60% higher Fe quota than cultures grown on ammonium (Maldonado and Price 1996). In addition, habitat and/or cell size (Sunda and Huntsman 1997) are the other important factors affecting trace metal quotas in marine algae. The laboratory culture studies showed that oceanic (smaller) isolates are able to survive under metal limitation due to their lower Fe′, Mn′ and Zn′ requirement and higher cell surface area to take up the metals when compared to neritic or larger species (Brand et al. 1983; Sunda and Huntsman 1992, 1995a, 1997).

External conditions can only explain part of the elemental composition variability in marine microalgae. Different physiological and biochemical characteristics among different species certainly are the other major factors causing the quota variability.

Trace Metal-Algae Studies

The research groups in the laboratories of Drs William G. Sunda and Francois M.M. Morel are the pioneers in applying the trace metal buffer system to study algal trace metal concentrations in marine microalgae culture. The results of these early studies may be found in the review papers (Morel et al. 1991, Sunda and Huntsman 1998c, Whitfield 2001). Morel et al. (1991) reported that the trace metal quotas of Fe, Mn, Zn, Co, Ni, and Cd mainly based on the model coastal diatom *Thalassiosira weissflogii* growing at 90% of the maximum growth rate were 6.7, 6.7, 4.2, 2.5, 1.7, and 1.7 μmol/mol C respectively, corresponding to 0.67, 0.67, 0.42, 0.25, 0.17, 0.17 mmol/mol P if C/P=100. Later on, the systematic detailed studies carried out by Sunda and Huntsman for the model algal species *Thalassiosira pseudonana*, *Thalassiosira oceanica*, *Thalassiosira weissflogii*, and *Emiliania huxleyi* (Sunda and Huntsman 1995a, b, c; 1996; 1997; 1998a, b; 2000) quantitatively revealed many important factors affecting the metal composition and metal uptake rate as discussed above.

The metal quotas obtained from the model species are certainly inadequate to reasonably estimate a representative value for the metal composition of the whole phytoplankton assemblages in the ocean. With the difficulties on choosing and culturing enough representative species, there had been no systematic culture studies focused on determining the trace metal elemental stoichiometry in various marine algal species until recently (Ho et al. 2003, 2004), which should be the most extensive study in terms of total species number used. All of the 15 species chosen in the study were grown under an identical and low Fe culture medium designed to mimic the trace metal concentration levels in natural seawater. The individual cellular metal concentrations of each species may be found in the paper. (Ho et al. 2003). It is noted that trace metal composition normalized to a major cellular element provides a cell volume independent concentration (also called quota), which presents a meaningful unit for comparison among different species. Carbon and phosphorus are the most often used denominators to calculate metal quotas in trace metal-algae studies. However, the extracellular cellulose or bio-inorganic calcium carbonate materials found in many dinoflagellates and coccolithophores species hamper the use of carbon-normalized unit among different species; also due to the traditional Redfield ratio (Redfield 1934, 1958), metal quota are usually normalized to P. Thus, phosphorus normalized elemental quotas are most often used by marine biologists and oceanographers (Broecker and Peng 1982; Bruland et al. 1991; Hecky et al. 1993; Geider and La Roche 2002; Li and Peng 2002). It should be pointed out that cellular P concentrations in marine microalgae may also vary plastically especially under extreme P limited condition. The studies compiled by Geider and La Roche (2002) showed that marine microalgae can exhibit large variations in the C/N/P ratios when grown under nutrient depleted conditions. The N/P ratio of some marine microalgae may vary from less than five to larger than 50 when the cells were grown under extreme N or P limited conditions. The critical N/P composition, representing the composition under N and P co-limitation condition, has been determined for several species, which ranged from 20 to 50 for the species tested (Geider and La Roche 2002). These N/P ratios under extreme nutrient depleted conditions are either far above or less than the Redfield ratio, suggesting that marine microalgae in the ocean in general are not under extreme N or P limited conditions. The relatively low variability for P content (normalized to cell volume) among the 15 species found by Ho et al. (2003) suggest that P normalized metal quotas are appropriate when the cells were grown under macronutrient replete condition.

Ho et al. (2003) found that there are relatively significant variabilities for all the metal quotas among the 15 species examined (Table 8.1). The ranges of the metal quotas from 10th to 90th percentile (excluding the lowest and highest data) are within a factor of 20 in general. In most of the species tested, the cellular quotas follow the order Fe>Mn>Zn>Cu>Co≈Cd, from the highest Fe quota of 7.5 mmol/mol P on average to the lowest Co and Cd quotas around 0.2 mmol/mol P on average. Considering the numerous essential roles Fe plays in algae, Fe is expected to have the highest quota among the trace metals in spite of the fact that Fe′ in natural seawater is extremely low. In contrast, the lowest Co and Cd quotas probably are due to the limited functions that Co and Cd can play in the eukaryotic species. Systematic metal quota differences were also revealed among different phyla (Fig. 8.2).

It is not surprising to find significant inter-species and inter-phyla differences for metal quotas in marine microalgae. Large variabilities for macronutrient element composition (C, N, and P) have been reported when various algal species were grown under identical nutrient replete condition (Geider and La Roche 2002), suggesting there is significant variability for the elemental composition among different species and clones. In addition to the interspecific variability, by reviewing numerous culture and field studies on the elemental composition of C, N, and P of marine microalgae and suspended particles, Geider and La Roche (2002) concluded that, depending highly on the availability of the major nutrients, the C/N/P

Table 8.1 Comparison of trace metal composition in natural plankton assemblages, suspended particulate materials, cultured algae, and deep sea water. All metal quotas are normalized to P. Numbers in the parentheses represent one standard deviation.

Reference	*Martin and Knauer 1973 (n=4)*[a]	*Martin et al. 1976 (n=6)*[b]	*Collier and Edmond 1984 (n=2)*[c]	*Kuss and Kremling 1999 (n=9)*[d]	*Cullen et al. 2003 w./wo.1 nM Fe*[e]	*Ho et al. 2003 (n=15)*	*Bruland and others*[f]
Sampling method	*Net*	*Net*	*Net*	*Centrifuge*	*Filtration*	*Culture*	*Deep water*
Sampling site	*Coastal*	*Pacific ocean*	*Pacific ocean*	*Atlantic ocean*	*Southern ocean*	*Laboratory*	*Pacific ocean*
Fe	7.4 (5.5)	3.6 (1.3)	4.6 (0.7)	4.6 (1.3)	N.A.	7.5 (5.3)	0.33
Zn	0.86 (0.63)	1.9 (1.2)	3.0 (1.3)	1.9 (0.7)	2.9/11	0.80 (0.52)	2.7
Mn	0.39 (0.21)	0.36 (0.11)	0.34 (0.04)	1.6 (0.2)	0.70/1.6	3.8 (2.4)	0.33
Cu	0.18 (0.10)	0.38 (0.06)	0.52 (0.05)	0.37 (0.06)	0.60/1.4	0.38 (0.35)	1.7
Ni	0.21 (0.16)	0.34 (0.14)	0.86 (0.17)	1.4 (0.4)	N.A.	N.A.	3.0
Co	N.A.	N.A.	N.A.	0.19 (0.02)	0.10/0.15	0.19 (0.13)	0.01
Cd	0.07 (0.02)	0.53 (0.08)	0.54 (0.10)	0.51 (0.09)	0.44/1.2	0.21 (0.22)	0.35

[a]Data are obtained from the group I raw data in Table 1 of Martin and Knauer (1973). Samples were collected in Monterey Bay, California, with 76 μm aperture phytoplankton net during blooming condition. Only data with Al content less than 100 μg/g (dry weight) are included. High Al concentration in the samples indicates that the data may be biased by the presence of alumino-silicate minerals (Bruland et al. 1991). The phosphorus concentration was obtained from Bruland et al. (1991).

[b]Data are obtained from Table 7-3 in Martin et al. (1976). Samples were collected in the oligotrophic water of North Pacific open ocean with 64 μm aperture plankton net. Only data with Al less than 100 μg/g are included, i.e. data from Station 69, 73, 75, 77, 78, and 85. The data of Station 54, 81, and 88 are not included as P concentrations were too low (81, 88) or some elements (esp. Fe) were abnormally high (54).

[c]Data are obtained from Table 3 in Collier and Edmond (1984). Samples were collected in the open ocean of North Pacific with 44 μm aperture net. Only data of MANOP C (Tow 1 and Tow 2) in the table are included. Other samples had either high Al contents (>100 μg/g) or no Al data reported.

[d]Data are obtained from Table 3 and 4 in Kuss and Kremling (1999). Only data of 'biogenic samples' are included in the table are included. Suspended particulate matters were collected in the surface water of the Atlantic open ocean with pumping and centrifuge method.

[e]Data are obtained from Table 3 and Table 2 in Cullen et al. (2003). Data with (left data) and without Fe added (right data) are both presented. The quotas obtained with 1 nM Fe added approximate the metal quotas without Fe limitation (Table 2). Suspended particulate matters were collected in the surface water of the Antartic ocean by pumping and memebrane filtration (0.45 μm). The data plotted in Figure 3 are the data with Fe added to exclude the influence of low growth rates caused by severe Fe limitation.

[f] Reference sources: Fe (Martin and Gordon 1988); Mn (Landing and Bruland 1980); Zn, Cu, Ni, and Cd (Bruland 1980); Co (Knauer et al. 1982).

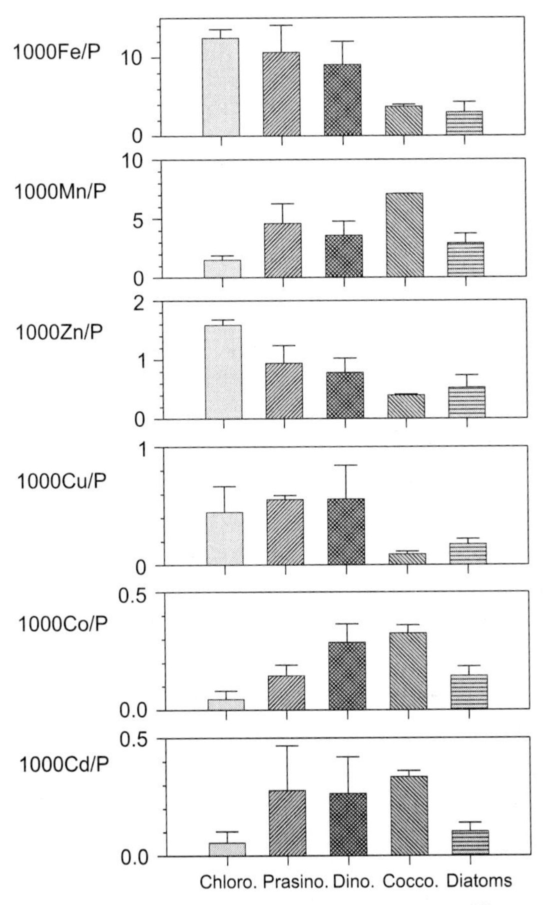

Fig. 8.2 Comparison of the average trace metal quotas among different phyla. For n > 2, the error bars represent 1 s.e.; for n = 2, the error bars represent 1 average deviation. Chlorophyceae (n = 3); Prasinophyceae (n = 3); Dinophyceae (n = 4); Prymesiophyceae (n = 2); Bacillariophyceae (n = 4). Figure is modified from the Fig. 4 of Ho et al. (2003).

ratios may range and vary dramatically in marine microalgae, though in general the elemental composition in most natural suspended materials collected in the ocean still follows the Redfield ratio. In spite of the significant variability found for the trace metal composition among different species or phyla, the possibility should not be excluded that the marine microalgae assemblage may still have a relatively constant trace metal composition as found in macronutrient composition in marine algae by Redfield. In fact, Redfield (1958) was aware that the elemental composition of plankton was uniform only in a 'statistical' sense, thus the significant variabilities of the composition among different species or phyla should exist.

In culture studies, to choose a truly representative species to stand for the entire marine microalgae population in the ocean is operationally impractical. The 15 eukaryotic species chosen in the study by Ho et al. (2003) certainly should not be considered as exact representatives of the marine microalgal assemblage in the oceans. However, since the 15 species have included the major eukaryotic phyla in the ocean, if the metal quotas in culture studies can approximately reflect the cellular quotas in nature, the distribution ranges of the metal quotas represented by the species may stand for the quota range of algal species in natural planktonic assemblages; the metal quota average obtained from the 15 species might then approximate the average found in nature.

Even though the M′ chosen in the study of Ho et al. (2003) were designed to mimic the conditions in natural seawater, whether the theoretical M′ value can reflect the conditions in nature are remained unanswered. The growth conditions like temperature or light intensity for the culture studies in the laboratory are also not exactly the same as the natural conditions in the field. Examining the metal quota variability affected by varying M′ and other major growth conditions would thus help us understand the variability extent of the quotas affected by the external and internal factors.

The question can be answered by the laboratory studies in which the quotas of a given trace metal in a given phytoplankton species are measured under a variety of growth conditions, and over a range of M′. Such detailed studies have already been performed by Sunda and Huntsman (1995b, 1998b, 2000). The studies of Sunda and Huntsman showed that the cellular trace metal quotas in marine phytoplankton are relatively well regulated over a certain range of bioavailable metal concentrations, a range that actually embraces the natural metal concentrations in the oceans but not severely limiting the algal growth. Using Fe as an example, when the bioavailable Fe concentration (Fe′) range roughly from 10 to 750 pM (Sunda and Huntsman 1995a), the cellular Fe quotas varied with Fe′ are listed as the following: Fe quotas of *E. huxleyi* varied from 11 to 24 μmol/mol C when Fe′ ranged from 10 to 750; Fe quotas of *Pelagomonas calceolata* varied from 8 to 10 μmol/mol C when Fe′ ranged from 11 to 580; Fe quotas of *Thalassiosira oceanica* varied from 5 to 34 μmol/mol C when Fe′ ranged from 10 to 760; Fe quotas of *Thalassiosira pseudonana* varied from 13 to 70 μmol/mol C when Fe′ ranged from 24 to 760; Fe quotas of *Thalassiosira weissflogii* varied from 11 to 31 μmol/mol C when Fe′ ranged from 28 to 750; Fe quotas of *Prorocentrum minimumi* varied from 11 to 37 μmol/mol C when Fe′ ranged from 23 to 750.

To summarize, an increase by factors of 20 to 75 in the unchelated Fe concentration (Fe′) only results cellular quota change by factors of 1.3 to 7 for all the algae examined, which included diatoms, coccolithophore, and dinaflagellate. Likewise, when Zn′ increased from 15 to 2454 pM, the Zn quotas only increased by factors of 3 to 10 for the algae examined (Sunda and Huntsman 1996). Similar results were also found for other trace metals. The detailed comparison for all the metals are compiled in the study of Ho et al. (2003).

In conclusion, bioavailable trace metal concentrations only modestly affect algal metal quota at the low concentration levels which are close to the natural concentrations. Trace metal quotas are generally regulated within a factor of 2 to 4 when the M′ varies over 1 order of magnitude. Similarly, the influence of other external factors, like metal-metal interaction, the light intensity, Fe limitation and so on, can rarely change the cellular quota of a given trace metal by more than a factor of 2 to 5 in the natural growth conditions (Sunda and Huntsman 1997; Cullen et al. 2003). Thus, it can be said that the quota variabilities affected by the major external factors are much smaller than the intrinsic variabilities that are seen among the 15 species, which span one or two orders of magnitude depending on the metal. Thus, the metal quotas reported from the culture study (Ho et al. 2003) mainly reflect the intrinsic biochemical requirements for the essential trace elements and the quotas determined in the culture study should be close to the metal quotas for the species in the ocean. The range of the metal quotas observed in the laboratory study can thus be taken as a first approximation of the range expected for the trace element composition of phytoplankton in the sea. The quota average of the culture study can be taken as a first approximation of the trace metal quotas of various species of phytoplankton. The average with one standard deviation of the metal quotas in the culture study of Ho et al. (2003) yields the following trace metal stoichiometry for marine microalgae:

$P_{1000}\ Fe_{7.5\pm5.3}\ Mn_{3.8\pm2.4}\ Zn_{0.80\pm0.52}\ Cu_{0.38\pm0.35}\ Co_{0.19\pm0.13}\ Cd_{0.21\pm0.22}.$

In other algal studies focusing on metal composition, the three model species of *Thalassiosira pseudonana*, *Thalassiosira weissflogii*, and *Emiliania huxleyi* have the most complete metal quota data. Two of the species *Thalassiosira weissflogii* and *Emiliania huxleyi* were already included in the study of Ho et al. (2003). The trace metal quotas of the two species obtained from Ho et al. (2003, 2004) are generally close to those reported by Sunda and Huntsman (Ho et al. 2003, 2004). The data in Ho et al. can thus well represent the value of the same species from other studies.

TRACE METAL COMPOSITION IN NATURAL PLANKTONIC ASSEMBLAGES

The Average and Variability of the Metal Composition

As discussed in the method section, while applying the metal data of suspended particles obtained in the field to represent the composition of plankton assemblages, it is essential to exclude the influence of abiogenic composition from the samples. Using Al and P concentrations as thresholds to evaluate the data obtained from field studies, the data sets complied in Table 8.1 can be considered to be reliable in representing the average metal composition for plankton assemblages in the sampling regions. The three earlier data sets all came from plankton tows samples, including the coastal sample (Martin and Knauer 1973) and two open ocean samples (Martin et al. 1976; Collier and Edmond 1984), collected with 76, 64, and 44 μm net apertures respectively. The data set of Kuss and Kremling (1999) came from pumping and centrifuge; compared to the data set of Cullen et al. (2003) which came from pumping plus membrane filtration. In spite of no available information of Al and Fe in the data set of Cullen et al. (2003), the data are included, as it is unlikely that the samples collected in the Antarctic region were severely influenced by lithogenic particles, where it is known to have low terrestrial dust input. In the study of Cullen et al. (2003), the metal quotas with and without Fe treated (adding 1 nM) are both included. To exclude the influence of the low growth rates due to the severe Fe limitation (Cullen et al. 2003), the metal quotas with Fe added are used for calculating the average. To visualize the difference among different data sets from different sampling methods, the data in Table 8.1 are presented in Fig. 8.3.

It should be noted that some of the average metal composition value in the data sets have large standard deviations (Table 8.1). This is partially due to the limited numbers of reliable data in each data set and partially due to the influence of abiogenic particles (Table 8.1). It can be justified by larger standard deviation for Fe and Mn, and smaller standard deviation for Co and Cd. As seen in Fig. 8.3, in spite of the fact that the data were obtained from different sampling methods and sampling regions, the variability of each metal composition from all the data sets are fairly constrictive. The metal composition follow the order Fe>Zn≈Mn>Cu>Co≈Cd, from the highest Fe quota of 7.4 mmol/mol P to the lowest Co and Cd quotas around 0.1 mmol/mol P, which are similar to the metal quota distribution trends observed in the culture studies (Ho et al. 2003). The Fe quotas are remarkably consistent with each other in the field datasets, ranging from 3.6 to 7.4 mmol/mol P. The highest Fe value, 7.4 mmol/mol P, is from the solo coastal

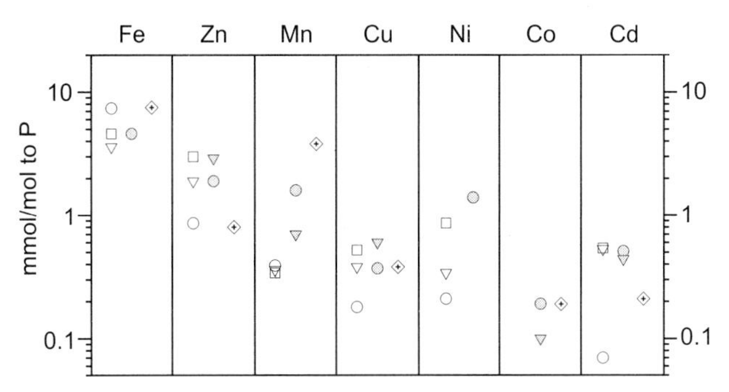

Fig. 8.3 Comparison of the average trace metal quotas among field and culture studies. All of the metals are normalized to P with the units of mmol/mol P. Open symbols represent the data obtained from plankton tow sampling method (Circle: Martin and Knauer 1973; Reverse triangle: Martin et al. 1976; Square: Collier and Edmond 1984); solid symbols represent the data obtained from centrifuge or filtration methods (Circle: Kuss and Kremling 1999; Reverse triangle: Cullen et al. 2003); the culture data is labeled with the crossed diamond symbol (Ho et al. 2003; 2004). Details for how the data were obtained are described in Table 8.1.

data (Martin and Knauer 1973). Neritic species are expected to have higher Fe quotas than oceanic species as culture studies have shown that neritic species have higher Fe demands to sustain maximum growth than oceanic species (Brand et al. 1983; Sunda and Huntsman 1995a). Overall, the Fe value among the field data sets only range by factors of 2. Similarly, the quotas of Zn and Cu in the data sets only vary approximately by factors of 3, ranging from 0.9 to 3.0 and from 0.2 to 0.6 mmol/mol P respectively. In both cases (Zn and Cu), the Zn and Cu value from the study of Martin and Knauer (1973) are the lowest among the data sets, using planktonic tow with a larger net aperture (76 μm) to collect plankton samples during a diatom bloom in coastal region.

Compared to Fe, Zn, and Cu, the variabilities of the Mn and Ni quotas are much larger, ranging from 0.36 to 1.6; and from 0.21 to 1.4 mmol/mol P respectively. Interestingly, the Mn quotas of the three data set that came from plankton nets are almost identical to one another, and both the Mn and Ni quotas from plankton nets are all significantly lower than the other 2 data sets obtained from pumping and filtration methods, suggesting the larger particles collected by plankton tows may contain lower Mn and Ni quotas than the particles collected by the pumping and centrifuge methods. The result obtained by analyzing the 15 Mn data from Ho et al. (2003) shows that the averaged Mn quota (with one standard error) of the four largest species, with cell volume more than 1000 $(\mu m)^3$, *Gymnodinium chlorophorum*, *Thoracosphaera heimii*, *Ditylum brightwellii*, and *Thalassiosira eccentrica*, is

1.8±0.2 mmol/mol P, which was much smaller than the average of the other 11 smaller species, 4.5± 0.6 mmol/mol P, indicating smaller phytoplankton may have higher Mn quotas or there are significant Mn adsorption on the cell surface. If the first situation is the case, samples collected by plankton nets would underestimate Mn quotas in phytoplankton assemblage; and the Mn quotas obtained from the samples collected by pumping and filtration should be more representative for total plankton. The Ni quotas determined in the samples collected by the plankton tows were also lower than the ones obtained from the centrifuge method. It is believed that prokaryotic algae may have kept using Ni and Co in the modern ocean and contain higher Ni and Co quotas due to the reducing conditions of early Proterozoic oceans that prokaryotic algae appeared from and the higher Ni and Co availability under anoxic conditions than other trace metals (Saito et al. 2002). Thus it is expected to see higher Ni quotas in the samples collected by the centrifuge method, which probably also included most of the prokaryotic algae (Fig. 8.3). Unfortunately, there were no Co data reported for the studies with plankton tow sampling methods. The two field Co data from pumping method were a close match for the data of the culture study, ranging from 0.1 to 0.19 mmol/mol P and remarkably close to the culture value (0.19), which is also the lowest average quota among the metals. All of the field studies listed in the Table 8.1, except Kuss and Kremling (1999), probably have undersampled the picoplankton and thus cyanobacteria. Due to their high total biomass in the open ocean, more future culture and field studies should focus on examining the trace metal composition of the bacteria.

The most noteworthy features in Fig. 8.3 is the fairly high and remarkably consistent Cd quotas in the open ocean samples collected from different sampling methods, ranging from 0.44 to 0.54 mmol/mol P, which is almost one order of magnitude higher than the coastal Cd value from Martin and Knauer (1973). These high Cd values from the oceanic samples, possibly reflect extensive substitution of Cd for Zn in the algae live in the Zn-poor open ocean and intensive uptake of cd by calcareous plankton organisms.

Compared to the culture data (Table 8.1), only with the notable exception of Mn, the field quotas for trace elements fall within the range of laboratory values (Fig. 8.3) and the individual field value of Fe, Zn, Cu, Co, and Cd are relatively close to one another among the different field studies and also close to the average of the culture values (Table 8.1 and Fig. 8.3).

The consistency of the average metal composition of the reliable field and culture studies raises a fundamental question about why the metal composition can be relatively constant in marine phytoplankton assemblage.

In spite of the fact that the interspecific variability is fairly large, ranging around one order of magnitude for most of the metals, the inter-phyla variability narrows the variability down to factors of 5 or even less after averaging the metal quotas among the different species from the same phyla (Fig. 8.2). The moderate inter-phyla metal variability can be smoothed further when averaging the metal contents from the whole algal community structure. Although M′ and other growth conditions are different in different regions, as discussed previously, the metal quotas are relatively well regulated by varying M′ and other growth factors under natural conditions. In addition, several lines of evidence suggest that marine microalgae that survive under environments with extremely low metal concentrations, develop various mechanisms to take up the essential trace metals, which would narrow down the metal quota differences for the algae grown under high and low M′ regions. For example, the cell surface/volume ratios are much smaller in the open oceans where the metal concentrations are extremely low and some open ocean species are able to take up organically bound Fe in natural seawater (Maldonado and Price 1999, Hutchins et al. 1999). However, it is possible that the generalities of the average metal composition may be lost under certain growth environments that are dominant by few algal species such as in the regions with *Emiliania huxleyi* blooming or red tide. To summarize, the average metal data with one standard deviation from the field studies yields the following trace metal stoichiometry:

$$P_{1000}\ Fe_{5.1\pm1.6}\ Mn_{0.68\pm0.54}\ Zn_{2.1\pm0.88}\ Cu_{0.41\pm0.16}\ Ni_{0.70\pm0.54}\ Co_{0.15\pm0.06}\ Cd_{0.42\pm0.20}$$

CONCLUSIONS

Extend the Redfield ratio?

The concept of the Redfield ratio may refer to the following two aspects: the first one is the consistency of the elemental composition of the plankton assemblage collected from different oceanic regions; the second one is the relationship between plankton composition and seawater composition. Can the Redfield ratio concept be usefully extended to the trace metals in terms of the two aspects? In this review, the intercomparison of the metal quotas between the culture and field datasets shows that the average metal quota from the field datasets obtained from various sampling methods and regions are relatively consistent to each other; and the average metal quotas measured in the field assemblages are also consistent with the average composition determined in culture studies. These striking similarities not only indicate that the metal composition obtained from laboratory culture studies can precisely reflect algal trace metal composition in the field but also support the data reported from the field studies listed in Table 8.1 may well represent the average metal composition for plankton assemblages. It can be said that it is very likely that there is constant trace metal composition for phytoplankton assemblages in the ocean. In this aspect, the traditional Redfield ratio may be extended to the trace metals. The overall trace metal stoichiometry obtained from the five field studies listed in Table 8.1 is:

$$P_{1000}\ Fe_{5.1\pm1.6}\ Mn_{0.68\pm0.54}\ Zn_{2.1\pm0.88}\ Cu_{0.41\pm0.16}\ Ni_{0.70\pm0.54}\ Co_{0.15\pm0.06}\ Cd_{0.42\pm0.20}$$

It appears more field studies are needed to confirm the consistency of the average metal quotas in the ocean. Comparing the metal composition in plankton with water metal/p ratios in the deep Pacific ocean water (Table 8.1), expressed as $P_{1000}\ Fe_{0.33}\ Mn_{0.33}\ Zn_{2.7}\ Cu_{1.7}\ Ni_{3.0}\ Co_{0.01}\ Cd_{0.35}$, the result shows that the composition for Cd and Zn are similar between the phytoplankton composition and the water composition, indicating that the two metals are mainly regulated by biological processes, taken up in the surface ocean by microalgae and regenerated in the deep oceans. Relatively, it is not surprising to obtain the low metal/P value in the deep water for Fe and Mn due to their property to be scavenged in the water column of the oceans. The higher Cu/P and Ni/P ratios in the deep water than the algal composition suggest that the sources of dissolved Cu and Ni in oceanic deep water are mainly from abiogenic materials instead of the decomposition of biogenic particles. Overall, the metal stoichiometry in marine microalgae provides a basis for modeling and examining how marine microalgae influence the relative distribution and the vertical transport of the trace metals in the ocean, especially for Cd and Zn.

Final Remark

Laboratory algal culture experiments observed that the essential metals at low concentrations can limit marine microalgal growth (Anderson and Morel 1982; Brand et al. 1983). Driven partially by the findings, John Martin and his colleagues eventually revealed and proved the importance of Fe on controlling algal productivity in the ocean and proposed the prominent 'The Iron Hypothesis' (Martin and Gordon 1988). Trace metal-algae study is a splendid example showing that marine microalgal culture study is a powerful analogue tool to understand the roles of marine microalgae play in nature.

ACKNOWLEDGMENTS

I thank Subba Rao Durvasula for his hearty encouragement while preparing this manuscript. I am deeply grateful for the valuable comments from the anonymous reviewer, which significantly enhanced the quality of this manuscript. I would like to thank Yuan-Hui Li and François M. M. Morel for their helpful comments on the manuscript.

REFERENCES

Anderson, M.A., F.M.M. Morel, and R.R.L. Guillard. 1978. Growth limitation of a coastal diatom by low zinc ion activity. Nature. 276: 70-71.

Anderson, M.A. and F.M.M. Morel. 1982. The influence of aqueous iron chemistry on the uptake of iron by the coastal diatom *Thalassiosira weissflogii.* Limnol. Oceanogr. 27: 789-813.

Boyd, P.W., A.J. Watson, C.S.B. Law, E.R. Abraham, T. Trull, R. Murdoch, D.C.E. Bakker, A.R. Bowie, K.O. Buesseler, H. Chang, M. Charette, P. Croot, K. Downing, R. Frew, M. Gall, M. Hadfield, J. Hall, M. Harvey, G. Jameson, J. LaRoche, M. Liddicoat, R. Ling, M.T. Maldonado, R.M. McKay, S. Nodder, S. Pickmere, R. Pridmore, S. Rintoul, K.K. Safi, P. Sutton, R. Strzepek, K. Tanneberger, S. Turner, A. Waite, and J. Zeldis. 2000. A mesoscale phytoplankton bloom in the polar southern ocean stimulated by Fe fertilization. Nature. 407: 695-702.

Brand, L.E. 1991. Minimum iron requirements of marine phytoplankton and the implications for the biogeochemical control of new production. Limnol. Oceanogr. 36: 1756-1771.

Brand, L.E., W.G. Sunda, and R.R.L. Guillard. 1983. Limitation of marine phytoplankton reproductive rates by zinc, manganese, and iron. Limnol. Oceanogr. 28: 1182-1198.

Brand, L.E., W.G. Sunda, and R.R.L. Guillard. 1986. Reduction of marine phytoplankton reproduction rates by copper and cadmium. J. Exp. Mar. Biol. Ecol. 96: 225-250.

Broecker, W.S. and T.-H. Peng. 1982. Tracers in the Sea. Lamont-Doherty Geological Observatory Columbia University, Palisades, New York, USA.

Bruland, K.W. 1980. Oceanographic distributions of cadmium, zinc, nickel, and copper in the north Pacific. Earth Planet. Sci. Lett. 47: 176-198.

Bruland, K.W. 1989. Complexation of zinc by natural organic ligands in the central North Pacific. Limnol. Oceanogr. 34: 269-285.

Bruland, K.W. 1992. Complexation of cadmium by natural organic ligands in the central North Pacific. Limnol. Oceanogr. 37: 1008-1017.

Bruland, K.W., J.R. Donat, and D.A. Hutchins. 1991. Interactive influences of bioactive trace metals on biological production in oceanic waters. Limnol. Oceanogr. 36: 1555-1577.

Coale, K.H., S.E. Fitzwater, R.M. Gordon, K.S. Johnson, and R.T. Barber. 1996. Control of community growth and export production by upwelled iron in the equatorial Pacific Ocean. Nature. 379: 621-624.

Collier, R. and J. Edmond. 1984. The trace element geochemistry of marine biogenic particulate matter. *Prog.* Oceanog. 13: 113-199.

Cullen, J.T. and R.M. Sherrell. 1999. Techniques for determination of trace metals in small samples of size-fractionated particulate matter: phytoplankton metals off central California. Mar. Chem. 67: 233-247.

Cullen, J.T., T.W. Lane, F.M.M. Morel, and R.M. Sherrell. 1999. Modulation of cadmium uptake in phytoplankton by seawater pCO_2. Nature. 402: 165-167.

Cullen, J.T., Z. Chase, K.H. Coale, S.E. Fitzwater, and R.M. Sherrell. 2003. Effect of iron limitation on the cadmium to phosphorus ratio of natural phytoplankton assemblages from the Southern Ocean. Limnol. Oceanogr. 48: 1079-1087.

Falkowski, P.G. 2000. Rationalizing elemental ratios in unicellular algae. J. Phycol. 36: 3-6.

Field, C.B., M.J. Behrenfeld, J.T. Randerson, and P. Falkowski. 1998. Primary production of the biosphere: Integrating terrestrial and oceanic components. Science. 281: 237-240.

Flemming, R.H. 1940. Composition of plankton and units for reporting populations and production. pp. 535-539 *In* Proceedings of the Sixth Pacific Science Congress of the Pacific Science Association. 1939. Uni. California Press, California, USA.

Geider, R.J. and J. La Roche. 2002. Redfield revisited: variability of C:N:P in marine microalgae and its biochemical basis. Eur. J. Phycol. 37: 1-17.

Guillard, R.R.L. 1975. Culture of phytoplankton for feeding marine invertebrates. pp 26-60. *In* Culture of Marine Invertebrate Animals., W.L. Smith, and M.H. Chanley [eds.], Plenum Press, New York, USA.

Harrison, G.I. and F.M.M. Morel, 1986. Response of the marine diatom *Thalassiosira weissflogii* to iron stress. *Limnol. Oceanogr.* 31: 989-997.

Hecky, R.E., P. Campbell, and L.L. Hendzel. 1993. The stoichiometry of carbon, nitrogen, and phosphorus in particulate matter of lakes and oceans. Limnol. Oceanogr. 38: 709-724.

Ho, T.-Y., A. Quigg, Z.V. Finkel, A.J. Milligan, K. Wyman, P.G. Falkowski, and F.M.M. Morel. 2003. The elemental composition of some marine phytoplankton. J. Phycol. 39: 1145-1159.

Ho, T.-Y., A. Quigg, Z.V. Finkel, A.J. Milligan, K. Wyman, P.G. Falkowski, and F.M.M. Morel. 2004. The elemental composition of some marine phytoplankton (Corrigendum). J. Phycol. 40: 227-227.

Ho, T.-Y., L. S. Wen, D.C. Lee, and C.F. You, (in prep.) Sources and abundance of trace metals in the suspended particles collected by size-fractionated plankton nets in the South China Sea.

Hudson, R.J.M. and F.M.M. Morel. 1989. Distinguishing between extracellular and intracellular iron in marine phytoplankton. Limnol. Oceanogr. 34: 1113-1120.

Hudson, R.J.M. and F.M.M. Morel. 1993. Trace metal transport by marine microorganisms - implications of metal coordination kinetics. Deep-Sea Res. I. 40: 129-150.

Hutchins, D.A., A.E. Witter, A. Butler, and G.W. Luther. 1999. Competition among marine phytoplankton for different chelated iron species. Nature. 400: 858-861.

Knauer, G.A., J.H. Martin, and R.M. Gordon. 1982. Cobalt in northeast Pacific waters. Nature. 297: 49-51.

Kuss, J. and K. Kremling. 1999. Spatial variability of particle associated trace elements in near-surface waters of the North Atlantic (30 degrees N/60 degrees W to 60 degrees N/2 degrees W), derived by large-volume sampling. Mar. Chem. 68: 71-86.

Landing, W.M. and K.W. Bruland. 1980. Manganese in the north Pacific. Earth Planet. Sci. Lett. 49: 45-56.

Lane, T.W. and F.M.M. Morel. 2000. A biological function for cadmium in marine diatoms. Proc. Natl. Acad. Sci. 97: 4627-4631.

Lee, J.G. and F.M.M. Morel. 1995. Replacement of zinc by cadmium in marine phytoplankton. Mar. Ecol. Prog. Ser. 127: 305-309.

Lee, J.G., S.B. Roberts, and F.M.M. Morel. 1995. Cadmium: A nutrient for the marine diatom *Thalassiosira weissflogii*. Limnol. Oceanogr. 40: 1056-1063.

Li, Y.H. and T.H. Peng. 2002. Latitudinal change of remineralization ratios in the oceans and its implication for nutrient cycles. Global Biogeochem. Cycles 16: 77-1-77-16.

Maldonado, M.T. and N.M. Price. 1996. Influence of N substrate on Fe requirements of marine centric diatoms. Mar. Ecol. Prog. Ser. 141: 161-172.

Maldonado, M.T. and N.M. Price. 1999. Utilization of iron bound to strong organic ligands by plankton communities in the subarctic Pacific Ocean. Deep-Sea. Res. II. 46: 2447-2473.

Martin, J.H. and R.M. Gordon. 1988. Northeast Pacific iron distributions in relation to phytoplankton productivity. Deep-sea Res. *I.* 35: 177-196.

Martin, J.H. and G.A. Knauer. 1973. The elemental composition of plankton. Geochim. Cosmochim. Acta. 37: 1639-1653.

Martin, J.H., K.W. Bruland, and W.W. Broenkow. 1976. Cadmium transport in the California current., pp. 159-184. *In* H.L. Windom and R.A. Duce [eds.], Marine Pollutant Transfer. Lexington Books (D.C. Health and Co.). Toronto, Canada.

Martin, J.H. and S.E. Fitzwater. 1988. Iron deficiency limits phytoplankton growth in the north-east Pacific subarctic. Nature. 331: 341-343.

Martin, J.H., K.H. Coale, K.S. Johnson, S.E. Fitzwater, R.M. Gordon, S.J. Tanner, C.N. Hunter, V.A. Elrod, J.L. Nowicki, T.L. Coley, R.T. Barber, S. Lindley, A.J. Watson, K. Van Scoy, C.S. Law, M.I. Liddicoat, R. Ling, T. Stanton, J. Stockel, C. Collins, A. Anderson, R. Bidigare, M. Ondrusek, M. Latasa, F.J. Millero, K. Lee, W. Yao, J.-Z. Zhang, G. Friederich, C. Sakamoto, F. Chavez, K. Buck, Z. Kolber, R. Greene, P.G. Falkowski, S.W. Chisholm, F. Hoge, R. Swift, J. Yungel, S. Turner, P. Nightingale, A. Hatton, P. Liss, and N. W. Tindale. 1994. Testing the iron hypothesis in ecosystems of the equatorial Pacific Ocean. Nature. 371: 123-129.

Moffett, J.W. 1995. The spatial and temporal variability of copper complexation by strong organic ligands in the Sargasso Sea. Deep-Sea *Res. I.* 42: 1273-1295.

Morel, F.M.M. and J.G. Hering. 1993. Principles and applications of aquatic chemistry. Wiley, New York, USA.

Morel, F.M.M. and R.J.M. Hudson. 1985. The geobiological cycle of trace elements in aquatic systems: Redfield revisited. pp 251-281. *In* W. Stumm, [ed.], Chemical Processes in Lakes. John Wiley, New York, USA.

Morel, F.M.M., R.J.M. Hudson, and N.M. Price. 1991. Limitation of productivity by trace metals in the sea. Limnol. Oceanogr. 36: 1742-1755.

Morel, F.M.M., J.R. Reinfelder, S.B. Roberts, C.P. Chamberlain, J.G. Lee, and D. Yee. 1994. Zinc and carbon co-limitation of marine-phytoplankton. Nature. 369: 740-742.

Morel, F.M.M., J.G. Rueter, D.M. Anderson, and R.R.L. Guillard. 1979. Aquil: a chemically defined phytoplankton culture medium for trace metal studies. J. Phycol. 15: 135-141.

Price, N.M., G.I. Harrison, J.G. Hering, R.J. Hudson, P.M.V. Nirel, B. Palenik, and F.M.M. Morel. 1988/89. Preparation and chemistry of the artificial algal culture medium Aquil. Biol. Oceanogr. 6: 443-461.

Price, N.M. and F.M.M. Morel. 1990. Cadmium and cobalt substitution for zinc in a marine diatom. Nature. 344: 658-660.

Redfield, A.C. 1934. On the proportions of organic derivatives in sea water and their relation to the composition of plankton. pp. 176-92. *In* R.J. Daniel [ed.], James Johnstone Memorial Volume. Liverpool University Press, UK.

Redfield, A.C. 1958. The biological control of the chemical factors in the environment. Am. Scientist. 46: 205-221.

Redfield, A.C., B.H. Ketchum, and F.A. Richards. 1963. The influence of organisms on the composition of sea-water. pp. 26-77. *In* M.N. Hill [ed.], The Sea., Interscience Publication New York USA.

Rue, E.L. and K.W. Bruland. 1995. Complexation of iron III by natural organic ligands in the Central north Pacific as determined by a new cometitive ligand equilibration/ adsorptive cathodic stripping voltammetric method. Mar. Chem. 50: 117-138.

Saito, M.A. and J.W. Moffett. 2001. Complexation of cobalt by natural organic ligands in the Sargasso Sea as determined by a new high-sensitivity electrochemical cobalt speciation method suitable for open ocean work. Mar. Chem. 75: 49-68.

Saito, M.A., J.W., Moffett, S.W. Chisholm, and J.B. Waterbury, 2002. Cobalt limitation and uptake in Prochlorococcus. Limnol. Oceanogr. 47: 1629-1636.

Sherrell, R.M. and E.A. Boyle. 1992. The trace metal composition of suspended particles in the oceanic water column near Bermuda. Earth Planet. *Sci. Lett.* 111: 155-174.

Sunda, W.G. and R.R.L. Guillard 1976. The relationship between cupric ion activity and the toxicity of copper to phytoplankton. J. Mar. Res. 34: 511-529.

Sunda, W.G. and S.A. Huntsman. 1983. Effect of competitive interactions between manganese and copper on cellular manganese and growth in estuarine and oceanic species of the diatom *Thalassiosira*. Limnol. Oceanogr. 28: 924-934.

Sunda, W.G. and S.A. Huntsman. 1992. Feedback interactions between zinc and phytoplankton in seawater. Limnol. Oceanogr. 37: 25-40.

Sunda, W.G. and S.A. Huntsman. 1995a. Iron uptake and growth limitation in oceanic and coastal phytoplankton. Mar. Chem. 50: 189-206.

Sunda, W.G. and S.A. Huntsman, 1995b. Cobalt and zinc interreplacement in marine phytoplankton: Biological and geochemical implications. Limnol. Oceanogr. 40: 1404-1417.

Sunda, W.G. and S.A. Huntsman. 1995c. Regulation of copper concentration in the oceanic nutricline by phytoplankton uptake and regeneration cycles. Limnol. Oceanogr. 40: 132-137.

Sunda, W.G. and S.A. Huntsman. 1996. Antagonisms between cadmium and zinc toxicity and manganese limitation in a coastal diatom. Limnol. Oceanogr. 41: 373-387.

Sunda, W.G. and S.A. Huntsman, 1997. Interrelated influence of iron, light and cell size on marine phytoplankton growth. Nature. 390: 389-392.

Sunda, W.G. and S.A. Huntsman, 1998a. Interactions among Cu^{2+}, Zn^{2+}, and Mn^{2+} in controlling cellular Mn, Zn, and growth rate in the coastal alga Chlamydomonas. Limnol. Oceanogr. 43: 1055-1064.

Sunda, W.G. and S.A. Huntsman. 1998b. Control of Cd concentrations in a coastal diatom by interactions among free ionic Cd, Zn, and Mn in seawater. Envir. Sci. Technol. 32: 2961-2968.

Sunda, W.G. and S.A. Huntsman. 1998c. Processes regulating cellular metal accumulation and physiological effects: phytoplankton as model systems. Sci. of the Total Envir. 219: 165-181.

Sunda, W.G. and S.A. Huntsman. 2000. Effect of Zn, Mn, and Fe on Cd accumulation in phytoplankton: Implications for oceanic Cd cycling. Limnol. Oceanogr. 45: 1501-1516.

Sunda, W.G., N.M., Price, and F.M.M. Morel 2004. Trace metal ion buffers and their use in culture studies. *In* R. Anderson [ed.], Algal Culturing Techniques., Academic Press. Burlington USA.

Westall, J.C., J.L. Zachary, and F.M.M. Morel 1986. MINEQL: a computer program for the calculation of the chemical equilibrium composition of aqueous systems. Report 86-01, Department of Chemistry, Oregon State University, Corvallis, OR, USA.

Whitfield, M. 2001. Interactions between phytoplankton and trace metals in the ocean. Adv. Mar. Biol. 41: 3-128.

Yee, D. and F.M.M. Morel. 1996. In vivo substitution of zinc by cobalt in carbonic anhydrase of a marine diatom. Limnol. Oceanogr. 41: 573-577.

9

Algal Cultures as a Tool to Study the Cycling of Dissolved Organic Nitrogen

Deborah A. Bronk[1] *and Kevin J. Flynn*[2]

[1] Department of Physical Sciences, The College of William and Mary, Virginia Institute of Marine Science, Route 1208; Greate Rd., Gloucester Point, VA 23062, USA
[2] Institute of Environmental Sustainability, University of Wales, Swansea SA2 8PP, UK

Abstract

The dissolved organic nitrogen (DON) pool is a large heterogeneous mixture of compounds including urea, amino acids, proteins, nucleic acids, amines, amides and humic substances, among others. Traditionally, the DON pool has been viewed as largely refractory and unimportant to planktonic nutrition. This view was supported by the persistence of relatively high DON concentrations even in environments where the inorganic nitrogen forms had been driven below the level of detection by planktonic uptake. Over the past decade, however, evidence has been accumulating that indicates the DON pool is a dynamic component of marine systems with relatively high rates of uptake and release being tightly coupled. Research into DON cycling is difficult, however, because of the complex chemical nature of the pool and the myriad biotic interactions that affect it. Cultures are therefore valuable tools to study DON cycling because they allow one to simplify a complex system by controlling more variables then are possible in field studies. The objectives of this chapter are to describe the analytical tools needed to study DON cycling in cultures, to review culture and field work related to DON production and bioavailability with a focus on more recent studies, to discuss the difficulties in extrapolating culture results to the field and to present recommendations for future research.

INTRODUCTION

Dissolved organic nitrogen (DON) is generally the largest pool of fixed nitrogen in most aquatic systems (reviewed in Bronk 2002). The pool consists of a wide range of organic compounds, such as urea, amino acids, combined amino acids, peptides, proteins, creatine and humic and fulvic acids. The exact composition of the pool, however, is unknown at any given time and likely changes over relatively short spatial and temporal scales. The fact that relatively high concentrations of DON persist even in oligotrophic environments believed to be nitrogen limited led to the traditional view that DON was largely refractory and therefore unimportant to phytoplankton nitrogen nutrition. Research conducted in the past decade or so, however, has challenged that belief and the DON pool is increasingly recognized as a dynamic component of aquatic nitrogen cycles. Research has shown that phytoplankton are both producers and consumers of DON. Much of what we know about algal-DON interactions has been learned through the use of cultures, particularly from work done in the 60s, 70s and 80s, and there are a number of reviews that describe DON research during this time period (Fogg 1966, Hellebust 1974, Flynn and Butler 1986, and Antia et al. 1991). The objectives of this chapter are to describe the analytical tools needed to study DON cycling, to review studies of DON production and bioavailability in cultures and in the field, to discuss the difficulties in extrapolating culture results to natural waters, and to present recommendations for promising areas of research in the future.

ANALYTICAL METHODS

Culture Media and Analytical Techniques

The ability to culture phytoplankton on artificial media greatly expanded the use of cultures in studies of phytoplankton physiology and growth (reviewed in Provasoli et al. 1957, Berges et al. 2001). One of the most widely used nutrient enrichment recipes is the 'f', or its half-strength counterpart, 'f/2' media developed by Guillard and Ryther (1962). Later media recipes were developed to improve autoclavability (Harrison et al. 1980) or for adaptation to specific studies such as trace metal physiology (Price et al. 1988/1989). Good reference sources on culturing techniques and protocols are Guillard (1975), Fogg and Thake (1987) and Hoffs (1999).

Types of Cultures and the Questions They can Address

Batch

The simplest culture system is a batch culture. Media is inoculated with a phytoplankton species and then the cells are left to grow under specific temperature and light conditions. To keep cells in suspension and to facilitate gas exchange, batch cultures are generally shaken, periodically swirled or aerated (reviewed in Guillard 1975). Batch cultures are ideal for asking questions relating to growth yield or growth stage, as well as to study the effect of light and temperature on cells. With respect to growth stages, batch cultures go through a lag phase, where cell numbers do not appreciably increase immediately after inoculation, exponential phase, where cell numbers increase exponentially and the culture is in steady state, stationary phase, where some resource is limiting the continued production of biomass, and, senescence, where cells begin to die.

Continuous

Continuous cultures, also known as chemostats, are a nutrient limited culture held in perpetual steady state through the constant addition of fresh media (reviewed in Smith and Waltman 1995). They are especially valuable because they are more representative of cells in the open ocean where biomass may be very low but growth rates of individual cells may be high due to high rates of nutrient regeneration (analogous to the dilution rate). As fresh media is added, an equal volume of old media, with the cells it contains, flows out of the culture chamber. The growth rate of the cells in the chemostat is controlled by the delivery rate of new media such that the growth rate equals the dilution rate (i.e. flow rate/volume of culture vessel). The biomass in the culture can be expanded by increasing the concentration of nutrients in the media at a given dilution rate. Chemostats may be considered ideal for addressing questions of nutrient limitation, including the effect of growth rate, temperature, or light on the limiting nutrient, as well as to compare the response of different species to a given set of growth conditions (Nagao and Miyazaki 2002). Chemostats, however, do not enable the ready examination of near-maximum growth rates (due to the risk of washout) and there are severe logistic constraints that limit their use when large volumes of cell suspension need to be removed for sampling. They also tend to be time consuming and temperamental to maintain.

Cage cultures

Cage cultures are batch or continuous culture systems where cells are grown within a dialysis bag or on one side of a filter or dialysis membrane. The membrane allows nutrients to diffuse into the culture area while simultaneously allowing excretory or release products to diffuse out (reviewed Sakshaug and Jensen 1978).

Axenic versus xenic cultures

Regardless of the type of culture system used to investigate questions of phytoplankton physiology and ecology, it is best to use axenic or pure cultures. In phytoplankton cultures, the main threat to axenicity is bacterial contamination. Bacteria may out-compete phytoplankton for substrates such that their presence can severely confound experimental results, particularly with respect to measuring DON release and uptake rates. If one is measuring DON release from phytoplankton, the presence of bacteria can result in recently released DON being taken up by the contaminant bacteria – in effect, the bacteria consume the experimental signal. If one is measuring algal DON uptake, the presence of contaminant bacteria can result in the appearance of uptake by the phytoplankton, when in fact no algal uptake occurred. There are a number of techniques that can be used to isolate axenic cultures including micro-pipetting and washing (Pringshein 1949), using antibiotics (Cottrell and Suttle 1993), or a combination of both (e.g. Nagai et al. 1998). Many phytoplankters, however, are extremely difficult to render and maintain axenic, particularly cultures of cyanobacteria and colonial forms (e.g. *Trichodesmium*).

Methods to Measure DON Concentration and Composition

DON is operationally defined as all nitrogen forms passing through a filter (generally 0.2 to 0.7 μm) that is not a form of dissolved inorganic nitrogen (DIN, ammonium, nitrate or nitrite). To measure bulk DON one must determine the concentration of total dissolved nitrogen (TDN) and then separately determine the concentration of ammonium and nitrate/nitrite. The final DON estimate, therefore, has the combined analytical error of three analyses. This makes it difficult to detect small but significant changes in concentrations of DON over time or between different environments. This lack of sensitivity is less of an issue in culture experiments because the biomass and substrate concentrations tend to be higher than those found in the environment and because substrate additions are in accordance with analytical detection limits.

There are three primary methods currently used to measure TDN concentrations: persulfate oxidation, ultraviolet (UV) oxidation, and high temperature combustion (Table 9.1, reviewed in Bronk et al. 2000 and Sharp et al. 2002). A number of methods also exist to measure concentrations of individual organic nitrogen compounds including urea, dissolved free amino acids (DFAA) and dissolved combined amino acids (DCAA), humic

Table 9.1 Methods for measuring concentrations of DON and individual organic compounds including dissolved primary amines (DPA) and dissolved free and combined amino acids (DFAA and DCAA)

Analyte	*Method*	*Reference*
Bulk DON concentrations	Persulfate oxidation	Menzel and Vaccaro 1964
		Valderamma 1981
	UV oxidation	Armstrong et al. 1966
	High temperature combustion	Hansell et al. 1993
Organic compounds:		
Urea	Monoximine	Price and Harrison 1987, Cozzi 2004
	Urease	McCarthy 1970
DPA	Flourometric opa	Parsons et al. 1984
	Fluorescamine	North 1975
DFAA	HPLC	Mopper and Lindroth 1982
DCAA	Liquid hydrolysis/HPLC	Parsons et al. 1984
	Vapor phase hydrolysis/HPLC	Keil and Kirchman 1991
Humic substances	XAD extraction	Hessen and Tranvik 1998
Nucleic acids (DNA, RNA)		Karl and Bailiff 1989, Sakano and
		Kamatani 1992, Siuda and Güde 1996
Proteins	Modified bicinchoninic acid	Smith et al. 1985, Nguyen and
Methylamines	Gas chromatography	Harvey 1994, Yang et al. 1993

substances, nucleic acids, proteins, and methylamines (Table 9.1), most of which exist at nanomolar to micromolar levels (summarized in Table 9.2). Concentrations of one of the most labile DON pools, free amino acids, are reported either as total dissolved primary amines (DPA), measured fluorometrically (North 1975, Parsons et al. 1984) or as individual DFAA, measured with HPLC (Mopper and Lindroth 1982). The simplest forms of

Table 9.2 Range of mean concentrations of organic nitrogen compounds found in nature (data from Bronk 2002)

Compound	*Range of mean (μM)*	*Reference*
Urea	0 to 13.0	reviewed in Bronk 2002
DFAA	0.001 to 0.7	reviewed in Bronk 2002
DCAA	0.15 to 4.20	reviewed in Bronk 2002
Humic substances	0.4 to 12.3	reviewed in Bronk 2002
Nucleic acids: DNA	0.05 to 80.6	Karl and Bailiff 1989
RNA	0.5 to 871	
Purine and pteridines	up to 12.6	Antia et al. 1991
Methylamines	0 to 0.62	Gibb et al. 1999

DCAA are dipeptides, with the most complex being soluble protein. To measure the nitrogen concentration within humic compounds, often the largest component of DON in aquatic systems, the humics are removed from solution by passing the sample through an XAD-8 resin (Aiken 1988). The TDN concentration in the eluate is then measured and the difference between the TDN concentration in the original sample and the TDN concentration of the eluate is assumed to be the concentration of humic substances.

Methods to Study DON Release and Bioavailability

To measure rates of DON release and uptake there are two main approaches – monitoring changes in ambient concentrations and tracer techniques. In the first, at its simplest, net flux out of or into biomass is determined by monitoring changes in concentration over time in the surrounding water. One permutation of this approach is the bioassay where a sample is spiked with a substrate to increase the magnitude of the fluxes (Flynn 1990, Carlsson et al. 1995, Seitzinger et al. 2002); if the spike is large, however, it can cause a significant perturbation to the system. The bioassay approach is ideally used in situations where only a single organism is present (i.e. axenic cultures), because in complex systems one cannot be sure biotic or abiotic transformations are not liberating nitrogen from DON as DIN, which is then the form incorporated by the phytoplankton. The most important biotic process is bacterial remineralization. Abiotic processes that can break down DON include photochemical decomposition (e.g. photochemical ammonification); under normal incubation conditions, however, cultures are generally not be exposed to the far ultraviolet range of light responsible for photochemical decomposition (reviewed in Mopper and Kieber 2002).

While useful in cultures, estimating flux rates by monitoring concentration changes is of limited use in the field because biomass is generally too low to produce a measurable change in substrate concentrations over the course of hours. The approach is also problematic if uptake and release are tightly coupled. For example, substantial uptake rates could be severely underestimated if regeneration rates similar in magnitude were co-occurring.

With the introduction of ^{15}N tracer techniques to phycology and oceanography (Neess et al. 1962, Dugdale and Goering 1967) it became possible to quantify small but significant nitrogen uptake rates. Isotopes also have the advantage that uptake rates can be measured even in situations where regeneration is co-occurring. There are also isotope dilution approaches that allow for the simultaneous measurement of uptake and regeneration rates of ammonium (Glibert et al. 1982), urea (Hansell and Goering 1989, Slawyk et al. 1990), and amino acids (Fuhrman 1987). Isotope approaches have the advantage that they are much more sensitive than monitoring concentration changes but they are also more costly and labor intensive. The number of compounds that can be studied is also limited to a relatively few compounds that have commercially available ^{15}N, ^{14}C, ^{13}C or ^{3}H labeled forms such as DFAA or urea (Fuhrman and Bell 1985, Hansell and Goering 1989, Wheeler and Kirchman 1986, Collos and Slawyk 1986, Bronk et al. 1998). Some organic substrates can be purchased in a number of different forms. For example, uptake of urea can be studied with ^{15}N-labeled urea (Lomas et al. 2002), dual-labeled urea (^{15}N and ^{13}C or ^{14}C; Hansell and Goering 1989, Bronk et al. 1998), urea labeled with radioactive ^{14}C (Price and Harrison 1988a, Lisa et al. 1995), or a sulfur analog of urea, ^{14}C-thiourea (Rees and Syrett 1979b); note that thiourea is toxic to at least one phytoplankter, *Emiliania huxleyi* (Palenik and Henson 1997).

Bioavailability of the DON pool as a whole is difficult to study because the pool is composed of a large number of compounds and the exact composition is unknown (reviewed in Antia et al. 1991, Benner 2002, Bronk 2002). This makes it especially difficult to determine which substrates to use in uptake studies. At first glance it may seem logical to add labeled forms of DON that are present at elevated concentration in seawater, such as glycine. In reality, however, the organic substances most often seen in seawater may only be present at elevated concentrations because their uptake rates are relatively low (Schell 1974, Flynn and Butler 1986, Flynn 1990). One approach is to let the phytoplankton community in natural waters produce a realistic ^{15}N-labeled DON tracer (Bronk et al. 1993a). In this approach a whole water sample is incubated with ^{15}N-labeled ammonium or nitrate and then the recently released $DO^{15}N$ is isolated (see below, Bronk and Glibert 1993a). More recently, Veuger et al. (2004) produced ^{15}N-labeled DON using a *Skeletonema costatum* culture, which was subsequently used as a tracer of DON uptake in the field. Other approaches used to study DON uptake use enzyme techniques where the hydrolysis of flourogenically tagged substrates are monitored (Pantoja et al. 1993; 1997, Pantoja and Lee 1999; Mulholland et al. 2002; Berg et al. 2002) or where enzyme activity is measured over time (e.g. Berges and Falkowski 1996).

Isotopic approaches have also been used to measure release of DON (Bronk and Glibert 1991) or individual DON compounds such as urea (Hansell and Goering 1989) and amino acids (Fuhrman 1987). To measure rates of release for the bulk DON pool, the pool is isolated at the end of a ^{15}N

incubation (reviewed in Bronk 2002). Isolating the pool can be accomplished using wet chemical techniques (Axler and Reuter 1986; Slawyk and Raimbault 1995; Bronk and Ward 1999, Veuger et al. 2004) or ion retardation resin (Bronk and Glibert 1991; Nagao and Miyazaki 1999; reviewed in Bronk 2002). If both the extracellular and intracellular DON pools are isolated, one can measure a release rate as well as an intracellular transformation rate of inorganic nitrogen to organic nitrogen within the cell (Bronk and Glibert 1991, Bronk 1999).

DOM Composition and Production in Culture and in Natural Waters

The mid-60s to mid-80s was a very active period in the study of algal nitrogen physiology in cultures, much of which involved the cycling of DON. Particularly common during this time were batch culture studies of nitrogen release where the accumulation of DON, or some specific organic compound, was measured over time. A number of studies demonstrate that there is appreciable production of organic matter in phytoplankton cultures and that organic matter accumulates during exponential growth with even more accumulating during the stationary phase (Fogg 1952, Allen 1956). For example, in one study the N-fixing cyanobacterium, *Calothrix scopulorum*, released an average of 40% of its assimilated nitrogen in culture under optimal growth conditions with the percentage increasing under suboptimal condition (Jones and Stewart 1969). In a study with continuous cultures, significant production of DON was documented in eight phytoplankton species with the release increasing when cells went from exponential to stationary growth (Newell et al. 1972). From studies such as these it was concluded by many that exponentially growing algae have low DON release rates but that at the end of exponential growth and the onset of senescence excretion of DON increases substantially. It is not always apparent from such studies, however, that the rate of release per cell changes, how the nature of limiting nutrient affects release (e.g. phosphorus, nitrogen, or silica), or to what extent other events associated with bloom termination (such as grazing or viral lysis) contribute to DON release.

While it is reasonable to expect nutrient stressed, senescent and damaged cells to leak organics (Fogg 1966) there has been an active debate over the question of whether healthy cells also do so (Fogg 1977, Sharp 1977). While DOC release from autotrophic organisms living in nutrient or trace metal-limited systems is easy to understand, explanations for why cells would release significant amounts of DON, are considerably more difficult to rationalize, particularly under the nitrogen-limited conditions that exist in many marine systems. The contention is that appreciable DON release only occurs from damaged or dying cells and that high rates of release in an apparently healthy population is an artifact of sample handling or filtration (Sharp 1977). More recently, however, there has been a growing consensus that extracellular release is a normal function of healthy cells (reviewed in Antia et al. 1991 and Bronk 2002). For example, the highest rates of DFAA excretion in diatoms is observed during exponential growth (Myklestad et al. 1989 and references therein). In another series of culture experiments, most of the DOM produced is again released during nutrient replete conditions (Biddanda and Benner 1997). In two clones of marine *Synechoccocus* spp. the highest rates of DON release, both absolute and as a percentage of gross nitrogen uptake, is observed during nitrogen sufficient growth, and the release rates decrease by over a factor of four when ammonium is depleted in the medium (i.e. they reached senescence; Bronk 1999). In cultures of *Scenedesmus quadricauda* and *Microcystis novacekii*, DON release is low in nitrogen-limited cultures but significantly higher in cultures that are nitrogen-replete (Nagao and Miyazaki 2002).

DON concentrations and composition in natural waters

Concentrations of bulk DON and individual organic nitrogen compounds vary widely across environments (Tables 9.2 and 9.3). A broad suite of biochemicals, including amino acids, proteins, urea, nucleic acids, and amino sugars, have been measured in seawater at low levels, but they represent only a small fraction of the total DON pool (Table 9.2, reviewed in Antia et al. 1991 and Benner 2002). The pool also includes other high molecular weight (HMW) dissolved compounds, such as humic substances. Humic substances are a class of organic compounds that are operationally defined based on their retention on hydrophobic resins, which can exist in

Table 9.3 Range of mean concentrations of DON in various marine and aquatic systems (from Bronk 2002 and Berman and Bronk 2003)

System	*Range of mean (μM)*	*Mean ± std (μM)*	*DON:TDN (%)*
Ocean – surface waters	0.8 to 11.0	5.8 ± 2.0	61.6 ± 32.9
Ocean – deep waters	1.4 to 9.8	4.3 ± 2.1	9.9 ± 2.6
Coastal/continental shelf	1.1 to 52.5	9.9 ± 8.1	65.3 ± 30.4
Estuarine	0.6 to 65.0	22.5 ± 17.5	68.9 ± 22.4
Rivers	2.9 to 90.0	34.7 ± 20.7	60.1 ± 23.5
Lakes	3.6 to 187.8	38.1 ± 34.7	~66.0 ± 15.3

relatively high concentrations in marine and aquatic systems (reviewed in Hessen and Tranvik 1998). The largest fraction of DON in aquatic systems, however, remains undefined (reviewed in Benner 2002).

When considering the composition of the DON pool, it is important to remember that the composition is greatly affected by uptake processes. For example, the composition of the DFAA pool in natural waters is largely composed of those compounds that bacteria and phytoplankton have the lowest affinity for (Flynn and Butler 1986). Amino acids such as glutamate, alanine, serine and glycine are often present at the highest concentrations, while others such as arginine, glutamine, and asparagine are often present at the lowest concentrations, in reflection of preferential uptake

DON release in natural waters

In culture, DON release is limited to direct release from algae, or possibly release due to viral infection. In the field, however, DON release (reviewed by Carlson 2002) is mediated by a number of processes including direct release from primary producers (Bronk 1999) and bacterioplankton (Ogawa et al. 2001), egestion, excretion, and sloppy feeding from micro- and mesozooplankton (Nagata and Kirchman 1991, Ward and Bronk 2001, Steinberg et al. 2002, 2004), viral lysis of bacterioplankton (Fuhrman 1992) and eukaryotic cells (Suttle 1994), and particle solubilization (Smith et al. 1992; summarized in Fig. 9.1). Here we will focus on the two release processes that are likely most important in natural waters: virus- and grazer-mediated release.

Viral infection is important to consider with respect to DON release because in the final stages of viral infection, the phage increase to such numbers that the cell actually bursts, resulting in the release of any dissolved organics present within the cell (Wommack and Colwell 2000). Viral infection and lysis is an ongoing process within plankton communities but the quantitative contribution of this process to rates of DON release is unknown. One aspect of viral ecology that has received attention is their role in the decline of blooms. A number of investigations have uncovered either direct (Nagasaki et al. 1994, Tarutani et al. 2000), or indirect (Sieburth et al. 1988, Milligan and Cosper 1994), evidence for the involvement of viral lysis in the large-scale mortality and cessation of monospecific microalgal blooms (reviewed by Wommack and Colwell 2000 and Weinbauer 2004). Viruses have also been implicated in the decline of blooms of the common coccolithophorid *Emiliania huxleyi* (Bratbak et al. 1993; Brussaard et al. 1996) and the diatom *Skeletonema* spp. (Bratbak et al. 1990), as well as harmful algal species such as *Aureococcus anophagefferens* (Gobler et al. 1997).

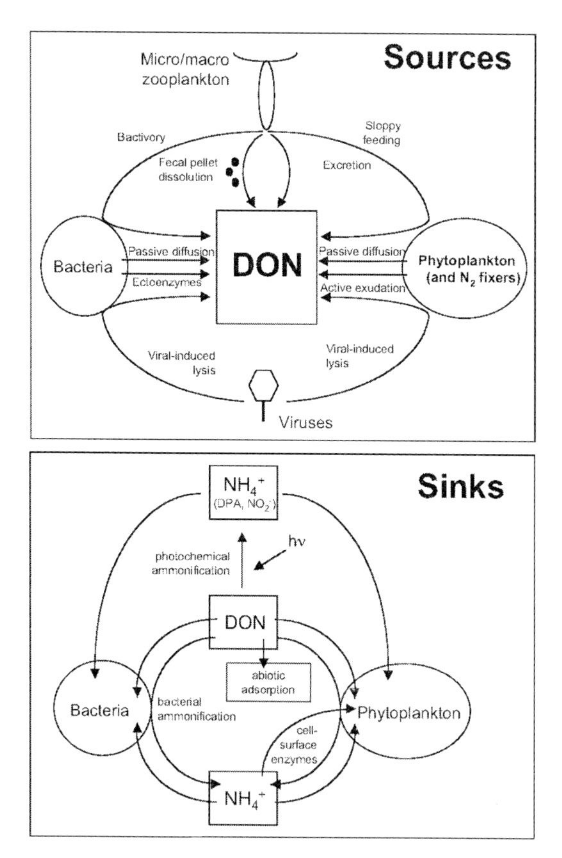

Fig. 9.1 Conceptual diagram of processes involved in dissolved organic nitrogen (DON) release (Sources) and processes involved in DON utilization (Sinks) in aquatic systems (modified from Bronk 2002)

Grazing by microzooplankton (e.g. flagellates and ciliates), on phytoplankton and bacteria, can release varying percentages of carbon, nitrogen or phosphorus, depending on prey type, via egestion and possibly diffusion (reviewed in Nagata 2000). This released DOM is composed of HMW and low MW (LMW) compounds (Taylor et al. 1985), DFAA (Flynn and Fielder 1989), and DCAA (Nagata and Kirchman 1991). DON is also released from micro- and macrozooplankton through excretion (Small et al. 1983; Steinberg et al. 2002) and fecal pellet dissolution (Lampitt et al. 1990, Jumars et al. 1989, Urban-Rich 1999). While there are few direct measurements, studies indicate excretion of DON by zooplankton can be a large proportion of the total nitrogen metabolized or ingested by the organism (Steinberg et al. 2000). Though ammonium is the primary nitrogen excretory product of many zooplankton, DON (i.e. urea and amino acids) is excreted at lower but significant levels (Bidigare 1983). Another important source of DON is sloppy feeding by mesozooplankton (i.e. copepods), where cells are not ingested whole but instead are broken during feeding thereby releasing intracellular dissolved compounds into the water column. A number of studies have demonstrated increased release of amino acids (Williams and Poulet 1986, Roman et al. 1988) or bulk DON (Hasagawa et al. 2000) due to zooplankton feeding.

DON BIOAVAILABILITY TO ALGAE IN CULTURE AND IN NATURAL WATERS

The study of DON bioavailability is especially suited to cultures because the pool is so complex in natural waters. Here we review uptake mechanisms used to access organic nitrogen and briefly survey the bioavailability of a suite of organic nitrogen compounds. We note that more detailed reviews of DON uptake can be found in Paul (1983), Flynn and Butler (1986), Antia et al. (1991) and Bronk (2002).

Uptake Mechanisms

Uptake of the smaller organic compounds, urea and amino acids, can occur by active transport or, if extracellular concentrations are extremely high (mM levels), through facilitated diffusion. For larger organic compounds such as peptides, humic acids, etc. direct uptake via transport proteins at the cytoplasmic membrane is not possible. Nitrogen associated with these larger compounds can be taken into the cell via pinocytosis or phagocytosis (cellular ingestion of substances within membrane bound vesicles) or after extracellular enzymatic cleavage of nitrogen from the larger compounds. Little is known about pino- or phagocytosis as a means of acquiring nitrogen, but uptake of a HMW (2000kDa) dextran by the dinoflagellate *Alexandrium catenella* has been documented (Legrand and Carlsson 1998).

The uptake mechanism that has received more attention of late is extracellular exzymatic breakdown and then subsequent adsorption of the free nitrogen (Fig. 9.2). A number of phytoplankton strains have been shown to have cell surface amine oxidase enzymes that can cleave amino groups from primary amines (Palenik and Morel 1990a) including *Chlamydomonas reinhardtii* (Langheinrich 1995) and *Gonyaulax polyedra* (Sankievicz and Colepicolo 1999). The resulting alpha-keto acids or aldehydes remain in the water column forming potential carbon sources for bacteria. This scenario illustrates that, though studies with ^{14}C-labeled organic compounds

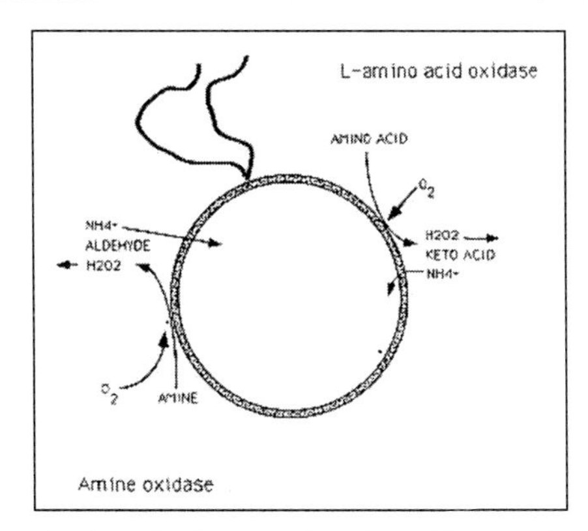

Fig. 9.2 Proposed mechanism of cell-surface L-amino acid and amine oxidases (from Palenik et al. 1988). The enzyme catalyzes the decomposition of an amine at the surface of a phytoplankton cell resulting in the release of ammonium (NH_4^+), a keto acid, and hydrogen peroxide (H_2O_2). The amine is shown labeled with ^{14}C to illustrate that in a tracer experiment, the amine will appear to be used by the bacterial fraction, while the N may actually being utilized by the phytoplankton.

generally show transfer of the label to the bacterial fraction, these results cannot necessarily be extrapolated to include utilization of the associated amino nitrogen. This type of event is analogous to the role of alkaline (marine species) and acidic (freshwater species) phosphatases that develop when algae are phosphorus-stressed. Mulholland et al. (1998) quantified extracellular amino acid oxidase activity in natural waters from a number of oceanic and estuarine systems using a fluorescent analog of lysine (Pantoja and Lee 1994), and found that oxidase activity is widespread. The presence of cell surface enzymes (Palenik et al. 1988, Palenik and Morel 1990b) also raises the possibility that phytoplankton may access nitrogen from larger moieties without taking up the entire molecule. Studies with axenic cultures of *Aureococcocus anophagefferens* found high rates of peptide hydrolysis suggesting that this harmful bloom former can access some components of the HMW DON pool as a nitrogen source (Berg et al. 2002). Studies with cultures of *Emiliania huxleyi*, however, found that, while short-chained aliphatic amines were readily taken up, longer chained aliphatic amines were not used as a nitrogen source (Palenik and Henson 1997).

Uptake of Individual Compounds

Urea

Urea is a small neutral molecule that can move into and out of cells by passive diffusion as well as by active transport; passive diffusion, however, is unlikely to be of significance at the low concentrations typical of natural waters. The degradation of urea is catalyzed intracellularly by two enzymes – urease and ATP:urea amidolyase; generally an organism will possess one but not both enzymes. Urease catalyzes the breakdown of urea to ammonium and carbonic acid (Collier et al. 1999). The enzyme itself is a constitutive nickel-ligated metaloprotein (Sakamoto and Bryant 2001); the ability of phytoplankters to use urea may have been underestimated in culture studies prior to the realization that nickel is required in the media for urea uptake via urease (Oliveria and Antia 1986). The enzyme ATP: urea amidolyase is a HMW protein that catalyzes two reactions with the end products being ammonium and bicarbonate. In contrast to urease, ATP: urea amidolyase appears to be inducible (Syrett and Leftley 1976). The ability to breakdown urea to ammonium appears to be common such that most freshwater and marine phytoplankton species tested grow efficiently on urea as the sole nitrogen source (reviewed in Antia et al. 1991).

Urea uptake kinetics have been studied in a number of species and are consistently found to be saturable and carrier-mediated (Table 9.4). A number of marine phytoplankton species have been shown to have a high affinity for urea with K_s values less than 1 μM (Table 9.4). In a comprehensive study of urea uptake by *Chlamydomonas reinhardtii*, uptake at concentrations less than 70 μM is mediated by a saturable transport mechanism with a K_s of 5.1 μM; at concentrations greater than 70 μM, urea transport into the cell takes place by passive diffusion (Hodson et al. 1975, Williams and Hodson 1977). These researchers also found that uptake only proceeds in the presence of light and oxygen.

Urea uptake is influenced by the availability of other nitrogen substrates. Many studies have shown that urea uptake is suppressed by the presence of ammonium (Lui and Roels 1970, Kirk and Kirk 1978a and b, Horrigan and McCarthy 1982, Lund 1987, Molloy 1987, Molloy and Syrett 1988a and b, Ricketts 1988) though not in all cases (Syrett and Leftley 1976). In contrast to ammonium, nitrate generally does not inhibit urea uptake, but the presence of urea has been shown to inhibit uptake of nitrate (McCarthy and Eppley 1972, Molloy and Syrett 1988b, Ricketts 1988) though, again, not in all cases (Kirk and Kirk 1978a and b). This is consistent with a common control of ammonium, nitrate and urea uptake centering on the intracellular concentrations of glutamine and 2-oxoglutarate and hence on the availability of intracellular ammonium and carbon skeletons typically produced during photosynthesis (Flynn 1991).

Table 9.4 Kinetic parameters for urea uptake in freshwater (FW) and marine (M) algae. Cultures were either nitrogen replete (R) or nitrogen depleted (D).

Phytoplankton species	V_{max} (h^{-1})	K_s (μM)	*Reference*	*FW/M*	*N status*
Chlorophytes					
Ankistrodesmus braunii Nägeli		0.8	Kirk and Kirk 1978c	FW	
Chlamydomonas reinhardtii Dangeard	0.67	5.1	Williams and Hodson 1977	FW	
Chlorella fusca Shihara et Krauss	ND	15.0	Syrett and Bekheet 1977	FW	R
	0.02	16.5	Bekheet and Syrett 1979	FW	R
Chlorella pyrenoidosa Chick	ND	0.6	Kirk and Kirk 1978c	FW	R
Eudorina elegans Ehrenberg	ND	0.1	Kirk and Kirk 1978c	FW	R
Golenkinia minutissima Iyengar et Balakrishnan	ND	0.7	Kirk and Kirk 1978c	FW	R
Gonium pectorale Müller	ND	0.1	Kirk and Kirk 1978c	FW	R
Pandorina morum Bory	ND	2.6	Kirk and Kirk 1978c	FW	R
Pleodorina californica Shaw	ND	0.5	Kirk and Kirk 1978c	FW	R
Scenedesmus obliquus Turpin	ND	0.8	Kirk and Kirk 1978c		R
Scenedesmus quadricauda Turpin	ND	1.2	Healey 1977	FW	R
Volvox carteri Iyengar	ND	0.9	Kirk and Kirk 1978c	FW	R
Cyanobacteria					
Psuedoanabaena catenata Lauterborn	ND	0.4	Healey 1977	FW	R
Diatoms					
Ditylum brightwellii Grunow	0.011	0.4	McCarthy 1972	M	R
Lauderia sp.	0.013	1.7	McCarthy 1972	M	R
Phaeodactylum tricornutum Bohlin	0.002–0.02	0.6–1.0	Rees and Syrett 1979a	M	D
Skeletonema costatum Greville	0.015	1.4	McCarthy 1972	M	R
Thalassiosira pseudonana Hustedt	0.008	0.4	McCarthy 1972	M	R
	0.08	0.5	Horrigan and McCarthy 1981	M	D
Thalassiosira weisflogii Grunow	0.024–0.03	0.5–1.7	McCarthy 1972	M	R
Dinoflagellates					
Alexandrium catenella	0.025 ± 0.008	28.4 ± 15.0	Collos et al. 2004	M	D
Prasinophytes					
Micromonas pusilla	0.054 ± 0.026	0.38 ± 0.07	Cochlan and Harrison 1991		

Amino acids

Microalgae can simultaneously take up a number of DFAA at natural concentrations and remove them to levels below detection (Lu and Stephens 1984). This capability appears to be wide spread, though there are differences in the ability to take up DFAA between phytoplankton groups (Flynn 1990). Uptake of amino acids occurs mainly through active transport, and there appears to be at least three transport systems, each of which facilitates the uptake of one of the three types of amino acids – acidic (negatively charged), neutral (uncharged), and basic (positively charged). Multiple amino acids have been shown to share the same porter system, which is based likely on the charge of a group of amino acids (Liu and Hellebust 1974b, North and Stephens 1972). Some amino acids are taken up but they can not be used for growth. For example, the toxic dinoflagellate *Alexandrium fundyense* can take up a range of amino acids but nitrogen derived from amino acids is not able to support significant growth (John and Flynn 1999).

It has been suggested that two intracellular pools occur in cells, a storage pool and a metabolic pool (Wheeler and Stephens 1977), although whether these are physically separate within the cell is not known. In general, amino acid uptake is subject to the regulatory process of transinhibition, a process in which the internal cellular concentration of the substrate limits its own transport across the plasma membrane. When intracellular pools of free amino acids are replete, then transport into the cell halts. In contrast to the uptake of inorganic nitrogen, the use of DFAA is promoted by nitrogen-stress and/or carbon-stress, because these conditions would adversely affect the size of the intracellular pool of free amino acids.

In general, amino acid uptake is controlled by a number of parameters including intracellular pools, nitrogen deprivation, light availability, and the presence of other nitrogen substrates. A number of studies have demonstrated that nitrogen deprivation increases amino acid uptake (North and Stephens, 1971, 1972, Wheeler et al. 1974, Flynn and Syrett 1985, 1986). For example, glycine uptake rates by *Platymonas* increase by a factor of ten when the cells become nitrogen depleted (North and Stephens 1971). Nitrogen-starvation, with or without carbon-starvation (i.e. incubation in darkness), is known to stimulate amino acid uptake (Flynn and Syrett 1986). Light deprivation can also induce amino acid uptake likely due to the need for carbon (Lewin and Hellebust 1975, 1976, 1978). The presence of ammonium has been shown to inhibit amino acid uptake (Flynn and Butler 1986), though not in all cases (North and Stephens 1971, Kirk and Kirk 1978c, Flynn and Syrett 1986). There are also studies that demonstrate instances when amino acids inhibit ammonium uptake (Flynn and Wright 1986) and nitrate uptake (Bilboa et al. 1981). Amino acids can also inhibit uptake of other amino acids. For example, lysine and arginine share the same porter system and can therefore inhibit each other (reviewed in Flynn and Syrett 1986).

Kinetic parameters have been determined for a number of different phytoplankton species (summarized in Table 9.5). Similar to urea, DFAA uptake appears to be saturable and carrier mediated. In more recent work, two axenic phytoplankton cultures were grown on eight different amino acids supplied at a concentration of 50 μM; *Thalassiosira pseudonana* is not able to grow on any of the amino acids but *Emiliania huxleyi* is able to grow on the neutral amino acids offered (Ietswaart et al. 1994).

Other organic compounds

A wide variety of other organic nitrogen compounds have been studied as a nitrogen source for algae (reviewed in Table 9.6 and Antia et al. 1991). The amino sugar, glucosamine is used as a nitrogen source by two diatoms (McLachlan and Craigie 1966) but a third diatom, *Phaeodactylum tricornutum*, is not able to grow on this substrate (Hayward 1965). The cryptomonad, *Hemiselmis virescens*, is able to grow on glucosamine, as well as on galactosamine (Antia and Chorney 1968). The dipeptide glycine-glycine is also a good nitrogen source for some phytoplankters (Turner 1979, Neilson and Larsson 1980).

Purines, pyrimidines, and pteridines have been documented to serve as nitrogen sources for a number of algal species (Table 9.6). Some purines and pteridines are primary excretory products that are the end products of nitrogen catabolism (Antia et al. 1991). The major purine bases found in nucleic acids are adenine and quanine and the major pyrimidine bases are thymine, cytosine, and uracil. Both adenine and quanine are able to serve as a sole nitrogen source for *Chlamydomonas reinherdtii* (Lisa et al. 1995) though the substrates appear to have been added at mM concentrations in this study. Kinetic experiments estimate K_s values of 3.3 and 3.2 μM for adenine and quanine respectively (Lisa et al. 1995). Berman and Chava (1999) also provide evidence that the purines guanine and hypoxanthine can serve as nitrogen sources for phytoplankton, and uptake appeared to be inducible and took place after a lag period. This study, however, also used relatively high substrate additions (100 μM) and non-axenic monocultures, raising the possibility that the nitrogen was taken up not in purine form but as a regeneration product of bacterial purine uptake. Pettersen and Kuntsen (1974) also show that guanine can be used as a nitrogen source (Pettersen

Table 9.5 Kinetic parameters for amino acid uptake for freshwater (FW) and marine (M) algae. Cultures were either nitrogen replete (R) or nitrogen depleted (D).

Phytoplankton species	*Type*	K_s (μM)	*Reference*	*FW/M*	*N status*
Alanine					
Nitzschia laevis Hustedt	Diatom	20	Lewin and Hellebust 1978	M	R
Arginine					
Chlamydomonas reinhardtii Dangeard	Chlorophyte	3.9	Kirk and Kirk 1978b	FW	R
Cyclotella cryptica Reiman	Diatom	3.0	Liu and Hellebust 1974 a, b	M	R
Nitzschia ovalis Arnott	Diatom	2.0	North and Stephens 1972	M	R
Volvox carteri Iyengar	Chlorophyte	3.8	Kirk and Kirk 1978a	FW	R
Aspartate					
Pleodorina californica Shaw	ND	0.5	Kirk and Kirk 1978c	FW	R
Scenedesmus obliquus Turpin	ND	0.8	Kirk and Kirk 1978c		R
Scenedesmus quadricauda Turpin	ND	1.2	Healey 1977	FW	R
Volvox carteri Iyengar	ND	0.9	Kirk and Kirk 1978c	FW	R
Glutamate					
Anabaena variabilis Kützing ex Gomont	Cyanobacteria	1.4–100	Chapman and Meeks 1983	FW	R
Cyclotella cryptica Reiman	Diatom	36	Liu and Hellebust 1974 a, b	M	R
Navicula angularis Grunow	Diatom	20	Lewin and Hellebust 1976	M	R
Navicula laevis Hustedt	Diatom	30	Lewin and Hellebust 1978	M	R
Navicula pavillardi Hustedt	Diatom	20	Lewin and Hellebust 1975	M	R
Glutamine					
Anabaena variabilis Kützing ex Gomont	Cyanobacteria	1.1–13.8	Chapman and Meeks 1983	FW	R
Glycine					
Gymnodinium breve Davis		110.0	Baden and Mende 1979	M	R
Phaeodactylum tricornutum Bohlin	Diatom	3.0	Lu and Stephens 1984	M	R
Tetraselmis subcordiformis Wille		19.0	North and Stephens 1971	M	D

Contd.

Table 9.5 Contd.

Phytoplankton species	*Type*	K_s (μM)	*Reference*	*FW/M*	*N status*
Guanine					
Phaeodactylum tricornutum Bohlin	Diatom	0.5	Shah and Syrett 1982	M	
Chlorella fusca Shihara et Krauss	Chlorophyte	0.1–1.0	Pedersen and Knutsen 1974	FW	R
Leucine					
Anabaena variabilis Kützing ex Gmont	Cyanobacteria	10.8	Thiel 1988	FW	R
Anacystis nidulans Richter	Cyanobacteria	125.0	Lee-Kaden and Simonis 1982	FW	R
Ankistrodesmus braunii (Nageli)	Chlorophyte	16	Kirk and Kirk 1978c	FW	R
Chlorella fusca Shihara et Krauss	Chlorophyte	2.5	Richards and Thurston 1980	FW	R
Pandorina morum Bory	Chlorophyte	52	Kirk and Kirk 1978c	FW	R
Scenedesmus obliquus (Turpin) Kruger	Chlorophyte	47	Kirk and Kirk 1978c	FW	R
Lycine					
Phaeodactylum tricornutum Bohlin	Diatom	2.3	Flynn and Syrett 1986	M	R
Phaeodactylum tricornutum Bohlin	Diatom	0.8	Flynn and Syrett 1986	M	D
Methionine					
Gymnodinium breve Davis		125.0	Baden and Mende 1979	M	R
Phenylalanine					
Chlorella fusca Shihara et Krauss	Chlorophyte	5.0	Pedersen and Knutsen 1974	FW	
Proline					
Cyclotella cryptica Reiman	Diatom	6.0	Liu and Hellebust 1974 a, b	M	R
Tyrosine					
Chlorella fusca Shihara et Krauss	Chlorophyte	0.4	Richards and Thurston 1980	FW	R
Valine					
Gymnodinium breve Davis		150.0	Baden and Mende 1979	M	R
Uric acid					
Platymonas concoutae Parke et Manton		1.5–3.4	Douglas 1983	M	

Table 9.6 Summary of culture studies that demonstrate whether a given freshwater (FW) or marine (M) phytoplankter can grow, Yes (Y), or is unable to grow, No (N), on a given substrate

Compound	*Organism*	*Use*	*FW/M*	*Reference*
Allantoic acid[1]	Chlorophyte	Y		Oliveira and Huynh 1990
	Chrysophyte	N		Oliveira and Huynh 1990
	Cyanobacterium	Y		Oliveira and Huynh 1990
	Diatoms (2)	N		Oliveira and Huynh 1990
	Dinoflagellate	N		Oliveira and Huynh 1990
	Eustigmatophyte	Y		Oliveira and Huynh 1990
	Prymnesiophytes (3)	Y		Oliveira and Huynh 1990
	Prymnesiophyte	N		Oliveira and Huynh 1990
Allantoin[1]	Chlorophycean algae (6)	Y	FW	Devi Prasad 1983
	Microalgae (12)	N	FW	Birdsey and Lynch 1962
	Microalgae (4 of 18)	Y	M	Antia et al. 1980
	Prasinophyte	N	M	Edge and Ricketts 1978
Amides				
Acetamide	Chlamydomonads (22 of 38)	Y	FW	Cain 1965
	Cyanobacterium	Y	Brackish	Kapp et al. 1975
	Microalgae (13 of 27)	Y		Neilson and Larsson 1980
	Aureococcus anophagefferens	Y	M	Berg et al. 2002
	Emiliania huxleyi	Y	M	Palenik and Henson 1997
	Pleurochrysis carterae	N	M	Palenik and Henson 1997
	Prymnesium parvum	N	M	Palenik and Henson 1997
	Thalassiosira pseudonana	Y	M	Palenik and Henson 1997
	Prorocentrum minimum	Y	M	Palenik and Henson 1997
Formamide	*Emiliania huxleyi*	Y	M	Palenik and Henson 1997
	Pleurochrysis carterae	Y	M	Palenik and Henson 1997
	Prymnesium parvum	N	M	Palenik and Henson 1997
	Thalassiosira pseudonana	N	M	Palenik and Henson 1997
	Prorocentrum minimum	N	M	Palenik and Henson 1997
Nicotinamide	*Phaeodactylum tricornutum*	Y		Hayward 1965
	Cryptomonad	N		Antia and Chorney 1968
Succinamide	Chlamydomonads (8)	Y	FW	Cain 1965
Amines				
Aniline	Cyanobacterium	N		Kapp et al. 1975
Diethylamine	Cyanobacterium	N		Kapp et al. 1975
	Diatom	Y		Wheeler and Hellebust 1981
Diethanolamine	Cyanobacterium	Y		Kapp et al. 1975
Ethylamine	Diatom	Y		Wheeler and Hellebust 1981
Methylamine	Cyanobacterium	N		Kapp et al. 1975
	Cryptomonad	N		Antia and Chorney 1968
	Diatom	Y		Wheeler and Hellebust 1981
p-aminobenzoic acid	Cryptomonad	N		Antia and Chorney 1968
Putrescine	Chlorophytes (2)	Y		Neilson and Larsson 1980
	Cryptomonad	N		Antia and Chorney 1968
	Cyanobacterium	Y		Kapp et al. 1975

Contd.

Table 9.6 Contd.

Compound	*Organism*	*Use*	*FW/M*	*Reference*
	Cyanobacterium	Y		Kapp et al. 1975
	Eustigmatophyte	Y		Neilson and Larsson 1980
	Rhodophyte	Y		Neilson and Larsson 1980
Amino sugars				
Glucosamine	*Aphanizomenon ovalisporum*	Y[2]	FW	Berman and Chava 1999
	Cryptomonad (achitinous)	Y		Antia and Chorney 1968
	Cyclotella spp.	Y[2]	FW	Berman and Chava 1999
	Diatoms (2) (chiton-forming)	Y		McLachlan and Craigie 1966
	Diatom (achitinous)	N		Hayward 1965
	Microalgae (26)	Y	M	Berland et al. 1976
	Pediastrum spp.	Y[1]	FW	Berman and Chava 1999
Combined amino acids				
Glycine-glycine	Chlorophytes (2)	Y	FW	Neilson and Larsson 1980
	Cryptomonad	N	M	Antia and Chorney 1968 Vieira and Klaveness 1986
	Microalgae (5)	N	FW	Antia and Chorney 1968 Vieira and Klaveness 1986
	Prymnesiophytes (3)	Y	M	Turner 1979
Nucleic acids				
Salmon sperm DNA	Cyanobacterium	Y		Kapp et al. 1975
Pteridines				
Pterin-6-carboxylic acid	Cyanobacterium	N		Kapp et al. 1975
Pterin	Cryptomonad	Y		Landymore and Antia 1978
Purine-Adenine	*Chlamydomonas reinhardtii*	T	FW	Lisa et al. 1995
	Chlamydomonads (33 of 38)	Y	FW	Cain 1965
	Chlorella protothecoides Krueger	Y		Rose and Casselton 1983
	Chlorella pyrenoidosa	Y		Ammann and Lynch 1964
	Cryptomonad	Poor	M	Antia and Chorney 1968
	Diatom	N	M	Shah and Syrett 1982
	Haematococcus–related species (6)	Poor		Droop 1961
	Prototheca zopfii Krueger	Y		Stacey and Casselton 1966
	Tetraselmis striata Butcher	Y		Edge and Ricketts 1978
	Emiliania huxleyi	Y	M	Palenik and Henson 1997
	Chlorophyta	Y		Sarcina and Casselton 1995
Purine-Guanine	*Aphanizomenon ovalisporum*			
	Chlamydomonas reinhardtii	T	FW	Lisa et al. 1995
	Chlorophytes (6)	Y	M	Shah and Syrett 1984
	Cryptomonad	Y	M	Antia and Chorney 1968
	Cyclotella spp.	Y[2]	FW	Berman and Chava 1999
	Diatom	Y	M	Shah and Syrett 1982

Contd.

Table 9.6 Contd.

Compound	*Organism*	*Use*	*FW/M*	*Reference*
	Diatom (3)	Y	M	Shah and Syrett 1984
	Pediastrum spp.	Y[2]	FW	Berman and Chava 1999
	Prasinophyte	Y	M	Gooday 1970
	Prasinophyte	Y	M	Edge and Ricketts 1978
	Prasinophyte (1)	Y	M	Shah and Syrett 1984
	Red algae (2)	N	M	Turner 1970
				Shah and Syrett 1984
	Supralittoral			
	Haematococcus (6)	Y	M	Droop 1961
Purine-Uric acid	Centric diatoms	Fair-Good	M	Guillard 1963
	Chlamydomonads (16 of 38)	Y	FW	Cain 1965
	Chlorophytes (5 of 8)	Y	FW	Birdsey and Lynch 1962
	Chlorophytes	Y	M	Turner 1979
	Cryptomonad	Y	M	Antia and Chorney 1968
	Cyanobacterium	Y	M	Van Baalen and Marler 1963
	Cyanobacterium	N	FW	Van Baalen 1965
	Diatoms (8)	Y	M	Fisher and Cowdell 1982
	Haematococcus –	Y		Droop 1961
	related species (5 of 6)			
	Microalgae (15 of 27)	Y		Neilson and Larsson 1980
	Prasinomonad	Y		Douglas 1983
	Prasinophytes (10)	Y	M	Turner 1979
	Prymnesiophytes (3)	Y	M	Turner 1979
	Supralittoral protests (4 of 5)	Y	M	Droop 1955
	Symbiotic alga	Y	M	Edge and Ricketts 1978
				Douglas 1983
Purine-Hypoxanthine	*Aphanizomenon ovalisporum*	Y[2]	FW	Berman and Chava 1999
	Cyclotella spp.	Y[2]	FW	
	Pediastrum spp.	Y[2]	FW	
	Emiliania huxleyi	Y	M	Palenik and Henson 1997
Purine-Xanthine	Chlorophytes (5 of 8)	Y	FW	Birdsey and Lynch 1962
	Chrysomonad	Y	M	Mahoney and McLaughlin 1977
	Cryptomonad	Y	M	Antia and Chorney 1968
	Cyanobacterium	Y	M	Kapp et al. 1975
	Dinoflagellate	Y	M	Mahoney and McLaughlin 1977
Pyrimidines-Uracil	Chlamydomonads (38)	N	FW	Cain 1965
	Cryptomonad	N	M	Antia and Chorney 1968
	Cyanobacterium	Poor	M	Kapp et al. 1975
	Haematococcus	Poor		Droop 1961
	(related spp. 3 of 6)			
	Prasinophyte	N	M	Edge and Ricketts 1978
	Rhodophyte	N		Turner 1970

[1] Ureides, a urea derivative.

[2] Non-axenic monocultures, substrate additions were 100 μM.

and Kuntsen 1974); surprisingly concentrations of ammonium up to 1 mM did not affect the rate of guanine uptake (Pettersen 1975). As in other organics, nitrogen deprivation increases the use of guanine (Shah and Syrett 1982, 1984). Rates of uptake of methylamine, urea, guanine, lysine and arginine increases as the periods of nitrogen deprivation increased. Uric acid is another good nitrogen source for growth for a number of phytoplankton species (Guillard 1963, Birdsey and Lynch 1962, Droop 1955, 1961, Cain 1965, Antia and Chorney 1968, Turner 1979, Neilson and Larsson 1980, Fisher and Cowdell 1982, Douglas 1983). Other purines, including adenine, guanine, hypoxanthine, and xanthine, can also be used as a nitrogen source (Antia et al. 1975, Devi Prasad 1983, Shah and Syrett 1984, Kapp et al. 1975).

Another group of organic compounds, humic substances, have traditionally been considered unavailable for assimilation due to their HMW and structural complexity. More recent studies of humic substances, however, indicate that they are not as refractory as once thought (Moran and Hodson 1994, Amon and Benner 1994, Gardner et al. 1996, See 2003). Recent work using ^{15}N-labeled estuarine humic substances, formed in the laboratory, indicates that the ability to take up humic-nitrogen is widespread in cultures of coastal phytoplankton. See et al. (submitted) surveyed 17 phytoplankton species isolated from coastal waters and found that all phytoplankters tested can take up humic-nitrogen in short-term (3 h) incubations. Time-course experiments, however, indicate that uptake of humic-nitrogen is not sustained, implying that once the fraction of bioavailable nitrogen associated with the humics is exhausted uptake ceases (See 2003, See et al. submitted).

DON uptake in natural waters

Observations of high and relatively constant concentrations of DON in the ocean, where production was considered at the time to be primarily limited by the availability of DIN, combined with field studies that showed organic substrates, primarily ^{14}C-lableled, were taken up by the small bacterial size fraction led to the dogma that DON was largely refractory in nature and therefore unimportant to phytoplankton nutrition (McCarthy et al. 1975). As a result, bacteria have traditionally been thought to be the primary users of DON in natural systems, and that direct uptake of DON into phytoplankton, without bacterial or photochemical remineralization, has been considered to be relatively minor. Evidence began to accumulate during the 1990s, however, that some organic compounds (i.e. urea) can be important to phytoplankton nutrition (Glibert and Garside 1992), that fluxes into and out of the DON pool are substantial (Bronk and Glibert 1993a, Bronk et al. 1994, Slawyk and Raimbault 1995), and that the nitrogen fraction of organic substrates need not follow the path of the carbon component of a molecule into the cell (Palenik and Morel 1990a). In natural waters DIN is still accepted as the most important form of nitrogen supporting algal growth, but it is now recognized that phytoplankton are not simply a source of DON that subsequently supports bacterial activity, but are a sink for DON as well. As we briefly review studies of DON uptake in natural waters here it is important to remember two things. First, uptake measurements in the field likely include bacterial, as well as, phytoplankton uptake. Second, once an organic substrate is added to an incubation in the field, it can be altered in a number of ways including bacterial or photochemical breakdown or abiotic adsorption such that the nitrogen form actually taken up can be very different than the one added.

Field studies of DON uptake have focused on urea and amino acid uptake because there were commercially available tracers. Urea has been the one exception to the believed bacterial dominance of DON utilization. In general, phytoplankton are believed to be the primary users of urea in marine systems (Price and Harrison 1988a), though some recent studies have questioned this belief (Tamminen and Irmisch 1996). For example, in the Thames Estuary (UK), the addition of a broad prokaryotic inhibitor decreases dark uptake rates of urea by 86 ± 25% suggesting that bacterial uptake of urea is substantial (Middelburg and Niewenhuize 2000). Lomas et al. (2002) reviewed urea uptake rates for over a decade in Chesapeake Bay and found that urea is consistently an important nitrogen source for the plankton community. Additional field studies have shown urea to be an important nitrogen source in the open ocean (Price and Harrison 1988b), the coastal ocean (Probyn et al. 1990, Veuger et al. 2004), the polar ocean (Cochlan and Bronk 2001), and lakes (Gu and Alexander 1993); more detailed reviews of urea uptake studies can be found in Antia et al. (1991) and Bronk (2002).

Bacteria are generally considered the primary users of DCAA and DFAA although autotrophic uptake of DFAA have been demonstrated on numerous occasions (reviewed in Flynn and Butler 1986, and Antia et al. 1991). Studies of cell-surface enzymes, however, suggest that phytoplankton use of DFAA may be greater than previously thought (see above, Palenik et al. 1988). For example, in a salt marsh phytoplankton community, addition of organic nitrogen, including glycine, glutamic acid, and an amino acid mixture, results in increased phytoplankton growth (Lewitus et al. 2000). Furthermore, in the Thames Estuary (UK), a study using a prokaryotic inhibitor indicated that approximately 51% of amino acid uptake is by autotrophs (Middelburg and Niewenhuize 2000).

Additional studies that measure uptake of other organic nitrogen compounds such as purines, pyrimidines, and amines show that though phytoplankton and bacteria can utilize these compounds, the uptake rates are quite low (reviewed in Antia et al. 1991). It must be remembered, however, that there is no reason to expect the use of any one form of DON to match that of DIN but that the combined use of a range of different DON forms can be highly significant even when studies of the use of any one form indicates only modest uptake.

Experiments in which natural humic substances, isolated from river water, are added to an assemblage of coastal phytoplankton reveal that growth and biomass formation are stimulated (Carlsson et al. 1993), though the mechanisms of the enhancement are unknown. The literature suggests that the nitrogen associated with humic substances can be removed by one of three mechanisms: through microbial activity (Müller-Wegener 1988), via excision by phytoplankton cell-surface enzymes (Fig. 9.2, Palenik and Morel 1990a) or through photodegradation to LMW compounds by exposure to ultraviolet radiation (reviewed in Mopper and Kieber 2002, summarized in Fig. 9.1).

Determining uptake rates for the DON pool as a whole, however, is problematic because it is so chemically complex. Bronk and Glibert (1993a) used ^{15}N-labeled DON produced *in situ* in Chesapeake Bay and found that during the decline of the spring bloom and during the summer, uptake rates of DON are comparable or higher than uptake rates of ammonium and nitrate. Although rates of DON uptake in the spring had a diel pattern and followed that of nitrate, both indications that autotrophic uptake dominated, the fate of the DON is unknown. Veuger et al. (2004) used ^{15}N-lableled DON produced in a *Skeletonema* culture to compare uptake of inorganic versus organic nitrogen uptake in a Danish fjord. They found that the organic forms (urea and DFAA) are important sources of nitrogen for the community and that DON other than urea and DFAA contributed to measured uptake.

The special case of blooms

Much of the recent work on DON utilization has been tied to the potential of DON as a nitrogen source to phytoplankton that can form harmful algal blooms (HABs, Paerl 1988). There is increasing evidence linking DON additions with the increase in HABs (Berg et al. 1997, Berman 1997, 2001, Carlsson et al. 1998). Dinoflagellates, in particular, are known to use organic nutrients, such as urea and DFAAs, both directly (Butler et al. 1979, Berg et al. 1997) and indirectly, via cell-surface enzymes (Palenik and Morel 1990a and b, Pantoja and Lee 1994, Mulholland et al. 1998). DON has been implicated in initiating the *A. anophagefferens* (brown tide) blooms in Long Island (NY, USA, Dzurica et al. 1989, Berg et al. 1997, LaRoche et al. 1997) with particular emphasis on urea utilization (Dzurica et al. 1989, Berg et al. 1997). Berg et al. (1997) found that 70% of the total nitrogen utilized during an *Aureococcus anophagefferens* bloom off Long Island is organic with the largest fraction contributed by urea. Later studies indicate that the ability to use HMW DON may give *Aureococcus anophagefferens* a competitive advantage over co-occurring species (Berg et al. 2003). High urea levels (greater than 1.5 μM) were also found to co-occur with dinoflagellate blooms in aquaculture ponds (Glibert and Terlizzi 1999). Typical of many dinoflagellates, *Karenia brevis* (formerly *Gymnodinium breve*), can take up a variety of organic nitrogen compounds (including amino acids) as nitrogen sources for growth (Steidinger et al. 1998; Bronk et al. in press). In cultures of *K.* Appl. Env. Microbiol., cell yields increase dramatically when glycine, leucine and aspartic acid are added (Shimizu et al. 1995).

What is not clear in most of these studies is whether the association between HABs and DON is direct or indirect. The formation of phytoplankton blooms is often associated with increases in DON concentrations. The decline of blooms are also known to be periods of significant DOM release from phytoplankton due to such processes as physiological stress or high grazing pressure (Carlson et al. 1994, Jenkinson and Biddanda 1995). DON accumulates during the year in temperate waters (Flynn and Butler 1986, Bronk et al. 1998) and HABs also become more frequent in summer months. Most HAB species are not in axenic culture, however, raising the possibility that any stimulation by the presence of DON is indirect because it may promote bacterial and other heterotrophic activities.

Problems Extrapolating Culture Results to the Field

With the wealth of culture data available on algal uptake of DON, can we answer the question – Do phytoplankton in natural waters fulfill a portion of their nitrogen needs by taking up dissolved organic compounds? The answer is a definitive – probably! Though uptake and release of a wide number of organic compounds has been documented in culture, it is much more difficult to address whether a given species actually uses that substrate in the environment. The answer to that question depends on many factors including the concentration of substrate in the environment, alternative sources of nitrogen that are available, and competition for the substrate by other phytoplankton species or bacteria. Here we review a number of reasons why it is difficult to extrapolate culture results to the field.

First, batch cultures generally are started with nutrient concentrations many times higher than those found in nature; for example, f/2 media has a starting nitrate concentration of 883 μM (Guillard 1975). While such studies indicate the potential for growth, the high concentrations used severely limit the ability to extrapolate results to natural waters. Second, cells in batch culture will eventually reach a state of extreme nutrient starvation, which likely has no counterpart in nature where regeneration possesses are ongoing. This is particularly problematic for extrapolating results from DON release studies to the field. Third, in batch cultures cells move through the phases of growth very quickly (h) relative to natural waters (d), which allows the cells little or no time to adapt to their changing environment. On the other hand, chemostats have potential for selecting for long-term adaptations that may not be representative of nature either.

Fourth, the practice of using non-axenic cultures or measuring uptake in the field using the GF/F filters (nominal pore size 0.7 μm) that are compatible with mass spectrometry results in both bacteria and phytoplankton being collected and analyzed. This practice severely limits our ability to understand algal nitrogen nutrition because results are confounded by possible bacterial uptake.

Fifth, there is the question of competition with bacteria for substrates. For algae to successfully compete with bacteria for an organic substrate, they must have transport systems with kinetic parameters that can function at the low concentrations of organic substrates found in the environment. Though kinetic parameters of a given species can be measured in the field under extreme bloom conditions, where a given species overwhelmingly dominates the biomass, most kinetic work must be done in cultures. One problem with measuring kinetic parameters in cultures is that the cultures are generally maintained at concentrations many times the concentrations seen in the field. The question is whether cells that have been grown under saturating nutrient conditions for many generations (often for years, even decades) will show the same uptake kinetic characteristics of the cells in the field where substrate concentrations are very low (Paul 1983).

Sixth, these are a number of processes that occur in natural waters that can alter organic substrates once they are added to an incubation (reviewed in Bronk 2002). Organic compounds can be broken down photochemically or through bacterial decomposition resulting in the release and subsequent uptake of ammonium or some other smaller labile moiety. Organics can also adsorb abiotically to particles such that they appear to be taken up.

Lastly, there is the issue of multiple nitrogen substrates. If a cell will use a given organic substrate as a sole nitrogen sources it does not necessarily mean it will use that substrate in the environment when other substrates are available. Furthermore, the inability of a cell to grow on a given substrate as its sole nitrogen source does not mean it does not use that substrate in the field. For example, glutamine and arginine are more readily catabolized than histidine and yet the DON source may still be important as an augment to total nitrogen nutrition (Antia et al. 1991).

SUGGESTIONS FOR FUTURE RESEARCH

Future research will in large measure be dictated, as it has in the past, by the development of new analytical techniques and approaches. There are many common practices that limit our ability to translate culture results to the field, as described above. Refining techniques and modifying protocols to circumvent these problems has the potential to dramatically advance our understanding of DON cycling in the environment. Some specific suggestions include curtailing the practice of conducting bioavailability studies with mM substrate additions if the goal of the study is to relate the findings to the environment. Expanded use of chemostats would allow cultures to be maintained at much lower nutrient and biomass concentrations thus making their results more applicable to natural waters. Greater attention to axenicity in cultures and more widespread use of techniques for separating bacteria from phytoplankton, such as flow cytometric sorting (Lipschultz 1995) has great potential to improve our understanding of algal nitrogen nutrition, particularly with respect to organic nitrogen forms that are highly desirable substrates for bacteria. Targeted studies on the affect of alternative nitrogen sources on DON uptake would be beneficial for deciphering the complex issue of substrate selection in the complex mix of organic nitrogen forms found in aquatic environments. Lastly, expanding our view from cultures and field incubation studies to define the role of the DON pool in the global carbon cycle remains one of the most challenging and important aspects of aquatic ecosystem research today. Progress in this area will require more complex mathematical models for the proper description of DON fluxes within ecosystem simulators. Currently these models typically involve a one-way flow of nitrogen from algae to bacteria, which is almost certainly incorrect. Even the most complex models of phytoplankton used in such simulators currently ignore a role for DON (Flynn, this volume). Culture work can have a pivotal role in improving models by providing important baseline controls for a range of variables. The challenge will be to design culture studies of algal growth using environmentally relevant species under quasi-natural conditions.

ACKNOWLEDGMENTS

The writing of this chapter was supported by NSF grant OCE-02218252 to DAB. This is contribution number 2677 of the Virginia Institute of Marine Science, The College of William and Mary.

REFERENCES

Aiken, G.R. 1988. A critical evaluation of the use of macroporous resins for the isolation of aquatic humic substances, In FH Frimmel, and RF Christman [eds.], Humic substances and than Role in the Environment. John Wiley and Sons, New York, USA.

Allen, M.B. 1956. Excretion of organic compounds by *Chlamydomonas*. Archiv. fur. Mikrobiologie. 24: 163-168.

Ammann, E.C. and V.H. Lynch. 1964. Purine metabolism by unicellular algae. II. Adenine, hypoxanthine and xanthine degradation by *Chlorella pyrenoidosa*. Biochimica et Biophysica Acta. 87: 370-379.

Amon, R.M.W. and R. Benner. 1994. Rapid cycling of high-molecular-weight dissolved organic matter in the ocean. Nature. 369: 549-551.

Antia, A.N.J., B.R. Berland, D.J. Bonin and S.Y. Maestrini. 1975. Comparative evaluation of certain organic and inorganic sources of nitrogen for phototrophic growth of marine microalgae. J. Exp. Mar. Biol. Ass. U.K. 55: 519-539.

Antia, J.N. and V. Chorney. 1968. Nature of the nitrogen compounds supporting phototrophic growth of the marine cryptomonad *Hemiselemis virescens*. J. Protozool. 15: 198-201.

Antia, N.J., B.R. Berland, D.J. Bonin and S.Y. Maestrini. 1980. Allantoin as nitrogen source for growth of marine benthic microalgae. Phycologia. 19: 103-109.

Antia, N.J., P.J. Harrison and L. Oliveira. 1991. Phycological Reviews: The role of dissolved organic nitrogen in phytoplankton nutrition, cell biology, and ecology. Phycologia. 30: 1-89.

Armstrong, F.A.J., P.M. Williams and J.D.H. Strickland. 1966. Photo-oxidation of organic matter in sea water by ultra-violet radiation, analytical and other applications. Nature. 211: 481-483.

Axler, R.P. and J.E. Reuter. 1986. A simple method for estimating the^{15}N content of DOM (DO^{15}N) in N cycling studies. Can. J. Fish. Aquat. Sci. 43: 130-133.

Baden, D.G. and T.J. Mende. 1979. Amino acid utilization by *Gymnodinium breve*. Phytochemistry. 18: 247-251.

Bekeheet, I.A. and P.J. Syrett. 1979. The uptake of urea by *Chlorella*. New Phytol. 82: 179-186.

Benner, R. 2002. Chemical composition and reactivity. 59-90. *In* D.A. Hansell, and C.A. Carlson [eds.]. Biogeochemistry of Marine Dissolved Organic Matter. Academic Press, San Diego, USA.

Berg, G.M., P.M. Glibert, M.W. Lomas and M.A. Burford. 1997. Organic nitrogen uptake and growth by the chrysophyte *Aureococcus anophagefferens* during a brown tide event. Mar. Biol. 129: 377-387.

Berg, G.M., D.J. Repeta and J. LaRoche. 2002. Dissolved organic nitrogen hydrolysis rates in axenic cultures of *Aureococcus anophagefferens* (Pelagophyceae): Comparison with heterotrophic bacteria. Appl. Env. Microb. 68: 401-404.

Berg, G.M., D.J. Repeta and J. LaRoche. 2003. The role of the picoeukaryote *Aureococcus anophagefferens* in cycling of marine high-molecular weight dissolved organic nitrogen. Limnol. Oceanog. 48: 1825-1830.

Berges, J.A. and P.G. Falkowski. 1996. Cell-associated proteolytic enzymes from marine phytoplankton. J. Phycol. 32: 566-574.

Berges, J.A., D.J. Franklin and P.J. Harrison. 2001. Evolution of an artificial seawater medium: improvements in enriched seawater, artificial water over the last two decades. J. Phycol. 37: 1138-1145.

Berland, B.R., D.J. Bonin, S.Y. Maestrini, M.L. Lizaraga-Partida and N.J. Antia. 1976. The nitrogen concentration requirement of glucosamine for supporting effective growth of marine microalgae. J. Mar. Biol. Assoc. U.K. 56: 629-637.

Berman, T. 1997. Dissolved organic nitrogen utilization by an *Aphanizomenon* bloom in Lake Kinneret. J Plankton Res. 19: 577-586.

Berman, T. 2001. The role of DON and the effect of N:P ratios on occurrence of cyanobacterial blooms: Implications from the outgrowth of *Aphanizomenon* in Lake Kinneret. Limnol. Oceanogr. 46: 443-447.

Berman, T. and S. Chava. 1999. Algal growth on organic compounds as nitrogen sources. J. Plankton Res. 21: 1423-1437.

Berman, T. and D.A. Bronk. 2003. Dissolved Organic Nitrogen: a dynamic participant in aquatic ecosystems. Aquatic Microb. Ecol. 31: 279-305.

Biddanada, B. and R. Benner. 1997. Carbon, nitrogen and carbohydrate fluxes during the production of particulate and dissolved organic matter by marine plankton. Limnol. Oceanogr. 42: 506-518.

Bidigare, R.B. 1983. Nitrogen excretion by marine zooplankton. pp. 385-410. *In* E.J. Carpenter, and D.G. Capone [eds.]. Nitrogen in the Marine Environment. Plenum Press, New York, USA.

Bilboa, M.M., J.M. Gabas and J.L. Serra. 1981. Inhibition of nitrite uptake by the diatom *Phaeodactylum tricornutum* by nitrate, ammonium, and some L-amino acids. Biochem. Soc. Trans. 9: 476-477.

Birdsey, E.C. and V.H. Lynch. 1962. Utilization of nitrogen compounds by unicellular algae. Science. 137: 763-764.

Bratbak, G., M. Heldal, S. Norland and T.F. Thingstad. 1990. Viruses as partners in spring bloom microbial trophodynamics. Appl. Env. Microbiol. 56: 1400-1405.

Bratbak, G., J.K. Egge and M. Heldal. 1993. Viral mortality of the marine alga *Emiliania huxleyi* (Haptophyceae) and termination of algal blooms. Mar. Ecol. Prog. Ser. 93: 39-48.

Bronk, D.A. 1999. Rates of NH_4^+ uptake, intracellular transformation, and dissolved organic nitrogen release in two clones of marine *Synechococcus* spp. J. Plankton Res. 21: 1337-1353.

Bronk, D.A. 2002. Dynamics of DON. pp. 153-249. *In* D.A. Hansell, and C.A. Carlson. [eds.]. Biogeochemistry of Marine Dissolved Organic Matter. Academic Press, San Diego, USA.

Bronk, D.A. and P.M. Glibert. 1991. A ^{15}N tracer method for the measurement of dissolved organic nitrogen release by phytoplankton. Mar. Ecol. Prog. Ser. 77: 171-182.

Bronk, D.A. and P.M. Glibert. 1993a. Application of a ^{15}N tracer method to the study of dissolved organic nitrogen uptake during spring and summer in Chesapeake Bay. Mar. Biol. 115: 501-508.

Bronk, D.A. and P.M. Glibert. 1993b. Contrasting patterns of dissolved organic nitrogen release by two size fractions of estuarine plankton during a period of rapid NH_4^+ consumption and NO_2^- production. Mar. Ecol. Prog. Ser. 96: 291-299.

Bronk, D.A. and B.B. Ward. 1999. Gross and net nitrogen uptake and DON release in the euphotic zone of Monterey Bay, California. Limnol. Oceanogr. 44: 573-585.

Bronk, D.A., P.M. Glibert and B.B. Ward. 1994. Nitrogen uptake, dissolved organic nitrogen release, and new production. Science. 265: 1843-1846.

Bronk, D.A., P.M. Glibert, T.C. Malone, S. Banahan and E. Sahlsten. 1998. Inorganic and organic nitrogen cycling in Chesapeake Bay: autotrophic versus heterotrophic processes and relationships to carbon flux. Aquat. Microb. Ecol. 15: 177-189.

Bronk, D.A., M. Lomas, P.M. Glibert, K.J. Schukert and M.P. Sanderson. 2000. Total dissolved nitrogen analysis: comparisons between the persulfate, UV and high temperature oxidation method. Mar. Chem. 69: 163-178.

Bronk, D.A., M.P. Sanderson, M.R. Mulholland, C.A. Heil and J.M. O'Neil. Organic and inorganic nitrogen uptake kinetics in field populations dominated by *Karenia brevis*. *In* Steidinger, K., G.A. Vargo and C.A. Heil. [eds]. The Proceedings of the 10th Conference on Harmful Algae, St. Petersburg, FL. (in press).

Brussaard, C.P.D., R.S. Kempers, A.J. Kop, R. Riegman and M. Heldal. 1996. Virus-like particles in a summer bloom of *Emiliania huxleyi* in the North Sea. Aquatic Microbial Ecology. 10: 105-113.

Butler, E.I., S. Knox and M.I. Liddicoat. 1979. The relationship between inorganic and organic nutrients in seawater. J. Mar. Biol. Ass. U.K. 59: 239-250.

Cain, B.J. 1965. Nitrogen utilization in 38 freshwater chlamydomonad algae. Can. J. Bot. 43: 1367-1378.

Carlson, C.A. 2002. Production and removal processes. pp. 91-151. *In* D.A. Hansell, and C.A. Carlson. [eds.]. Biogeochemistry of Marine Dissolved Organic Matter. Academic Press, Amsterdam, the Netherlands

Carlson, C.A., H.W. Ducklow and A.F. Michaels. 1994. Annual flux of dissolved organic carbon from the euphotic zone in the northwestern Sargasso Sea. Nature. 371: 405-408.

Carlsson, P., A.Z. Segatto and E. Granéli. 1993. Nitrogen bound to humic matter of terrestrial origin - a nitrogen pool for coastal phytoplankton? Mar. Ecol. Prog. Ser. 97: 105-116.

Carlsson, P., E. Granéli, P. Tester and L. Boni. 1995. Influences of riverine humic substances on bacteria, protozoa, phytoplankton, and copepods in a coastal plankton community. Mar. Ecol. Prog. Ser. 127: 213-221.

Carlsson, P., H. Edling and C. Bechemin. 1998. Interactions between a marine dinoflagellate (*Alexandrium catenella*) and a bacterial community utilizing riverine humic substances. Aquat. Microb. Ecol. 16: 65-80.

Chapman, J.S. and J.C. Meeks. 1983. Glutamine and glutamate transport by *Anabaena variabilis*. J. Bacteriol. 156: 122-129.

Cochlan, W.P. and P.J. Harrison. 1991. Kinetics of nitrogen (nitrate, ammonium, and urea) uptake by the picoflagellate *Micromonas pusilla* (Prasinophyceae). J. Exp. Mar. Biol. Ecol. 153: 129-141.

Cochlan, W.P. and D.A. Bronk. 2001. Nitrogen uptake kinetics in the Ross Sea, Antarctica. Deep-Sea Res. II. 48: 4127-4153.

Collier, J.L., B. Brahamsla and B. Palenik. 1999. The marine cyanobacterium, *Synechococcus sp.* WH7805, requires urease to utilize urea as a nitrogen source: molecular genetic and biochemical analysis of the enzyme. Microbiol. 145: 447-454.

Collos, Y. and G. Slawyk. 1986. ^{13}C and ^{15}N uptake by marine phytoplankton - IV. Uptake ratios and the contribution of nitrate to the productivity of Antarctic waters (Indian Ocean sector). Deep-Sea Res. 33: 1039-1051.

Collos, Y., C. Gagne, M. Laabir, A. Vaquer, P. Cecchi and P. Souchu. 2004. Nitrogenous nutrition of *Alexandrium catenella* (Dinophyceae) in cultures and in Thau Lagoon, Southern France. J. Phycol. 40: 96-103.

Cottrell, M. and C. Suttle. 1993. Production of axenic cultures of *Micromonas pusilla* (Prasinophyceae) using antibiotics. J. Phycol. 29: 385-387.

Cozzi, S. 2004. A new application of the diacetyl monoxime method to the automated determination of dissolved urea in seawater. Mar. Biol. 145: 843-848.

Devi Prasad, P.V. 1983. Hypoxanthine and allantoin as nitrogen sources for the growth of some freshwater green algae. New Phytol. 93: 575-580.

Douglas, A.E. 1983. Uric acid utilization in *Platymonas convolutae* and symbiotic *Convoluta roscoffensis*. J. Mar. Biol. Assoc. U.K. 63: 435-447.

Droop, M.R. 1955. Some new supra-littoral protista. J. Mar. Biol. Assoc. U.K. 34: 233-245.

Droop, M.R. 1961. *Haematococcus pluvialis* and its allies. III. Organic nutrition. Revue Algologique. 4: 247-259.

Dugdale, R.C. and J.J. Goering. 1967. Uptake of new and regenerated forms of nitrogen in primary productivity. Limnol. Oceanogr. 12: 196-206.

Dzurica, S., C. Lee and E.M. Cosper. 1989. Role of environmental variables, specifically organic compounds and micronutrients, in the growth of the chrysophyte *Aureococcus anophagefferens*. pp. 229-252. *In* E.M., Cosper, V.M. Bricelj and E.J. Carpenter. [eds.]. Novel Phytoplankton Blooms. Spriner-Verlag, Berlin, Germany.

Edge, P.A. and T.R. Ricketts. 1978. Some notes on the growth and nutrition of *Platymonas striata* Butcher (Prasinophyceae). Nova Hedwigia. 29: 676-682.

Fisher, N.S. and R.A. Cowdell. 1982. Growth of marine planktonic diatoms on inorganic and organic nitrogen. Mar. Biol. 72: 147-155.

Flynn, K.J. 1990. Composition of intracellular and extracellular pools of amino acids, and amino acid utilization of microalgae of different sizes. J. Exp. Mar. Biol. Ecol. 139: 151-166.

Flynn, K.J. 1991. Algal carbon-nitrogen metabolism: a biochemical basis for modeling the interactions between nitrate and ammonium uptake. J. Plankton Res. 13: 373-387.

Flynn, K.J. and P.J. Syrett. 1985. Development of the ability to take up L-lysine by the diatom *Phaeodactylum tricornutum*. Mar. Biol. 89: 317-127.

Flynn, K.J. and I. Butler. 1986. Nitrogen sources for the growth of microalgae: role of dissolved free amino acids. Mar. Ecol. Prog. Ser. 34: 281-304.

Flynn, K.J. and P.J. Syrett. 1986. Utilization of L-lysine and L-arginine by the diatom *Phaeodactylum tricornutum*. Mar. Biol. 90: 159-163.

Flynn, K.J. and C.R.N. Wright. 1986. The simultaneous assimilation of ammonium and L-arginine by the marine diatom *Phaeodactylum tricornutum* Bohlin. J. Exp. Mar. Biol. Ecol. 95: 257-269.

Flynn, K.J. and J. Fielder. 1989. Changes in intracellular and extracellular amino acids during the predation of the chlorophyte *Dunaliella primolecta* by the heterotrophic dinoflagellate *Oxyrrhis marina* and the use of the glutamine/glutamate ratio as an indicator of nutrient status in mixed populations. Mar. Ecol. Prog. Ser. 53: 117-127.

Fogg, G.E. 1952. The production to extracellular nitrogenous substances by blue-green alga. Proc. R. Soc. Lond. Ser. 139: 373-397.

Fogg, G.E. 1966. The extracellular products of algae. Oceanogr. Mar. Biol. Ann. Rev. 4: 195-212.

Fogg, G.E. 1977. Excretion of organic matter by phytoplankton. Limnol. Oceanog. 22: 576-577.

Fogg, G.E. and B. Thake. 1987. Algal cultures and phytoplankton ecology, University of Wisconsin Press, Madison, WI, USA.

Fuhrman, J. 1987. Close coupling between release and uptake of dissolved free amino acids in seawater studied by an isotope dilution approach. Mar. Ecol. Prog. Ser. 37: 45-52.

Fuhrman, J. 1992. Bacterioplankton roles in cycling of organic matter: The microbial loop, Vol. Plenum Press, New York, USA.

Fuhrman, J.A. and T.M. Bell. 1985. Biological considerations in the measurement of dissolved free amino acids in seawater and implications for chemical and microbiological studies. Mar. Ecol. Prog. Ser. 25: 13-21.

Gardner, W., R. Benner, R. Amon, J. Cotner, J. Caveletto and J. Johnson. 1996. Effects of high molecular weight dissolved organic matter on the nitrogen dynamics on the Mississippi River plume. Mar. Ecol. Prog. Ser. 133: 287-297.

Gibb, S.W., R.F.C. Mantoura, P.S. Liss and R.G. Barlow. 1999. Distributions and biogeochemistries of methylamines and ammonium in the Arabian Sea. Deep-Sea Res. II. 46: 593-615.

Glibert, P.M. and C. Garside. 1992. Diel variability in nitrogenous nutrient uptake by phytoplankton in the Chesapeake Bay. J. Plankton Res. 14: 271-288.

Glibert, P.M. and D.E. Terlizzi. 1999. Cooccurrence of elevated urea levels and dinoflagellate blooms in temperate estuarine aquaculture ponds. Appl. Environ. Microbiol. 65: 5594-5596.

Glibert, P.M., F. Lipschultz, J.J. McCarthy and M.A. Altabet. 1982. Isotope dilution models of uptake and remineralization of ammonium by marine plankton. Limnol. Oceanogr. 27: 639-650.

Gobler, C.J., D.A. Hutchins, N.S. Fisher, E.M. Cosper and S.A. Sañudo-Wilhelmy. 1997. Release and bioavailability of C, N, P, Se, and Fe following viral lysis of a marine chrysophyte. Limnol. Oceanogr. 42: 1492-1504.

Gooday, G.W. 1970. A physiological comparison of the symbiotic alga *Platymonas convolute* and its free-living relatives. J. Mar. Biol. Assoc. U.K. 50: 199-208.

Gu, B. and V. Alexander. 1993. Dissolved nitrogen uptake by a cyanobacterial bloom (*Anabaena flos-aquae*) in a subarctic lake. Appl. Env. Microbiol. 59(2): 422-430.

Guillard, R.R.L. 1963. Organic sources of nitrogen for marine centric diatoms. pp. 93-104. *In:* C.H. Oppenheimer, [ed]. Symposium on Marine Microbiology. Thomas, Springfield, IL. USA.

Guillard, R.R.L. 1975. Culture of phytoplankton for feeding marine invertebrates. pp. 20-60. *In* W.L. Smith, and M.H. Chanley [eds.]. Culture of Marine Invertebrate Animals. Plenum Press, New York, USA.

Guillard, R.R.L. and J.J. Ryther. 1962. Studies on marine planktonic diatoms, I. *Cyclotella nana* Hustedt and *Detonula confervacae* (Cleve) Gran. Can. J. Microbiol. 8: 229-239.

Hansell, D.A. and J.J. Goering. 1989. A method for estimating uptake and production rates for urea in seawater using ^{14}C urea and ^{15}N urea. Can. J. Fish. Aquat. Sci. 46: 198-202.

Hansell, D.A., P.M. Williams and B.B. Ward. 1993. Measurements of DOC and DON in the Southern California Bight using oxidation by high temperature combustion. Deep-Sea Res. 40: 219-234.

Harrison, P.J., R.E. Waters and F.J.R. Taylor. 1980. A broad spectrum artificial seawater medium for coastal and open ocean phytoplankton. J. Phycol. 16: 28-35.

Hasegawa, T., I. Koike and H. Mukai. 2000. Dissolved organic nitrogen dynamics in coastal waters and the effect of copepods. J. Exp. Mar. Biol. Ecol. 244: 219-238.

Hayward, J. 1965. Studies on the growth of *Phaeodactylum tricornutum* (Bohlin). I. the effect of certain organic nitrogenous substances on growth. Physioligica Plantarum. 18: 201-207.

Healey, F.P. 1977. Ammonium and urea uptake by some freshwater algae. Can. J. Bot. 55: 61-69.

Hellebust, J.A. 1974. Extracellular products. 838-863. *In* W.P.D. Stewart, [ed.]. Algal physiology and biochemistry. University of California Press, Berkeley, USA.

Hessen, D.O. and L.J. Tranvik. 1998. Aquatic Humic Substance. Ecology and Biogeochemistry, Vol. Springer, Berlin, Germany.

Hodson, R.C., S.K. Williams and W.J. Davidson. 1975. Metabolic control of urea catabolism in *Chlamydomonas reinhardtii* and *Chlorella pyrenoidosa*. J. Bacteriol. 121: 143-158.

Hoffs, F.H. 1999. Plankton Culture Manual, Vol. Florida Aqua Farms Inc.

Horrigan, S.G. and J.J. McCarthy. 1981. Urea uptake by phytoplankton at various stages of nutrient depletion. J. Plankton Res. 3: 403-413.

Horrigan, S.G. and J.J. McCarthy. 1982. Phytoplankton uptake of ammonium and urea during growth of oxidised forms of nitrogen. J. Plankton Res. 4: 379-389.

Ietswaart, T., P.J. Schneider and R.A. Prins. 1994. Utilization of organic nitrogen sources by two phytoplankton species and a bacterial isolate in pure and mixed cultures. Appl. Env. Microbiol. 60: 1554-1560.

Jenkinson, I.R. and B.A. Biddanda. 1995. Bulk-phase viscoelastic properties of seawater: relationship with plankton components. J. Plankton Res. 17: 2251-2274.

John, E.H. and K.J. Flynn. 1999. Amino acid uptake by the toxic dinoflagellae *Alexandrium fundyense*. Mar. Biol. 133: 11-19.

Jones, K. and W.D.P. Stewart. 1969. Nitrogen turnover in marine and brackish habitats. III. The production of extracellular nitrogen by *Calothrix scopulorum*. J. Mar. Biol. Ass. U.K. 49: 475-488.

Jumars, P.A., D.L. Penry, J.A. Baross, M.J. Perry and B.W. Frost. 1989. Closing the microbial loop: dissolved carbon pathway to heterotrophic bacteria from incomplete ingestion, digestion, and absorption in animals. Deep-Sea Res. 36: 483-495.

Kapp, R., S.E. Stevens and J.L. Fox. 1975. A survey of available nitrogen sources for the growth of blue-green alga, *Agmenellum quadruplicatum*. Arch. Microbiol. 104: 135-138.

Karl, D.M. and M.D. Bailiff. 1989. The measurement and distribution of dissolved nucleic acids in aquatic environments. Limnol. Oceanogr. 34: 543-558.

Keil, R.G. and D.L. Kirchman. 1991. Dissolved combined amino acids in marine waters as determined by a vapor-phase hydrolysis method. Mar. Chem. 33: 243-259.

Kirk, D.L. and M.M. Kirk. 1978a. Carrier-mediated uptake of arginine and urea by *Volvox carterif. nagariensis*. Plant Physiol. 61: 549-555.

Kirk, D.L. and M.M. Kirk. 1978b. Carrier-mediated uptake of arginine and urea by *Chlamydomonas reinhardtii*. Plant Physiol. 61: 556-560.

Kirk, D.L. and M.M. Kirk. 1978c. Amino acids and urea uptake in ten species of chlorophyta. J. Phycology. 14: 198-203.

Lampitt, R.S., T. Noji and B. von Bodungen. 1990. What happens to zooplankton faecal pellets? Implications for material flux. Mar. Biol. 104: 15-23.

Landymore, A.F. and N.J. Antia. 1978. White-light-promoted degradation of leucopterin and related pteridines dissolved in seawater, with evidence for involvement of complexation from major divalent cations of seawater. Mar. Chem. 6: 309-325.

Langheinrich, U. 1995. Plasma membrane-associated aminopeptidase activities of *Chlamydomonas reinhardtii* and their biochemical characterization. Biochimica et Biophysica Acta. 12491: 45-57.

LaRoche, J., R. Nuzzi, R. Waters, K. Wyman, P.G. Falkowski and D.W.R. Wallace. 1997. Brown Tide blooms in Long Island's coastal waters linked to interannual variability in groundwater flow. Global Change Biol. 3: 397-410.

Lee-Kaden, J. and W. Simonis. 1982. Amino acid uptake and energy coupling dependent on photosynthesis in *Anacystis nidulans*. J. Bacteriol. 151: 229-236.

Legrand, C. and P. Carlsson. 1998. Uptake of high molecular weight dextran by the dinoflagellate *Alexandrium catenella*. Aquat. Microb. Ecol. 16: 81-86.

Lewin, J. and J.A. Hellebust. 1975. Heterotrophic nutrition of the marine pennate diatom *Navicula pavillardi* Hustedt. Can. J. Microbiol. 21: 1335-1342.

Lewin, J. and J.A. Hellebust. 1976. Heterotrophic nutrition of the marine pennate diatom *Nitzschia angularis* var. *affinis*. Mar. Biol. 36: 313-320.

Lewin, J. and J.A. Hellebust. 1978. Utilization of glutamate and glucose for heterotrophic growth by the marine pennate diatom *Nitzschia laevis*. Mar. Biol. 47: 1-7.

Lewitus, A.J., E.T. Koepfler and R.J. Pigg. 2000. Use of dissolved organic nitrogen by a salt marsh phytoplankton bloom community. Arch. Hydrobiol. Spec. Issues Adv. Limnol. 55: 441-456.

Lipschultz, F. 1995. Nitrogen-specific uptake rates of marine phytoplankton isolated from natural populations of particles by flow cytometry. Mar. Ecol. Prog. Ser. 123: 245-258.

Lisa, T., P. Piedras, J. Cardenas and M. Pineda. 1995. Utilization of adenine and guanine as nitrogen sources by *Chlamydomonas reinhardtii* cells. Plant Cell Env. 18: 583-588.

Liu, M.S. and J.A. Hellebust. 1974a. Utilization of amino acids as nitrogen sources, and their effects on nitrate reductase in the marine diatom *Cyclotella cryptica*. Canadian J. Microbiol. 20: 1119-1125.

Liu, M.S. and J.A. Hellebust. 1974b. Uptake of amino acids by the marine centric diatom *Cyclotella cryptica*. Canadian J. Microbiol. 20: 1109-1118.

Lomas, M.W., T.M. Trice, P.M. Glibert, D.A. Bronk and J.J. McCarthy. 2002. Temporal and spatial dynamics of urea concentrations in Chesapeake Bay: Biological versus physical forcing. Estuaries. 25: 469-482.

Lu, M. and G.C. Stephens. 1984. Demonstration of net influx of free amino acids in *Phaeodactylum tricornutum* using high performance liquid chromatography. J. Phycol. 20: 584-589.

Lui, N.S.T. and O.A. Roels. 1970. Nitrogen metabolism of aquatic organisms. I. The assimilation and formation of urea in *Ochromonas malhamensis*. Arch. Biochem. Biophysics. 139: 269-277.

Lund, B.A. 1987. Mutual interference of ammonium, nitrate, and urea on uptake of ^{15}N sources by the marine diatom *Skeletonema costatum* (Grev.) Cleve. J. Exp. Mar. Biol. Ecol. 113: 167-180.

Mahoney, J.B. and J.J.A. MacLaughlin. 1977. The association of phytoflagellate blooms in lower New York bay with hypertrophication. J. exp. mar. Biol. Ecol. 28: 53-65.

McCarthy, J.J. 1970. A urease method for urea in seawater. Limnol. Oceanogr. 15: 309-313.

McCarthy, J.J. 1972. The uptake of urea by natural populations of marine phytoplankton. Limnol. Oceanogr. 17: 738-748.

McCarthy, J.J. and R.W. Eppley. 1972. A comparison of chemical, isotopic, and methods for measuring assimilation of marine phytoplankton. Limnol. Oceanogr.17: 371-382.

McCarthy, J.J., W.R. Taylor and J.L. Taft. 1975. The dynamics of nitrogen and phosphorus cycling in the open waters of the Chesapeake Bay. pp. 664-681. *In* T.M. Church, [ed.]. Marine Chemistry in the Coastal Environment, Vol. 18. ACS Symposium Series, Washington, DC, USA.

McLachlan, J. and J.S. Craigie. 1966. Chitan bifres in *Cyclotella cryptica* and the growth of *C. cryptica* and *Thalassiosira fluviatilis*. pp. 511-517. *In* H. Barnes, (ed). Some Contemporary Studies in Marine Science. Allen and Unwin, London, UK.

Menzel, D.W. and R.F. Vaccaro. 1964. The measurement of dissolved organic and particulate carbon in seawater. Limnol. Oceanogr. 9: 138-142.

Middelburg, J.J. and J. Nieuwenhuize. 2000. Nitrogen uptake by heterotrophic bacteria and phytoplankton in the nitrate-rich Thames estuary. Mar. Ecol. Prog. Ser. 203: 13-21.

Milligan, K.L.D. and E.M. Cosper. 1994. Isolation of virus capable of lysing the brown tide microalga, *Aureococcus anophagefferens*. Science Wash. 266: 805-807.

Molloy, C.J. 1987. Interactions in the assimilation of nitrogen compounds by unicellular algae. University College, Swansea, Wales.

Molloy, C.J. and P.J. Syrett. 1988a. Interrelationships between uptake of urea and uptake of ammonium by microalgae. J. exp. mar. Biol. and Ecol. 118: 85-95.

Molloy, C.J. and P.J. Syrett. 1988b. Effect of light on N deprivation on inhibition of nitrate uptake by urea in microalgae. J. exp. mar. Biol. and Ecol. 118: 97-101.

Mopper, K. and P. Lindroth. 1982. Diel and depth variations in dissolved free amino acids and ammonium in the Baltic Sea determined by shipboard HPLC analysis. Limnol. Oceanogr. 27: 366-347.

Mopper, K. and D.J. Kieber. 2002. Photochemistry and the cycling of carbon, sulfur, nitrogen and phosphorus. 455-489. *In* Hansell, D.A. and C.A. Carlson [eds.]. Biogeochemistry of Marine Dissolved Organic Matter. Academic Press, Amsterdam, the Netherlands.

Moran, M.A. and R.E. Hodson. 1994. Support of bacterioplankton production by dissolved humic substances from three marine environments. Mar. Ecol. Prog. Ser. 110: 241-247.

Mulholland, M.R., P.M. Glibert, G.M. Berg, L. Van Heukelem, S. Pantoja and C. Lee. 1998. Extracellular amino acid oxidation by microplankton: a cross-system comparison. Aquat. Microb. Ecol. 15: 141-152.

Mulholland, M.R., C.J. Gobler and C. Lee. 2002. Peptide hydrolysis, amino acid oxidation, and nitrogen uptake in communities seasonally dominated by *Aureococcus anophagefferens*. Limnol. Oceanogr. 47: 1094-1108.

Müller-Wegener, U. 1988. Interaction of humic substances with biota. pp. 179-192. *In* F.H. Frimmel, and R.F. Christman [eds.]. Humic substances and their role in the environment. John Wiley and Sons Limited, New York, USA.

Myklestad, S., O. Holm-Hansen, K.M. Varum and B.E. Volcani. 1989. Rate of release of extracellular amino acids and carbohydrates from the marine diatom *Chaetoceros affinis*. J. Plankton Res. 11: 763-773.

Nagai, S., I. Imai and T. Manabe. 1998. A simple and quick technique for establishing axenic cultures of the centric diatom *Coscinodiscus wailesii* Gran. J. Plankton. Res. 20: 1417-1420.

Nagao, F. and T. Miyazaki. 1999. A modified ^{15}N tracer method and new calculation for estimating release of dissolved organic nitrogen by freshwater planktonic algae. Aquat. Microb. Ecol. 16: 309-314.

Nagao, F. and T. Miyazaki. 2002. Release of dissolved organic nitrogen from *Scenedesmus quaricauda* (Chlorophyta) and *Microcystis novcekii* (Cyanobacteria). Aquat. Microb. Ecol. 27: 275-284.

Nagasaki, K., M. Ando, S. Itakura, I. Imai and Y. Ishida. 1994. Viral mortality in the final stage of *Heterosigma akashiwo* (Raphidophyceae) red tide. J. Plankton Res. 16: 1595-1599.

Nagata, T. 2000. Production mechanisms of dissolved organic matter. pp. 121-152. *In* D.L. Kirchman, (ed). Microbial Ecology of the Oceans. Wiley, New York, USA.

Nagata, T. and D.L. Kirchman. 1991. Release of dissolved free and combined amino acids by bacterivorous marine flagellates. Limnol. Oceanog. 36: 433-443.

Neess, J.C., R.C. Dugdale, V.A. Dugdale and J. Goering. 1962. Nitrogen metabolism in lakes. I. Measurement of nitrogen fixation with ^{15}N. Limnol. Oceanog. 7: 163-169.

Neilson, A.H. and T. Larsson. 1980. The utilization of organic nitrogen for growth of algae: physiological aspects. Physiol. Plant. 48: 542-553.

Newell, B.S., G. Dalmont and B.R. Grant. 1972. The excretion of organic nitrogen by marine algae in batch and continuous culture. Can. J. Bot. 50: 2605-2611.

Nguyen, R.T. and H.R. Harvey. 1994. A rapid micro-scale method for the extraction and analysis of protein in marine samples. Mar. Chem. 45: 1-14.

North, B.B. 1975. Primary amines in California coastal waters. Utilization by phytoplankton. Limnol. Oceanogr. 20: 20-27.

North, B.B. and G.C. Stephens. 1971. Uptake and assimilation of amino acids by *Platymonas*. II. Increased uptake in nitrogen-deficient cells. Biol. Bull. 140: 242-254.

North, B.B. and G.C. Stephens. 1972. Amino acid transport in *Nitzschia ovalis* Arnott. J. Phycol. 8: 54-68.

Ogawa, H., Y. Amagai, I. Koike, K. Keiser and R. Benner. 2001. Production of refractory dissolved organic matter by bacteria. Science. 292: 917-920.

Oliveira, L. and A. Antia. 1986. Some observations on the urea-degrading enzyme of the diatom *Cyclotella cryptica* and the role of nickel in its production. J. Plank. Res. 8: 235-242.

Oliveira, L. and H. Huynh. 1990. Phototrophic growth of microalgae with allantoic acid or hypoxanthine serving as nitrogen source, implications for purine-N utilization. Canadian J. Fish. Aquatic Sci. 47: 351-356.

Paerl, H.W. 1988. Nuisance phytoplankton blooms in coastal, estuarine, and inland waters. Limnol. Oceanogr. 33: 823-847.

Palenik, B. and F.M.M. Morel. 1990a. Comparison of cell-surface L-amino acid oxidases form several marine phytoplankton. Mar. Ecol. Prog. Ser. 59: 195-201.

Palenik, B. and F.M.M. Morel. 1990b. Amino acid utilization by marine phytoplankton: A novel mechanism. Limnol. Oceanogr. 35: 260-269.

Palenik, B. and S.E. Henson. 1997. The use of amides and other organic nitrogen sources by the phytoplankton *Emiliania huxleyi*. Limnol. Oceanogr. 42: 1544-1551.

Palenik, B., D.J. Kieber and F.M.M. Morel. 1988. Dissolved organic nitrogen use by phytoplankton: the role of cell-surface enzymes. Biol. Oceanogr. 6: 347-354.

Pantoja, S. and C. Lee. 1994. Cell-surface oxidation of amino acids in seawater. Limnol. Oceanogr. 39: 1718-1726.

Pantoja, S. and C. Lee. 1999. Peptide decomposition by extracellular hydrolysis in coastal seawater and salt marsh sediment. Mar. Chem. 63: 273-291.

Pantoja, S., C. Lee, J.F. Marecek and B.P. Palenik. 1993. Synthesis and use of fluorescent molecular probes for measuring cell-surface enzymatic oxidation of amino acids and amines in seawater. Anal. Biochem. 211: 210-218.

Pantoja, S., C. Lee and J.F. Marecek. 1997. Hydrolysis of peptides in seawater and sediment. Mar. Chem. 57: 25-40.

Parsons, T.R., Y. Maita and C. Lalli. 1984. A manual of chemical and biological methods for seawater analysis, Pergamon Press, Oxford, UK.

Paul, J.H. 1983. Uptake of organic nitrogen. 275-308. *In* Carpenter, E.J. and D.G. Capone [eds.]. Nitrogen in the Marine Environment. Plenum Press, New York, USA.

Pedersen, A.-G. and G. Knutsen. 1974. Uptake of L-phenylalanine in cynchronous *Chlorella fusca*. Characterization of the uptake system. Pysiologia Plantarum. 32: 294-300.

Pettersen, R. 1975. Control by ammonium of intercompartmental guanine transport in *Chlorella fusca*. Zeitschrift fur Pflanzenphysiologie. 76: 213-223.

Pettersen, R. and G. Knutsen. 1974. Uptake of guanine by synchronized *Chlorella fusca*. Characterization of the transport system in autospores. Archives fur Microbiol. 96: 233-246.

Price, N. and P. Harrison. 1987. Comparison of methods for the analysis of dissolved urea in seawater. Mar. Biol. 94: 307-317.

Price, N.M., G.I. Harrison, J.G. Hering, R.J. Hudson, P.M.V. Nirel, B. Palenik and F.M.M. Morel. 1988/1989. Preparation and chemistry of the artificial algal culture medium Aquil. Biol. Oceanogr. 6: 443-461.

Price, N.M. and P.J. Harrison. 1988a. Uptake of urea C and urea N by the coastal marine diatom *Thalassiosira pseudonana*. Limnol. Oceanogr. 33: 528-537.

Price, N.M. and P.J. Harrison. 1988b. Urea uptake by Sargasso Sea phytoplankton: saturated and *in situ* uptake rates. Deep-Sea Res. 35: 1579-1593.

Pringsheim, E.G. 1949. Pure Culture of Algae, Cambridge University Press, Cambridge, UK.

Probyn, T.A., H.N. Waldron and A.G. James. 1990. Size-fractionated measurements of nitrogen uptake in aged upwelled waters: Implications for pelagic food webs. Limnol. Oceanogr. 35: 202-210.

Provasoli, L., J.J.A. McLaughlin and M.R. Droop. 1957. The development of artificial media for marine algae. Arch Mikrobiol. 25: 392-428.

Rees, T.A.V. and P.J. Syrett. 1979a. The uptake of urea by the diatom, *Phaeodactylum*. New Phytol. 82: 169-178.

Rees, T.A.V. and P.J. Syrett. 1979b. Mechanisms for urea uptake by the diatom *Phaeodactylum tricornutum*: the uptake of thiourea. New Phytol. 83: 37-48.

Richards, L. and C.F. Thurston. 1980. Uptake of Leucine and tryosine and their intracellular pools in *Chlorella fusca* var. *vacuolata*. J. Gen. Microbiol. 121: 39-47.

Ricketts, T.R. 1988. Homeostatisin nitrogenous uptake/assimilation by the green alga *Platymonas (Tetraselmis) stiata* (Prasinophyceae). Annals Bot. 61: 451-458.

Roman, M.R., H. Ducklow, J. Fuhrman, C. Garside, P. Glibert, T. Malone and G.B. McManus. 1988. Production, consumption, and nutrient cycling in a laboratory mesocosm. Mar. Ecol. Prog. Ser. 42: 39-52.

Rose, P.K. and P.J. Casselton. 1983. Purine catabolism in *Prototheca zopfi*. British Phycological J. 18: 210.

Sakamoto, T. and D.A. Bryant. 2001. Requirement of nickel as an essential micronutrient for the utilization of urea in the marine cyanobacterium *Synechococcus* sp. PCC 7002. Microbes Environ. 16: 177-184.

Sakano, S. and A. Kamatani. 1992. Determination of dissolved nucleic acids in seawater by the fluorescence dye, ethidium bromide. Mar. Chem. 37: 239-255.

Sakshaug, E. and A. Jensen. 1978. The use of cage cultures in studies of the biochemistry and ecology of marine phytoplankton. Mar. Biol. Ann. Rev. 16: 81-106.

Sankievicz, D. and P. Colepicolo. 1999. A new member of the leucyl aminopeptidase family purified and identified from a marine unicellular algae. Biochem. Biophys. Res. Commun. 262: 557-561.

Sarcina, M. and P.J. Casselton. 1995. Degradation of adenine by *Protheca zopfii* (Chlorophyta). J. Phycol. 31: 575-576.

Schell, D.M. 1974. Uptake and regeneration of free amino acids in marine waters of southeast Alaska. Limnol. Oceanogr. 19: 260-270.

See, J.H. 2003. Availability of humic nitrogen to phytoplankton. Ph.D., The College of William and Mary, Williamsburg.

See, J.H., D.A. Bronk and A.J. Lewitus. Uptake of *Spartina*-derived humic nitrogen by estuarine phytoplankton. (submitted)

Seitzinger, S.P., R.W. Sanders and R. Styles. 2002. Bioavailability of DON from natural and anthropogenic sources to estuarine plankton. Limnol. Oceanogr. 47: 353-366.

Shah, N. and P.J. Syrett. 1982. Uptake of guanine by the diatom *Phaeodactylum tricornutum*. J. Phycol. 18: 579-587.

Shah, N. and P.J. Syrett. 1984. Uptake of guanine and hypoxanthine by marine microalgae. J. Mar. Biol. Assoc. U.K. 64: 545-556.

Sharp, J.H. 1977. Excretion of organic matter by phytoplankton: Do healthy cells do it? Limnol. Oceanogr. 22: 381-399.

Sharp, J.H., K.R. Rinker, K.B. Savidge, J. Abell, J.Y. Benaim, D.A. Bronk, D.J. Burdige, G. Cauwet, W. Chen, M.D. Doval, D. Hansell, C. Hopkinson, G. Kattner, N. Kaumeyer, K.J. McGlathery, J. Merriam, J. Morley, K. Nagel, H. Ogawa, C. Pollard, P. Raimbault, S. Seitzinger, G. Spyres, F. Tirendi, T.W. Walsh and C.S. Wong. 2002. A preliminary methods comparison for measurement of dissolved organic nitrogen in seawater. Mar. Chem. 78: 171-184.

Shimizu, Y., N. Watanabe and G. Wrensford. 1995. Biosynthesis of brevetoxins and heterotrophic metabolism in *Gymnodinium breve*. *In* P., Lassus, G. Arzul, E. Erard, P. Gentien and C. Marcaillou [eds.]. Harmful Marine Algal Blooms. Lavoisier, Intercept, Ltd. Paris, France.

Sieburth, J.M., P.W. Johnson and P.E. Hargraves. 1988. Ultrastructure and ecology of *Aureococcus anophagefferens* new genus new species Chrysophyceae the dominant picoplankter during a bloom in Narragansett Bay Rhode Island USA summer 1985. J. Phycol. 24: 416-425.

Siuda, W. and H. Güde. 1996. Determination of dissolved deoxyribonucleic acid concentration in lake water. Aquatic Microb. Ecol. 11: 193-202.

Slawyk, G. and P. Raimbault. 1995. Simple procedure for simultaneous recovery of dissolved inorganic and organic nitrogen in ^{15}N-tracer experiments and improving the isotopic mass balance. Mar. Ecol. Prog. Ser. 124: 289-299.

Slawyk, G., P. Raimbault and S.L. Helquen. 1990. Recovery of urea nitrogen from seawater for measurement of ^{15}N abundance in urea regeneration studies using the isotope dilution approach. Mar. Chem., 30: 343-362.

Small, L.F., S.W. Fowler, S.A. Morre and J. La Rosa. 1983. Dissolved and fecal pellet carbon and nitrogen release by zooplankton in tropical waters. Deep Sea Res. 30: 1199-1220.

Smith, D.C., M. Simon, A.L. Alldredge and F. Azam. 1992. Intense hydrolytic enzyme activity on marine aggregates and implications for rapid particle dissolution. Nature. 359: 139-141.

Smith, H.L. and P. Waltman. 1995. The Theory of the Chemostat. Dynamics of Microbial Competition, Cambridge University Press, Cambridge, UK.

Smith, P.K., R.I. Krohn, T.I. Hermanson, A.K. Mallia, F.H. Gartner, M.D. Provenzano, E.K. Fugimoto, N.M. Goeke, B. Olson and D.C. Sklenk. 1985. Measurement of protein using bicinchonic acid. Analyt. Biochem. 150: 76-85.

Stacey, J.L. and P.J. Casselton. 1966. Utilization of adenine but not nitrate as nitrogen source by *Prototheca zopfi*. Nature. 211

Steidinger, K.A., G.A. Vargo, P.A. Tester and C.R. Tomas. 1998. Bloom dynamics and physiology of *Gymnodinium breve* with emphasis on the Gulf of Mexico. pp. 133-153. *In*: D.M., Anderson, A.D. Cembella and G.M. Hallegraeff (eds) Physiological Ecology of Harmful Algal Blooms. Springer-Verlag.

Steinberg, D.K., C.A. Carlson, N.R. Bates, S.A. Goldthwait, L.P. Madin and A.F. Michaels. 2000. Zooplankton vertical migration and the active transport of dissolved organic and inorganic carbon in the Sargasso Sea. Deep-Sea Res. I. 47: 137-158.

Steinberg, D.K., S.A. Goldthwait and D.A. Hansell. 2002. Zooplankton vertical migration and the active transport of dissolved organic and inorganic nitrogen in the Sargasso Sea. Deep Sea Res. I. 49: 1445-1461.

Steinberg, D.K., N.B. Nelson, C.A. Carlson and A.C. Prusak. 2004. Production of chromophoric dissolved organic matter (CDOM) in the open ocean by zooplankton and the colonial cyanobacterium *Trichodesmium* spp. Mar. Ecol. Prog. Ser. 267: 45-64.

Suttle, C.A. 1994. The significance of viruses to mortality in aquatic microbial communities. Microb. Ecol. 28: 237-243.

Syrett, P.J. and I.A. Bekeheet. 1977. The uptake of thiourea by *Chlorella*. New Phytol. 79: 291-297.

Syrett, P.J. and J.W. Leftley. 1976. Nitrate and urea assimilation in algae. pp. 221-234. *In* Sunderland, N. (ed). Perspectives in Experimental Biology. Vol. 2 Botany. Pergamon, Oxford, UK.

Tamminen, T. and A. Irmisch. 1996. Urea uptake kinetics of a midsummer planktonic community on the SW coast of Finland. Mar. Ecol. Prog. Ser. 130: 201-211.

Tarutani, K., K. Nagasaki and M. Yamaguchi. 2000. Viral impact on total abundance and clonal composition of the harmful bloom-forming phytoplankton *Heterosigma akashiwo*. Appl. Env. Microb. 66: 4916-4920.

Taylor, G.T., R. Iturriaga and C.W. Sullivan. 1985. Interactions of bacterivorous grazers and heterotrophic bacteria with dissolved organic matter. Mar. Ecol. Prog. Ser. 23: 129-141.

Thiel, T. 1988. Transport of leucine in the cyanobacterium *Anabaena variabilis*. Arch. Microbiol. 149: 466-470.

Turner, M.F. 1970. A note on the nutrition of *Phodella*. British Phycological J. 5: 15-18.

Turner, M.F. 1979. Nutrition of some marine microalgae with special reference to vitamin requirements and utilization of nitrogen and carbon sources. J. Mar. Biol. Assoc. U.K. 59: 535-552.

Urban-Rich, J. 1999. Release of dissolved organic carbon from copepod fecal pellets in the Greenland Sea. J. Exp. Mar. Biol. Ecol. 232: 107-124.

Valderrama, J.C. 1981. The simultaneous analysis of total nitrogen and total phosphorus in natural waters. Mar. Chem. 10: 109-122.

Van Baalen, C. 1965. The photooxidation of uric acid by *Anacystis nidulans*. Plant Physiol. 40: 368-371.

Van Baalen, C. and J.E. Marler. 1963. Characteristics of marine blue-green algae with uric acid as anitrogen source. J. Gen. Microbiol. 32: 457-463.

Veuger, B., J.J. Middelburg, H.T.S. Boschker, J. Nieuwenhuize, P. van Rijswijk, E.J. Rocchelle-Newall and N. Navarro. 2004. Microbial uptake of dissolved organic and inorganic nitrogen in Randers Fjord. Est. Coast. Shelf Sci. 61: 507-515.

Vieira, A.A.H. and D. Klaveness. 1986. The utilization of organic nitrogen compounds as sole nitrogen source by some freshwater phytoplankters. Nordic J. Bot. 6: 93-97.

Ward, B.B. and D.A. Bronk. 2001. Net nitrogen uptake and DON release in surface waters: Importance of trophic interactions implied from size fractionation experiments. Mar. Ecol. Prog. Ser. 219: 11-24.

Weinbauer, M.G. 2004. Ecology of prokaryotic viruses. FEMS Microbiol. Rev. 28: 127-181.

Wheeler, P.A. and J.A. Hellebust. 1981. Uptake and concentration of alkylamines by a marine diatom. Plant. Physiol. 67: 367-372.

Wheeler, P.A. and D.L. Kirchman. 1986. Utilization of inorganic and organic nitrogen by bacteria in marine systems. Limnol. Oceanogr. 31: 998-1009.

Wheeler, P.A., B.B. North and G.C. Stephens. 1974. Amino acid uptake by marine phytoplankters. Limnol. Oceanogr. 19: 249-259.

Wheeler, P.A. and G.C. Stephens. 1977. Metabolic segregation of intracellular free amino acids in *Platymonas* (Chlorophyta). J. Phycol. 13: 193-197.

Williams, R. and S.A. Poulet. 1986. Relationships between the zooplankton, phytoplankton, particulate matter, and dissolved free amino acids in the Celtic Sea. Mar. Biol. 90: 279-284.

Williams, S.K. and R.C. Hodson. 1977. Transport of urea at low concentrations in *Chlamydomonas reinhardtii*. J. Bacteriol. 130: 266-273.

Wommack, K.E. and R.R. Colwell. 2000. Virioplankton: Viruses in aquatic ecosystems. Microbiol. Mol. Biol. Rev. 64(1): 69-114.

Yang, X.-H., C. Lee and M.I. Scranton. 1993. Determination of nanomolar concentrations of individual dissolved low molecular weight amines and organic acids n seawater. Anal. Chem. 65: 572-576.

10

Osmotrophy in Marine Microalgae

Alan J. Lewitus[1]

[1]Belle W. Baruch Institute for Marine and Coastal Sciences, University of South Carolina, and Marine Resources Research Institute, South Carolina Department of Natural Resources Hollings Marine Laboratory, 331 Fort Johnson Road, Charleston, South Carolina, USA 29412

Abstract

The widespread ability of microalgae to use dissolved organic compounds has been long recognized from culture studies, but the ecological relevance of the process traditionally has been a point of contention. In recent years, research focus on this issue has increased, due in part to a growing appreciation for the lability of dissolved organic matter (DOM) pools, and for the nutritional versatility of microalgae in natural communities. Evidence for the latter has come primarily from demonstrations of mixotrophic nutrition via phagotrophy, but more recently, evidence for the ecological importance of mixotrophic use of DOM also has accumulated. A review of microalgal culture research into the physiological ecology of osmotrophic nutrition is presented, revisiting early unresolved debates on the ecological importance of this pathway, exploring the complications in laboratory tests of this process that confound ecological application, and addressing recent discoveries that have stimulated newfound interest in microalgal heterotrophy.

INTRODUCTION

The ecological relevance of microalgal heterotrophy (defined as the use of exogenous organic substrates as metabolic carbon source, Margulis et al. 1990) is historically one of the most debated issues in marine science. Use of microalgal cultures to examine this topic can serve as a classic illustration of the complexities involved in extrapolating laboratory findings to natural ecosystem function. The ability of microalgae to assimilate organic compounds is nearly ubiquitous under appropriate laboratory conditions. However, whether the conditions necessary for expression of this pathway in culture are environmentally relevant or exceptional (i.e. a laboratory artifact) is still unresolved when applied to the trophic dynamics of most marine ecosystems.

As discussed below, the case is now strong that dissolved organic substrates are used by microalgae in certain environments, for example hypereutrophic ponds, sea ice, or estuarine sediment porewater. There is growing recognition that dissolved organic nitrogen (DON) loading can promote harmful algal blooms (HABs) through direct use. The bioavailability of dissolved organic matter (DOM) is considered higher than previously thought, based on the discoveries of the 'microbial loop' (and other sources of organic-rich microenvironments), UV photolysis of macromolecules, and realization that some components of humic matter (high molecular weight fractions) are highly labile. Also, it is now well accepted that microalgal phagotrophy (colloquially termed 'mixotrophy') is widespread and can contribute significantly to overall community protist grazing.

Despite these advances, inclusion of microalgal osmotrophic pathways in marine food web models is lacking, although an emphasis on mixotrophy vis-à-vis algal phagotrophy has emerged (Thingstad et al. 1996, Baretta-Bekker et al. 1998, Stickney et al. 2000, Quintana et al. 2002, Anderson et al. 2003, Zhang et al. 2003). Wetzel (1994) pointed out the need to incorporate 'multiple resource utilization' pathways into microbial loop models, stressing the importance of autotrophic and heterotrophic competition for carbon and nitrogen resources. Although bacteria-phytoplankton competition for inorganic nutrients is popularly accepted (Laws et al. 1985, Wheeler and Kirchman 1986, Caron 1994, Kirchman 1994, Kirchman and Wheeler 1998), similar considerations for organic nutrients and carbon have been slow to follow (Flynn and Butler 1986, Antia et al. 1991).

This chapter addresses the role of microalgal culture experiments in advancing knowledge of the physiological ecology of marine microalgal heterotrophy. It will emphasize microalgal use of dissolved organic carbon (DOC) rather than DON, which has been reviewed recently (Bronk 2002, Berman and Bronk 2003), dissolved organic phosphorus (DOP), or particulate carbon (microalgal phagotrophy), also reviewed recently (Jones 1997, Raven 1997, Stoecker 1998, 1999). Some reference to freshwater species is included, but the focus is on marine systems. The chapter emphasizes ongoing gaps in our understanding of this issue, with a recommendation to integrate this pathway into testable hypotheses and models, and adopt previous approaches and findings using modern techniques and diverse cultures.

A HISTORICAL PERSPECTIVE

From the recognition that soil extract was requisite or stimulatory to the growth of some microalgae (Provasoli et al. 1957, Morrill and Loeblich 1979), to the multiple demonstrations of DOM enhancement of growth in the light or support of dark (chemoheterotrophic) growth, the widespread ability of microalgae to use organic growth substrates has long been realized (reviewed in Danforth 1962, Lewin 1963, Van Baalan and Pulich 1973, Droop, 1974, Neilson and Lewin 1974, Ukeles and Rose 1976, Hellebust and Lewin 1977, Sepers 1977, Antia 1980, Bonin and Maestrini 1981). As the number of examined species increased, it became apparent that the cellular mechanisms, regulation, and capacity for DOM uptake varied greatly with substrate, algal species, and environmental condition.

In a classic review on heterotrophic nutrition in diatoms, Hellebust and Lewin (1977) described the group's remarkable variability in metabolic pathways for organic substrate uptake and their environmental regulation. For example, some species could grow chemoheterotrophically, while others whose growth was stimulated by organic enrichment were obligate phototrophs. Light could stimulate DOM uptake or inhibit it. Some diatoms had inducible transport systems and others constitutive. Several specific uptake systems for DOM were documented in this group that varied widely with species. The authors surmised that "among those species studied so far, no two have responded to organic substances in quite the same way. This high degree of specificity is most remarkable, and is probably good indirect evidence for the evolutionary adaptation of species to various specific environmental habitats and situations." Furthermore, Hellebust and Lewin (1977) pointed to the fact that several diatoms contained transport systems specific only to substrates that are commonly abundant (e.g. glucose, lactate and glutamate) as strong support for "the assumption that these species have evolved highly sophisticated and ecologically significant mechanisms for facultative heterotrophy."

Although these conclusions argued for an ecological role for this process in diatoms, Hellebust and Lewin (1977) were not suggesting that diatoms were primary users of DOM at the typically low substrate concentrations found in most aquatic environments, but rather suggested that these algae might be competitive in organically enriched environments such as those associated with sediments or in hypereutrophic systems such as sewage treatment ponds. As was recognized over a decade before, bacteria appeared to have much more efficient DOM uptake systems than those phytoplankton species tested. For example, Michaelis-Menten uptake kinetics consistently revealed microalgal half-saturation constants higher than in bacteria, and higher than the concentrations normally found in natural waters (Hobbie and Crawford 1969, Hellebust and Lewin 1977, Martinez and Azam 1993, Vallino et al. 1996). A seminal study (Wright and Hobbie 1965, Hobbie and Wright 1965), comparing uptake of glucose and acetate by *Chlamydomonas* sp. Ehrenberg and a bacterial isolate demonstrated greater substrate affinity by the bacterium. Also, whereas bacterial uptake was biphasic, indicating a high-affinity (i.e. active transport) and low-affinity (i.e. diffusion) stage, only a simple diffusion-like mechanism of uptake was suggested by the microalgal kinetic curves. In this and other laboratory studies, the DOM concentrations necessary to support dark growth or stimulate low-light growth were much higher than those typically measured in marine or estuarine waters (Barber 1968, Williams 1970, Bennett and Hobbie 1972, Antia et al 1975, Hellebust and Lewin 1977, Admiraal and Peletier 1979, Flynn and Butler 1986, Mulholland et al 1988, Lewitus and Caron 1991a,b, Martinez and Azam 1993). Wright and Hobbie's (1965) findings had a profound influence on perception of this issue, and introduced healthy skepticism to extrapolations of culture findings on microalgal heterotrophy to nature.

In contrast to Wright and Hobbie's (1965) results with *Chlamydomonas* sp., several studies since have indicated complex rather than simple DOM uptake kinetics in the same *Chlamydomonas* species (Bennett and Hobbie 1972) and other microalgae (Hellebust 1970, Lylis and Trainor 1973, Liu and Hellebust 1974, Sheath and Hellebust 1974, Bollman and Robinson 1977, 1985, Sepers 1977, Chapman and Meeks 1983, Flynn and Butler 1986). Hellebust (1970) suggested that the amino acid transport systems involved in the diatom, *Melosira nummuloides* C. Agard, were "as complex as those reported for any other microorganism", including active uptake at low substrate concentrations, a relatively long duration of active uptake (> 1 h), and a transport constant for arginine (0.77 μM) comparable to many bacteria.

As mentioned, enzyme kinetic studies on diatoms generally have demonstrated relatively higher half-saturation constants than substrate concentrations typically associated with marine environments, with some exceptions where transport constants were < 1 μM (Bunt 1969, Hellebust 1970, Shah and Syrett 1982, Flynn and Syrett 1986a). As heterotrophic capabilities are widespread in pennate, but not centric diatoms, the case was presented that uptake systems in the former are adapted to organic-rich conditions associated with benthic or epiphytic environments (Wheeler et al. 1974, Lee et al. 1975, Ukeles and Rose 1976, Hellebust and Lewin 1977, Admiraal and Peletier 1979, Pip and Robinson 1982, Admiraal 1984, Flynn and Butler 1986). Nilsson and Sundbäck (1996) used microautoradiography to determine pronounced uptake of ^{3}H-labeled amino acids by benthic diatoms in sandy sediment porewaters. The potential for microalgal heterotrophy in sediment porewaters is further addressed in the 'Sediments' section below.

The traditional thresholds used to evaluate the ecological relevance of laboratory-derived microalgal DOM uptake constants need a readjustment given the recent findings indicating higher DOM lability in sediment porewater and other aquatic systems (Geller 1985, Kieber et al. 1989, Bronk and Glibert 1993, Wetzel et al. 1995, Lindell et al. 1995, 1996, Moran and Zepp 1997, Hopkinson et al. 1998, Seitzinger and Sanders 1999, Stepanaukas et al. 1999, 2002, Cole 1999, Seitzinger et al. 2002, Berg et al. 2003). However, the argument that bacteria have a relatively greater affinity for DOM uptake than microalgae based on the preponderance of kinetic-based comparisons is more difficult to counter. As discussed in the 'Taxonomic variability' section, culture studies on the DOM uptake properties of microalgae include very few of the smaller nanoflagellates or picoplankton. Theoretically, these types would be expected to have a relatively high affinity for substrate uptake (i.e. smaller cell surface-to-volume ratio). Furthermore, in marine and especially freshwater systems, several cases exist where investigators demonstrated significant uptake of DOM in direct competition with bacteria (Table 10.1). The use of half-saturation constants as criteria for uptake capability under natural conditions is problematic on several grounds. These include the critical effects of incubation time (Saks and Kahn 1979, Admiraal 1984, Admiraal et al. 1984) and other experimental conditions (see 'Considerations in culture experiments' section below), and the need to simultaneously consider the maximum substrate velocity (V_{max}) in order to encompass competitive advantages associated with luxury uptake and storage capacity (Neilson and Lewin 1974, Tilman and Kilham 1976, Saks and Kahn 1979, Healey 1980).

Culture studies on the ecological role and physiological properties of microalgal heterotrophy have dwindled since the research boom on these topics in the 1960's and 1970's, but an interest in DON use has resurged, primarily based on supposition of its importance in relieving N limitation (reviewed in Flynn and Butler 1986, Antia et al. 1991, Bronk 2002, Berman and Bronk 2003). Flynn and Butler (1986) and Antia et al. (1991) suggested a reassessment of the importance of microalgal use of dissolved free amino acids (DFAA), citing several new or revisited bases; e.g. the ecological importance of pico- and nanophytoplankton, the use of appropriate culture maintenance conditions, the need to include AA mixtures in laboratory

Table 10.1 Tracer studies demonstrating DOC uptake by microalgae in mixed communities

Environment	*Labeled Substrate*	*Taxa*	*Method*	*Reference*
Lake	^{14}C-acetate, glucose	Small flagellates, epiphytes	Size-fractionation	Allen 1971
Lake	^{14}C-acetate, glucose	Cyanobacteria (4 species)	Microautoradiography	Saunders 1972
Ice-covered lake	^{14}C-acetate, glucose	Diatoms, cryptophytes	Antibiotics	Maeda and Ichimura 1973
Salt marsh	^{14}C-AA mix, glucose, glycerol, mannitol, urea	Epiphytic diatoms	Size-fractionation	Lee et al. 1973
Lake	^{3}H-acetate, AA mix, glucose, glycollate	Chlorophytes, cyanobacteria	Microautoradiography	Pollingher and Berman 1976
Lake	^{14}C-glucose	Cryptophytes, diatoms	Microautoradiography	McKinley 1977
Coastal marine	^{14}C-glycine	Diatoms, dinoflagellates	Microautoradiography	Wheeler et al. 1977
Freshwater pond	^{3}H-acetate	*Scenedesmus*	Size-fractionation	Vincent 1980
Lake	^{14}C-acetate	Chlorophytes	Microautoradiography	Vincent and Goldman 1980
Lake	^{14}C-fructose, glucose, sucrose	Epiphytic microalgae	Microautoradiography	Pip and Robinson 1982
Lake	^{14}C-, ^{3}H-glucose	Diatoms, cyanobacteria	Microautoradiography	Moll 1984
Polar sea ice	^{14}C-serine	Diatoms	Microautoradiography	Palmisano et al. 1985
Polar marine planktonic, benthic, and sea ice	^{3}H-AA mix, glucose, glutamate, glycine, leucine	Diatoms (7 species)	Single cell micromanipulation	Rivkin and Putt 1987
Lake	^{14}C-AA mix	*Oscillatoria rubescens*	Microautoradiography	Feuillade et al. 1988
Lake	^{3}H-AA mix	*Oscillatoria rubescens*	Microautoradiography	Bourdier et al. 1989
Stream	^{3}H-AA mix	Periphytic *Chlamydomonas*	Microautoradiography	Amblard et al. 1990
Coastal, open ocean	^{14}C-AA mix, ^{3}H-AA mix, glucose, mannitol	Picoplankton, diatoms	Microautoradiography	Paerl 1991
Marine microbial mats	^{3}H-AA mix, acetate, glucose	Cyanobacteria, diatoms	Microautoradiography	Paerl et al. 1993
Freshwater fish ponds	*N*-acetyl-D-[1-^{3}H] glucosamine	Cyanobacteria, diatoms	Microautoradiography	Nedoma et al. 1994
Brackish-to-marine sediment porewater	^{3}H-AA mix	Benthic diatoms, cyanobacteria, flagellates	Microautoradiography	Nilsson and Sundbäck 1996
Estuarine tidal creek	^{14}Glycine	Eukaryotes	Antibiotics	Lewitus et al. 2000
Open ocean	^{35}S-methionine	*Prochlorococcus*	Flow cytometry	Zubkov et al. 2003
Estuarine tidal creek	FITC-dextran	*Kryptoperidinium foliaceum*, *Scrippsiella* sp.	Laser confocal microscopy	This chapter

experiments, critical consideration of light conditions, and recognition of microalgal cell surface enzymes are addressed below. These and other rationale for the need to reconsider the ecological importance of microalgal DOC use are addressed below.

TAXONOMIC VARIABILITY

Culture Bias (Lack of Information on Flagellates and Picophytoplankton)

Notwithstanding the important work of Antia's laboratory on cryptophytes and chrysophytes (reviewed in Antia 1980), it has been argued that our understanding of the physiological parameters of marine microalgal DOM use is based traditionally on diatoms, a group associated with spring blooms and with tendencies toward relatively high photosynthetic and NO_3^- uptake capacities (Flynn and Butler 1986, Berg et al. 1997, 2003, Smayda 1997, Doblin et al. 1999, Bronk 2002, Fan et al. 2003). Relatively little is known of the heterotrophic properties of the types of microalgae most commonly found in environments suboptimal for support of photosynthesis (e.g. the summer communities in temperate estuaries). These include flagellates (e.g. dinoflagellates) and pico-sized cyanobacteria, prochlorophytes, or eukaryotes. The relative lack of information on flagellates and picophytoplankton is partly due to the difficulty in maintaining cultures under standard protocols, and in establishing axenic cultures (Granéli et al. 1997, John and Flynn 1999, Zubkov et al. 2003). As standard protocols traditionally involved mineral media and moderate-to-high growth irradiance, culture collections in the past were probably biased against microalgal species with high heterotrophic capabilities (Bennett and Hobbie 1972, Antia et al. 1975, Flynn and Butler 1986). Although present day culture collections have greatly expanded in taxonomic diversity, the examination of heterotrophic pathways in phytoflagellates and picophytoplankton in defined laboratory studies has been extended to relatively few species.

Flagellates

Lewitus and Kana (1994) tested the effects of glucose, glycerol or acetate addition on the growth rates of 16 phytoplankton species isolated from a subestuary of Chesapeake Bay (salinity 7-12 PSU) during late winter/early spring, a nanoflagellate-dominated community. Substrate additions as low as 1 µM significantly stimulated growth in 9 of the isolates (Table 10.2). The growth rate of one of the isolates, *Nephroselmis* sp. F. Stein, a 4-µm diameter

Table 10.2 Effect of organic compound additions on growth rate of axenic cultures freshly isolated from a Chesapeake Bay subestuary (11 PSU). Acetate, glycerol, or glucose were added at ≤ 10 μM; irradiance = 12 μE m^{-2} s^{-1}. HP designates isolated at Horn Point Laboratory; CCMP designates strains deposited at Provasoli-Guillard National Center for Culture of Marine Phytoplankton.

Class	*Species*	*Diameter (μm)*	*Significant Increase in Specific Growth Rate*
Bacillariophyceae	*Phaeodactylum* sp. HP9101	2.5 (at mid-region)	no
Chlorophyceae	*Chlorella* sp. HP9001	2	yes
	Ankistrodesmus sp. HP9101	1.5 (at mid-region)	yes
Cryptophyceae	*Chroomonas* sp. HP9001	3	no
	Chroomonas sp. HP9004	7	no
	Hemiselmis sp. HP9001	3.5	yes
	Storeatula major HP9001	8	yes
Dinophyceae	*Karlodinium micrum* HP9001	8	yes
	Prorocentrum minimum HP9001	8	yes
Euglenophyceae	*Eutreptia* sp. HP9101	9	no
Pedinellophyceae	*Pseudopedinella pyriforme* HP9001	9	yes
Prasinophyceae	*Nephroselmis* sp. HP9001	4	yes
	Pyramimonas sp. HP9001	5	no
Prymnesiophyceae	*Diacronema* sp. HP9101	3	yes
	(CCMP 1610)	3	Yes
	unidentified (CCMP 1611)	3	no
Raphidophyceae	*Heterosigma akashiwo* HP9001	9	no

prasinophyte, was stimulated > 7-fold in axenic cultures (Lewitus and Kana 1994, 1995). Glucose addition to the prasinophyte resulted in loss of chloroplast material, consistent with catabolite repression of chloroplast development (e.g. as found in *Euglena gracilis* G.A. Klebs, some strains of *Chlorella* M. Beijerinck, and *Poterioochromonas malhamensis* (Pringsheim) Peterfi; Pringsheim 1952, Myers and Graham 1956, App and Jagendorf 1963, Shihira-Ishikawa and Hase 1964, Harris and Kirk 1969, Monroy and Schwartzbach 1984, Lewitus and Caron 1991a). Thus, *Nephroselmis* sp. appeared to favor heterotrophic over autotrophic nutrition. Very small prasinophytes are common in oceanic and estuarine waters (Biegala et al. 2003, Rodriguez et al. 2003, Worden et al. 2004, Lewitus pers. observ.), yet few physiological studies with these picoeukaryotes exist. Because standard enrichment protocols for phytoplankton usually include mineral media (which may not support optimal growth for some species, Lewitus and Kana 1994), and because chloroplast loss may be mistaken for obligate heterotrophy, the prevalence of this *Nephroselmis* species or other flagellates with similar physiological characteristics is unknown.

Chrysophytes, cryptophytes, and dinoflagellates are known to be nutritionally versatile (Antia 1980, Gaines and Elbrächter 1987). Kristiansen (1990) stated "It is doubtful if any entirely photoautotrophic chrysophytes exist". More work has been done on freshwater than marine chrysophytes, although phagotrophy is widespread throughout this group (Sanders and Porter 1988). As mentioned, a relatively higher capacity for heterotrophic than photoautotrophic growth has been demonstrated in the freshwater chrysophyte, *Poterioochromonas malhamensis*, but a similar preference for heterotrophy has not been discovered in marine species. As with diatoms, growth enhancement by DOC substrate addition at high concentrations has been demonstrated for several marine flagellates (Rahat and Jahn 1965, Droop and McGill 1966, Pintner and Provasoli 1968, Antia et al. 1969, Cheng and Antia 1970, Ukeles and Rose 1976, Morrill and Loeblich 1979). In these studies, glycerol was predominantly the favored substrate, but the requirement for high concentrations to support growth (Lewitus and Caron 1991b) argues against an active uptake mechanism (Antia 1980). However, studies on enzyme kinetics for glycerol uptake are lacking for these taxa.

Over the last decade, a major advance was made in understanding the importance of cell surface enzymes for DOM use by phytoflagellates. Palenik and Morel (1990a,b, 1991) documented the occurrence of cell surface amine oxidases in *Prymnesium parvum* N. Carter (Prymnesiophyceae), *Pleurochrysis carterae* (Braarud and Fagerland) Christensen (Prymnesiophyceae), and *Amphidinium operculatum* Claperede and Lachmann (Dinophyceae), with half-saturation constants < 0.5 μM for amino acids and primary amines. N uptake by this mechanism does not require direct assimilation of DON, but rather involves extracellular DON degradation and uptake of released ammonium. The process has since been shown to be an important route of phytoplankton N acquisition in some natural communities, including those dominated by *Aureococcus anophagefferens* Hargraves et Sieburth (Pelagophyceae) or *Trichodesmium* C.G. Ehrenberg ex Gomont (Cyanophyceae); Pantoja and Lee (1994), Mulholland et al. (1998, 2003).

Cell surface or extracellular enzymatic breakdown of complex macromolecules by microalgae has received attention recently. Martinez and Azam (1993) demonstrated cell surface aminopeptidase activity in several marine *Synechococcus* Nägeli strains, and suggested that proteolysis was important to the N nutrition of these cyanobacteria. Stoecker and Gustafson (2003) measured cell surface leucine aminopeptidase activity in several dinoflagellate species, and speculated that the mechanism was used to enhance amino acid availability at the cell surface. The authors found a correlation between aminopeptidase activity and dinoflagellate abundance in Chesapeake Bay samples, and concluded that bloom populations might contribute significantly to overall microbial proteolytic activity. Cell surface peptide hydrolysis also was shown in axenic cultures of *Aureococcus anophagefferens* (Berg et al. 2002, 2003, Mulholland et al. 2002b), and Berg et al. (2003) hypothesized that this was a mechanism used to scavenge N from HMW DON in groundwater.

Also recently, heterotrophic and phototrophic protists have been considered major contributors to glucosaminidase and glucosidase activities. β-glucosaminidase activity was developed as a tool to measure protist bacterivorous activity, based on its role in degrading bacterial cell wall peptidoglycan (Vrba et al. 1993, 1996). The assay has been applied using both cell extracts (vacuolar enzyme) and intact cells (extracellular enzyme); Vrba et al. (1996). Sherr and Sherr (1999), using cell extracts, found that several marine microalgal species exhibited β-glucosaminidase activity comparable to heterotrophic flagellates. Vrba et al. (1997) found an association between low-affinity β-N-acetylglucosaminidase activity and diatom biomass in a freshwater eutrophic reservoir, and suggested ectoenzyme production by diatoms as a possible explanation. A significant contribution of microalgae to extracellular α- and β-glucosidase activity in freshwater systems has been proposed recently (Sabater et al. 2003, Vrba et al. 2004).

Picophytoplankton

Based on their small size (i.e. higher surface area:volume ratio), coccoid cyanobacteria such as *Synechococcus* may be expected to have relatively high capabilities for taking up DOM. Two decades ago, the evidence for heterotrophy in this and other marine and freshwater cyanobacteria under natural conditions was weak, but Smith (1982) suggested that understanding the natural relevance of the process was biased by the lack of pure cultures, the use of unnatural culture conditions, and a limited knowledge of allelochemical interactions. Paerl (1991) demonstrated uptake of an amino acid (AA) mixture at trace concentrations by axenic isolates of *Synechococcus*. Uptake of ^{3}H-labeled AAs occurred in the dark, but was light-stimulated. Furthermore, DCMU only partially reduced the light-stimulated AA uptake but completely inhibited photosynthetic ^{14}C uptake, suggesting direct photostimulation of heterotrophic C use. When combined with microautoradiographic evidence of AA uptake in picophytoplankton from diverse N-limited oceanic communities, the author concluded that the process was widespread in this group, and proposed that DON use may explain high picoplankton primary production in DIN-limited waters.

The existence of viable chloroplast-containing cells from aphotic oceanic depths has been recognized for decades, including samples from the deep-sea (Bernard 1963, Fournier 1966, 1970, 1971, Hamilton et al. 1968, Malone et al. 1973, Platt et al. 1983). Anderson (1982) and Lewitus and Broenkow (1985) documented deep-sea fluorescence maxima associated with oxygen minimum zones in the Eastern Tropical North Pacific, and Broenkow et al. (1985) suggested that these peaks were produced by cyanobacteria, based on *in situ* fluorometric spectral characteristics. The existence of a metabolically active chroococcoid cyanobacterial community in aphotic/anoxic waters was later demonstrated in the Baltic Sea (Detmer et al. 1993). The authors suggested a possible role of heterotrophy in survival or growth at these depths, based on observations of high growth potential under aphotic/anoxic conditions. More recent studies in the Eastern Tropical North Pacific and in the Arabian Sea also confirmed that subphotic oxygen minimum zone pigment layers were dominated by the cyanobacterium, *Prochlorococcus marinus* Chisholm et al. 2001, with some contribution from *Synechococcus* in the former region (Goericke et al. 2000). The authors questioned whether, at the ambient low irradiances, strictly photoautotrophic nutrition could sustain *Prochlorococcus* population growth in the face of protozoan grazing pressure, and suggested photoheterotrophy as a reasonable alternative. Zubkov et al. (2003) demonstrated that *Prochlorococcus* and *Synechococcus* in Arabian Sea waters incorporated ^{35}S-methionine (at ca. natural concentrations) at rates comparable to heterotrophic bacteria. However, the estimated contribution of *Prochlorococcus* to daily turnover of the methionine pool was much greater than that of *Synechococcus*. Worden et al. (2004) hypothesized different ecological niches for these cyanobacteria genera in oceanic environments, based on nutrient preferences. *Synechococcus* and picoeukaryotes tend to co-exist and have relatively high NO_3 uptake capabilities, while *Prochlorococcus*, which may not be capable of using NO_3 (Moore et al. 2002), may instead use regenerated nutrients such as NH_4 or DON. Tests of this intriguing hypothesis are limited by the availability of axenic *Prochlorococcus* cultures. As Zubkov et al. (2003) pointed out, Rippka et al. (2000) included methionine in the medium used to establish the first axenic culture of *Prochlorococcus*.

Harmful Algae

The preponderance of recent literature on DOM use by phytoplankton has come from studies on harmful algae (reviews by Paerl 1988, 1997, Carlsson and Granéli 1998, Anderson et al. 2002). Much of the attention has been triggered by the hypothetical relationship between anthropogenically related increased nutrient loading and an associated global increase in the prevalence and distribution of HABs (Smayda 1989, 1990, Hallegraeff 1993, Paerl 1997, Glibert et al. 2001, Anderson et al. 2002). A key proposed link is that harmful species may have a greater relative capability to use DON, a constituent of anthropogenic pollution. Rationale for categorically distinguishing harmful bloom species from phytoplankton typically viewed as supportive of higher trophic levels include the traditional view that the former cannot adequately compete for inorganic nutrients at low concentrations, and thus have evolved mechanisms for competing in alternative nutrient environments (discussed in Smayda 1997, with recognition that this may be over-generalized). These can include DIN-rich conditions that allow increased biomass of all phytoplankton, which may favor species with allelopathic capability (e.g. chemical deterrence of competitors or predators, chemical auto-stimulation), or DOM-rich conditions which would favor species capable of metabolizing exogenous organic substrates. Although recent experimental tests of DOM uptake have emphasized the role of DON in bloom promotion under DIN-limited conditions (Maestrini et al. 1999, Bronk 2002, Berman and Bronk 2003), the corresponding value of DOC as a promoting factor is not as clearly argued and has lagged in experimental examination. As discussed in the following sections, some recent studies suggest that explanations of HAB proliferation warrant consideration of heterotrophic pathways.

Dinoflagellates

Dinoflagellates have been long suspected of having relatively high osmotrophic capabilities. Many species are not amenable to culturing using standard mineral media, and are difficult to grow axenically (Provasoli et al. 1957, Pant and Fogg 1976, Morrill and Loeblich 1979, Granéli et al. 1997, Doblin et al. 1999, John and Flynn 1999). Phagotrophic mixotrophs are prevalent in this group, and phagotrophic mechanisms are diverse (Gaines and Elbrächter 1987, Schnepf and Elbrächter 1992, Bockstahler and Coats 1993, Lewitus et al. 1999). Furthermore, dinoflagellate blooms have been associated with DOM-rich waters (Prakash and Rashid 1968, Mahoney and McLaughlin 1977, Granéli and Moreira 1990, Burkholder et al. 1997, Carlsson and Granéli 1998, Anderson et al. 2002). Stoecker and Gustafson (2003) demonstrated a high potential for dinoflagellates to use macromolecules by cell surface enzymatic breakdown (see above). Another HMW DOM pool that may be widely used by dinoflagellates is humic matter.

Dinoflagellate blooms are commonly associated with rain-driven runoff of humic-rich water (Prakash and Rashid 1968, Glover 1978, Therriault et al. 1985, Granéli et al. 1989, Lara et al. 1993, Carlsson et al. 1993, Carlsson and Granéli 1993, 1998). A stimulatory effect of DOM on dinoflagellates has been attributed to enhanced trace metal availability (i.e. chelation), direct use of DOM as C or N source, or uptake of bacteria-mediated DOM degradation products (Anderson and Morel 1982, Granéli et al. 1985, 1986, Sunda and Huntsman 1995, Doblin et al. 1999). The bioavailability of humic matter has been a contentious issue, but recent data indicate that a much greater fraction of this pool is labile than previously thought, especially the higher molecular weight fractions (Carlsson et al. 1998, Stepanaukas et al. 1999, 2000, Bronk 2002, Stolte et al. 2002, Berman and Bronk 2003).

Some culture experiments have supported the relatively high ability of dinoflagellates to use HMW DOM associated with humic matter. The addition of riverine humic substances to N-limited axenic and nonaxenic cultures of *Alexandrium catenella* (Whedon and Kofoid) E. Balech stimulated growth, which was attributed to direct uptake of DON (Carlsson et al. 1998). In exploring possible uptake mechanisms, the authors demonstrated direct uptake of HMW fluorescently labeled dextran, a polysaccharide. A companion study by Legrand and Carlsson (1998) demonstrated uptake of HMW (2000 kDa) but not lower molecular weight (20 kDa) dextrans. The use of HMW DOM (e.g. colloids) may be through pinocytosis or phagocytosis (Klut et al. 1987, Carlsson et al. 1998, Legrand and Carlsson 1998). Stolte et al. (2002) found that *Alexandrium tamarense* (Lebour) E. Balech used HMW DOM as a N source, and also suggested the possibility of phagocytotic colloidal uptake. The direct uptake of HMW DOM has been proposed as a 'niche' for phototrophic and heterotrophic flagellates in direct competition with bacteria (Sherr 1988, Marchant and Scott 1993, Tranvik et al. 1993, Legrand and Carlsson 1998, Schuster et al. 1998).

Other dinoflagellate blooms potentially stimulated by humic matter recently were discovered in South Carolina tidal creeks. In 1999, bloom initiation of *Kryptoperidinium foliaceum* (Stein) Lindemann in North Inlet, a high-salinity salt marsh estuary, followed rain-driven runoff events, and over the bloom period, *K. foliaceum* abundance varied inversely with DOC, DON, and DOP concentrations, and positively with dissolved inorganic carbon concentrations, suggesting high respiratory activity (Lewitus et al. 2001; *K. foliaceum* referred to as a *Scrippsiella* species, but later renamed by Wolny et al. in press). *K. foliaceum* bloom populations were tested for dextran uptake capability. Fluorescein isothiocyanate (FITC) labeled HMW (2000 kDa) dextran was taken up rapidly by *K. foliaceum* bloom populations, but lower molecular weight dextran (3 kDa) was not (Figs. 10.1 and 10.2). Pre-bloom (10^3 cell ml^{-1}) and peak bloom populations (6×10^4 cell ml^{-1}) ingested dextran rapidly (intracellular fluorescence was detected in > 50% of cells after 30 min). Uptake was observed in up to 74% and 67% of cells from pre-bloom and peak-bloom periods, respectively. Pronounced and rapid uptake of HMW dextran also was observed in bloom isolates of *K. foliaceum*, in cultures of *Scrippsiella trochoidea* (Stein) Loeblich III and *K. foliaceum*, and in natural populations of *K. foliaceum* and an unidentified *Scrippsiella* sp. (Table 10.3). Blooms of *K. foliaceum* and *Scrippsiella* sp. were commonly observed in South Carolina estuarine tidal creeks in recent years. In these same experiments, dextran uptake was not observed in monocultures of the following species isolated from South Carolina estuaries: the chrysophyte, *Ochromonas* sp. Wyssotski SCAEL970626, and the raphidophytes, *Fibrocapsa japonica* S. Toriumi and H. Takano SCC1-01, *Chattonella subsalsa* B. Biecheler CAAE 1662, and *Heterosigma akashiwo* (Y. Hada) Y. Hada ex Y. Hara et M. Chihara CAAE 1663 (data not shown). The use of colloidal humic substances by phagotrophy implies that both organic sources of N and C are assimilated, raising hypothesized roles of the process in relieving nutrient

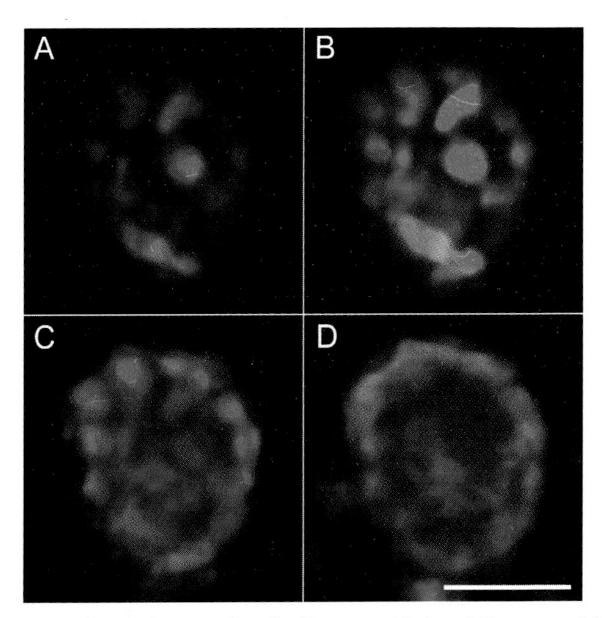

Fig. 10.1 Laser confocal micrographs of a *Kryptoperidinium foliaceum* cell from a natural bloom sample at successive 0.5-µm depth intervals from "A" to "B" to "C" to "D". The red is chlorophyll fluorescence from numerous chloroplasts, and green is (FITC) fluorescence from ingested stained dextran. Scale bar is 10 µm.

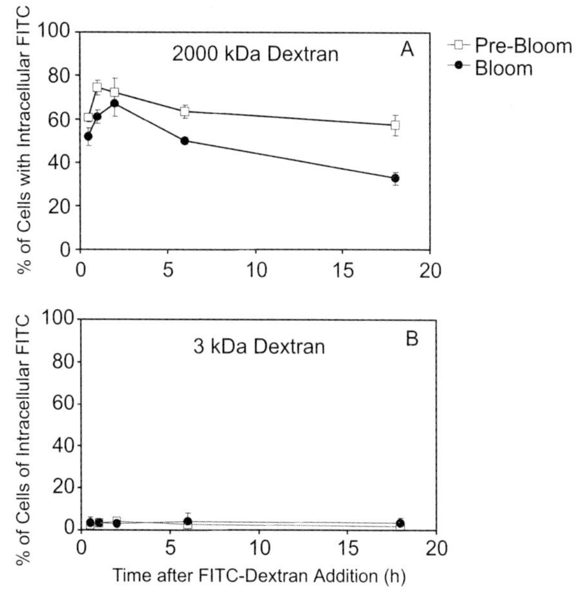

Fig. 10.2 Time-course of fluorescein isothiocyanate (FITC) labeled dextran uptake by *Kryptoperidinium foliaceum* populations collected during a pre-bloom (open squares) or bloom (closed circles) stage, showing % of cells with intracellular FITC fluorescence. (A) uptake of 2000 kDa dextran, (B) uptake of 3 kDa dextran. Mean and standard deviation (error bars) of triplicate samples.

Table 10.3 Percent of population estimated to take up FITC-labeled high molecular weight (2000 kDa) dextran based on observations of intracellular FITC fluorescence

Population	*Antibiotics*	*Irradiance Condition*	*Incubation Time (h)*	*% Cells with FITC*
Kryptoperidinium foliaceum isolate	No	Continuous light	2	38 ± 6
	No	Continuous darkness	2	57 ± 4
	No	12L:12D	2	34 ± 6
	Yes	12L:12D	2	39 ± 10
Scrippsiella trochoidea CCMP1331	No	Continuous light	2	26 ± 5
	No	Continuous darkness	2	37 ± 1
	No	12L:12D	2	41 ± 10
	Yes	12L:12D	2	31 ± 2
K. foliaceum UTEX1688	No	12L:12D	0.5	35
K. foliaceum (bloom)			0.5	25
Scrippsiella sp. (bloom)			0.5	33

and/or light limitation. Adequate tests of these hypotheses under defined laboratory conditions are not yet possible for *K. foliaceum* and *Scrippsiella* sp. from South Carolina waters because isolates have not been amenable to long-term culturing.

Aureococcus

Blooms of the 'brown tide' pelagophyte, *Aureococcus anophagefferens*, have recurred periodically in northeastern U.S. estuaries since first observed in 1985. Hypotheses on factors triggering bloom formation by this picoeukaryote have included consideration of heterotrophic pathways. Dzurica et al. (1989) demonstrated a relatively high ability of nonaxenic cultures to take up glucose and glutamate in comparison with several other microalgae. A potential role of DOM was reinforced by findings that bloom populations had greater growth responses and uptake abilities with DON substrates than with NO_3 (Berg et al. 1997). In fact, the low NO_3 utilization capability resembled properties of a closely related pelagophyte, *Aureoumbra lagunensis* D.A. Stockwell et al., the 'Texas brown tide' (DeYoe and Suttle 1994). Using axenic cultures, Berg et al. (2002) found that *A. anophagefferens* was able to take up a variety of DON compounds (urea, acetamide, aminopepetidase, chitobiose), with exceptionally high rates of aminopeptide hydrolysis that were greater than those of two of three heterotrophic bacteria tested. Berg et al. (2003) tested culture responses to HMW DON isolated from sediment pore water, and found that *A. anophagefferens* was capable of using a large fraction of this N pool. The protein fraction was most preferentially used, through cell surface enzymatic peptide hydrolysis.

A high capability for DON uptake from many sources is consistent with LaRoche et al.'s (1997) hypothesis that bloom formation is inversely related to DIN-rich groundwater flow; i.e. that blooms formed during drought years when restricted groundwater input resulted in relatively high water column DON:DIN. However, Berg et al.'s (2003) finding that porewater DON was an exceptionally favorable source for growth adds a complication to the hypothesis in the potential link to groundwater DON. Whereas a high capability for DON use (including urea) may be instrumental in supporting high *A. anophagefferens* biomass, formation and sustenance of monospecific blooms may relate to DOC uptake. Blooms are associated with relatively high DOM C:N ratios and low irradiances (Milligan and Cosper 1997, Breuer et al. 1999, Gobler and Sañudo-Wilhelmy 2003). Gobler and Sañudo-Wilhelmy (2001) found that *A. anophagefferens* bloom populations were generally stimulated by glucose, but not urea, additions, and hypothesized that labile exudates from mixed phytoplankton blooms may select for monospecific blooms, based on DOC use under low light availability. The relative roles of DON and DOC in stimulating *A. anophagefferens* blooms therefore may be a function of bloom stage. Tests of this hypothesis will rely, in large part, on culture studies to compare the physiological parameters of DOC and DON assimilation and metabolic function.

APPLICATION OF CULTURE RESULTS TO MICROALGAL HETEROTROPHY IN SPECIFIC ENVIRONMENTS

Hypereutrophic Environments

The association of phytoplankton species exhibiting high heterotrophic abilities under laboratory conditions with their distribution in exceptionally organic-rich waters has been recognized for some time. For example, Ryther (1954) tested the effects of organic and inorganic nutrients on the growth of two small chlorophyte isolates (*Nannochloris atomas* Butcher, *Stichococcus* sp. Nägeli) that formed blooms in association with duck farm waste input to Great South Bay and Moriches Bay, New York. These chlorophytes grew equally as well on DON substrates as on DIN. Mahoney and McLaughlin (1977) examined growth responses to organic N, P and C compounds in three flagellates that dominated blooms in a hypereutrophic section of Lower New York Bay that was influenced by sewage effluent. Isolates of the dinoflagellates, *Prorocentrum micans* Ehrenberg and *Heterocapsa rotundata* (Lohmann) G. Hansen (formerly *Katodinium rotundatum* (Lohmann) Loeblich III), and the raphidophyte, *Heterosigma akashiwo* (Y. Hada) Y. Hada ex Y. Hara et M. Chihara (formerly *Olisthodiscus luteus* N. Carter) displayed capabilities of using most of the substrates. The authors argued that the lowest C concentrations tested (5 mg L^{-1}) were reasonable estimates of levels in hypereutrophic systems. Iwasaki (1979) showed a growth enhancement effect of organic substrate additions to axenic cultures of several dinoflagellates and raphidophytes, and proposed that this pathway may contribute to their growth or survival in the organic-rich environments where they typically bloom (e.g. the Inland Sea of Japan, pearl oyster culture grounds, and waters impacted by pulp waste). Neilson and Larsson (1980) also found a relatively high ability of isolates from polluted environments to use DON compared to isolates typically found in unpolluted areas or to strains obtained from culture collections. Chemoheterotrophic growth and DOM-stimulated low light growth were demonstrated in axenic cultures of most of the diatom isolates from muddy sediments of the Eems-Dollard estuary influenced by organic enrichment from wastewater discharge, but not isolates from organic-poor sandflats (Admiraal and Peletier 1979, Admiraal 1984).

Sediments

The potential importance of microphytobenthos to estuarine primary production is now widely recognized (Pinckney and Zingmark 1993, MacIntyre et al. 1996). Porewaters are known to be a significant source of inorganic and organic nutrients, derived from groundwater and reminerilization of sinking particulate organic matter (Henrichs and Farrington 1979, Jørgensen 1982, Nilsson and Sundbäck 1996, Paerl 1997, Burdige and Zheng 1998, Hopkinson et al. 1998, Middleboe et al. 1998, Berg et al. 2003). Light availability at the interface between sediment and water column is widely variable and often limited by turbidity via sediment resuspension (see 'Light' section). The repercussions of this relatively high nutrient, low/fluctuating light environment on microphytobenthic photosynthetic properties have been studied recently (MacIntyre and Cullen 1996, MacIntyre et al. 2000), but recent studies into the potential contribution of DOC uptake by this group are lacking. The need to revisit the ecological importance of this pathway is further justified by the newfound recognition that DOC in sediment porewaters can be highly labile (Hopkinson et al. 1998, Berg et al. 2003). In fact, based on the recognized importance of DOM photolysis in enhancing bioavailability, Cole (1999) proposed that refractory DOC accumulating in aphotic layer sediments may become labile when exposed to light (e.g. during resuspension, tidal scouring, or bioturbation). Lee et al. (1975) and Hellebust and Lewin (1977) suggested the importance of microbenthic organism interaction (i.e. microenvironments) in DOC flow (see 'Microenvironment' section). Diatom production of flocculent layers of heteropolysaccharides may enhance DOC availability directly or following heterotrophic microbial degradation (Decho 1990, Antia et al. 1991, Arnosti et al. 1994, Underwood and Smith 1998, Staats et al. 2000). Consideration that a labile concentrated pool of DOC exists in low light environments has implications not only to potential heterotrophy of benthic microalgae, but also to overlying phytoplankton that can take advantage of access to this pool during light limiting periods. For example, in shallow waters, certain dinoflagellates may be able to access porewater DOC and/or diatom DOC exudates by exposure to resuspended sediments or through active downward migration (Horstmann 1980, Larsen 1987, Kondo et al. 1990, Nilsson and Sundbäck 1996, Smayda 1997).

The presence of non-pigmented diatoms in benthic habitat is evidence that osmotrophy is a natural process in benthic diatoms (Lewin and Lewin 1967, Admiraal 1984). As mentioned in the 'Historical perspective' section, several studies have shown organic substrate stimulation of pennate diatom growth in limiting light that may typify the water/sediment interface. For example, Cooksey and Chansang (1976) found growth stimulation by several DOC compounds of three *Amphora* spp. Ehrenberg ex Kützing at 2% of the saturating light level. Saks and Kahn (1979) demonstrated growth stimulation of a salt marsh diatom, *Cylindrotheca closterium* (Ehrenberg) Lewin and Reimann, by low concentrations of various organic substrates in the light and dark, and concluded that, in the short term (1 h) but not a longer term (10 d), the diatom outcompeted a co-occurring bacterium, *Aeromonas* sp. Stanier, for uptake. Lewitus et al. (2000) demonstrated a significant increase in this same diatom species in response to glycine addition in 72-h natural community bioassays from a South Carolina salt marsh, but the stimulatory effect did not occur in the presence of antibiotics, suggesting that stimulation was mediated by bacterial breakdown of glycine. However, uptake of a mixture of ^{3}H-labeled amino acids was demonstrated by microautoradiography to be common in microphytobenthic diatoms in a sandy bay in Sweden, including *C. closterium* (Nilsson and Sundbäck 1996). Nilsson and Sundbäck (1996) also observed uptake in benthic cyanobacteria (e.g. *Microcrocis* sp. Richter, *Phormidium* sp. Kützing, *Merismopedia* sp. Meyen) and phytoflagellates. The potential for porewater DOC uptake in these groups and others (e.g. green algae) that may be constantly in contact with marine sediment porewater or intermittently exposed (e.g. diurnal or tidally-driven vertical migration, sediment resuspension, cyst cycles) rarely has been examined experimentally using cultures (see 'Culture bias' section).

Microenvironments

The concentration of organic matter can be elevated within microenvironments, and consideration that microalgal heterotrophy may occur at these locations has a sound theoretical basis (Riley 1963, Sloan and Strickland 1966, White 1974, Lee et al. 1975, Hellebust and Lewin 1977). One implication of this reasoning is that DOM concentration estimates from bottle samples may not be an appropriate benchmark for bioavailable DOM. Considerations of elevated DOM bioavailability in microenvironments can include a wide range of circumstances involving microbial community interactions, including excretory, grazing or cell lysing processes associated with macroaggregates (marine snow, microbial mats, flocculent masses), phytoplankton blooms, 'microbial loop' food webs or even microenvironments surrounding individual micro- or macroorganisms.

Uptake of algal exudates by microalgae may explain species shifts in community dynamics. As mentioned, Gobler and Sañudo-Wilhelmy (2001, 2003) showed that glucose additions to mixed bloom assemblages selected for the growth of *Aureococcus anophagefferens*, and the authors hypothesized that heterotrophic use of bloom exudates by this species under limiting light may be instrumental in formation of monospecific blooms. Oligotrophic oceanic colonies of the diazotrophic cyanobacterium, *Trichodesmium*, contain mixtures of cells with and without N_2-fixing capabilities (Bergman and Carpenter 1991, Lin et al. 1998), and both uptake and release of DON have been shown in natural populations and cultures (Capone et al. 1994, Glibert and Bronk 1994, Mulholland and Capone 1999, 2000, Mulholland et al. 1999, Berman-Frank et al. 2001). Mulholland et al. (2001) confirmed the ability of *Trichodesmium* cultures to fix N_2 in the presence of 1 μM NH_4, glutamate, or glutamine, but found nitrogenase inhibition in response to 10 μM additions. The authors concluded that simultaneous uptake of N_2 and combined N was probable under ambient nutrient conditions. Based on alkaline phosphatase measurements in natural and culture populations, Mulholland et al. (2002a) concluded that DOP could be an important source of P for the cyanobacterium. DON release by *Trichodesmium* colonies was proposed as an important N source for associated microorganisms, including microalgae (Paerl et al. 1989, Paerl 1991, Capone et al. 1994, Glibert and Bronk 1994, Mulholland et al. 1998). Recent studies have proposed a role of *Trichodesmium*-produced DON as N source for supporting blooms of the dinoflagellate, *Karenia brevis* (C. C. Davis) G. Hansen and Ø. Moestrup (Heil et al. 2001, Lenes et al. 2001, Lester et al. 2001, Vargo et al. 2001).

Lytic processes may lead to release of labile DOM through algicidal viruses or bacteria, autolysis of bacteria or algae, or 'sloppy feeding' by zooplankton (Fuhrman and Suttle 1993, Strom et al. 1997, Agusti et al. 1998, Ferrier-Pagés et al. 1998, Fuhrman 1999, Bronk 2002). Strom et al. (1997), using laboratory experiments with a variety of zooplankton and phytoplankton types, showed that grazing was the dominant route of ambient DOC supply relative to algal excretion. Glucose is an important photosynthetic product of microalgae, and its release during bloom senescence, grazing, or other forms of lysis may contribute to its high ambient concentrations in estuarine and marine systems (Ittekot et al. 1981, Rich et al. 1996, Skoog and Benner 1997, Biersmith and Benner 1998). The widespread ability of microalgae to use this C source may be related to its abundance as a lytic product.

A strong case can be made for microalgal use of DOC-rich microenvironments based on chemotactic capabilities, although again this process has only been studied in a few marine species. Levandowsky and Hauser (1978) reviewed evidence of chemosensory responses of the heterotrophic dinoflagellate, *Crypthecodinium cohnii* (Seligo) Chatton *in* Grassé, to several DOC compounds. Cooksey and Cooksey (1988) demonstrated chemotaxy to sugars in the benthic diatoms, *Amphora* spp. Willey and Waterbury (1989) showed chemotactic behavior of *Synechococcus* to nanomolar levels of glycine and alanine. Lee et al. (1999) demonstrated chemotaxy to amino acids in a small (3 μm) cryptophyte (*Chroomonas* sp. Hansgirg HP9001, Table 10.2) when grown with glycine, but not with NO_3. The authors hypothesized that this process was inducible in response to NO_3 limitation.

CONSIDERATIONS IN CULTURE EXPERIMENTS

Strain Variability and Genetic Bias

Numerous examples exist of microalgal species that exhibit strain differences in heterotrophic capabilities. Bronk (2002) suggested that conflicting results on DON use by cultures of *Emiliania huxleyi* (Lohmann) Hay and Mohler (Prymnesiophyceae) may be attributed to strain variability. Yamada et al. (1983) found that one isolate of the diatom, *Skeletonema costatum* (Greville) Cleve, grew equally well with arginine or NO_3 as sole N source, but another isolate failed to grow on arginine. In these and other examples (Palmer and Togasaki 1971, Carpenter et al. 1972, Ukeles and Rose 1976, Flynn and Butler 1986, Gaines and Elbrächter, 1987, Fruend et al. 1993), it is difficult to differentiate between true strain variability and 'genetic biasing' that can occur during culture maintenance. As mentioned, the standard protocols for maintaining microalgae typically involve mineral medium and adequate light, while heterotrophic bacteria are routinely cultured on organic-rich medium and without light. Therefore, the possibility that cultures become genetically altered after long-term maintenance should be seriously considered, and comparisons between microalgal and bacterial DOM uptake capabilities using such cultures may include this bias. Some possible examples of loss of heterotrophic potential over culture maintenance time include glucose uptake capability by the diatom, *Cyclotella cryptica* Reimann, Lewin and Guillard (White 1974), and urea uptake by the diatom, *Skeletonema costatum* (Carpenter et al. 1972). Also, the growth of *Storeatula major* D.R.A. Hill (Cryptophyceae) was enhanced by glucose addition in experiments with freshly isolated cultures (Lewitus and Kana 1994, Table 10.2). In contrast, this same isolate was not stimulated by a range of glucose concentrations (10 μM to 10 mM) after a decade of maintenance on f/2 medium (supplied with NO_3 and PO_4) and moderate light (Willis and Lewitus, unpub. data). The contrasting situation in which photosynthetic ability was lost by long-term maintenance in organic-rich medium was hypothesized by Jones et al. (1996) for the freshwater chrysophyte, *Poterioochromonas malhamensis*.

Choice of Substrates

Culture experiments examining microalgal organic substrate uptake and physiological response must be critically evaluated based on the type of substrate used. As is the case with diatoms (see 'Historical Perspective' section), high variability for preferred substrates exists in all taxa (Flynn and Butler 1986, Antia et al. 1991, Berg et al. 1997, Berman and Bronk 2003). Conclusions based on relatively high or low uptake of tested substrate(s) need to consider that niche differences in substrate use may be the rule in natural microalgal communities. Lylis and Trainor (1973) argued that experiments demonstrating a relatively low ability of diatom cultures to use acetate should not draw general conclusions that DOC uptake is not ecologically important in this group because other substrates such as glucose have been shown to be more favorably used. The use of DON substrates as C or N source can depend on the role of the substrate in photosynthetic metabolism. For example, some amino acids can be more readily used for C biosynthate than others, and this has been related to the substrate-specific requirement of light (i.e. photosynthate) for uptake (Liu and Hellebust 1976, Flynn and Butler 1986, Flynn and Syrett 1986b). The ease by which organic C compounds can be stored or their potential as respiratory sources also may influence uptake capability, and therefore evaluation of organic substrate effect based on short-term growth responses is not always an accurate indicator of assimilation (Pearce and Carr 1969, Bennett and Hobbie 1972, Neilsen and Lewin 1974, Hellebust and Lewin 1977, Baden and Mende 1978, Richardson and Fogg 1982, Krupka and Feuillade 1988).

Mixtures of substrate additions may cause greatly different physiological responses than single compound additions. Flynn and Butler (1986) pointed out that, when inorganic and organic N are both available, amino acid use can be more efficient than when DON is supplied alone because the need for deamination and subsequent assimilation of NH_4 is reduced. An interdependency between DOC and DON has been demonstrated where enzymes involved in the uptake of one class are induced by the presence of the other. For example, glucose or other DOC substrates have been shown to stimulate amino acid uptake, either by induction of intracellular or cell surface enzymes, or by increasing biosynthetic needs resulting from higher respiration rates (Sauer et al. 1983, Sauer 1984, Berg et al. 1997, Mullholland et al. 1998). Several amino acid uptake systems are present in symbiotic *Chlorella* spp. M. Beijerinck, some of which can be induced by glucose even when growing photosynthetically (Cho and Komor 1985, Komor et al. 1988). Lewin and Hellebust (1976) showed that chemoheterotrophic growth of *Nitzschia angularis* var. *affinis* (Grun.) Perag. on glucose required the presence of amino acids. Growth of some cyanobacteria on glucose was found to be enhanced by amino acids (Smith 1982). Other examples exist where the presence of one DOC compound causes induction of another (e.g. purines induce adenine and guanine uptake by *Chlamydomonas reinhardtii* P.A. Dangeard; Lisa et al. 1995) or the same compound will cause induction of its own uptake system (Neilsen and Lewin 1974, Saks and Kahn 1979, Combres et al. 1994). Transinhibition can also occur where, for example, one substrate inhibits the uptake system for another. Several examples have been cited where DON uptake is inhibited by DIN and vice-versa, but transinhibition of one AA for another also has been demonstrated (Flynn and Butler 1986, Ricketts 1988, Antia et al. 1991, Lisa et al. 1995, Mulholland et al. 1998). The numerous cases of inducible DOM uptake systems imply that lags in enzyme synthesis should be an important consideration in experimental design and interpretation. For example, after transfer from light to dark, DOM uptake systems may require an acclimation period to deplete intracellular AA pools, ATP, cAMP, or other photosynthetic products (North and Stephens 1972, Lylis and Trainor 1973, Neilson and Lewin 1974, Lewin and Hellebust 1975, 1976, 1978, Bollman and Robinson 1977, Sepers 1977, Admiraal 1984, Flynn and Butler 1986, Flynn and Syrett 1986a, Antia et al. 1991).

Importance of Light Conditions and the Relationship Between Heterotrophy and Phototrophy

Heterotrophy refers to the use of organic substrates as C sources. However, environments most often cited as potential sites for microalgal DOC use involve light (i.e. energy) limitation rather than DIC limitation. Although DIC limitation is not as commonly addressed, it has theoretical and empirical bases for occurring, for example during blooms, in microbial mats, or in shallow turbid estuaries (Fogg 1965, Mahoney and MacLaughlin 1977, Bonin and Maestrini 1981, Admiraal et al. 1982, Bollman and Robinson

1985, Raven and Johnston 1991, Nilsson and Sundbäck 1996, Miao and Wu 2002). The use of organic substrates as alternative C sources may pertain in these cases, and this possibility has been understudied. Light is a source of energy and therefore with respect to its function in relieving light limitation, DOC may most accurately be considered a source of energy alternative or supplemental to light (Bonin and Maestrini 1981, Lewitus and Kana 1994, 1995, Berg et al. 1997).

Based on the close interregulation of osmotrophic and photosynthetic pathways, physiological responses of microalgae to DOM use have strict dependence on light conditions; e.g. irradiance level, light/dark cycle, spectral characteristics. Several studies have shown that, when grown under limiting light conditions, DOC use can enhance microalgal growth, but the physiological repercussions of this pathway on photosynthetic metabolism can vary greatly, and include repression of synthesis of photosynthetic machinery, and reduction of, no effect on, or enhancement of photosynthetic performance and potential (Wiessner and Gaffron 1964, White 1974, Schwelitz et al. 1978, Ogawa and Aiba 1981, Saks 1983, Lewitus and Caron 1991a.b, Lewitus et al. 1991, Lewitus and Kana 1994, 1995, Kovacs et al. 2000). Under severe light limitation, the allocation of biosynthate to light harvesting machinery may be restricted such that relief of this energy limitation by DOC respiration may lead to enhanced pigmentation (Lewitus and Kana 1994). That biomass production from photoheterotrophic nutrition can exceed photoautotrophic or chemoheterotrophic nutrition at limiting light levels has been recognized for some time, and is the basis for highly active research efforts on mixotrophic cultivation in biotechnology and aquaculture applications (Day and Tsavalos 1996, Day et al. 1999, Wood et al. 1999, Garci et al. 2000, Lee 2001, Xie et al. 2001, Lebeau and Roberts 2003, Xu et al. 2004).

SUMMARY

The widespread capability of microalgae to assimilate dissolved organic substrates in culture experiments has stimulated curiosity on the process's ecological relevance for decades. Arguments for microalgal DOM uptake in the face of bacterial competition traditionally failed on kinetic grounds, and the relatively high substrate concentrations needed to solicit dark growth or low-light growth enhancement in many studies is still validly criticized. One fallout of this criticism was a loss of momentum in laboratory research on the physiological responses and environmental regulation of marine microalgal osmotrophy, so effectively gained in the 1960's, 1970's, and into the 1980's. Our understanding of microalgal ecology has evolved dramatically over the last two decades, and we now recognize the importance of microenvironments such as those associated with the 'microbial loop', and the widespread significance of small phytoplankton to primary production and phagotrophic microalgae ('mixotrophs') to community grazing. With advances in DOM analytical capabilities and recognition of the important influence of photolysis, the consequence of relatively high DOM lability to microbial food web dynamics has been an active research focus. An important subtopic is the consideration that DON from runoff, atmospheric deposition, or groundwater may be more available to marine microalgae in relief of N limitation. To this end, cell surface or extracellular amino acid oxidases and more recently proteolytic enzymes have received attention in ecological contexts. At the forefront of this research is consideration that many harmful algal blooms may be stimulated by this 'new' DON input. With the advent of these new paradigms, the use of cultures to acquire basic knowledge of microalgal physiology under alternative nutritional states is needed to reevaluate the ecological role of microalgal heterotrophy.

ACKNOWLEDGEMENTS

I am grateful to Krista DeMattio, Megan Heidenreich, and Bonnie Willis for isolating *Kryptoperidinium foliaceum* and *Scrippsiella* sp., and conducting dextran experiments. Thanks also to Patrick Brown for help on the laser confocal miscroscope. These experiments were funded by U.S. ECOHAB Program (sponsored by NOAA/NSF/EPA/NASA/ONR) grant NA16OP2796, NOAA grants NA06OA0675 and NA86RG0052 (SC Sea Grant Consortium grant). Contribution 1386 of University of South Carolina's Belle W. Baruch Institute for Marine and Coastal Sciences, Contribution 546 of South Carolina Department of Natural Resources Marine Resources Research Institute, and ECOHAB Contribution 115.

REFERENCES

Admiraal, W. 1984. The ecology of estuarine sediment-inhabiting diatoms. pp. 265-322 *In* F.E. Round and D.J. Chapman [eds.]. Progress in Phycology Research. Biopress Ltd. Bristol, England.

Admiraal, W., R.W.P.M. Laane and H. Peletier. 1984. Participation of diatoms in the amino acid cycle of coastal waters; uptake and excretion in cultures. Mar. Ecol. Prog. Ser. 15: 303-306.

Admiraal, W. and H. Peletier. 1979. Influence of organic compounds and light limitation on the growth rate of estuarine benthic diatoms. Br. Phycol. J. 14: 197-206.

Admiraal, W., H. Peletier and H. Zomer. 1982. Observations and experiments on the population dynamics of epipelic diatoms from an estuarine mudflat. Estuar. Coast. Shelf Sci. 14: 471-487.

Agusti, S., M.P. Satta, M.P. Mura and E. Benavent. 1998. Dissolved esterase activity as a tracer of phytoplankton lysis: Evidence of high phytoplankton lysis rates in the northeastern Mediterranean. Limnol. Oceanogr. 43: 1836-1849.

Allen, H.L. 1971. Dissolved organic carbon utilization in size-fractionated algal and bacterial commumities. Int. Revue ges. Hydrobiol. 56: 731-749.

Amblard, C., P. Couture and G. Bourdier. 1990. Effects of a pulp and paper mill effluent on the structure and metabolism of periphytic algae in experimental streams. Aquat. Toxicol. 18: 137-162.

Anderson, D.M., P.M. Glibert and J.M. Burkholder. 2002. Harmful algal blooms and eutrophication: nutrient sources, composition, and consequences. Estuaries 25: 704-726.

Anderson, D.M. and F.M.M. Morel. 1982. The influence of aqueous iron chemistry on the uptake of iron by the coastal diatom *Thalassiosira weissflogii*. Limnol. Oceanogr. 27: 789-813.

Anderson, J.J. 1982. The nitrite-oxygen interface at the top of the oxygen minimum zone in the eastern tropical North Pacific. Deep-Sea Res. 29: 1193-1201.

Anderson, J.T., R.R. Hood and X. Zhang. 2003. Quantification of *Pfiesteria piscicida* growth and encystment parameters using a numerical model. Mar. Ecol. Prog. Ser. 246: 105-113.

Antia, N.J. 1980. Nutritional physiology and biochemistry of marine cryptomonads and chrysomonads. pp. 67-115. *In* M. Levandowsky and S.H. Hutner [eds.]. Biochemistry and Physiology of Protozoa. Academic Press, New York, USA.

Antia, N.J., B.R. Berland, D.J. Bonin and S.Y. Maestrini. 1975. Comparative evaluation of certain organic and inorganic sources of nitrogen for phototrophic growth of marine microalgae. J. Mar. Biol. Ass. U.K. 55: 519-539.

Antia, N.J., J.Y. Cheng and F.J.R. Taylor. 1969. The heterotrophic growth of a marine photosynthetic cryptomonad (*Chroomonas salina*). Proc. Intl. Seaweed Symp. 6: 17-29.

Antia, N.J., P.J. Harrison and L. Oliveira. 1991. The role of dissolved organic nitrogen in phytoplankton nutrition, cell biology and ecology. Phycologia 30: 1-89.

App, A.A. and A.T. Jagendorf. 1963. Repression of chloroplast development in *Euglena gracilis* by substrates. J. Protozool. 10: 340-343.

Arnosti, C., D.J. Repeta and N.V. Blough. 1994. Rapid bacterial degradation of polysaccharides in anoxic marine systems. Geochem. Geophys. Acta 58: 2639-2652.

Baden, D.G. and T.J. Mende. 1978. Amino acid utilization by *Gymnodinium breve*. Phytochem. 18: 247-251.

Barber, R.T. 1968. Dissolved organic carbon from deep sea waters resists oxidation. Nature 220: 274-275.

Baretta-Bekker, J.G., J.W. Baretta, A.S. Hansen and B. Riemann. 1998. An improved model of carbon and nutrient dynamics in the microbial food web in marine enclosures. Aquat. Microb. Ecol. 14: 91-108.

Bennett, M.E. and J.E. Hobbie. 1972. The uptake of glucose by *Chlamydomonas* sp. J. Phycol. 8: 392-398.

Berg, G.M., P.M. Glibert, M.W. Lomas and M. Burford. 1997. Organic nitrogen uptake and growth by the chrysophyte *Aureococcus anophagefferens* during a brown tide event. Mar. Biol. 129: 377-387.

Berg, G.M., D.J. Repeta and J. LaRoche. 2002. Dissolved organic nitrogen hydrolysis rates in axenic cultures of *Aureococcus anophagefferens* (Pelagophyceae): Comparison with heterotrophic bacteria. Appl. Environ. Microbiol. 68: 401-404.

Berg, G.M., D.J. Repeta and J. LaRoche. 2003. The role of the picoeukaryote *Aureococcus anophagefferens* in cycling of marine high-molecular weight dissolved organic matter. Limnol. Oceanogr. 48: 1825-1830.

Bergman, B. and E.J. Carpenter. 1991. Nitrogenase confined to randomly distributed trichomes in the marine cyanobacterium *Trichodesmium thiebautii*. J. Phycol. 27: 158-165.

Berman, T. and D.A. Bronk. 2003. Dissolved organic nitrogen: a dynamic participant in aquatic ecosystems. Aquat. Microb. Ecol. 31: 279-305.

Berman-Frank, I., P. Lundgren, Y. Chen, H. Kupper, Z. Kolber, B. Bergmann and P.G. Falkowski. 2001. Segragation of nitrogen fixation and oxygenic photosynthesis in the marine cyanobacteria *Trichodesmium*. Science 294: 1534-1539.

Bernard, F. 1963. Density of flagellates and Myxophyceae in the heterotrophic layers related to environment. pp. 215-228. *In* C.H. Oppenheimer [ed.]. Marine Microbiology. Thomas, Springfield, Illinois, USA.

Biegala, I.C., F. Not, D. Vaulot and N. Simon. 2003. Quantitative assessment of picoeukaryotes in the natural environment by using taxon-specific oligonucleotide probes in association with tyramide signal amplification-fluorescence *in situ* hybridization and flow cytometry. Appl. Environ. Microb. 69: 5519-5529.

Biersmith, A. and R. Benner. 1998. Carbohydrates in phytoplankton and freshly produced dissolved organic matter. Mar. Chem. 63: 131-144.

Bockstahler, K.R. and D.W. Coats. 1993. Spatial and temporal aspects of mixotrophy in Chesapeake Bay dinoflagellates. J. Euk. Microbiol. 40: 49-60.

Bollman, R.C. and G.G.C. Robinson. 1977. The kinetics of organic acid uptake by three Chlorophyta in axenic culture. J. Phycol. 13: 1-5.

Bollman, R.C. and G.G.C. Robinson. 1985. Heterotrophic potential of the green alga, *Ankistrodesmus braunii* (Naeg.). Can. J. Microbiol. 31: 549-554.

Bonin, D.J. and S.Y. Maestrini. 1981. Importance of organic nutrients for phytoplankton growth in natural environments: Implications for algal species succession. Can. Bull. Fish. Aquat. Sci. 210: 279-291.

Bourdier, G., J. Bohatier, M. Fueillade and J. Fueillade. 1989. Amino acid incorporation by a natural population of *Oscillatoria rubescens*. A microautoradiographic study. FEMS Microbiol. Ecol. 62: 185-190.

Breuer E., S.A. Sañudo-Wilhelmy and R.C. Aller. 1999. Trace metals and dissolved organic carbon in an estuary with restricted river flow and a brown tide bloom. Estuaries 22: 603-615.

Broenkow, W.W., A.J. Lewitus and M.A. Yarbrough. 1985. Spectral observations of pigment fluorescence in intermediate depth waters of the North Pacific. J. Mar. Res. 43: 875-891.

Bronk, D.A. 2002. Dynamics of DON. pp. 153-247. *In* D.A. Hansell and C.A. Carlson [eds.]. Biogeochemistry of Marine Dissolved Organic Matter. Academic Press, San Diego, USA.

Bronk, D.A. and P.M. Glibert. 1993. Application of a ^{15}N tracer method to the study of dissolved organic nitrogen uptake during spring and summer in Chesapeake Bay. Mar. Biol. 115: 501-508.

Bunt, J.S. 1969. Observations on photoheterotrophy in a marine diatom. J. Phycol. 5: 37-42.

Burdige, D.J. and S. Zheng. 1998. The biogeochemical cycling of dissolved organic nitrogen in estuarine sediments. Limnol. Oceanogr. 43: 1796-1813.

Burkholder, J.M., M.A. Mallin, H.B. Glasgow Jr., L.M. Larsen, M.R. McIver, G.C. Shank, N. Deamer-Melia, D.S. Briley, J. Springer, B.W. Touchette and E.K. Hannon. 1997. Impacts to a coastal river and estuary from rupture of a large swine waste holding lagoon. J. Environ. Qual. 26: 1451-1466.

Capone, D.G., M.D. Ferrier and E.J. Carpenter. 1994. Amino acid cycling in colonies of the planktonic marine cyanobacterium, *Trichodesmium thiebautii*. Appl. Environ. Microbiol. 60: 3989-3995.

Carlsson, P., H. Edling and C. Béchemin. 1998. Interactions between a marine dinoflagellate (*Alexandrium catanella*) and a bacterial community utilizing riverine humic substances. Aquat. Microb. Ecol. 16: 65-80.

Carlsson, P. and E. Granéli. 1993. Availability of humic bound nitrogen for coastal phytoplankton. Estuar. Coast. Shelf Sci. 36: 433-447.

Carlsson, P. and E. Granéli. 1998. Utilization of dissolved organic matter (DOM) by phytoplankton, including harmful species. pp. 509-524. *In* D.M. Anderson, A.D. Cembella and G.M. Hallegraeff [eds.]. Physiological Ecology of Harmful Algal Blooms. Springer, Berlin, Germany.

Carlsson, P., A.Z. Segatto and E. Granéli. 1993. Nitrogen bound to humic matter of terrestrial origin - a nitrogen pool for coastal phytoplankton? Mar. Ecol. Prog. Ser. 97: 105-116.

Caron, D.A. 1994. Inorganic nutrients, bacteria, and the microbial loop. Microb. Ecol. 28: 295-298.

Carpenter, E.J., C.C. Remsen and B.W. Schroeder. 1972. Comparison of laboratory and *in situ* measurements of urea decomposition by a marine diatom. J. Exp. Mar. Biol. Ecol. 8: 259-264.

Chapman, J.S. and J.C. Meeks. 1983. Glutamine and glutamate transport by *Anabaena variabilis*. J. Bact. 156: 122-129.

Cheng, J.Y. and N.J. Antia. 1970. Enhancement by glycerol of phototrophic growth of marine planktonic algae and its significance to ecology of glycerol pollution. J. Fish. Res. Bd. Canada 27: 335-346.

Chisholm, S.W., S.L. Frankel, R. Goericke, R.J. Olson, B. Palenik, J.B. Waterbury, L. West-Johnsrud and E.R. Zettler. 2001. Validation List No. 79 Int. J. Syst. Evol. Microbiol. 51: 263-265.

Cho, B.-H. and E. Komor. 1985. The amino acid transport systems of the autotrophically grown green alga *Chlorella*. Biochim. Biophys. Acta 821: 384-392.

Cole, J.J. 1999. Aquatic microbiology for ecosystem scientists: new and recycled paradigms in ecological microbiology. Ecosystems 2: 215-225.

Combres, C., G. Laliberté, J. Sevrin Reyssac and J. de la Noüe. 1994. Effect of acetate on growth and ammonium uptake in the microalga *Scenedesmus obliquus*. Physiol. Plant. 91: 729-734.

Cooksey, K.E. and H. Chansang. 1976. Isolation and physiological studies on three isolates of *Amphora* (Bacillariophyceae). J. Phycol. 12: 455-460.

Cooksey, B. and K.E. Cooksey. 1988. Chemical signal response in diatoms of the genus *Amphora*. J. Cell. Sci. 91: 523-529.

Danforth, W.F. 1962. Substrate assimilation and heterotrophy. pp. 99-123. *In* R.A. Lewin [ed.]. Physiology and Biochemistry of Algae. Academic Press, New York, USA.

Day, J.G., E.E. Benson and R.A. Fleck. 1999. *In vitro* culture and conservation of microalgae: Applications for aquaculture, biotechnology and environmental research. In Vitro Cell. Dev. Biol.: Plant 35: 127-136.

Day, J.G. and A.J. Tsavalos. 1996. An investigation of the heterotrophic culture of the green alga *Tetraselmis*. J. Appl. Phycol. 8: 73-77.

Decho, A.W. 1990. Microbial exopolymer secretions in ocean environments: their role(s) in food webs and marine processes. Oceanogr. Mar. Biol. Ann. Rev. 28: 73-153.

Detmer, A.E., H.C. Giesenhagen, V.M. Trenkel, H. Auf dem Venne and F.J. Jochem. 1993. Phototrophic and heterotrophic pico- and nanoplankton in anoxic depths of the central Baltic Sea. Mar. Ecol. Prog. Ser. 99: 197-203.

DeYoe, H.R. and C.A. Suttle. 1994. The inability of the Texas "brown tide" alga to use nitrate and the role of nitrogen in the initiation of a persistent bloom of this organism. J. Phycol. 30: 800-806.

Doblin, M.A., S.I. Blackburn and G.M. Hallegraeff. 1999. Growth and biomass stimulation of the toxic dinoflagellate *Gymnodinium catenatum* (Graham) by dissolved organic substances. J. Exp. Mar. Biol. Ecol. 236: 33-47.

Droop, M.R. 1974. Heterotrophy of carbon. pp. 530-559. *In* W.D.P. Stewart [ed.]. Algal Physiology and Biochemistry. Blackwell, Oxford, UK.

Droop, M.R. and S. McGill. 1966. The carbon nutrition of some algae: the inability to utilize glycolic acid for growth. J. Mar. Biol. Assoc. U.K. 46: 679-684.

Dzurica, S., C. Lee, E.M. Cosper and E.J. Carpenter. 1989. Role of environmental variables, specifically organic compounds and micronutrients, in the growth of the chrysophyte *Aureococcus anophagefferens*. pp. 229-252. *In* E.M. Cosper, V.M. Bricelj and E.J. Carpenter [eds.]. Novel Phytoplankton Blooms: Causes and Impacts of Recurrent Brown Tides and other Unusual Blooms. Springer-Verlag, New York, USA.

Fan, C., P.M. Glibert, J. Alexander and M.W. Lomas. 2003. Characterization of urease activity in three marine phytoplankton species, *Aureococcus anophagefferens*, *Prorocentrum minimum*, and *Thalassiosira weissflogii*. Mar. Biol. 142: 949-958.

Ferrier-Pagés, C., M. Karner and F. Rassoulzadegan. 1998. Release of dissolved amino acids by flagellates and ciliates grazing on bacteria. Oceanol. Acta 21: 485-494.

Feuillade, M., J. Bohatier, G. Bourdier, Ph. Dufour, J. Feudillade and H. Krupka. 1988. Amino acid uptake by a natural population of *Oscillatoria rubescens* in relation to uptake by bacterioplankton. Arch. Hydrobiol. 113: 345-358.

Flynn, K.J. and I. Butler. 1986. Nitrogen sources for the growth of marine microalgae: role of dissolved free amino acids. Mar. Ecol. Prog. Ser. 34: 281-304.

Flynn, K.J. and P.J. Syrett. 1986a. Characteristics of the uptake system for L-lysine and L-arginine in *Phaeodactylum tricornutum*. Mar. Biol. 90: 151-158.

Flynn, K.J. and P.J. Syrett. 1986b. Utilization of L-lysine and L-arginine by the diatom *Phaeodactylum tricornutum*. Mar. Biol. 90: 159-163.

Fogg, G.E. 1965. Algal Cultures and Phytoplankton Ecology. University of Wisconsin Press, Madison and Milwaukee, USA.

Fournier, R.O. 1966. North Atlantic Deep-Sea fertility. Science 153: 1250-1252.

Fournier, R.O. 1970. Studies on pigmented microorganisms from aphotic marine environments. Limnol. Oceanogr. 15: 675-682.

Fournier, R.O. 1971. Studies on pigmented microorganisms from aphotic marine environments. II. North Atlantic distribution. Limnol. Oceanogr. 16: 952-961.

Fruend, C., E. Romem and A.F. Post. 1993. Ecological physiology of an assembly of photosynthetic microalgae in wastewater oxidation ponds. Wat. Sci. Tech. 27: 143-149.

Fuhrman, J.A. 1999. Marine viruses and their biogeochemical and ecological effects. Nature 399: 541-548.

Fuhrman, J.A. and C.A. Suttle. 1993. Viruses in marine planktonic systems. Oceanography 6: 50-62.

Gaines, G. and M. Elbrächter. 1987. Heterotrophic nutrition. pp. 224-268. *In* F.J.R. Taylor [ed.]. The Biology of Dinoflagellates. Blackwell Scientific, Oxford, UK.

Garci, M.C.C., J.M.F. Sevilla, F.G.A. Fernandez, E.M. Grima and F.G. Camacho. 2000. Mixotrophic growth of *Phaeodactylum tricornutum* on glycerol: growth rate and fatty acid profile. J. Appl. Phycol. 12: 239-248.

Geller, A. 1985. Light-induced conversion of refractory, high molecular weight lake water constituents. Schweiz. Z. Hydrol. 47: 21-26.

Glibert, P.M. and D.A. Bronk. 1994. Release of dissolved organic nitrogen by marine diazotrophic cyanobacteria, *Trichodesmium* spp. Appl. Environ. Microbiol. 60: 3996-4000.

Glibert, P.M., R. Magnien, M.W. Lomas, J. Alexander, C. Fan, E. Haramoto, M. Trice and T.M. Kana. 2001. Harmful algal blooms in the Chesapeake and coastal bays of Maryland, USA: Comparison of 1997, 1998, and 1999 events. Estuaries 24: 875-883.

Glover, H, 1978. Iron in marine coastal waters; seasonal variation and its apparent correlation with a dinoflagellate bloom. Limnol. Oceanogr. 23: 534-7.

Gobler, C.J. and S.A. Sañudo-Wilhelmy. 2001. Effects of organic carbon, organic nitrogen, inorganic nutrients, and iron additions on the growth of phytoplankton and bacteria during a brown tide bloom. Mar. Ecol. Prog. Ser. 209: 19-34.

Gobler, C.J. and S.A. Sañudo-Wilhelmy. 2003. Cycling of colloidal organic carbon and nitrogen during an estuarine phytoplankton bloom. Limnol. Oceanogr. 48: 2314-2320.

Granéli, E., L. Edler, D. Gedziorowska and U. Nyman. 1985. Influence of humic and fulvic acids on *Prorocentrum minimum* (Pav.) Schiller. pp. 201-206. *In* D.M. Anderson, A.W. White and D.G. Baden [eds.]. Toxic Dinoflagellates. Elsevier, New York, USA.

Granéli, E., H. Persson and L. Edler. 1986. Connections between trace metals, chelators and Red Tide blooms in the Laholm Bay, SE Kattegat – An experimental approach. Mar. Environ. Res. 18: 61-78.

Granéli, E., P. Olson, B. Sundstrom and L. Edler. 1989. *In situ* studies of the effects of humic acids on dinoflagellates and diatoms. pp. 209-212. *In* T. Okaichi, D.M. Anderson and T. Nemoto [eds.]. Red Tides: Biology, Environmental Science and Toxicology. Elsevier, New York, USA.

Granéli, E. and M.O. Moreira. 1990. Effects of river water of different origin on the growth of marine dinoflagellates and diatoms in laboratory culture. J. Exp. Mar. Biol. Ecol. 136: 89-106.

Granéli, E., D.M. Anderson, P. Carlsson and Y. Maestrini. 1997. Light and dark carbon uptake by *Dinophysis* species in comparison to other photosynthetic and heterotrophic dinoflagellates. Aquat. Microb. Ecol. 13: 177-186.

Goericke, R., R.J. Olson and A. Shalapyonok. 2000. A novel niche for *Prochlorococcus* sp. *In* low-light suboxic environments in the Arabian Sea and the Eastern Tropical North Pacific. Deep-Sea Res. 47: 1183-1205.

Hallegraeff, G.M. 1993. A review of harmful algal blooms and their apparent global increase. *Phycologia* 32: 79-99.

Hamilton, R.D., O. Holm-Hansen and J.D.H. Strickland. 1968. Notes on the occurrence of living microscopic organisms in deep water. Deep-Sea Res. 15: 651-656.

Harris, R.C. and J.T.O. Kirk. 1969. Control of chloroplast formation in *Euglena gracilis*: antagonism between carbon and nitrogen sources. Biochem. J. 113: 195-205.

Healey, F.P. 1980. Slope of the monod equation as an indicator of advantage in nutrient competition. Microb. Ecol. 5: 281-286.

Heil, C.A., G.A. Vargo, D.N. Spence, M.B. Neely, R. Merkt, K.M. Lester and J.J. Walsh. 2001. Nutrient stoichiometry of a *Gymnodinium breve* bloom: What limits blooms in oligotrophic environments? pp. 165-168 *In* G.M. Hallegraeff, S. Blackburn, C. Bolch and R. Lewis [eds.]. Harmful Algal Blooms 2000. IOC UNESCO, Paris, France.

Hellebust, J.A. 1970. The uptake and utilization of organic substances by marine phytoplankters. pp. 223-256. *In* D.W. Hood [ed.]. Symposium on Organic Matter in Natural Waters. University of Alaska, Fairbanks, Alaska, USA.

Hellebust, J.A. and J. Lewin. 1977. Heterotrophic nutrition. pp. 169-197. *In* D. Werner [ed.]. The Biology of Diatoms. Univ. California Press, Berkeley, California, USA.

Henrichs, S.M. and J.W. Farrington. 1979. Amino acids in interstitial waters of marine sediments. Nature 279: 319-322.

Hobbie, J.E. and C.C. Crawford. 1969. Bacterial uptake of organic substrate: new methods of study and application to eutrophication. Verh. Internat. Verein. Limnol. 17: 725-730.

Hobbie, J.E. and R.T. Wright. 1965. Competition between planktonic bacteria and algae for organic solutes. Mem. 1st Ital. Idrobiol. Suppl. 18: 175-183.

Hopkinson, C.S., A.E. Giblin, R.H. Garritt, J. Tucker and M.A.J. Hullar. 1998. Influence of the benthos on growth of planktonic estuarine bacteria. Aquat. Microb. Ecol. 16: 109-118.

Horstmann, U. 1980. Observations on the peculiar diurnal migrations of a red tide Dinophyceae in tropical shallow waters. J. Phycol. 16: 481-485.

Ittekot, V., U. Brockmann, W. Michaelis and E.T. Degens. 1981. Dissolved free and combined carbohydrates during a phytoplankton bloom in the northern North Sea. Mar. Ecol. Prog. Ser. 4: 299-305.

Iwasaki, H. 1979. Physiological ecology of red tide flagellates. pp. 357-393. *In* M. Levandowsky and S. Hutner [eds.]. Biochemistry and Physiology of Protozoa. Academic Press, New York, USA.

John, E.H. and K.J. Flynn. 1999. Amino acid uptake by the toxic dinoflagellate *Alexandrium fundyense*. Mar. Biol. 133: 11-19.

Jones, H.L.J. 1997. A classification of mixotrophic protists based on their behavior. Freshwat. Biol. 37: 35-43.

Jones, H.L.J., C. Cockell and L.J. Rothschild. 1996. Intraspecies variation in *Poterioochromonas* due to long-term culturing conditions. Phycologist 43: 24.

Jørgensen, N.O.G. 1982. Heterotrophic assimilation and occurrence of dissolved free amino acids in a shallow estuary. Mar. Ecol. Prog. Ser. 8: 145-159.

Kieber, D.J., J. McDaniel and K. Mopper. 1989. Photochemical source of biological substrates in sea water: Implications for carbon cycling. Nature 341: 637-639.

Kirchman, D.L. 1994. The uptake of inorganic nutrients by heterotrophic bacteria. Microb. Ecol. 28: 255-271.

Kirchman, D.L. and P.A. Wheeler. 1998. Uptake of ammonium and nitrate by heterotrophic bacteria and phytoplankton in the sub-Arctic Pacific. Deep-Sea Res. 45: 347-365.

Klut, M.E., T. Bialputra and N.J. Antia. 1987. Some observations on the structure and function of the dinoflagellate pusule. Can J. Bot. 65: 736-744.

Komor, E., B.-H. Cho and M. Kraus. 1988. The occurrence of the glucose-inducible transport systems for glucose, proline, and arginine in different species of *Chlorella*. Bot. Acta 101: 321-326.

Kondo, K., Y. Seike and Y. Date. 1990. Red tides in the brackish Lake Nakanoumi (III). The stimulative effects of organic substances in the interstitial water of bottom sediments and in the excreta from *Skeletonema costatum* on the growth of *Prorocentrum minimum*. Bull. Plank. Soc. Japan 37: 34-47.

Kovacs, L., W. Wiessner, M. Kis, F. Nagy, D. Mende and S. Detemer. 2000. Short- and long-term redox regulation of photosynthetic light energy distribution and photosystem stoichiometry by acetate metabolism in the green alga, *Chlamydobotrys stellata*. Photosynth. Res. 65: 231-247.

Kristiansen, J. 1990. Phylum Chrysophyta. pp. 438-453. *In* L. Margulis, J.O. Corliss, M. Melkonian and D.J. Chapman [eds.]. Handbook of Protoctista. Jones and Bartlett, Boston, USA.

Krupka, H.M. and M. Feuillade. 1988. Amino acids as a nitrogen source for growth of *Oscillatoria rubescens* D. C. Arch. Hydrobiol. 112: 125-142.

Lara, R.J., U. Hubberten and G. Kattner. 1993. Contribution of humic substances to the dissolved nitrogen pool in the Greenland Sea. Mar. Chem. 41: 327-336.

LaRoche, J., R. Nuzzi, R. Waters, K. Wyman, P.G. Falkowski and D.W.R. Wallace. 1997. Brown tide blooms in Long Island's coastal waters linked to interannual variability in groundwater flow. Global Change Biol. 3: 101-114.

Larsen, J. 1987. Algal studies of the Danish Wadden Sea. IV. A taxonomic study of the interstitial euglenoid flagellates. Nord. J. Bot. 7: 589-607.

Laws, E.A., W.G. Harrison and G.R. DiTullio. 1985. A comparison of nitrogen assimilation rates based on ^{15}N uptake and autotrophic protein synthesis. Deep-Sea Res. 32: 85-95.

Lee, E.S., A.J. Lewitus and R.K. Zimmer. 1999. Chemoreception in a marine cryptophyte: Behavioral plasticity in response to amino acids and nitrate. Limnol. Oceanogr. 44: 1571-1574.

Lee, J.J., M.E. McEnery, E.M. Kennedy and H. Rubin. 1973. Educiing the functional relationships among the diatom assemblages within sublittoral salt marsh epiphytic communities. Bull. Ecol. Res. Comm. 17: 387-397.

Lee, J.J., M.E. McEnery, E.M. Kennedy and H. Rubin. 1975. A nutritional analysis of a sublittoral diatom assemblage epiphytic on *Enteromorpha* from a Long Island salt marsh. J. Phycol. 11: 14-19.

Lee, Y.K. 2001. Microalgal mass culture systems and methods: Their limitation and potential. J. Appl. Phycol. 13: 307-315.

Lebeau, T. and J.M. Robert. 2003. Diatom cultivation and biotechnologically relevant products. Part I: Cultivation at various scales. Appl. Microbiol. Biotechnol. 60: 612-623.

Legrand, C. and P. Carlsson. 1998. Uptake of high molecular weight dextran by the dinoflagellate *Alexandrium catenella*. Aquat. Microb. Ecol. 16: 81-86.

Lenes, J.M., B.P. Darrow, C. Cattrall, C.A. Heil, M. Callahan, G.A. Vargo and R.H. Byrne. 2001. Iron fertilization and the *Trichodesmium* response on the West Florida shelf. Limnol. Oceanogr. 46: 1261-1277.

Lester, K.M., R. Merkt, C.A. Heil, G.A. Vargo, M.B. Neely, D.N. Spence, L. Melahn and J.J. Walsh. 2001. Evolution of a *Gymnodinium breve* (Gymnodiniales, Dinophyceae) red tide bloom on the West Florida shelf: relationship with organic nitrogen and phosphorus. pp. 161-164. *In* G.M. Hallegraeff, S. Blackburn, C. Bolch and R. Lewis [eds.]. Harmful Algal Blooms 2000. IOC UNESCO, Paris, France.

Levandowsky, M. and D.C.R. Hauser. 1978. Chemosensory responses of swimming algae and protozoa. pp. 145-210. *In* G.H. Bairne and J.F. Danielli [eds.]. International Review of Cytology. Academic Press, New York, USA.

Lewin, J.C. 1963. Heterotrophy in marine diatoms. pp. 229-235. *In* C.H. Oppenheimer [ed.]. Marine Microbiology. C.C. Thomas, Springfield, Illinois, USA.

Lewin, J. and J.A. Hellebust. 1975. Heterotrophic nutrition of the marine pennate diatom *Navicula pavillardi* Hustedt. Can. J. Microbiol. 21: 1335-1342.

Lewin, J. and J.A. Hellebust. 1976. Heterotrophic nutrition of the marine pennate diatom *Nitzschia angularis* var. *affinis* (Grun.) Perag. Mar. Biol. 36: 313-320.

Lewin, J. and J.A. Hellebust. 1978. Utilization of glutamate and glucose for heterotrophic growth by the marine pinnate diatom *Nitzschia laevis*. Mar. Biol. 47: 1-7.

Lewin, J. and R.A. Lewin. 1967. Culture and nutrition of some apochlorotic diatoms of the genus *Nitzschia*. J. Gen. Microbiol. 46: 361-367.

Lewitus, A.J. and W.W. Broenkow. 1985. Intermediate depth pigment maxima in oxygen minimum zones. Deep-Sea Res. 32: 1101-1115.

Lewitus A.J. and D.A. Caron. 1991a. Physiological responses of phytoflagellates to dissolved organic substrate additions. 1. Dominant role of heterotrophic nutrition in *Poterioochromonas malhamensis* (Chrysophyceae). Plant Cell Physiol. 32: 671-680.

Lewitus, A.J. and D.A. Caron. 1991b. Physiological responses of phytoflagellates to dissolved organic substrate additions. 2. Dominant role of autotrophic nutrition in *Pyrenomonas salina* (Cryptophyceae). Plant Cell Physiol. 32: 791-801.

Lewitus, A.J., D.A. Caron and K.R. Miller. 1991. Effects of light and glycerol on the organization of the photosynthetic apparatus in the facultative heterotroph *Pyrenomonas salina* (Cryptophyceae). J. Phycol. 27: 578-587.

Lewitus, A.J., H.B. Glasgow and J.M. Burkholder. 1999. Kleptoplastidy in the toxic dinoflagellate, *Pfiesteria piscicida* (Dinophyceae). J. Phycol. 35: 303-12.

Lewitus, A.J., K.C. Hayes, S.G. Gransden, H.B. Glasgow, Jr., J.M. Burkholder, P.M. Glibert and S.L. Morton. 2001. Ecological characterization of a widespread *Scrippsiella* red tide in South Carolina estuaries: a newly observed phenomenon. pp. 129-132. *In* G.M. Hallegraeff, S. Blackburn, C. Bolch and R. Lewis [eds.]. Harmful Algal Blooms 2000. IOC UNESCO, Paris, France.

Lewitus, A.J. and T.M. Kana. 1994. Responses of estuarine phytoplankton to exogenous glucose: Stimulation versus inhibition of photosynthesis and respiration. Limnol. Oceanog. 39: 182-189.

Lewitus, A.J. and T.M. Kana. 1995. Light respiration in six estuarine phytoplankton clones: contrasts under autotrophic and mixotrophic growth conditions. J. Phycol. 31: 754-761.

Lewitus, A.J., E.T. Koepfler and R. Pigg. 2000. Use of dissolved organic nitrogen by a salt marsh phytoplankton bloom community. Arch. Hydrobiol. Spec. Issues Adv. Limnol. 55: 441-56.

Lin, S., S. Henze, P. Lundgren, B. Bergman and E.J. Carpenter. 1998. Whole cell immunolocalization of nitrogenase in the marine diazotrophic cyanobacterium *Trichodesmium*. Appl. Envir. Microbiol. 64: 3052-3058.

Lindell, M.J., W. Granéli and L.J. Tranvik. 1995. Enhanced bacterial growth in response to photochemical transformation of dissolved organic matter. Limnol. Oceanogr. 40: 195-199.

Lindell, M.J., H.W. Granéli and L.J. Tranvik. 1996. Effects of sunlight on bacterial growth in lakes of different humic content. Aquat. Microb. Ecol. 11: 135-141.

Lisa, T.A., P. Piedras, J. Cárdenas and M. Pineda. 1995. Utilization of adenine and guanine as nitrogen sources by *Chlamydomonas reinhardtii* cells. Plant, Cell Environ. 18: 583-588.

Liu, M.S. and J.A. Hellebust. 1974. Uptake of amino acids by the marine centric diatom *Cyclotella cryptica.* Can. J. Microbiol. 20: 1109-1118.

Liu, M.S. and J.A. Hellebust. 1976. Regulation of proline metabolism in the marine centric diatom *Cyclotella cryptica*. Can. J. Bot. 54: 949-959.

Lylis, J.C. and F.R. Trainor. 1973. The heterotrophic capabilities of *Cyclotella meneghiniana.* J. Phycol. 9: 365-369.

MacIntyre, H.L. and J.J. Cullen. 1996. Primary production by suspended and benthic microalgae in a turbid estuary: time-scales of variability in San Antonio Bay, Texas. Mar. Ecol. Prog. Ser. 145: 245-268.

MacIntyre. H.L., D.C. Geider and D.C. Miller. 1996. Microphytobenthos: the ecological role of the "secret garden" of unvegetated shallow-water marine habitats. I. Distribution, abundance and primary production. Estuaries 19: 186-201.

MacIntyre, H.L., T.M. Kana and R.J. Geider. 2000. The effect of water motion on short-term rates of photosynthesis by marine phytoplankton. Trends in Plant Sci. 5: 12-17.

Maeda, O. and S.-E. Ichimura. 1973. On the high density of a phytoplankton population found in a lake under ice. Int. Revue des. Hydrobiol. 58: 673-685.

Maestrini, S.Y., M. Balode, C. Béchemin and I. Purina. 1999. Nitrogenous organic substances as potential nitrogen sources, for summer phytoplankton in the Gulf of Riga, eastern Baltic Sea. Plankton Biol. Ecol. 46: 8-17.

Mahoney, J.B. and J.J.A. McLaughlin. 1977. The association of phytoflagellate blooms in Lower New York Bay with hypereutrophication. J. Exp. Mar. Biol. Ecol. 28: 53-65.

Malone, T.C., C. Garside, R. Anderson and O.A. Roels. 1973. The possible occurrence of photosynthetic microorganisms in deep-sea sediments of the North Atlantic. J. Phycol. 9: 482-488.

Marchant, H.J. and F.J. Scott. 1993. Uptake of sub-micrometre particles and dissolved organic material by Antarctic choanoflagellates. Mar. Ecol. Prog. Ser. 92: 59-64.

Margulis, L., J.O. Corliss, M. Melkonian and D.J. Chapman [eds.] 1990. Handbook of Protoctista. Jones and Bartlett Publishers, Boston, USA.

Martinez, J. and F. Azam. 1993. Aminopeptidase activity in marine chroococcoid cyanobacteria. Appl. Environ. Microbiol. 59: 3701-3707.

McKinley, K.R. 1977. Light-mediated uptake of ^{3}H-glucose in a small hard-water lake. Ecology 58: 1356-1365.

Miao, X.L. and Q.Y. Wu. 2002. Inorganic carbon utilization in some marine phytoplankton species. Acta Botanica Sin. 44: 395-399.

Middleboe, M., N. Kroer, N.O.G. Jørgensen and D. Paluski. 1998. Influence of sediment on pelagic carbon and nitrogen turnover in a shallow Danish estuary. Aquat. Microb. Ecol. 14: 81-90.

Milligan, A.J. and E.M. Cosper. 1997. Growth and photosynthesis of the 'brown tide' microalga *Aureococcus anophagefferens* in subsaturating constant and fluctuating irradiance. Mar. Ecol. Prog. Ser. 153: 67-75.

Moll, R. 1984. Heterotrophy by phytoplankton and bacteria in Lake Michigan. Verh. Internat. Verein. Limnol. 22: 431-434.

Monroy, A.F. and S.D. Schwartzbach. 1984. Catabolite repression of chloroplast development in *Euglena*. Proc. Natl. Acad. Sci. USA 81: 2786-2790.

Moore, L.R., A.F. Post, G. Rocap and S.W. Chisholm. 2002. Utilization of different nitrogen sources by the marine cyanobacteria *Prochlorococcus* and *Synechococcus*. Limnol. Oceanogr. 47: 989-996.

Moran, M.A. and R.G. Zepp. 1997. Role of photoreactions in the formation of biologically labile compounds from dissolved organic matter. Limnol. Oceanogr. 42: 1307-1316.

Morrill, L.C. and A.R. Loeblich III. 1979. An investigation of heterotrophic and photoheterotrophic capabilities in marine Pyrrhophyta. Phycologia 18: 394-404.

Mulholland, M.R. and D.G. Capone. 1999. N_2 fixation, N uptake and N metabolism in natural and cultured populations of *Trichodesmium* spp. Mar. Ecol. Prog. Ser. 188: 33-49.

Mulholland, M.R. and D.G. Capone. 2000. The nitrogen physiology of the marine N_2-fixing cyanobacteria *Trichodesmium* spp. Trends Plant Sci. 5: 148-153.

Mulholland, M.R., P.M. Glibert, G.M. Berg, L. Van Heukelem, S. Pantoja and C. Lee. 1998. Extracellular amino acid oxidation by microplankton: a cross-ecosystem comparison. Aquat. Microb. Ecol. 15: 141-152.

Mulholland, M.R., K. Ohki and D.G. Capone. 1999. N utilization and metabolism relative to patterns of N_2 fixation in cultures of *Trichodesmium* NIBB1067. J. Phycol. 35: 977-988.

Mulholland, M.R., K. Ohki and D.G. Capone. 2001. Nutrient controls on nitrogen uptake and metabolism by natural populations and cultures of *Trichodesmium* (cyanobacteria). J. Phycol. 37: 1001-1009.

Mulholland, M.R., S. Floge, E.J. Carpenter and D.G. Capone. 2002a. Phosphorus dynamics in culture and natural populations of *Trichodesmium* spp. Mar. Ecol. Prog. Ser. 239: 45-55.

Mulholland, M.R., C.J. Gobler and C. Lee. 2002b. Peptide hydrolysis, amino acid oxidation and N uptake in communities seasonally dominated by *Aureococcus anophagefferens*. Limnol. Oceanogr. 47: 1094-1108.

Mulholland, M.R., C. Lee and P.M. Glibert. 2003. Extracellular enzyme activity and uptake of carbon and nitrogen along an estuarine salinity and nutrient gradient. Mar. Ecol. Prog. Ser. 258: 3-17.

Myers, J. and J. Graham. 1956. The role of photosynthesis in the physiology of *Ochromonas*. J. Cell. Comp. Phys. 47: 397-414.

Nedoma, J., J. Vrba, J. Hejzlar, K. Šimek and V. Straškrabová. 1994. *N*-acetylglucosamine dynamics in freshwater environments: Concentration of amino sugars, extracellular enzyme activities, and microbial uptake. Limnol. Oceanogr. 39: 1088-1100.

Neilson, A.H. and R.A. Lewin. 1974. The uptake and utilization of organic carbon by algae: a essay in comparative biochemistry. Phycologia 13: 227-264.

Neilson, A.H. and T. Larsson. 1980. The utilization of organic nitrogen for growth of algae: physiological aspects. Physiol. Plant. 48: 542-553.

Nilsson, C. and K. Sundbäck. 1996. Amino acid uptake in natural microphytobenthic assemblages studied by microautoradiography. Hydrobiologia 332: 119-129.

North, B.B. and G.C. Stephens. 1972. Amino acid transport in *Nitzschia ovalis* Arnott. J. Phycol. 8: 64-68.

Ogawa, T. and S. Aiba. 1981. Bioenergetic analysis of mixotrophic growth in *Chlorella vulgaris* and *Scenedesmus acutus*. Biotechnol. Bioenerg. 23: 1121-1132.

Paerl, H.W. 1988. Nuisance phytoplankton blooms in coastal, estuarine and inland waters. Limnol. Oceanogr. 33: 823-847.

Paerl, H.W. 1991. Ecophysiological and trophic implications of light-stimulated amino acid utilization in marine picoplankton. Appl. Environ. Microbiol. 57: 473-479.

Paerl, H.W. 1997. Coastal eutrophication and harmful algal blooms: Importance of atmospheric deposition and groundwater as "new" nitrogen and other nutrient sources.

Paerl, H.W., B.M. Bebout and L.E. Prufert. 1989. Bacterial associations with marine *Oscillatoria* sp. (*Trichodesmium* sp.) populations: ecophysiological implications. J. Phycol. 25: 773-784.

Paerl, H.W., B.M. Bebout, S.B. Joye and D.J. Des Marais. 1993. Microscale characterization of dissolved organic matter production and uptake in marine microbial mat communities. Limnol. Oceanogr. 38: 1150-1161.

Palenik, B. and F.M.M. Morel. 1990a. Comparison of cell-surface L-amino acid oxidases from several marine phytoplankton. Mar. Ecol. Prog. Ser. 59: 195-201.

Palenik, B. and F.M.M. Morel. 1990b. Amino acid utilization by marine phytoplankton: A novel mechanism. Limnol. Oceanogr. 35: 260-269.

Palenik, B. and F.M.M. Morel. 1991. Amine oxidases of marine phytoplankton. Appl. Environ. Microbiol. 57: 2440-2443.

Palmer, E.G. and R.K. Togasaki. 1971. Acetate metabolism by an obligate phototrophic strain of *Pandora morum*. J. Protozool. 18: 640-644.

Palmisano, A.C., S.T. Kottmeier, R.L. Moe and C.W. Sullivan. 1985. Sea ice microbial communities. IV. The effect of light perturbation on microalgae at the ice-seawater interface in McMurdo Sound, Antarctica. Mar. Ecol. Prog. Ser. 21: 37-45.

Pant, A. and G.E. Fogg. 1976. Uptake of glycollate by *Skeletoneme costatum* (Grev.) Cleve in bacterized culture. J. Exp. Mar. Biol. Ecol. 22: 227-264.

Pantoja, S. and C. Lee. 1994. Cell-surface oxidation of amino acids in seawater. Limnol. Oceanogr. 39: 1718-1726.

Pearce, J. and N.G. Carr. 1969. The incorporation and metabolism of glucose by *Anabaena variabilis*. J. Gen. Microbiol. 54: 451-462.

Pinckney, J and R.G. Zingmark. 1993. Biomass and production of benthic microalgal communities in estuarine habitats. Estuaries 16: 887-897.

Pintner, I.J. and L. Provasoli. 1968. Heterotrophy in subdued light of 3 *Chrysochromulina* species. Bull. Misaki Mar. Biol. Inst., Kyoto Univ. 12: 25-31.

Pip, E. and G.G.C. Robinson. 1982. A study of the seasonal dynamics of three phycoperiphytic communities using nuclear track autoradiography. II. Organic carbon uptake. Arch. Hydrobiol. 96: 47-64.

Platt, T., D.V. Subba Rao, J.C. Smith, W.K. Li, B. Irwin, E.P.W. Horne and D.D. Sameoto. 1983. Photosynthetically-competent phytoplankton from the aphotic zone of the deep ocean. Mar. Ecol. Prog. Ser. 10: 105-110.

Pollingher, U. and T. Berman. 1976. Autoradiographic screening for potential heterotrophc in natural algal populations of Lake Kinneret. Microb. Ecol. 2: 252-260.

Prakash, A. and M.A. Rashid. 1968. Influence of humic substances on the growth of marine phytoplankton: dinoflagellates. Limnol. Oceanogr. 13: 598-606.

Pringsheim, E.G. 1952. On the nutrition of *Ochromonas*. Q.J. Microscop. Sci. 93: 71-96.

Provasoli, L., J.J.A. McLaughlin and M.R. Droop. 1957. The development of artificial media for marine algae. Arch Mikrobiol. 25: 392-428.

Quintana, X.D., F.A. Comín and R. Moreno-Amich. 2002. Biomass-size spectra in aquatic communities in shallow fluctuating Mediterranean salt marshes (Empordà wetlands, NE Spain). J. Plankton Res. 24: 1149-1161.

Rahat, M. and T.L. Jahn. 1965. Growth of *Prymnesium parvum* in the dark; note on the ichthyotoxin formation. J. Protozool. 12: 246-250.

Raven, J.A. 1997. Phagotrophy in phototrophs. Limnol. Oceanogr. 42: 198-205.

Raven, J.A. and A.M. Johnston. 1991. Carbon assimilation mechanisms: Implications for intensive culture of seaweeds. pp. 151-166. *In* G. Garcia Reina and M. Pedersen [eds.]. Seaweed Cellular Biotechnology, Physiology and Intensive Cultivation. Universidad las Palmas Gvan Canaria, Spain.

Rich, J.H., H.W. Ducklow and D.L. Kirchman. 1996. Concentrations and uptake of neutral monosaccharides along 140° W in the equatorial Pacific: contribution of glucose to heterotrophic bacterial activity and DOM flux. Limnol. Oceanogr. 41: 595-604.

Richardson, K. and G.E. Fogg. 1982. The role of dissolved organic material in the nutrition and survival of marine dinoflagellates. Phycologia 18: 378-382.

Ricketts, T.R. 1988. Homeostasis in nitrogenous uptake/assimilation by the green alga *Platymonas* (*Tetraselmis*) *striata* (Prasinophyceae). Ann. Bot. 61: 451-458.

Riley, G.A. 1963. Organic aggregates in sea water and the dynamics of their formation and utilization. Limnol. Oceanogr. 35: 455-463.

Rippka, R., T. Coursin, W. Hess, C. Lichtle, D.J. Scanlan, K.A. Palinska, I. Iteman, F. Partensky, J. Houmard and M. Herdman. 2000. *Prochlorococcus marinus* Chisholm et al. 1992 subsp. *Pastoris* subsp. Nov. strain PCC 9511, the first axenic chlorophyll a(2)/b(2)-containing cyanobacterium (Oxyphoto-bacteria). Int. J. Syst. Evol. Microbiol. 50: 1833-1847.

Rivkin, R.B. and M. Putt. 1987. Heterotrophy and photoheterotrophy by Antarctic microalgae: Light-dependent incorporation of amino acids and glucose. J. Phycol. 23: 442-452.

Rodriguez, F., Y. Pazos, J. Maneiro and M. Zapata. 2003. Temporal variation in phytoplankton assemblages and pigment composition at a fixed station of the Ria of Pontevedra (NW Spain). Estuar. Coast. Shelf Sci. 58: 499-515.

Ryther, J.H. 1954. The ecology of phytoplankton blooms in Moriches Bay and Great South Bay, Long Island, New York. Biol. Bull. 106: 198-206.

Sabater, S., E. Vilalta, A. Gaudes, H. Guasch, I. Munoz and A. Romani. 2003. Ecological implications of mass growth of benthic cyanobacteria in rivers. Aquat. Microb. Ecol. 32: 175-184.

Saks, N.M. 1983. Primary production and heterotrophy of a pennate and a centric salt marsh diatom. Mar. Biol. 76: 241-246.

Saks, N.M. and E.G. Kahn. 1979. Substrate competition between a salt marsh diatom and a bacterial population. J. Phycol. 15: 17-21.

Sanders, R.W. and K.G. Porter. 1988. Phagotrophic phytoflagellates. Adv. Microb. Ecol. 10: 167-192.

Sauer, N. 1984. A general amino-acid permease is inducible in *Chlorella vulgaris*. Planta 161: 425-431.

Sauer, N., E. Komor and W. Tanner. 1983. Regulation and characterization of two inducible amino-acid transport systems in *Chlorella vulgaris*. Planta 159: 404-410.

Saunders, G.W. 1972. Potential heterotrophy in a natural population of *Oscillatoria agardhii* var. *isothrix* Skuja. Limnol. Oceanogr. 17: 704-711.

Schnepf, E. and M. Elbrächter. 1992. Nutritional strategies in dinoflagellates. A review with emphasis on cell biological aspects. Eur. J. Protistol. 28: 3-24.

Schuster, S., J.M. Arrieta and G.J. Herndl. 1998. Adsorption of dissolved free amino acids on colloidal DOM enhances colloidal DOM utilization but reduces amino acid uptake by orders of magnitude in marine bacterioplankton. Mar. Ecol. Prog. Ser. 166: 99-108.

Schwelitz, F.D., P.L. Cisneros, J.A. Jagielo, J.L. Comer and K.A. Butterfield. 1978. The relationship of fixed carbon and nitrogen sources to the greening process in *Euglena gracilis* strain Z. J. Protozool. 25: 257-261.

Seitzinger, S.P. and R.W. Sanders. 1999. Atmospheric inputs of dissolved organic nitrogen stimulate coastal bacteria and algae. Limnol. Oceanogr. 44: 721-730.

Seitzinger, S.P., R.W. Sanders and R.V. Stiles. 2002. Bioavailability of DON from natural and anthropogenic sources to estuarine plankton. Limnol. Oceanogr. 47: 353-366.

Sepers, A.B.J. 1977. The utilization of dissolved organic compounds in aquatic environments. Hydrobiologia 52: 39-54.

Shah, N. and P.J. Syrett. 1982. Uptake of guanine by the diatom *Phaeodactylum tricornutum*. J. Phycol. 18: 579-587.

Sheath, R.G. and J.A. Hellebust. 1974. Glucose transport systames and growth characteristics of *Bracteacoccus minor*. J. Phycol. 10: 34-41

Sherr, E.B. 1988. Direct use of high molecular weight polysaccharide by heterotrophic flagellates. Nature 335: 348-351.

Sherr, E.B. and B.F. Sherr. 1999. β-glucosaminidase activity in marine microbes. FEMS Microbiol. Ecol. 28: 111-119.

Shihira-Ishikawa, I. and E. Hase. 1964. Nutritional control of cell pigmentation in *Chlorella protothecoides* with special reference to the degeneration of the chloroplast induced by glucose. Plant Cell Physiol. 5: 227-240.

Skoog, A. and R. Benner. 1997. Aldoses in various size fractions of marine organic matter: implications for carbon cycling. Limnol. Oceanogr. 42: 1803-1813.

Sloan, P.R. and J.D.H. Strickland. 1966. Heterotrophy of four marine phytoplankters at a low substrate concentration. J. Phycol. 2: 29-32.

Smayda, T.J. 1989. Primary production and the global epidemic of phytoplankton blooms in the sea: a linkage? pp. 449-484. *In* E.M. Cosper, V.M. Bricelj and E.J. Carpenter [eds.]. Novel Phytoplankton Blooms: Causes and Impacts of Recurrent Brown Tides and other Unusual Blooms. Springer-Verlag, New York, USA.

Smayda, T.J. 1990. Novel and nuisance phytoplankton blooms in the sea: Evidence for a global epidemic. pp. 29-40. *In* E. Granéli, B. Sundström, L. Edler and D.M. Anderson [eds.] 1990. Toxic Marine Phytoplankton. Elsevier Science, New York, USA.

Smayda, T.J. 1997. Harmful algal blooms: Their ecophysiology and general relevance to phytoplankton blooms in the sea. Limnol. Oceanogr. 42: 1137-1153.

Smith, A.J. Modes of cyanobacterial carbon metabolism. pp. 47-85. *In* N.G. Carr and B.A. Whitton [eds.]. The Biology of Cyanobacteria. Blackwell Scientific, Oxford, UK.

Staats, N., L.J. Stal and L.R. Mure. 2000. Exopolysaccharide production by the epipelic diatom *Cylindrotheca closterium*: effects of nutrient conditions. J. Exp. Mar. Biol. Ecol. 249: 13-27.

Stepanauskas, R., H. Edling and L. Tranvik. 1999. Differential dissolved organic nitrogen availability and bacterial aminopeptidase activity in limnic and marine waters. Microb. Ecol. 38: 264-272.

Stepanauskas, R., N.O.G. Jørgensen, O.R. Eigaard, A. Žvikas, L.J. Tranvik and L. Leonardson. 2002. Summer inputs of riverine nutrients to the Baltic Sea: Bioavailability and eutrophication relevance. Ecol. Monographs 72: 579-597.

Stepanauskas, R, H. Laudon and N.O.G. Jørgensen. 2000. High DON bioavailability in boreal streams during a spring flood. Limnol. Oceanogr. 45: 1298-1307.

Stickney, H.L., R.R. Hood and D.K. Stoecker. 2000. The impact of mixotrophy on planktonic marine ecosystems. Ecol. Model. 125: 203-230.

Stoecker, D.K. 1998. Conceptual models of mixotrophy in planktonic protists and some ecological and evolutionary implications. Eur. J. Protistol. 34: 281-290.

Stoecker, D.K. 1999. Mixotrophy among dinoflagellates. J. Eukaryot. Microbiol. 46: 397-401.

Stoecker, D.K. and D.E. Gustafson Jr. 2003. Cell-surface proteolytic activity of photosynthetic dinoflagellates. Aquat. Microb. Ecol. 30: 175-183.

Stolte, W., R. Panosso, L-Å. Gisselson and E. Granéli. 2002. Utilization efficiency of nitrogen associated with riverine dissolved organic carbon (>1 kDa) by two toxin-producing phytoplankton species. Aquat. Microb. Ecol. 29: 97-105.

Strom, S.L., R. Benner, S. Ziegler and M.J. Dagg. 1997. Planktonic grazers are a potentially important source of marine dissolved organic carbon. Limnol. Oceanogr. 42: 1364-1374.

Sunda, W.G. and S.A. Huntsman. 1995. Iron uptake and growth limitation in oceanic and coastal phytoplankton. Mar. Chem. 50: 189-206.

Therriault, J.C., J. Painchaud and M. Levasseur. 1985. Factors controlling the occurrence of *Protogonyaulax tamarensis* and shellfish toxicity in the St. Lawrence Esturay: freshwater runoff and the stability of the water column. pp. 141-146. *In* D.M. Anderson, A.M. White and D.G. Baden [eds.]. Toxic Dinoflagellates. Elsevier, New York, USA.

Thingstad, T.F., H. Havskum, K. Garde and B. Riemann. 1996. On the strategy of "eating your competitor": A mathematical analysis of algal mixotrophy. Ecology 77: 2108-2118.

Tilman, D. and P. Kilham 1976. Sinking in freshwater phytoplankton: Some ecological implications of cell nutrient status and physical mixing processes. Limnol. Oceanogr. 21: 409-417.

Tranvik, L.J., E.B. Sherr and B.F. Sherr. 1993. Uptake and utilization of 'colloidal DOM' by heterotrophic flagellates in seawater. Mar. Ecol. Prog. Ser. 38: 251-258.

Ukeles, R. and W.E. Rose. 1976. Observations on organic carbon utilization by photosynthetic marine microalgae. Mar. Biol. 37: 11-28.

Underwood, G.J.C. and D.J. Smith. 1998. Predicting epipelic diatom exopolymer concentrations in intertidal sediments from sediment chlorophyll *a*. Microb. Ecol. 35: 116-125.

Vallino, J.J., C.S. Hopkinson and J.E. Hobbie. 1996. Modeling bacterial utilization of dissolved organic matter: Optimization replaces Monad growth kinetics. Limnol. Oceanogr. 41: 1591-1609.

Van Baalan, C. and W.M. Pulich. 1973. Heterotrophic growth of the microalgae. CRC Crit. Rev. Microbiol. 2: 229-255.

Vargo, G.A., C.A. Heil, D. Spence, M.B. Neely, R. Merkt, K. Lester, R.H. Weisberg, J.J. Walsh and K. Fanning. 2001. The hydrographic regime, nutrient requirements, and transport of a *Gymnodinium breve* Davis red tide on the West Florida shelf. pp. 157-

160. *In* G.M. Hallegraeff, S. Blackburn, C. Bolch and R. Lewis [eds.]. Harmful Algal Blooms 2000. IOC UNESCO, Paris, France.

Vincent, W.F. 1980. The physiological ecology of a *Scenedesmus* population in the hypolimnion of a heterotrophic pond. II. Heterotrophy. Br. Phycol. J. 15: 35-41.

Vincent, W.F. and C.R. Goldman. 1980. Evdience for algal heterotrophy in Lake Tahoe, California-Nevada. Limnol. Oceanogr. 25: 89-99.

Vrba, J., C. Callier, T. Bittl, K. Šimek, R. Bertoni, P. Filandr, P. Hartman, J. Hejzlar, M. Macek and J. Nedoma. 2004. Are bacteria the major producers of extracellular glycolytic enzymes in aquatic environments? Int. Rev. Hydrobiol. 89: 102-117.

Vrba, J., J. Kofronová-Bobková, J. Pernthaler, K. Šimek, M. Macek and R. Psenner. 1997. Extracellular, low-affinity β-*N*-acetylglucosaminidases linked to the dynamics of diatoms and crustaceans in freshwater systems of different trophic degree. Int. Rev. Ges. Hydrobiol. 82: 277-286.

Vrba, J., K. Šimek, J. Nedoma and P. Hartman. 1993. 4-Methyl-umbelliferyl-β-*N*-acetylglucosaminide hydrolysis by a high-affinity enzyme, a putative marker of protozoan bacterivory. Appl. Environ. Microbiol. 59: 3091-3101.

Vrba, J., K. Šimek, J. Pernthaler and R. Psenner. 1996. Evaluation of extracellular, high-affinity β-*N*-acetylglucosaminidase measurements from freshwater lakes: an enzyme assay to estimate protistan grazing on bacteria and picocyanobacteria. Microb. Ecol. 32: 81-99.

Wetzel, R.L. 1994. Modeling the microbial loop – an estuarine modeler's perspective. Microb. Ecol. 28: 331-334.

Wetzel, R.G., P.G. Hatcher, and T.S. Bianchi. 1995. Natural photolysis by ultraviolet irradiance of recalcitrant dissolved organic matter to simple substrates for rapid bacterial metabolism. Limnol. Oceanogr. 40: 1369-1380.

Wheeler, P.A. and D.L. Kirchman. 1986. Utilization of inorganic and organic nitrogen by bacteria in marine systems. Limnol. Oceanogr. 31: 998-1009.

Wheeler, P., B. North, M. Littler and G. Stephens. 1977. Uptake of glycine by natural phytoplankton communities. Limnol. Oceanogr. 22: 900-910.

Wheeler, P.A., B.B. North and G.C. Stephens. 1974. Amino acid uptake by marine phytoplankters. Limnol. Oceanogr. 19: 249-259.

White, A.W. 1974. Uptake of organic compounds by two facultatively heterotrophic marine centric diatoms. J. Phycol. 10: 433-438.

Wiessner, W. and H. Gaffron. 1964. Role of photosynthesis in the light-induced assimilation of acetate by *Chlamydobotris.* Nature 201: 725-726.

Willey, J.M. and J.B. Waterbury. 1989. Chemotaxis towards nitrogenous compounds by swimming strains of marine *Synechococcus* spp. Appl. Environ. Microbiol. 55: 1888-1894.

Williams, P.J.LeB. 1970. Heterotrophic utilization of dissolved organic compounds in the sea. I. Size distribution of population and relationship between respiration and incorporation of growth substances. J. Mar. Biol. Ass. U.K. 50: 859-870.

Wolny, J.L., J.W. Kempton and A.J. Lewitus. 2004. Taxonomic re-evaluation of a South Carolina 'red tide' dinoflagellate indicates placement in the genus *Kryptoperidinium* Proc. 10th International Conference on Harmful Algal Blooms (in press).

Wood, B.J.B., P.H.K. Grimson, J.B. German and M. Turner. 1999. Photoheterotrophy in the production of phytoplankton organisms. J. Biotechnol. 70: 175-183.

Worden, A.Z., J.K. Nolan and B. Palenik. 2004. Assessing the dynamics and ecology of marine picophytoplankton: The importance of the eukaryotic component. Limnol. Oceanogr. 49: 168-179.

Wright, R.T. and J.E. Hobbie. 1965. The uptake of organic solutes by planktonic bacteria and algae. Ocean Sci. Ocean. Eng. 1: 116-127.

Xie, J.L., Y.X. Zhang, Y.G. Li and Y.H. Wang. 2001. Mixotrophic cultivation of *Platymonas subcordiformis*. J. Appl. Phycol. 13: 343-347.

Xu, F., H.H. Hu, W. Cong, Z.L. Cai and F. Ouyang. 2004. Growth characteristics and eicosapentaenoic acid production by *Nannochloropsis* sp. in mixotrophic conditions. Biotechnol. Lett. 26: 51-53.

Yamada, M., Y. Arai, A. Tsuruta and Y. Yoshida. 1983. Utilization of organic nitrogenous compounds as nitrogen sources by marine phytoplankton. Bull. Jap. Soc. Scient. Fish. 49: 1445-1448.

Zhang, X.S., J.T. Anderson and R.R. Hood. 2003. Modeling *Pfiesteria piscicida* population dynamics: a new approach for tracking size and mass in mixotrophic species. Mar. Ecol. Prog. Ser. 256: 29-44.

Zubkov, M.V., B.M. Fuchs, G.A. Tarran, P.H. Burkill and R. Amann. 2003. High rate of uptake of organic nitrogen compounds by *Prochlorococcus* cyanobacteria as a key to their dominance in oligotrophic oceanic waters. Appl. Environ. Microbiol. 69: 1299-1304.

11

Role of the Cell Cycle in the Metabolism of Marine Microalgae

Jacco C. Kromkamp[1] and Pascal Claquin[2]

[1]Centre for Estuarine and Marine Ecology, Netherlands Institute of Ecology, PO Box 140, 4400 AC Yerseke, The Netherlands

[2]Laboratoire de Biologie et de Biotechnologies Marines Université de Caen Basse-Normandie Esplanade de la paix, 14032 Caen Cedex, France

Abstract

In this chapter we summarize the effect of the cell cycle on nutrient uptake and metabolism and the interaction with light and the naturally occurring light-dark cycle. Different methods exist to examine the cell cycle, but the most preferred method is flow cytometric analysis of the DNA content of algal cells.

Cell cycle checkpoints 'check' whether conditions are met to proceed to the next phase. If conditions are not met, cells stay in that particular phase. The naturally occurring light-dark cycle can induce synchronized growth in picocyanobacteria, green microalgae and dinoflagellates, but it has not been shown in diatoms. The cue for synchronization differs: in some species it is the dark/light ('dawn') transition whereas in other species it is the 'dusk' transition. The occurrence of cell cycle checkpoints can be investigated by limiting or starving cells for a particular nutrient or for light, although these methods do not always yield the same answer, for reasons not clear yet. Starvation for light or nitrogen causes most algal cells to accumulate in the G1-phase, although diatoms and *Euglena* also seem to have a 2nd light control point in the G2+M-phase. Diatoms have a major checkpoint in the G2+M-phase: as a consequence slow growing cells which are nutrient or light limited (Si-limitation excepted) will become heavily silicified. A 2nd Si-control point seem to be present in some diatoms in the G1-phase. Hardly any information is available for nitrogen or phosphorus, but a phosphorus-checkpoint in the S-phase for diatoms and picocyanobacteria has been described, although *Prochlorocccus* can also accumulate in the G2+M-phase during P-starvation, suggesting a checkpoint in these phases as well. As nutrient uptake often seems to be confined to a particular cell cycle phase we speculate about the role of (partly) synchronized algal communities in phytoplankton species diversity.

THE CELL CYCLE: INTRODUCTION

During normal cell division algal cells go through a series of steps that together make up the cell cycle. Microscopically two phases can be easily identified: the interphase, during which the algae grow, followed by cell division. The eukaryotic cell cycle is best studied, especially in yeast and mammalian cell, as a result of direct link with cancer research, and the genes involved in controlling the cell cycle are highly conserved from simple unicellular cells such as yeast to complex metazoans (Nurse 2000). During the M-phase nuclear division (mitosis) is followed by cell division (cytokinesis) and during this latter phase the formation of two daughter cells by cell division takes place. The interphase consists of three steps: the G1-phase (G stands for gap), during which most of the growth takes place, followed by the S-phase (S stands for synthesis) in which DNA replication takes place. The S-phase is followed by a second gap phase, G2, in which the cells prepare themselves for mitosis (Fig. 11.1). Although no new DNA is synthesized in either G1 or G2, repair of damaged DNA can take place in the gap phases. Although the length of all the phases is variable, the largest variability is found in G1, the major growth phase.

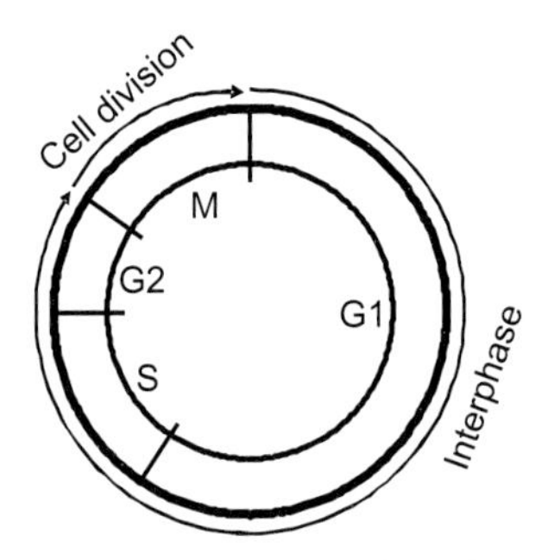

Fig. 11.1 The cell cycle showing the 4 typical phases in eukaryotic cells

The cell cycle is tightly regulated by a complex interplay of a family of proteins, cyclins, which are activated by cyclin dependent protein kinases (cdk's). The activated complexes phosphorylate specific target proteins. Every phase of the cell cycle has its own family of cyclin/cdk complexes for triggering different processes in the cell cycle. G1 cyclins belong to the cyclin D family, but others cyclins (like cyclin E) can be found as well in the G1-

phase. Cyclins A and E can be found in the S -phase, whereas cyclins B seem to be restricted to the M-phase. The cdk family is also highly specific and only bind to a specific cyclin (like cdk4 binds to cyclin D) (Nagyova et al. 2003, De Veylder et al. 2003, Nurse 2000) or http://www.biocarta.com/genes/index.asp). Breakdown of a cyclin terminates the activity of the corresponding cdk, leading to varying cyclin concentrations during the cell cycle. The activity of the cdks can also be regulated by inhibitory proteins (e.g. p19) which cause conformational changes by inhibitory phosphorylation by cdk-dependent kinases (e.g. wee1) or activation of the ckd/cyclin complex by phosphatase (e.g. Cdc25).

Prokaryotic organisms have a different way of DNA replication, and they can often divide faster than they replicate their genome. This is made possible because replication forks can start before the previous round of DNA duplication is ended. Nevertheless, cyanobacteria and prochlorophytes proceed through cell cycle phases similar to that of eukaryotic algae (Vaulot 1994), although it has been shown that several strains of *Synechococcus* and *Prochorococcus* have more than two copies of the genome (Vaulot 1994, Binder and Chisholm 1995b), and that the copy numbers of one, two, three, four, five ...) do not follow that *E. coli* model which predicts that the copy nr follows 2^n. This suggests that DNA synthesis is loosely coupled to the cell division cycle. Nevertheless, slow growing prokaryotic phototrophs often have a defined S-phase and hardly any cells with more than two copies of the genome (Binder and Chisholm 1995b)

Cell Cycle Checkpoints

The cyclin/ckd complexes also play an import role in the cell cycle checkpoints (Murray and Hunt 1993). These checkpoints play a vital role in the coordination of the cell cycle and prevent entering into the next phase when the previous phase has not been completed (Fig. 11.2). The G1-checkpoint (also called restriction point or Start) is present in all eukaryotic cells and controls whether cells are able to enter the DNA-synthesis phase. When cells pass this restriction point they are committed to DNA duplication and cell division. Cells need to exceed a critical size before they can pass the G1/S-checkpoint (De Veylder et al. 2003), and as diatoms need to exceed a certain size before they can divide the same is likely to be true for algae as well. This checkpoint also senses DNA-damage. The G2/M checkpoint serves to ensure that the S-phase has been completed and that no damaged DNA is present before the cells are allowed to proceed into the M-phase. For more information on the molecular mechanisms involved in the way these

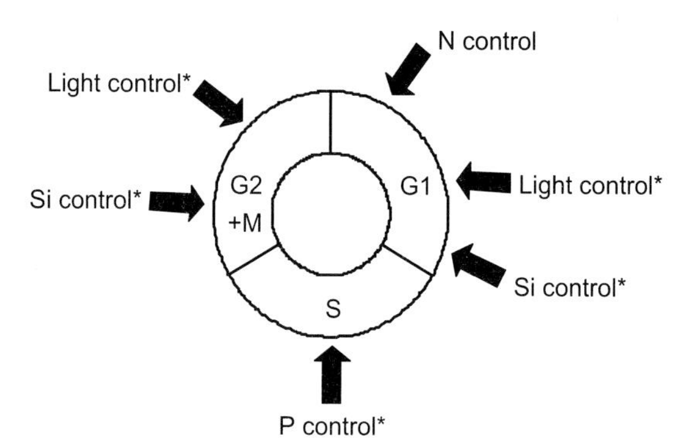

Fig. 11.2 Principal environmental checkpoints observed in the cell cycle of eukaryotic microalgae measured using flow cytometric (FCM) determination of the cell cycle stages. FCM analyses cannot separate the G2 and M-phase; *specific arrest points in the diatoms cell cycle. The light control checkpoint in the G2+M-phase is also described for *Euglena* sp. (Zachleder and Van Der Ende 1992).

checkpoints work we refer to reviews by De Veylder et al. (2003) and Lisby and Rothstein (2004).

Environmental factors influence progression of algal cells through the cell cycle, of which light and nutrient availability are the most important ones (Prezelin 1992). Limitation or starvation for light or a particular nutrient can be used to investigate in which phase of the cell cycle the cells will encounter a checkpoint because if a certain phase contains such a checkpoint, cells will accumulate in this particular phase. Nutrient deprivation, light and temperature conditions can entail cell cycle arrests (Brzezinski et al. 1990, Olson et al. 1986, Vaulot et al. 1986, 1987), or an increase of the duration of specific phases (Claquin et al. 2002, Olson et al. 1986). As mentioned by Pascual and Caswell (1997), this leads to a conceptual view of the cell cycle known as the transition point hypothesis, in which an environmental factor has no effect on cell progression beyond a certain point in the cycle (Spudich and Sager 1980, Vaulot et al. 1986).

Deficiencies in nutrients or in light cause arrest of cells in the G1 phase in the majority of microalgae groups (Spudich and Sager 1980, Vaulot et al. 1986, Zachleder and van der Ende 1992). Diatoms and Euglenophytes appear to have a different regulation of the cell cycle, arrest of the cell cycle in G1 and G2 phases was shown for diatoms (Brzezinski et al. 1990, Olson et al. 1986, Vaulot et al. 1986, 1987). *Euglena gracilis* seem to have light restriction points downstream of G1 as well (Hagiwara et al. 2001, 2002). However, Jacquet et al. (2001) showed that different strains of *Prochlorococcus* are affected by light both in S or G1 phases and Parpais et al. (1996), observed in *Prochlorococcus* spp. an arrest of cells in the S phase under phosphorus limitation as well.

Before discussing the influence of cell cycle on nutrient metabolism, first the effect of light is mentioned as photoautotrophs do not only depend on light as their only or main source of energy, but because algae do not only experience variations in light intensity but also in the time they are exposed to light. As a result, the total daily light dose is dependent on the light intensity as well as the length of the light period (photophase). However, first we will describe how the different stages in the cell cycle can be measured.

Cell Cycle Measurements

In order to study the cell cycle, it is necessary to determine the proportion of cells in the different cell cycle phases. Flow cytometric analysis appears to be the most efficient method to study the cell cycle as it allows a rapid analysis of a large population. The cellular DNA can be stained by various fluorochromes. Common fluorochromes specific for DNA such as ethidium bromide and propidium iodide emit photons at wavelengths which interfere with the red chlorophyll fluorescence and require chlorophyll extraction prior to DNA staining (Jochem and Meyerdierks 1999, Olson and Chisholm 1986, Van Bleijswijk and Veldhuis 1995). Other type of DNA dyes are excitable by ultraviolet (UV), like DAPI and Hoechst, which emit at blue wavelengths and therefore do not interfere with chlorophyll autofluorescence, and because of this it was the preferred dye for some time (Binder and Chisholm 1995a, Lemaire et al. 1999, Vaulot et al. 1987, Vaulot and Partensky 1992). However, a high power UV laser is required for flow cytometry using this dye, which is not available on most benchtop flow cytometers suitable for shipboard experiments (Jochem and Meyerdierks 1999). A recent generation of blue light excitable DNA stains emits in the green and thus do not overlap the spectrum of chlorophyll fluorescence emission, simplifying DNA labeling of microalgae. These dyes for DNA are PicoGreen, SYTOX (Veldhuis et al. 1997) or YOYO-1 (Jochem and Meyerdierks 1999). After flow cytometric analysis the DNA distribution obtained can be deconvoluted into the different populations of G1, S and G2+M phases (because the DNA content in the G2 and M-phase is the same, a flow cytometric analysis of stained DNA cannot distinguish between cells that are in G2 or in the M-phase).

Other methods can be used to determine the cell cycle activity like microscopic observation, [^{3}H]thymidine labelling or molecular markers.

Indeed, by using microscopy the proportion of cells in mitosis and in the interphase can be estimated after labelling with simple DNA stains as carnine acid (Saburova and Polikarpov 2003), or DNA fluorochrome staining, [^{3}H]thymidine nucleotides incorporation can be used to monitor DNA replication and the detection of gene expression by Northern-Blots can give information about the cell cycle stage (Planchais et al. 2000).

Methods of Synchronization Cells and of Measurements of Cell Cycle Activity

A synchronous cell culture is characterized by a high proportion of cells proceeding to the same event of the cell cycle at the same time. Cells in unsynchronized populations progress at different rates through the cell cycle phases. In order to obtain a synchronized population several techniques can be used but the general feature is usually the same. The principle consists in reversibly arresting cells at a certain definite stage. When all cells have accumulated at this particular stage, removal of the blockade will restart their progression through the cell cycle in parallel, at least for a couple of divisions. Temperature shocks, inhibitor treatments, nutrient starvation, photoperiod variations can be used to synchronize unicellular algae. A light–dark cycle, sometimes associated with temperature cycles or nutrient starvation, is commonly used (Chisholm 1981).

Studies about the cell cycle were mostly performed on few genera of green algae, like *Chlorella, Chlamydomonas, Dunaliella, Scenedesmus* (Lemaire et al. 1999, Senger and Bishop 1967, Spudich and Sager 1980, Strasser et al. 1999, Wegmann and Metzner 1971) and in some other groups of microalgae, like *Euglena* (Euglenophyta) (Winter and Brandt 1986), *Emiliania huxleyi, Pavlova lutheri* (Haptophyta), some Dinophyta (*Gonyaulax, Amphidinium*) or various diatoms (*Cylindrotheca fusiformis, Skeletonema costatum, Navicula*) (Chisholm 1981). Most species can be synchronized by using a light-dark cycle, however, the patterns shown by diatoms grown on light and dark cycles appear to be fundamentally different from those of other groups of microalgae and growth is frequently unsynchronized. Several methods were thus developed to synchronize diatoms. Silicon starvation coupled with light stress was used to synchronize diatoms, particularly in the species *Cylindrotheca fusiformis* (Busby and Lewin 1967, Darley et al. 1976, Hildebrand et al. 1998, Kröger and Wetherbee 2000, Sullivan 1979). Recently, methods which avoided the use of silicon starvation and light and dark stresses were developed. The synchronization of the population of *C. fusiformis* was performed using cell cycle inhibitors as nocodazole (methyl

(5-[2-thienylcarbonyl]-1H-benzimidazol-2-yl) carbamate) or aphidicolin (Claquin et al. 2004). Nocodazole is an antimitotic agent that disrupts microtubules (Luduena and Roach 1991, Vasquez et al. 1997), arresting the cell cycle at the G2/M phase. Aphidicolin inhibits eukaryotic and viral DNA replication by blocking DNA polymerases (Cheng and Kuchta 1993, Spadari et al. 1985) and blocks cells in the end of the G1 phase (Planchais et al. 2000). After washing, these cell cycle arrests are reversible. By using these cell cycle inhibitors, the cells were synchronized under a light-dark cycle or under continuous light (Claquin et al. 2004). The additional benefits of these methods are that they allow to separate the processes related to the light and dark cycle from those related to the cell cycle. Planchais et al. (2000) made a short review of the cell cycle inhibitors mainly used in plant cell cycle studies. Some of them could be tested on the different microalgae groups.

Others methods based on the behavior of algae to investigate synchronization can also be found in the literature: for example, Wong and Whiteley (1996) described a method based on the motility of the heterotrophic dinoflagellate *Crypthecodinium cohii* as a function of the cell cycle stages. Cells in the G1 phase were motile, whereas cells in S and G2+M phases were non-motile because they lost their flagella. By collecting the swimming cells they obtained only cells in the G1 phase which could then be used to follow synchronized division of the population.

Light: Influence of Light Intensity and Light Period on Growth and Photosynthesis

Photoautrophs depend on light for cell growth, and the length of the G1-phase is dependent on the ambient irradiance and thus on the rate of photosynthesis (Zachleder and van den Ende 1992, Fig. 11.3). When algal cells go beyond the G1/S checkpoint the progression through the cell cycle becomes independent of the light intensity and the cells can divide in the dark. As a result of this dependence on light it is not surprising that light-dark periods will influence the cell cycle and many algal species will be entrained by a light-dark period, although it is in many cases unclear whether the 'dawn' or 'dusk' transition acts as a zeitgeber. It has been shown for field populations of both *Prochlorococcus* (Vaulot et al, 1987) and *Synechococcus* (Vaulot et al, 1996) that the cell cycle is highly synchronized to the diel cycle. Also, in most dinoflagellates the diel cycle entrains cell division, giving maximum growth rates of one per day. Van Dolah and Leighfield (1999) showed for the red tide dinoflagellate *Gymnodinium breve* that the dark/light transition ('dawn') is the cue which entrains the cells. In

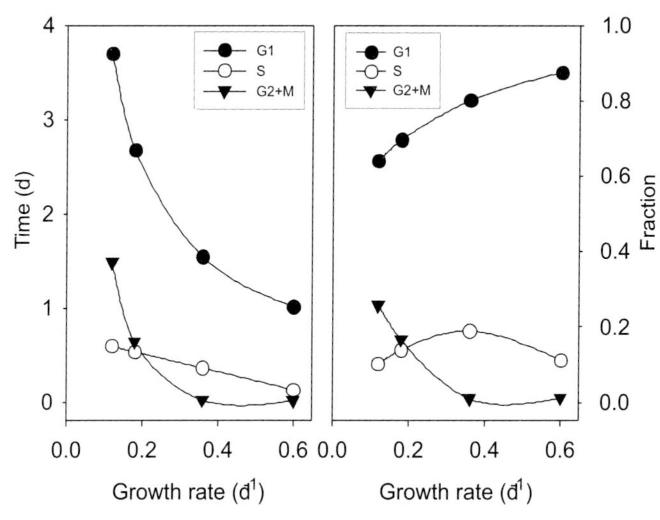

Fig. 11.3 Effect of irradiance on the absolute length of the cell cycle phases (left) and on the relative proportion of the different stages of the cell cycle (right)

most of the photophase, the cells were in the dark and a delay of the L/D-transition caused a similar delay in the onset of the S and G2+M phases. Shifting the L/D ('dusk') transition forward did not affect the timing of the cell cycle. Natural populations of the same species showed the same timing to the diel cycle, but estimated growth rates were approx. 0.29 divisions per day, indicating that the cells would experience 3-4 diel cycles before they could pass the G1/S checkpoint. The situation in *Euglena*, which will arrest in G1, S or G2+M when transferred to continuous darkness, is different (Hagiwara et al. 2001). In this case the L/D-transition ('dusk') is the Zeitgeber entraining the cell cycle (Fig. 11.4). If cells are entrained by the naturally occurring light dark cycle, this can induce variability in physiological processes: using synchronized cultures of the green alga *Scenedesmus obliquus* Strasser et al. (1999) showed that relative photosystem II (PS II) antenna size, maximum PS II efficiency and photochemical quenching coefficient were influenced by the stage of the cell cycle, although no information about the cell cycle stage was measured. Similar results were reported by Kaftan et al. (1999). Winter and Brandt (1986) showed cell cycle dependent changes in antenna size in *Euglena gracilis.* Claquin et al. (2004) used nocodazol to synchronize cultures of the marine diatom *Cylindrotheca fusiformis* and studied variation in photosynthesis and respiration as function of the cell cycle stage. They observed that the maximum rate of PS II electron transport and oxygen evolution were lowest when the percentage

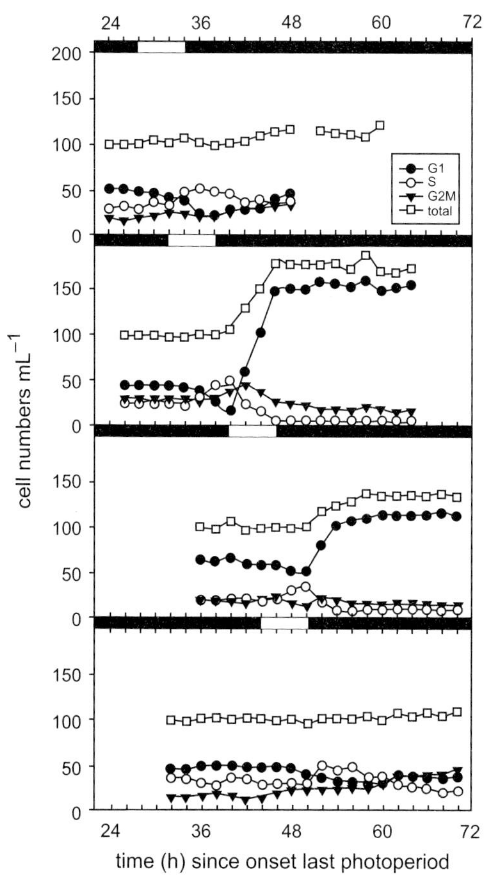

Fig. 11.4 Changes in the number of cells of *Euglena gracilis* in different phases of the cell cycle. The cells were grown in a 14:8 LD-cycle and than transferred to darkness. The light break around subjective dusk (B) produced the biggest increase in cell numbers, followed by a light break in midnight (C), midday, whereas the "dawn" transition (D) did not produce a response at all. From Hagiwara et al (2002).

G2+M cells were highest (late dark, early morning period), and that it increased in the G1-phase (Fig. 11.5). That the photosynthetic activity would by highest in the G1-phase could be expected because this phase is the major growth phase, which thus needs an efficient rate of photosynthesis. The rate of total oxygen uptake followed the pattern in photosynthesis. These authors also demonstrate that light stimulated the rate of oxygen uptake, which was attributed partly to the Mehler reaction. The light stimulated rate of oxygen uptake as a fraction of the total rate of oxygen uptake in the light was, however, independent of the cell cycle.

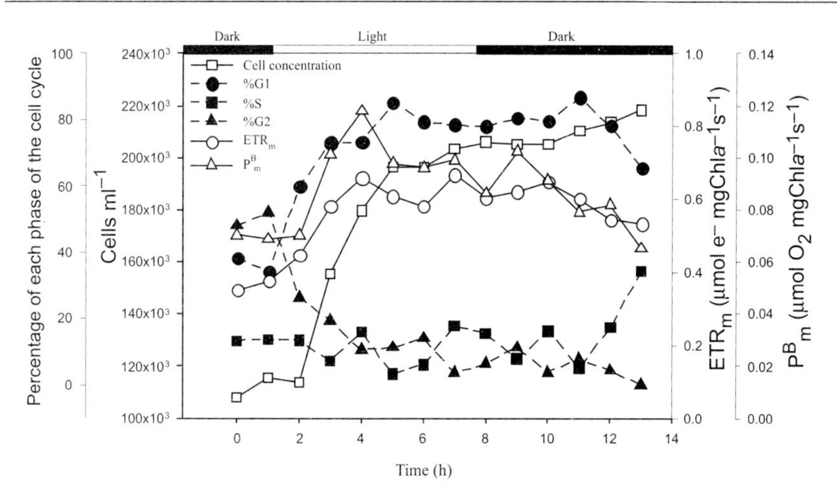

Fig. 11.5 Changes in cell number, cell cycle stage and photosynthesis in synchronized cultures of the diatom *Cylindrotheca fusiformis*. ETR: PSII electron transport rate, P^B_m: maximum rate of oxygen production (photosynthetic capacity). From Claquin et al. (2004).

Influence of Light/Dark Cycle on Nutrient Metabolism

As shown in the previous section on checkpoints, some nutrients are only taken up during a specific stage of the cell cycle. This was explored by Pascual and Caswell (1997) who investigated the effect of cell cycle dependent nutrient uptake using mathematical model. In their chemostat model, nutrients were only taken up during a particular phase of the cell cycle, and the cells could only proceed through the cell cycle after conditions satisfying the transition point have been met. Despite the fact that their model implicitly assumed that the cells were growing synchronously, no light-dark cycles were given. When the limiting nutrient was supplied continually in their model, stable conditions were reached at higher dilution rates, whereas at low dilution rates strong oscillations in cell numbers were observed. At a constant dilution rate oscillations occurred at relative high nutrient concentration, whereas at low nutrient concentration stable conditions developed. When a variable nutrient input with a regular interval was simulated, both periodic and aperiodic responses were observed (Fig. 11.6). When they refined their model to include nutrient storage, oscillations in cell number were again observed with frequencies different from the forcing. As under natural conditions the light-dark cycle might entrain phytoplankton species inducing synchronized growth, this also may lead to phasing in nutrient uptake kinetics. On the other hand Olson and Chisholm (1983) showed that nutrient forcing can override the

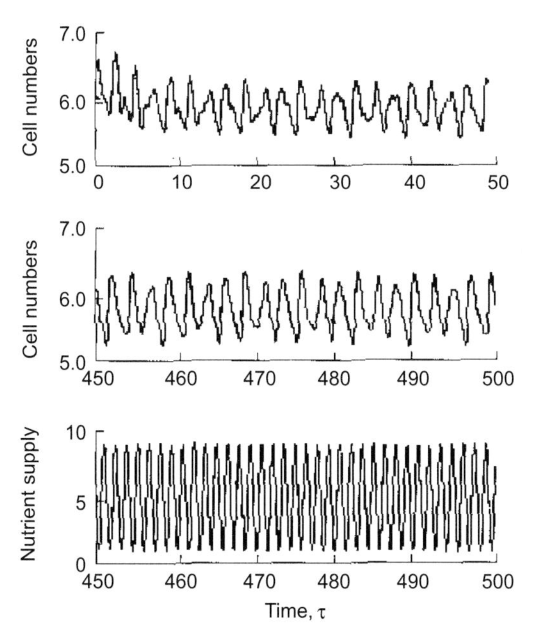

Fig. 11.6 Fluctuations in cell number under a periodic nutrient supply. A: transients, B: long-term dynamic. C: nutrient supply. Notice the aperiodic fluctuation in cell number in B, despite the fact that the nutrients were supplied with regular interval. From (Pascual and Caswell 1997).

light-dark cycle and drive cell division. Very complex oscillatory patterns may be the outcome, which are not phased to either the LD or nutrient input cycles, making it difficult to drawn conclusions about the forcing functions behind the changing cell numbers. More research in this field is needed.

Cell Cycle and P-uptake and P-content

Only a few papers on the effect of the cell cycle on phosphorus metabolism have been published. As phosphorus is an essential element not only in the energy metabolism but also as important structural component of DNA, a shortage of P may be expected to lead to a checkpoint in the DNA synthesis phase. However, Parpais et al. (1996) showed for different strains of the photosynthetic prokaryote *Prochlorococcus* an increase in the number of cells that were arrested in the S phase as well as in the G2+M phase during P-starvation, apart from one of the North Atlantic strains tested, which only showed an increase in the percentage of S-phase arrested cells. Interestingly, the S-phase arrested cells of Mediterranean strain CCMP 1378 did not resume growth after addition of phosphate, whereas cells arrested in the G1 or G2+M-phase did. An increase in the number of cells arrested in the S-phase was only demonstrated recently for the first time for a eukaryotic alga, the diatom *Thalassiosira pseudonana* (Table 11.1, Claquin 2002). However, when cells of this diatom were grown in steady state P-limited continuous cultures, the percentage of cells in the S-phase decreased whereas the percentage of cells in the G2+M phase increased when the growth rate decreased (i.e. when the intensity of the limitation increased) (Fig. 11.7). This increase in percentage of cells in the G2+M-phase with lower growth rates was observed also for nitrogen- and light-limited cells. Although limitation and starvation experiments generally, but not always lead to the same conclusion (Vaulot 1994), these results indeed show that starvation can lead to different behavior than limitation, an observation

Table 11.1 Percentage of cells in each phase of cell cycle in the diatom *Cylindrotheca fusiformis* after 6 and 24 as a function of Si, N or P starvation and under unlimited conditions (Claquin 2002)

	%G1	*%S*	*%G2+M*
Unlimited cells	67.5	6.4	26.1
Si starved cells 6 h	39.8	7.0	53.2
Si starved cells 24 h	39.7	8.5	51.7
P starved cells 6 h	65.8	5.5	28.7
P starved cells 24 h	17.9	66.2	15.9
N starved cells 6 h	68.8	10.6	20.6
N starved cells 24 h	88.1	6.0	5.9

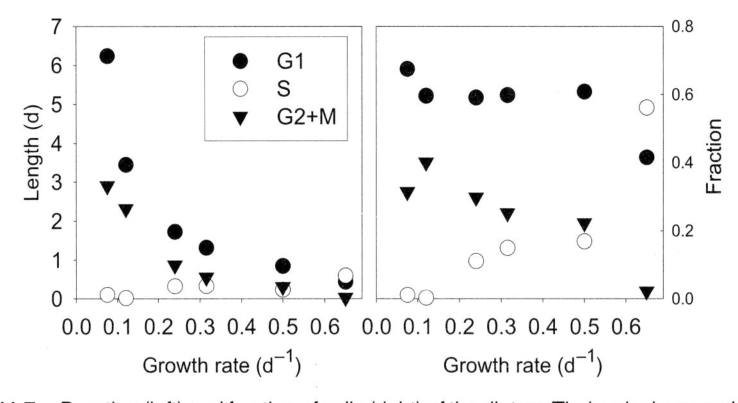

Fig. 11.7 Duration (left) and fraction of cells (right) of the diatom *Thalassiosira pseudonana* in different phases of the cell cycle when grown in steady state P-limited continuous cultures. Notice the large increase in percentage of G2+M-phase cells and the decrease in percentage of S-phase cells with low growth rates. Data from Claquin et al (2002).

also made by Brzezinski et al. (1990) for the Si-limited cells of the diatom *Thalassiosira weisflogii*, which arrests in G1 and G2 or M after Si-starvation, but does not change the length of the M-phase during Si-limitation.

The results of the cultures studies on *Prochloroccus* strains by Parpais et al. (1996) suggested that P-limitation could only be induced with N:P-ratios exceeding 50, making it difficult to actually have P-limited *Prochloroccus* strains in nature. Vaulot et al (1996) studied the addition of low concentrations of phosphate to *Synechococcus* in the surface waters of the Mediterranean waters in summer. The cells were highly synchronized to the diel cycle with most cells in G1 during the day and DNA synthesis starting around 16.00 with a maximum of cells in the G2+M-phase around 21:00 h. Addition of P or N+P, but not of N, shortened the G1-phase and decreased the % of cells in G2+M around 18:00 and increased the number of cells in G2+M around noon the next day, demonstrating that the *Synechococcus* were indeed P-limited, and dose response curves suggested that the limitation was stronger at the end of June than in the middle of July, thus demonstrating that some oligotrophic regions maybe P-limited rather than N-limited.

Cell Cycle and Si-uptake and Si-content

The main particularity of diatoms is the silicified cell wall called a frustule, which essentially consists of hydrated amorphous silica. The silicon metabolism consequently plays a fundamental role in diatoms (see Martin-Jézéquel et al. 2000, for a review), vegetative division cannot occur without formation of the valves of the daughter cells and cell growth cannot occur without girdle band formation (Gordon and Drum 1994, Pickett-Heaps et al. 1990, Volcani 1981). Silicic acid is mainly present in the seawater at pH 8.0 in the following two forms: as $Si(OH)_4$, which is the prevailing form representing 97% of dissolved silica; and as $SiO(OH)_3^-$ that composes the remaining 3% (Ingri 1978, Stumm and Morgan 1981). Most diatom species consume the prevailing form ($Si(OH)_4$) (Del Amo and Brzezinski 1999), however, it was shown that *Phaeodactylum tricornutum* could also use the minority form (Riedel and Nelson 1985, Del Amo and Brzezinski 1999). However, it is important to note that *Phaeodactylum tricornutum* is a special species which can grow and perform divisions in the absence of silicic acid (Round et al. 1990).

The assimilation of silicic acid is done by the means of active sodium dependent transport in marine diatoms (Paasche 1973a, b, Azam et al. 1974, Bhattacharyya and Volcani 1980) and it may be coupled with potassium in freshwater species (Sullivan 1976). This transport uses sodium/silicic acid symporters (Bhattacharyya and Volcani 1980, Hildebrand et al. 1997). Hildebrand et al. (1997) characterized these silicon transporters in *Cylindrotheca fusiformis*, and they proved the existence of five genes (Silicic acid transporter (*SIT*) genes) coding for these transporters (Hildebrand et al. 1998). The presence of these genes could be generalized for other pennate or centric, marine or freshwater diatom species (Hildebrand et al. 1998). Hildebrand et al. (1998) observed that the rate of mRNA expression of *SIT* genes was multiplied four-fold just before the maximal phase of the synthesis of a frustule (Martin-Jézéquel et al. 2000). Silicon uptake and deposition appear to be associated with the formation of new siliceous valves just prior to cell division, and thus mainly seem to be confined to the G2 and M period between cytokinesis and daughter cell separation (Hildebrand 2000, Sullivan 1986, Sullivan and Volcani 1981). The silicon metabolism in diatoms is then strictly linked to the cell cycle (Martin-Jézéquel et al. 2000). However, some species can assimilate silicon before this phase and stock it (Chisholm et al. 1978, Brzezinski and Conley 1994). The girdle bands are formed after the valves and can represent a significant part of the total biogenic silica of the cell in some species (Round 1972). Chiappino and Volcani (1977) showed that the formation of the first and often of the second girdle bands follow the formation of new valves in *Navicula pelliculosa,* whereas the formation of other bands can proceed till the last moment before the mitosis. The Si-uptake does not require photosynthetic energy, but does require energy from respiration (Raven 1983, Sullivan 1976, Sullivan 1980), consequently the silicic acid uptake can take place in the dark as well as in the light (Chisholm 1981, Martin-Jézéquel et al. 2000).

The models of Brzezinski (1992) and Flynn and Martin-Jézéquel (2000) based on a large amount of data about the relation of the cell cycle and the silicon metabolism (Blank et al. 1986, Blank and Sullivan 1979, Brzezinski and Conley 1994, Darley and Volcani 1969, Okita and Volcani 1978, 1980, Sullivan 1976, 1977) suggested that the increase of the length of the G2 phase should have entailed an increase of the silicate uptake, this hypothesis developed by Martin-Jézéquel et al. (2000) was confirmed by Claquin et al. (2002). They observed in continuous cultures of the marine diatom *Thalassiosira pseudonana* that the increase of the length of the G2 phase, which was due to a decrease of the growth rate under stronger light, nitrogen or phosphorus limitation, entailed an augmentation of the silification of the cells. Thus it appears that the cellular silicon content and the frustule thickness (i.e. BSi per cell surface) are regulated by the total amount of Si taken up which is directly driven by the length of the cell cycle (i.e. by the growth rate). Hence, the silicon content variations seem not to be linked to the type of the limitation but to the intensity of the limitation. This model is not appropriate under silicon limitation; in this case the cellular amount of BSi decreases with lower growth rate (Martin-Jézéquel et al. 2000).

As mention before, the strict link between the silicon metabolism and the cell cycle associated with a low respiratory energy requirement independent of photosynthesis (Löbel et al. 1996, Martin-Jézéquel et al. 2000, Raven 1983) allows to explain the uncoupling between the carbon and silicon metabolisms (Claquin 2002, Claquin and Martin-Jézéquel 2005). This uncoupling controls the Si/C or Si/N ratio variations frequently observed in connection with growth conditions changes due to light intensity, temperature or nutrient limitation (Brzezinski 1985, Claquin et al. 2002, Davis 1976, Harrison et al. 1976, 1977, Paasche 1980).

Cell Cycle and DIN-uptake and N-content

Nitrogen is a major element in the control of microalgae growth. Microalgae can assimilate both forms of inorganic nitrogen, nitrate and ammonium and several algae can take up organic nitrogen in the form of urea or amino acids. Nitrogen and C-metabolism are tightly coupled because protein synthesis needs both, and as such it is logical that nitrogen metabolism depends on photosynthesis. As the reduction of nitrate via nitrite to ammonia is light dependent, nitrate uptake takes place in the light only (Dortch 1990) and this can cause patterns of diel variation in nitrogen metabolism which is thus connected to photosynthesis (Turpin 1991, Turpin et al. 1988, Vergera et al. 1998). Influence of the cell cycle on nitrogen metabolism is shown by the fact that the few algae investigated arrest mainly in the G1-phase (Hildebrand and Dahlin 2000, Olson et al. 1986, Vaulot et al. 1987), like light-limited algae, although several diatoms and cyanobacteria also show an arrestpoint in G2+M (Vaulot, 1994). Hildebrand and Dahlin (2000) showed in *Cylindrotheca fusiformis* a relationship between the nitrate transporter (NAT) mRNA levels and the cell cycle. They observed high levels of NAT mRNA in early G1 phase, a decrease during the remainder of G1, then an increase during the S and into the G2 phase, and finally a decrease after the M phase. Olson et al. (1986) observed in the diatom *Thalassiosira weissflogii* and in the coccolithophore *Hymenomonas carterae* an exclusive lengthening of the G1 phase by nitrogen limitation. They concluded that the G1 phase is more nitrogen dependent than the other phases, but in *Thalassiosira pseudonana* a major nitrogen-dependent segment in the G2+M phase was shown (Claquin et al. 2002).

Claquin et al. (2002) showed that even if the nitrate uptake is linked to the cell cycle, the length phase variations do not necessarily affect the global metabolism of nitrogen because they observed that the cellular nitrogen content did not change significantly with growth rate under either light or phosphorus limitation. Numerous studies proved that the nitrogen uptake is uncoupled from assimilation (Boyd and Gradmann 1999, Collos 1982, Collos and Slawyk 1976, Raimbault and Mingazzini 1987) and the assimilation of nitrogen depends on the regulation of different processes mainly linked with the carbon metabolism (Cresswell and Syrett 1979, Dortch et al. 1979, Huppe and Turpin 1994, Turpin 1991).

CONCLUSIONS

In this overview we summarized the effects of the influence of a particular cell cycle stage on the physiological behavior of different microalgae and how the cell cycle is influenced by changes in the environment. Limitation for light of a particular nutrient will lengthen the duration of all cycle stages, but especially the G1-phase, the major growth phase, seems most affected. Due to the periodic nature of light, the cell cycle of algae is often entrained by the natural occurring light-dark cycle: picocyanobacteria like *Synechococcus* sp or *Prochloroccocus*, dinoflagellates and some green algae seem to synchronize their cell cycle to the light-dark cycle, whereas diatoms do not seem to do this. A result of this entrainment to the L:D-cycle is that the maximum growth rate will be maximally one division per day. At present it is difficult to generalize because synchronized growth has not been investigated in many species and it is therefore impossible to say if all green algae or all cyanobacteria show synchronized cell division as a result of the natural occurring light-dark cycle. Also with respect to the 'cue' causing the synchronization variability between groups exist: the dinoflagellate *Gymnodinium* uses the dawn transitions as zeitgeber (van Dolah and Leighfield 1999), whereas *Euglena* uses dusk as zeitgeber (Hagiwara et al. 2001).

The role of the various checkpoints in the cell needs further investigations. If a particular nutrient is taken up only during the G1 phase, as seems the case for nitrogen, than the G1/S cell cycle checkpoints determines whether a cell will proceed to the next phase. In unsynchronized cultures this will mean that only a part of the cells is actually involved in the uptake of the nutrient, and determination of nutrient uptake kinetics will then lead to underestimates of the true kinetic parameters. If nutrients are taken up during a particular phase only, this can also potentially lead to large oscillations in cell number, and combined with nutrient storage the oscillations might have a periodicity different from the forcing, as shown by a modelling exercise by Pascual and Caswell (1997). In our view this mechanism can also explain part of the diversity of the phytoplankton: if two populations of different algae are synchronized by the L:D cycle but take up the same limiting nutrient in a different phase of the cell cycle, competition for the same nutrient will not lead to exclusion because the competition for the limiting nutrient is temporary separated. Alternatively, competition for a limiting nutrient between synchronized species like green microalgae and unsynchronized species like diatoms might also lead to stable coexistence if the synchronized species has a higher affinity for the nutrient because of the difference in timing of nutrient consumption: only during the time in which the green algae are in the cell cycle phase in which nutrient uptake takes place are the diatoms in disadvantage. Thus, the effect of competition on the coexistence depends on the nutrient availability, whether the limiting nutrient is taken up only during a particular phase of the cell cycle and the degree of synchronization. Unfortunately only little experimental evidence is available in the literature to test these ideas. More work on different algae where nutrients are only taken up in a particular phase of the cell and on the influence cell cycle checkpoints is needed. Also, the molecular mechanisms behind the cell cycle checkpoints need further investigation.

ACKNOWLEDGEMENTS

We like to thank the van-Gogh exchange programme. This is publication 3596 of the NIOO-CEME.

REFERENCES

Azam, F., B.B. Hemmingsen and B.E. Volcani. 1974. Role of silicon in diatom metabolism. V. Silicic acid transport and metabolism in the heterotrophic diatom *Nitzschia alba*. Arch. Microbiol. 97: 103-114.

Battacharyya, P. and B. E. Volcani. 1980. Sodium-dependent silicate transport in the apochlorotic marine diatom *Nitzschia alba*. Proc. Natl. Acad. Sci. USA. 77: 6368-6390.

Binder, B.J. and S.W. Chisholm. 1995a. Cell cycle regulation in marine Synechococcus sp. strains. Appl. Environm. Microbiol. 61: 708-717.

Binder, B.J. and S.W. Chisholm. 1995b. Cell-Cycle Regulation in Marine Synechococcus Sp Strains. Appl. Environm. Microbiol 61: 708-717.

Blank, G.S., D.H. Robinson and C.W. Sullivan. 1986. Diatom mineralization of silicic acid. VIII. Metabolic requirements and the timing of protein synthesis. J. Phycol. 22: 382-389.

Blank, G.S. and C.W. Sullivan. 1979. Diatom mineralization of silicic acid. III. Si(OH)4 binding abd light dependent transport in *Nitzschia angularis*. Arch. Microbiol. 123: 157-164.

Boyd, C.N. and D. Gradmann. 1999. Electrophysiology of the marine diatom Coscinodiscus wailesii - III. Uptake of nitrate and ammonium. J. Exp. Bot. 50: 461-467.

Brzezinski, M.A. 1985. The Si:C:N ratio of the marine diatoms : interspecific variability and the effect of some environmental variables. J. Phycol. 21: 347-357.

Brzezinski, M.A. 1992. Cell-cycle effects on the kinetics of silicic acid uptake and resource competition among diatoms. J. Plankton Res. 14: 1511-1539.

Brzezinski, M.A. and D.J. Conley. 1994. Silicon deposition during the cell cycle of *Thalassiosira weisflogii* (Bacillariophycae) determined using rhodamine 123 and propidium iodide staining. J. Phycol. 30: 45-55.

Brzezinski, M.A., R.J. Olson and S. W. Chisholm. 1990. Silicon availability and cell-cycle progression in marine diatoms. Mar. Ecol.Progr. Ser. 67: 83-96.

Busby, W.F. and J.C. Lewin. 1967. Silicate uptake and silica shell formation by synchronously dividing cells of the diatom *Navicula pelliculosa* (Breb) Hilse. J. Phycol. 3: 127-131.

Cheng, C.H. and R.D. Kuchta. 1993. DNA Polymerase-Epsilon - Aphidicolin inhibition and the Relationship between polymerase and exonuclease activity. Biochemistry 32: 8568-8574.

Chiappino, M. L., and B. E. Volcani. 1977. Studies on the biochemistry and fine structure of silica shell formation in diatoms. VIII. Sequential cell wall development in the pennate *Navicula pelliculosa*. Protoplasma 93: 205-221.

Chisholm, S.W. 1981. Temporal patterns of cell division in unicellular algae, p. 150-181. *In* T. Platt [ed.], Physiological bases of phytoplankton ecology. Canadian Bulletin of Fisheries and Aquatic Sciences.

Chisholm, S.W., F. Azam and R.W. Eppley. 1978. Silicic acid incorporation in marine diatoms on light:dark cycles: use as an essay for phased cell division. Limnol. Oceanogr. 23: 518-529.

Claquin, P. 2002. Régulations des métabolismes du carbone et du silicium au cours de la division et de la croissance cellulaires des diatomées. Conséquences sur la production phytoplanctonique. PhD-thesis, Université de Bretagne Occidentale, Brest, France.

Claquin, P., J.C. Kromkamp and V. Martin-Jézéquel. 2004. Relationship between photosynthetic metabolism and cell cycle in a synchronized culture of the marine alga *Cylindrotheca fusiformis* (Bacilliarophyceae). Eur. J. Phycol. 39: 33-41.

Claquin, P. and V. Martin-Jézéquel. 2005. Regulations of the Si and C uptake and of the soluble free Si pool in a synchronized culture of *Cylindrotheca fusiformis* (Bacillariophyceae): Effects on the Si/C ratio. Mar. Biol. 146: 877-886.

Claquin, P.V. Martin-Jézéquel, J.C. Kromkamp, M.J.W. Veldhuis, and G.W. Kraay. 2002. Uncoupling of silicon compared to carbon and nitrogen metabolism, and role of the cell cycle in continuous cultures of *Thalassiosira pseudonana* (Bacillariophyceae) under light, nitrogen and phosphorus control. J. Phycol. 38: 922-930.

Collos, Y. 1982. Transient situations in nitrate assimilation by marine diatoms. III. Short-term uncoupling of nitrate uptake and reduction. 62: 285-295.

Collos Y. and G. Slawyk. 1976. Significance of cellular nitrate content cultures. Mar. Biol. 34: 27-32.

Creswell, R.C. and P.J. Syrett. 1979. Ammonium inhibition of nitrate uptake by the diatom Phaeodactylum tricornutum. Plant Sci. Lett. 14: 321-325.

Darley, W.M., C.W. Sullivan and B.E. Volcani. 1976. Studies on the biochemistry and fine structure of silica shell formation in diatoms, division cycle and chemical composition of *Navicula pelliculosa* during light-dark synchronized growth. Planta. 130: 159-167.

Darley, W.M. and B.E. Volcani. 1969. Role of silicon in diatom metabolism. A silicon requirement for Deoxyribonucleic acid synthesis in the diatom *Cylindrotheca fusiformis* Reimann and Lewin. Exp. Cell Res. 58: 334-342.

Davis, C.O. 1976. Continuous culture of marine diatoms under silicate limitation II Effect of light intensity on growth and nutrient uptake of *Skeletonema costatum*. J. Phycol. 12: 291-300.

de Veylder, L., J. Doubès and D. Inze. 2003. Plant Cell cycle transitions. Curr. Opin. Plant Biol. 6: 536-543.

Del Amo, Y. and M.A. Brzezinski. 1999. The chemical form of dissolved Si taken up by marine diatoms. J. Phycol. 35: 1162-1170.

Dortch, Q. 1990. The interaction between ammonium and nitrate uptake in phytoplankton. Mar. Ecol. Progr. Ser. 61, 183-201.

Dortch, Q, S.I. Ahmed and T.T. Packard. 1979. Nitrate reductase and glutamate dehydrogenase activities in Skeletonema costatum as measures of nitrogen assimilation rates: J. Plankton Res. 1: 169-186.

Flynn, K. J. and V. Martin-Jezequel.2000. Modelling Si-N-limited growth of diatoms. J. Plankton Res. 22: 447-472.

Gordon, R. and R.W. Drum. 1994. The chemical basis of diatom morphogenis. Int. Rev. Cytol. 150: 243-372.

Hagiwara, S., A. Bolige, Y.L. Zhang, M. Takahashi, A. Yamagishi and K. Goto. 2002. Circadian gating of photoinduction of commitment to cell-cycle transitions in relation to photoperiodic control of cell reproduction in Euglena. Photochem. Photobiol. 76: 105-115.

Hagiwara, S., M. Takahashi, A. Yamagishi, Y. Zhang and K. Goto. 2001. Novel findings regarding photoinduced comments of G1-, S and G2-phase cells to cell-cycle transitions in darknes and dark-induced G1-, S- and G2-phase arrests in *Euglena*. Photochem. Photobiol 74: 726-733.

Harrison, P.J., H.L. Conway and R.C. Dugdale. 1976. Marine diatoms grown in chemostats under silicate or ammonium limitation. I Cellular chemical composition and steady-state growth kinetics of *Skeletonema costatum*. Mar. Biol. 35: 177-186.

Harrison, P.J., H.L. Conway, R.W. Holmes and C.O. Davis. 1977. Marine diatoms grown in chemostats under silicate or ammonium limitation III Cellular Chemical composition and morphology of *Chaetoceros debilis*, *Skeletonema costatum* and *Thalassiosira gravida*. Mar. Biol 43: 19-31.

Hildebrand, M. 2000. Silicic acid transport and its control during cell wall silicification in diatoms, p. 171-188. *In* E. Bauerlein [ed.], Biomineralization of Nano-and Micro-Structures. Wiley-VCH. Weinheim, Germany.

Hildebrand, M.K. Dahlin, 2000. Nitrate transporter genes from the diatom Cylindrotheca fusiformis (Bacillariophyceae) : mRNA levels controlled by nitrogen source and by the cell cycle. J. Phycol. 36: 702-713.

Hildebrand, M., K. Dahlin and B.E. Volcani. 1998. Characterization of a silicon transporter gene family in *Cylindrotheca fusiformis*: sequences, expression analysis, and identification of homologs in other diatoms. Mol. Gen. Genet. 260: 480-486.

Hildebrand, M., B.E. Volcani, W. Gassmann, and J.I. Schroeder. 1997. A gene family of silicon transporters. Nature 385: 688-689

Huppe, H.C. and D.H. Turpin. 1994. Integration of carbon and nitrogen metabolism in plant and algal cells. Ann. Rev. Plant Physiol. Plant Mol. Biol. 45: 577-600.

Ingri, N. 1978. Aqueous silicic acid, silicates and silicate complexes. p. 5-31. *In* G. Bendz and I. Lindqvist [eds.], Biochemistry of Silicon and Regulated Problems. Plenum Press, London, U.K.

Jacquet, S., F. Partensky, D. Marie, R. Cassoti and D. Vaulot. 2001. Cell cycle regulation by light in *Prochlorococcus* strains. Appl. Environ.Microbiol. 67: 782-790.

Jochem, F.J. and D. Meyerdierks. 1999. Cytometric measurement of the DNA cell cycle in the presence of chlorophyll autofluorescence in marine eukaryotic phytoplankton by the blue-light excited dye YOYO-1. Mar. Ecol.Progr.Ser. 185: 301-307.

Kaftan, D., T. Meszaros, J. Whitmarsh and L. Nedbal. 1999. Characterization of photosystem II activity and heterogeneity during the cell cycle of the green alga Scenedesmus quadricauda. Plant Physiol. 120: 433-441.

Kröger, N. and R. Wetherbee. 2000. Pleuralins are involved in theca differentiation in the diatom Cylindrotheca fusiformis. Protist 151: 263-273.

Lemaire, S.D., M. Hours, C. Gerard-Hirne, A. Trouabal, O. Roche and J.P. Jacquot. 1999. Analysis of light/dark synchronization of cell-wall-less *Chlamydomonas reinhardtii* (Chlorophyta) cells by flow cytometry. Eur. J. Phycol. 34: 279-286.

Lisby, M. and R. Rothstein. 2004. DNA damage checkpoints and repair centers. Current opinion in Plant Biology 16: 328-334.

Löbel, K.D., J.K. West and L.L. Hench. 1996. Computational model for protein-mediated biomineralization of the diatom frustule. Mar. Biol. 126: 353-360.

Luduena, R.F. and M.C. Roach. 1991. Tubulin sulfhydryl-groups as probes and targets for antimitotic and antimicrotubule agents. Pharmacology and Therapeutics 49: 133-152.

Martin-Jézéquel, V., M. Hildebrand and M.A. Brzezinski. 2000. Silicon metabolism in diatoms: implications for growth. J. Phycol. 36: 821-840.

Murray, A. and T. Hunt. 1993. The Cell Cycle: An Introduction. Oxford University Press, UK.

Nagyova, B., M. Slaninova, E. Galova and D. Vlcek. 2003. Cell cycle checkpoints: from yeast to algae. Biologia 58: 617-626.

Nurse, P. 2000. A long twentieth century of the cell cycle and beyond. Cell 100: 71-78.

Okita, T.W. and B.E. Volcani. 1978. Role of silicon in diatom metabolism. IX. Differential synthesis of DNA polymerases and DNA-binding proteins during silicate starvation and recovery in *Cylindrotheca fusiformis*. Biochim. Biophys. Acta 519: 76-86.

Okita, T.W. and B.E. Volcani. 1980. Rôle of silicon in diatom metabolism X Polypeptide labelling patterns during the cell cycle, silicate starvation and recovery in *Cylindrotheca fusiformis*. Exp. Cell Res. 125: 471-481.

Olson, R.J. and S.W. Chisholm. 1983. Effects of photocycles and periodic ammonium supply on three marine phytoplankton species. I. Cell division patterns. J. Phycol. 19: 522-528.

Olson, R.J. and S.W. Chisholm. 1986. Effects of light and nitrogen limitation on the cell cycle of the dinoflagellate *Amphidinium carteri*. J. Plankton Res. 8: 785-793.

Olson, R.J., D. Vaulot and S.W. Chisholm. 1986. Effects of environmental stresses on the cell cycle of two marine phytoplankton species. Plant Physiol. 80: 918-925.

Paasche, E. 1973a. Silicon and the ecology of marine plankton diatoms. I. *Thalassiosira pseudonana* (*Cyclotella nana*) growth in a chemostat with silicate as limiting nutrient. Mar. Biol. 19: 117-126.

Paasche, E. 1973b. Silicon and the ecology of marine plankton diatoms. II. Silicate uptake kinetics in five diatom species. Mar. Biol. 19: 262-269.

Paasche, E. 1980. Silicon content of five marine plankton diatom species measured with a rapid filter method. Limnol. Oceanogr. 25: 474-480.

Parpais, J., D. Marie, F. Partensky, P. Morin and D. Vaulot. 1996. Effect of phosphorus starvation on the cell cycle of the photosynthetic prokaryote *Prochlorococcus* spp. Mar. Ecol. Progr. Ser. 132: 265-274.

Pascual, M. and H. Caswell. 1997. From the cell cycle to population cycles in phytoplankton-nutrient interactions. Ecology. 78: 897-912.

Pickett-Heaps, J., A.-M.M. Schmid and L.A. Edgar. 1990. The cell biology of diatom valve formation. Progr. Phycol. Res. 7: 1-168.

Planchais, S., N. Glab, D. Inzé and C. Bergounioux. 2000. Chemical inhibitors: a tool for plant cell cycle studies. FEBS Lett. 476: 78-83.

Prezelin, B. 1992. Diel periodicity in phytoplankton productivity. Hydrobiolgia 238: 1-35.

Raimbault, P. and M. Mingazzini. 1987. Diurnal variations of intracellular nitrate storage by marine diatoms: effects of nutritional state. J. Exp. Mar. Biol. Ecol. 112: 217-232.

Raven, J.A. 1983. The transport and function of silicon in plants. Biological Revue 58: 178-207.

Riedel, G. and D.M. Nelson. 1985. Silicon uptake by algae with no known Si requirement.2. Strong Ph-dependence of uptake kinetic-parameters in *Phaeodactylum-tricornutum* (Bacillariophyceae). J. Phycol. 21: 168-171.

Round, F.E. 1972. The formation of girdle, intercalary bands and septa in diatoms. Nova Hedwiga 23: 449-463.

Saburova, M.A. and I.G. Polikarpov. 2003. Diatom activity within soft sediments: behavioural and physiological processes. Mar. Ecol. Progr. Ser. 251: 115-126.

Senger, H., and N.I. Bishop. 1967. Quantum yield of photosynthesis in synchronous *Scenedesmus* cultures. Nature. 214: 140-142.

Spadari, S. 1985. Control of cell division by aphidicolin without adverse effects upon resting cells. Arzneimittel-Forschung. 35: 1108-1116.

Spadari, S., F. Focher, F. Sala, G. Ciarrocchi, G. Koch, A. Falaschi and G. Pedralinoy. 1985. Control of cell-division by aphidicolin without adverse-effects upon resting cells. Arnzneimittel-Forschung 35: 1108-1116.

Spudich, J.L. and R. Sager. 1980. Regulation of the *Chlamydomonas* cell cycle by light and dark. J. Cell Biol. 85: 136-145.

Strasser, B.J., H. Dau, I. Heinze and H. Senger. 1999. Comparaison of light induced and cell cycle dependent changes in the photosynthetic apparatus: A fluorescence induction study on the green alga *Scenedesmus obliquus*. Photosynth. Res. 60: 217-227.

Stumm, W. and J.J. Morgan. 1981. Aquatic Chemistry. 2nd ed. Wiley. New York, U.S.A.

Sullivan, C.W. 1976. Diatom mineralization of silicic acid. I. $Si(OH)_4$ transport characteristics in *Navicula pelliculosa*. J. Phycol. 12: 390-396.

Sullivan, C.W. 1977. Diatom mineralization of silicic acid. II. $Si(OH)_4$ transport rate during the cell cycle of *Navicula pelliculosa*. J. Phycol. 13: 86-91.

Sullivan, C.W. 1979. Diatom mineralization of silicic acid IV Kinetics of soluble Si Pool formation in exponentially growing and synchronized *Navicula Pelliculosa*. J. Phycol. 15: 210-216.

Sullivan, C.W. 1980. Diatom mineralization of silicic acid. V. Energetic and macromolecular requirements for $Si(OH)_4$ mineralization events during the cell cycle of *Navicula pelliculosa*. J. Phycol. 16: 321-328.

Sullivan, C.W. 1986. Silicification by diatoms, pp. 59-89. *In* D. Evered and M. O'Connor [eds.]. Silicon Biochemistry. Ciba foundation Symposium 121, Wiley Interscience. Chicester, U.K.

Sullivan, C.W. and B.E. Volcani. 1981. Silicon in the cellular metabolism of diatoms, pp. 15-42. *In* T. L. Simpson and B. E. Volcani [eds.], Silicon and siliceous structures in biological systems. Springer Verlag. New York, U.S.A.

Turpin, D.H. 1991. Effects of inorganic N availability on algal. Photosynthesis and carbon metabolism. J. Phycol. 27: 14-20.

Turpin, D.H., I.R. Elrifi, D.G. Birch, H.G.Weger and J.J. Holmes. 1988. Interactions between photosynthesis, respiration, and nitrogen assimilation in microalgae. Can. J. Bot. 66: 2083-2097.

Van Bleijswijk, J.D.L. and M.J.W. Veldhuis. 1995. *In situ* gross growth rate of *Emiliana huxleyi* in enclosures with different phosphate loading revealed by diel changes in DNA content. Mar. Ecol. Progr. Ser. 121: 271-277.

van Dolah, F.M., and T.A. Leighfield. 1999. Diel phasing of the cell-cycle in the Florida red tide Dinoflagellate, *Gymnodinium breve.* J. Phycol. 35: 1404-1411.

Vasquez, R.J., B. Howell, A.M.C. Yvon, P. Wadsworth and L. Cassimeris. 1997. Nanomolar concentrations of nocodazole alter microtubule dynamic instability in vivo and in vitro. Molecular Biology of the Cell 8: 973-985.

Vaulot, D. 1994. The cell cycle of phytoplankton. Molecular ecology of aquatic microbes. NATO ASI.

Vaulot, D., N. LeBot, D. Marie and E. Fukai. 1996. Effect of phosphorus on the *Synechococcus* cell cycle in surface Mediterranean waters during summer. Appl. Environm. Microbiol. 62: 2527-2533.

Vaulot, D., R.J. Olson and S.W. Chisholm. 1986. Light and dark control of the cell cycle in two marine phytoplankton species. Exp.Cell Res. 167: 38-52.

Vaulot, D., R.J. Olson S. Merkel, and S.W. Chisholm. 1987. Cell-cycle response to nutrient starvation in two phytoplankton species *Thalassiosira weissflogii* and *Hymenomonas carterae.* Mar. Biol. 95: 625-630.

Vaulot, D. and F. Partensky. 1992. Cell cycle distribution of prochlorophytes in the north western Mediterranean Sea. Deep Sea Res. 39: 727-742.

Veldhuis, M.J.W., T.L. Cucci and M. E. Sieracki. 1997. Cellular DNA content of marine phytoplankton using two new fluorochromes : Taxonomic and ecological implications. J. Phycol. 33: 527-541.

Vergera, J.J., J.A. Berges and P.G. Falkowski. 1998. Diel periodicity of nitrate reductase activity and protein levels in the marine diatom *Thalassiosira weissflogii* (Bacillariophyceae) J. Phycol. 34: 952-961.

Volcani, B.E. Cell wall formation in diatoms : morphogenesis and biochemistry, pp. 157-200. *In* T.L. Simpson and B.E. Volcani [eds.]. 1981. Silicon and Siliceous Structures in Biological Systems. Springer Verlag, New York, U.S.A.

Wegmann, K. and H. Metzner. 1971. Synchronization of *Dunaliella* cultures. Arch. Mikrobiol. 78: 360-367.

Winter, J. and P. Brandt. 1986. Stage-specific state I-state transitions during the cell cycle of *Euglena gracilis.* Plant Physiol. 81: 548-552.

Wong, J.T.Y. and A. Whiteley. 1996. An improved method of cell cycle synchronisation for the heterotrophic dinoflagellate Crypthecodinium cohnii Biecheler analyzed by flow cytometry. J. Exp. Mar. Biol. Ecol. 197: 91-99.

Zachleder, V. and H. van der Ende. 1992. Cell cycle events in the green alga *Chlamydomonas eugametos* and their control by environmental factors. J. Cell Sci. 102: 469-474.

12

Nutritional Value of Microalgae and Applications

John K. Volkman and Malcolm R. Brown

CSIRO Marine Research and Aquafin CRC, GPO Box 1538, Hobart, Tasmania 7001, Australia

Abstract

Microalgae provide a well balanced mixture of nutrients to organisms higher in the food web. Indeed life in the oceans is ultimately dependent on the organic matter produced by microalgae through photosynthesis. These aquatic food-chains are mimicked in aquaculture where microalgae are used as live feeds for a variety of animals including molluscs, larval stages of crustaceans and some fish species, and as food for zooplankton which in turn are used as live feeds for other animals. Microalgae contain high, but variable, percentages of the key macronutrients: typically 25 to 40% protein (of DW), 5 to 30% carbohydrate and 10 to 30% lipid. The proportions of each nutrient can be modified by careful selection of growth conditions or by harvesting the microalgae at different growth stages. All species have similar amino acid compositions and are rich in the essential amino acids. The sugar composition of polysaccharides is more variable, but most have high proportions of glucose (21–87%). Most microalgae are rich sources of one or both of the essential polyunsaturated fatty acids eicosapentaenoic acid 20:5(n-3) and docosahexaenoic acid 22:6(n-3). These fatty acids are essential for the growth of marine fish larvae. Green microalgae tend to be poor sources of these polyunsaturated fatty acids: prasinophytes have moderate amounts of 22:6(n-3) (4-10%), whereas chlorophytes are deficient in both acids (0-3%). Microalgae also contain significant quantities of micronutrients and antioxidants such as the vitamins ascorbic acid (1–16 mg g^{-1} DW), riboflavin (20–40 $\mu g\ g^{-1}$) and tocopherols as well as different carotenoids and a surprising variety of novel lipids. Details of the nutritional values of the different classes of microalgae, based on their biochemical composition, are discussed in this review plus comments on the applications of microalgae to aquaculture nutrition. Uses of microalgae also extend well beyond aquaculture to include commercial sources of carotenoid pigments, single cell oils having a high content of polyunsaturated fatty acids, dietary supplements such as '*Spirulina*' tablets, biodiesel and many others.

INTRODUCTION

Microalgae have an important role to play in aquaculture. They are the primary food source for larval and juvenile bivalves, and for the larvae of some crustaceans and fish species. Microalgae are also used to enhance the biochemical composition of zooplankton such as rotifers and copepods that are used as food for late-larval and juvenile crustaceans and fish (Reitan et al. 1997). Commercially important marine animals that rely on cultured algae as food for part of their life cycle include edible oysters (e.g. *Crassostrea gigas* and *Ostrea edulis*), pearl oyster (*Pinctada maxima*), abalone (*Haliotis* spp.), and shrimp (*Penaeus* spp.). Juvenile fish such as Atlantic halibut (*Hippoglossus hippoglossus*) and barramundi (*Lates calcarifer*) are fed on commercially reared zooplankton, before being introduced to a pelleted diet.

Size is an important criterion for choosing appropriate microalgal species for aquaculture. Larval animals are usually fed microalgae in the nanoplankton size range (2–20 μm). Shrimp larvae can use larger species such as diatom chains (e.g. *Skeletonema* spp. which can be up to 60 μm long), whereas larger 'sticky' pennate diatoms such as *Nitzschia* spp. and *Navicula* spp. (> 20 μm) are readily grazed by abalone. The usefulness of a microalga for aquaculture also depends on its nutritional quality and on how easy it is to culture which in turn is reflected in its tolerance to temperature, salinity and light conditions, especially if grown in outdoor tanks or ponds (Jeffrey et al. 1992). It is perhaps not surprising that aquaculturists worldwide have chosen to work with a limited number of microalgal species that grow well even though the biochemical composition of these species may not be ideally matched to the nutritional requirements of aquatic animals.

Information on the gross composition, amino acids, lipid classes, fatty acids, sugars and vitamins of many microalgal species is available in the literature, although most studies do not provide data for all biochemical classes. An exception is the work of Brown et al. (1997) which presents biochemical data for about 40 species from seven algal classes to assess how composition relates to differences in the nutritional value of the species. Representatives of most of the major algal taxa were examined: diatoms, haptophytes, prasinophytes, chlorophytes, eustigmatophytes, cryptomonads and a rhodophyte.

METHODS

Algal Culture

Microalgae are grown in mass culture for aquaculture operations, but most biochemical data in the literature have been derived from small scale cultures. These data can provide a useful guide as long as it is appreciated that culturing conditions can markedly affect the relative proportions of the biochemical classes, and to a lesser extent affect the distribution of compounds within a biochemical class. Important considerations are the choice of culture medium, light intensity (e.g. 70–80 μmol photons m^{-2} s^{-1} white fluorescent light is often used), light:dark cycle, temperature, extent of aeration, CO_2 supplementation, pH and stage of culture at both the inoculum and at harvest. A completely chemically defined medium is preferred. A common one is the f/2 medium of Guillard and Ryther (1962) (see Guillard 1975 for a review of culture media), the widespread use of which has dramatically advanced algal aquaculture (Wikfors and Ohno 2001).

Procedures for Biochemical Analyses

Sensitive and specific analytical methods are required to measure the nutrient concentrations in the small quantities of microalgal samples typically available from small-scale culturing (≤ 20 mg DW). Cells are usually harvested by centrifugation or by filtering onto glass fiber filters. Cells should be washed with sufficient volumes of 0.5 M ammonium formate (e.g. 3 × 5 mL for filters) to remove residual salts from the seawater medium prior to drying for dry weight determinations. Many microalgae are rapidly lysed on harvest with release of lipolytic enzymes which can result in a high proportion of free fatty acids in the extract. Some authors recommend boiling cells before analysis to inactivate these enzymes (Budge and Parrish 1999). Note that samples should contain a large enough mass of total lipids (> 4% DW) to overcome the absorption of polar lipids on glassware (Bergen et al. 2000).

For amino acids, algal samples are hydrolysed in hydrolysis-tubes at 110°C for between 12 to 48 h with either 6 M HCl (Enright et al. 1986b, Daume et al. 2003) or 4 M methanesulphonic acid containing 0.02% tryptamine (MSA reagent) (Brown 1991). Hydrolysates from 6 M HCl can be freeze-dried, and derivatised directly prior to analysis, though this process gives a poor recovery of tryptophan. Residual lipids and carbohydrate can interfere with the derivatisation process (Cohen et al. 1988). The MSA reagent provides a good recovery of tryptophan, though the hydrolysate must be passed through a cation-exchange resin (AG 50W-X8; Bio-Rad, Ca, USA) to remove the reagent (Lazarus 1973, Brown 1991). However, an advantage is that this process separates amino acids from lipids and carbohydrates, producing a purified amino acid fraction for derivatisation and analysis. Common reagents used for amino acid derivatisation include phenylisothiocyanate (Bidlingmeyer et al. 1984) and 9-fluorenylmethoxycarbonyl chloride (Daume et al. 2003). The anhydroamino acid residues can be summed to give an estimate of the protein content.

Most studies of lipids in microalgae use some variation of the original extraction method of Bligh and Dyer (1959) designed for analysis of fish lipids which uses successive extractions with a chloroform-methanol-water mixture (1:2:0.8; v/v/v), although good yields can also be obtained with other solvents such as ethanol and hexane-ethanol (1:2.5) (Molina Grima et al. 1994). Higher contents of methanol have been shown to give better yields of the more-polar lipids such as phospholipids (Smedes and Askland 1999). The resultant extracts are combined, and separated into chloroform and aqueous-methanol layers by the addition of water and chloroform. The chloroform layers are then concentrated by rotary evaporation to provide a total lipid extract. Note that there can be some loss to the aqueous phase or water-solvent interface of more polar lipid classes that contain water-soluble moieties such as sugars. Lipid classes can be separated into biochemical classes by HPLC (Kato et al. 1996, Silversand and Haux 1997, Nordbäch et al. 1998), or by open column or thin-layer chromatography (Christie 1982). The use of Ag^+-impregnated supports can be very useful in separating lipids with different degrees of unsaturation (Dobson et al. 1995). Quantitative data can be determined by thin-layer chromatography with flame ionization detection (Iatroscan TLC-FID) (Volkman and Nichols 1991, Parrish et al. 1992, Bergen et al. 2000) or using a light-scattering detector with HPLC (Christie 1998, Nordbäch et al. 1998). Total lipids are often transesterified or saponified to produce total fatty acids which are analyzed as methyl esters by capillary GC using polar and non-polar capillary columns and GC-mass spectrometry (Volkman et al. 1989).

The aqueous-methanol layers from a lipid extraction contain mono- and oligo-saccharides, and these can be assayed for carbohydrate by the phenol-sulphuric acid method (Dubois et al. 1956). The residues remaining after lipid extraction contain polysaccharide which can be hydrolysed with 0.5 M H_2SO_4 at 100°C and analyzed for carbohydrate (Dubois et al.

1956). Constituent sugars may be converted to alditol acetate derivatives (Blakeney et al. 1983) and quantified by gas chromatography (Whyte 1987, Brown 1991). Total cell carbohydrate is determined as the sum of the mono-, oligosaccharide and polysaccharide carbohydrate. Alternatively, the method of Dubois et al. (1956) can be applied directly to the cell pellet or freeze-dried biomass to estimate total carbohydrate (Brown et al. 1993a).

A range of specific and sensitive assays are available for analysis of vitamins in microalgae. For example, ascorbic acid, α-tocopherol, riboflavin and thiamin can be assayed by reverse phase HPLC with fluorimetric detection (Brown et al. 1999), whereas biotin and other B group vitamins are assayed by microbiological methods (Seguineau et al. 1996). GC methods used for sterol analysis are also suitable for the analysis of lipid-soluble vitamins such as α-tocopherol (reviewed by Rupérez et al. 2001). Note that α-tocopherol elutes near cholesterol on non-polar GC phases.

Biochemical Composition

Proximate Composition

The proximate composition, i.e. the relative percentages of protein, carbohydrate, lipid and mineral (or ash) can vary substantially between microalgae species. Under standard conditions (i.e. where nutrients are not limited), microalgae typically contain between 10 to 40% of DW as protein, 10 to 30% as lipid, 5 to 30% as carbohydrate and 10 to 40% as ash (Whyte 1987, Renaud et al. 1999, Knuckey et al. 2002) (Table 12.1). Due to this variability it is difficult to categorize algal classes based on proximate composition alone. Nevertheless, cryptophytes are generally richer in protein than other microalgae (McCausland et al. 1999, Renaud et al. 1999), whereas diatoms have higher concentrations of ash (Brown and Jeffrey 1995, Renaud et al. 1999) (Table 12.1).

The reported percentages of proximate components in microalgae generally sum to less than 100%. In two independent studies that examined a total of 28 species or strains Renaud et al. (1999) found an average value of 77% (range: 63 to 92%), whereas Knuckey et al. (2002) found an average value of 96% (range: 78 to 117%). Aside from potential analytical error, reasons for these discrepancies relate to: 1) other nutrient components unaccounted for when using a specific assay, and 2) methodologies that under/over-estimate nutrients within certain algal species. With respect to the former, residual moisture can account for up to 8% of proximate weight in microalgae after freeze-drying (Table 4 in Rebolloso Fuentes et al. 2000).

Table 12.1 Proximate composition of microalgal species and classes, based on logarithmic phase cultures. Data are expressed as % of DW showing the range of values from selected studies, with mean values shown in parentheses. n.a. denotes not analyzed.

Algal species/Class studied	*Protein*	*Carbohydrate*	*Lipid*	*Ash*	*Reference*
Data from individual studies:					
Chlorophytes					
D. tertiolecta	39	28	23	18	McCausland et al. (1999)
Dunalliella sp.	64	13	23	12	Thomas et al. (1984)
Chlorella spp. (CS-247 and CS-195), *Stichococcus* sp.	15-23 (19)	6-16 (12)	9-17 (14)	n.a.	Brown and Jeffrey (1992)
Prasinophytes					
Tetraselmis sp. CS-362	30	26	16	17	Brown et al. (1998a)
Tetraselmis spp. (NT18 and TEQL01), *Nephroselmis* sp.	26-32 (30)	9-14 (11)	10-14 (12)	11-17 (14)	Renaud et al. (1999)
Cryptophytes					
Rhodomonas salina	59	19	19	10	McCausland et al. (1999)
Cryptomonas sp., *Rhodomonas* sp.	29-47 (38)	4-9 (6)	19-22 (20)	15-17 (16)	Renaud et al. (1999)
Diatoms					
Amphora coffeaformis, Chaetoceros sp., *Fragilaria pinnata, Nitzschia* sp., *N.* cf. *frustulum, Skeletonema* sp., *S. costatum*, unidentified chain diatom	22-37 (26)	4-15 (7)	13-20 (15)	21-39 (30)	Renaud et al. (1999)
Attheya septentrionalis, Entomoneis sp., *Extubocellulus spinifera, Minidiscus* spp., *Thalassiosira oceanica, T. pseudonana, Nitzschia* spp., *Papiliocellulus simplex*	12-38 (24)	9-30 (19)	14-42 (25)	9-43 (29)	Knuckey et al. (2002)
T. pseudonana, Chaetoceros sp., *C. calcitrans*	21-23 (22)	8-17 (12)	10-16 (12)	27-38 (33)	Whyte (1987)
Cylindrotheca fusiformis, Navicula jeffreyi, Nitzschia closterium, Lauderia annulata	16-38 (29)	5-12 (8)	18-20 (19)	9-35 (20)	Brown and Jeffrey (1995)

Contd.

Table 12.1 Contd.

Algal species/Class studied	*Protein*	*Carbohydrate*	*Lipid*	*Ash*	*Reference*
Skeletonema sp., *S. costatum*					
Rhodophytes					
Rhodosorus sp.	32	20	5	16	Renaud et al. (1999)
Porphyridium cruentum	34	32	7	20	Rebolloso Fuentes et al. (2000)
Eustigmatophytes					
Nannochloropsis-like sp.	17	23	26	16	Brown et al. (1998a)
Nannochloropsis salina	41	14	21	8	Thomas et al. (1984)
Prymnesiophytes					
Pavlova pinguis	33	41	19	12	Brown et al. (1998a)
Isochrysis sp. NT14, 3 unidentified spp. (NT19, CS-260 and QLD-B2)	24-32 (28)	7-16 (12)	13-23 (18)	13-20 (16)	Renaud et al. (1999)
Isochrysis sp. (T.ISO), *P. lutheri*	22-29 (25)	12-27 (19)	32-37 (34)	4-5 (5)	Knuckey et al. (2002)
Isochyrsis sp. (T.ISO), *I. galbana*	28-33 (31)	8-9 (9)	21-25 (23)	9-12 (10)	Whyte (1987)
Summary Data for Classes:					
Chlorophytes	15-64 (32)	6-28 (15)	9-23 (18)	12-18 (15)	
Prasinophytes	26-32 (30)	9-26 (16)	10-16 (13)	11-17 (15)	
Cryptophytes	29-59 (42)	4-19 (11)	19-22 (20)	10-17 (14)	
Diatoms	12-38 (25)	5-30 (12)	10-42 (17)	9-43 (28)	
Rhodophytes	32-34 (33)	20-32 (26)	5-7 (6)	16-20 (18)	
Eustigmatophytes	17-41 (29)	14-23 (19)	21-26 (24)	8-16 (12)	
Prymnesiophytes	22-33 (29)	7-41 (16)	13-37 (23)	4-20 (12)	

Crude fiber, not normally accounted for in standard carbohydrate assays using acid hydrolysis, may be associated with cell wall structures (e.g. cellulose matrix, chitan and pectic polysaccharides in scales surrounding cells; Vesk et al. 1990) and can comprise more than 10% of total carbohydrate (Parsons et al. 1961). Some microalgae also contain a few percent of polymeric non-extractable lipidic 'algaenans' (Gelin et al. 1997, 1999).

The choice of methodologies can also lead to errors. Lipid can be difficult to extract from microalgae with tough cell walls (e.g. *Nannochloris* sp., *Nannochloropsis* spp., *Tetraselmis* spp. and *Chlorella* spp.) and hence can be underestimated (Volkman et al. 1989). Carbohydrate assays are usually based on colorimetric assays with glucose as the standard (e.g. phenol-sulfuric acid method; Dubois et al. 1956) and are accurate if glucose is the predominant sugar, but may under- or over-estimate true carbohydrate if there are high proportions of other sugars that are either more or less reactive than glucose. Protein can be estimated by summation of individual amino acids from cell hydrolysates and is the most accurate method (Brown 1991). However, less specific colorimetric assays (Clayton et al. 1988) generally give comparable results (M.R. Brown, unpublished data). 'Crude protein' is often reported as total % N × 6.25, but this method can overestimate protein by up to 30% if cells have particularly high concentrations of non-protein N (e.g. from nucleic acids, hexosamines and amides) (Lee and Picard 1982).

Variation in proximate compositions also occurs within the same species under different culture conditions, for example light, temperature and nutrient status (Harrison et al. 1990, Thompson et al. 1992) though not in a consistent way across all species. Knowledge of how species respond under different environments allows the production of nutrients of interest to be optimized. For example, when nitrate is limiting the levels of carbohydrate double at the expense of protein in species such as *Isochrysis* sp. (T.ISO) and *Chaetoceros calcitrans* (Harrison et al. 1990, Brown et al. 1993b), whereas high light intensity stimulates the accumulation of carbohydrate in *Thalassiosira pseudonana* (Brown et al. 1996). In general, microalgae are richer in polar lipids during logarithmic phase, but accumulate triacylglycerols during stationary phase (Dunstan et al. 1993).

The molar N:P ratio of microalgae and cyanobacteria can be quite variable in nutrient-limited cells, ranging from less than five when phosphate is greatly in excess to greater than 100 when inorganic N is greatly in excess (Geider and La Roche 2002). Under nutrient-replete growth conditions the cellular N:P ratio is more restricted, typically ranging from 5 to 19, with most observations below the Redfield ratio of 16. Lowest values of N:P are associated with nitrate- and phosphate-replete conditions in both field samples and in culture (Geider and La Roche 2002). The C:N ratio is also quite variable with the average C:N ratios of nutrient-replete phytoplankton cultures and oceanic particulate matter slightly greater than the Redfield ratio of 6.6. The use of N and P in the cosmopolitan marine species *Emiliania huxleyi* has been extensively studied by Riegman et al. (2000). This work showed that *E. huxleyi* has an exceptional P assimilation capability. Its uptake rate of nitrate in the dark is 70% lower than in the light, while N-limited cells are smaller than P-limited cells and contain 50% less organic and inorganic carbon. As a consequence of its high affinity for inorganic phosphate, and the presence of two different types of alkaline phosphatase, *E. huxleyi* is able to perform well in ecosystems where productivity is controlled by P availability.

Amino acids

Most of the amino acids within microalgae are incorporated as protein. The content of free amino acids can increase significantly during the stationary phase (e.g. from to 2 to 40 times their concentration during exponential phase; Flynn 1990); an analysis of five species found concentrations ranging from 3 to 12% of total DW (Derrien et al. 1998). While the majority of the free amino acids are the same as those incorporated in protein, microalgae also contain other amino acids as intermediary metabolites, e.g. ornithine, 4-aminobutryic acid and hyroxy-proline and collectively they can constitute up to 1 to 2% of the DW (Brown 1991, Derrien et al. 1998).

Many microalgae algae also contain trace amounts of mycosporine-like derivatives of amino acids that efficiently absorb UV-light (Jeffrey et al. 1999). These compounds are transferred through the food chain and may have a UV-protective role in natural ecosystems (Newman et al. 2000, Tartarotti et al. 2004). Accurate quantitative data are lacking on their concentration in algae, though surface bloom-forming dinoflagellates, cryptomonads, haptophytes and raphidophytes seem to have the highest concentrations (Jeffrey et al. 1999). Dense laboratory cultures of *Phaeocystis antarctica* contained one derivative, mycosporine-glycine:valine in concentrations of up to 500 $\mu g\ ml^{-1}$ of culture, though data were not expressed on a DW basis (Newman et al. 2000).

Most amino acid analyses of microalgae have assessed profiles from cell hydrolysates, thus assessing total (i.e. free plus protein-bound) amino acids. In analyses of 37 microalgae, Brown and co-workers found that the amino acid compositions were remarkably similar (Brown 1991, Brown et al. 1997) (Table 12.2). Aspartate and glutamate were present in highest concentrations (typically 8 to 12% of total amino acids); cystine, methionine, tryptophan and histidine were in lowest concentrations (typically 0.5 to 3%) and average values of other amino acids ranged from 4 to 8%. Also, there were no class-specific differences in composition although species from the genus *Tetraselmis* (*T. suecica* and *T. chui*) contained twice the arginine content of other species (Brown 1991). Other studies have shown that the amino acid compositions of microalgae are relatively unaffected by light and nutrient conditions (Brown et al. 1993a,b, Daume et al. 2003).

Table 12.2 The amino acid composition (weight % of total amino acids) of microalgae, showing range and mean data from 37 species. Data compiled from Brown (1991), Brown and Jeffrey (1992), Volkman et al. (1993), Dunstan et al. (1994), and M. R. Brown unpublished. Data are compared to those of oyster larvae, *Crassostrea gigas.*

Amino acid	*Range*	*Mean ± s.d.*	Crassostrea gigas
Essential:			
arginine	4.8 - 15.3	7.3 ± 2.1	7.7
histidine	1.4 - 2.7	2.0 ± 0.3	1.7
isoleucine	3.4 - 5.9	4.6 ± 0.6	4.5
leucine	6.5 - 10.4	8.1 ± 1.0	7.0
lysine	5.0 - 8.6	6.2 ± 0.8	8.4
methionine	1.4 - 3.2	2.2 ± 0.4	1.3
phenylalanine	4.5 - 7.7	6.2 ± 0.8	5.3
threonine	4.0 - 7.3	5.3 ± 0.7	5.2
tryptophan	0.4 - 3.1	1.6 ± 0.5	1.6
valine	4.4 - 6.8	5.8 ± 0.6	5.5
Non-Essential:			
alanine	5.3 - 9.1	7.6 ± 0.9	5.7
aspartate	7.5 - 10.9	9.1 ± 0.9	10.2
cystine	0.4 - 2.0	0.8 ± 0.4	0.7
glutamate	9.5 - 12.6	10.9 ± 0.8	12.5
glycine	5.2 - 7.7	6.0 ± 0.5	9.3
proline	4.1 - 13.3	6.1 ± 1.9	4.5
serine	4.5 - 8.4	5.7 ± 0.8	5.1
tyrosine	3.3 - 6.9	4.8 ± 0.8	4.6

Sugars and carbohydrates

Carbohydrates in microalgae are distributed between polymeric fractions and the soluble fractions of simple sugars, i.e. the mono-, di- and oligo-saccharides. Most carbohydrate within the polymeric fraction is readily-hydrolysable polysaccharide, though some microalgae can have significant amounts of fibrous material such as cellulose (e.g. *Tetraselmis* spp.; Parsons et al. 1961) or chitan (e.g. *Thalassiosira* spp.; Falk et al. 1966). Polysaccharide typically comprises 80 to 95% of the total carbohydrate in microalgae, excluding the fibre (Whyte 1987, Brown 1991).

Several studies have examined the constituent sugar composition of algal polysaccharide following acid hydrolysis to characterize the polysaccharide types and determine phylogenetic differences in profiles (Chu et al. 1982, Whyte 1987, Brown 1991). The sugar composition of microalgae varies appreciably. Glucose is the predominant sugar, ranging from 20 to 90% of total sugars, and its percentage increases in most microalgae, except for diatoms, with culture age (Chu et al. 1982, Whyte 1987). These observations are consistent with glucan (polysaccharides rich in glucose) being the major food reserves in microalgae (Handa and Yanagi 1969). In a survey of 16 microalgae, Brown (1991) found galactose (1 to 20% of polysaccharide sugars) and mannose (2 to 46%) were also common, with arabinose, fucose, rhamnose, ribose and xylose found in varying proportions (0 to 17%). Haptophytes contained more arabinose, on average, than microalgae from other classes, whereas diatoms had higher percentages of galactose and mannose than most other species.

Diatoms can also excrete up to 50% of their polysaccharide into the culture medium during the stationary phase (Mykelstad 1974). Red microalgae are encapsulated within a sulfated polysaccharide rich in xylose, glucose and galactose, and which dissolves into the growth medium (Geresh and Arad 1991). The prasinophyte *Prasinococcus capsulatus* forms a copious, and unique, sulfated and carboxylated polyanionic polysaccharide named capsulan. This polysaccharide capsule is presumed to originate in the Golgi body and is secreted through a crown of 10 pores in the cell wall, the 'decapore' (Sieburth et al. 1999). Large scale production of the red microalga *Porphyridium* is being investigated as a source of polysaccharides for the cosmetic, drugs and health food markets (Singh et al. 2000).

Lipids

Lipids are an important source of storage energy for microalgae. They also function as insulators of delicate internal organs as membrane constituents, and they provide hormones and other bioactive components that regulate cell biochemistry. Lipids also have a vital role in tolerance to several physiological stressors in a variety of organisms including cyanobacteria (Singh et al. 2002). The major lipid classes in microalgae are triacylglycerols (TAG; an energy store), phospholipids (e.g. phosphatidylcholine PC, phosphatidylglycerol PG, and phosphatidylethanolamine PE) and glycolipids (e.g. monogalactosyldiacylglycerol MGDG, digalactosyldiacylglycerol DGDG, and sulfoquinovosyldiacylglycerol SQDG) (Okuyama et al. 1992, Budge and Parrish 1999). Betaine lipids have been reported in only a few species (Eichenberger 1993, Eichenberger et al. 1996), but it may be that they are more widely distributed. Diacylglyceryltrimethylhomoserine (DGTS) and hydroxymethyl-N,N,N-trimethyl-β-alanine (DGTA) occur in the cryptophyte *Chroomonas salina* (Eichenberger et al. 1996). DGTA, diacylglyceryl carboxyhydroxymethylcholine (DGCC) and diacylglyceryl glucuronide (DGGA) occur as major lipid components in some haptophytes such as *Pavlova* (Eichenberger and Gribi 1997). Both DGTA and DGTS are produced in the chrysophyte *Ochromonas* and the cryptomonad *Cryptomonas*; but in the presence of either DGTS or DGTA, the phospholipid PC is often not produced (Eichenberger 1993). All microalgae contain sterols (Volkman 1986, Volkman et al. 1998) and these can occur either as the free sterol (most common) or as a fatty acid ester, glycoside or sulfate. The distributions can be exceedingly simple (e.g. eustigmatophytes usually contain just cholesterol with traces of other sterols) to highly complex (dinoflagellates can contain 20 or more 4-desmethyl and 4-methyl sterols) (Volkman 1986). A few microalgae also contain unusual lipids as discussed later.

Fatty acids

Fatty acids are denoted by a short-hand nomenclature x:y(n-z) where x is the number of carbon atoms, y is the number of double bonds and z is the position of the double bond closest to the methyl (ω) end of the molecule. All double bonds are assumed to be *cis* (Z) and methylene-interrupted unless otherwise stated. Thus the polyunsaturated fatty acid (PUFA) arachidonic acid (AA) is denoted by 20:4(n-6), eicosapentaenoic acid (EPA) is denoted by 20:5(n-3), and docosahexaenoic acid (DHA) is denoted by 22:6(n-3). In the older literature these are often referred to as 20:4ω6, 20:5ω3 and 22:6ω3 respectively, hence the common expression 'omega-3' and 'omega-6' fatty acids.

EPA and DHA are essential dietary components for a variety of mollusc, shrimp and fish larvae, and may also be essential for other marine animals (Langdon and Waldock 1981, Castell et al. 1986, Volkman et al. 1992, Sargent et al. 1997). AA may be important for fish larvae (Tocher and Sargent 1984). Good sources of AA include diatoms, eustigmatophytes and rhodophytes (Fig. 12.1).

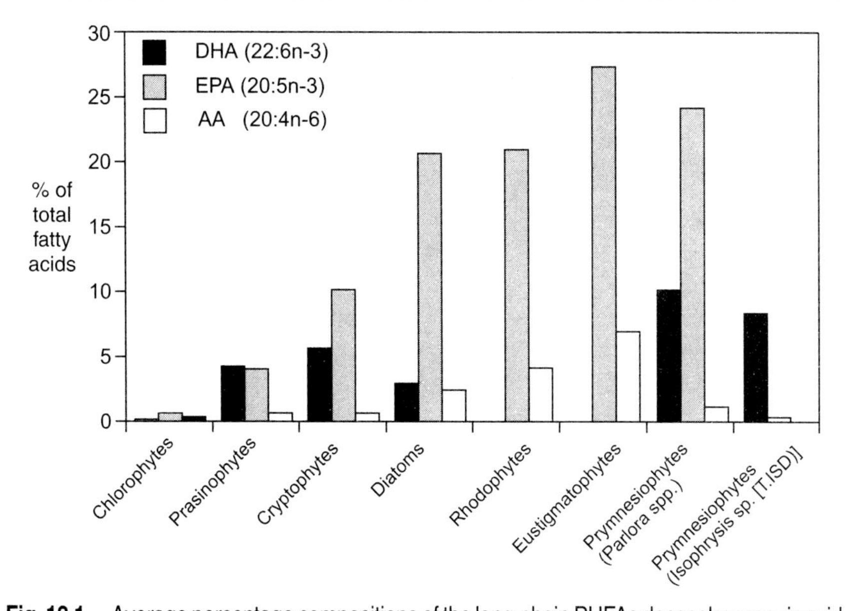

Fig. 12.1 Average percentage compositions of the long-chain PUFAs docosahexaenoic acid (DHA; 22:6n-3), eicosapentaenoic acid (EPA; 20:5n-3) and arachidonic acid (AA, 20:4n-6) of microalgae commonly used in aquaculture. Data compiled from over 40 species analysed at CSIRO Marine Research.

Microalgae contain a great variety of fatty acids and their distributions are often distinctly different between algal classes (Volkman et al. 1989). Diatoms, eustigmatophytes, cryptomonads, rhodophytes and some haptophytes (e.g. *Pavlova* species) are excellent sources of EPA (from 7–34% of total fatty acids) (Volkman et al. 1989, 1991, 1993, Dunstan et al. 1992, 1994). Many representatives from these classes have been used successfully as food in larval culture (reviewed by Brown et al. 1997). For example the diatom *Skeletonema costatum* provides a good diet for molluscs, due to its high proportion of PUFA such as EPA (Berge et al. 1995)

Many cryptomonads and haptophytes are relatively rich in DHA (0.2–11%), whereas eustigmatophytes, rhodophytes and diatoms contain small amounts of AA (0–4%). Chlorophytes are deficient in both C_{20} and C_{22} PUFAs, although some species have small amounts of 20:5(n-3) (up to 3.2%). Chlorophytes generally have low nutritional value and are not suitable as a single species diet (Brown et al. 1997). Prasinophytes contain significant proportions of C_{20} and C_{22} PUFAs (but rarely both) and *Tetraselmis* species have been used successfully for shrimp and mollusc culture (Wikfors et al. 1996, Brown et al. 1997).

The C_{18} PUFAs, 18:2(n-6) and/or 18:3(n-3) are essential for many freshwater fish (Castell et al. 1986, Brett and Müller-Navarra 1997). The dietary C_{18} PUFA can be elongated and desaturated to form longer chain PUFA more efficiently by freshwater fish than by marine fish, although dietary EPA and DHA, when available, are preferentially incorporated into fish tissues. Significant levels of 18:2(n-6) and 18:3(n-3) are found in most microalgal groups, except for the diatoms and eustigmatophytes which contain very low levels (Volkman et al. 1989, 1993, Dunstan et al. 1994). The major C_{18} PUFA in prasinophytes and haptophytes is 18:4(n-3) (Volkman et al. 1989, Dunstan et al. 1992). The importance of these polyunsaturated C_{18} PUFA in marine finfish diets is not known.

A common feature of all microalgal fatty acid distributions is a high proportion of the saturated C_{16} fatty acid palmitic acid (16:0). Diatoms and haptophytes can also contain significant amounts of 14:0, and diatoms contain very high proportions of palmitoleic acid [16:1(n-7)] which is often the major fatty acid present. These acids are not usually considered as specific nutritional factors, but they are readily utilized for energy (Thompson et al. 1993). Haptophytes generally contain the highest proportions of saturated fats (ca. 33%), followed by diatoms and eustigmatophytes (ca. 27%), chlorophytes and prasinophytes (ca. 23%) and cryptomonads (ca. 18%) (Volkman et al. 1989, 1991, 1993; Dunstan et al. 1992, 1994).

An unusual feature of the fatty acids of some dinoflagellates is the presence of very long-chain C_{28} highly unsaturated fatty acids [28:7(n-6)] and [28:8(n-3)] (Mansour et al. 1999, Leblond and Chapman 2000). The biological function of these fatty acids is yet to be elucidated, but it is known that they are associated with phospholipids and hence may be membrane constituents synthesized in the cytoplasm (Leblond and Chapman 2000). The effects of these unusual fatty acids being present in a diet are unknown.

The pathways by which polyunsaturated fatty acids are biosynthesized by microalgae combine features found in both plants and animals (Moreno et al. 1979). Thus, in the n-6 pathway, 18:2 is desaturated to 18:3(n-6), elongated to 20:3(n-6), and subsequently desaturated to 20:4 (n-6) and then to 20:5(n-3). In the n-3 pathway, 18:2 is first desaturated to 18:3(n-3) which is then sequentially converted, apparently by the same enzymatic sequence of the n-6 pathway to 18:4(n-3), 20:4(n-3) and 20:5(n-3) (Shiran et al. 1996). Specific aspects of the pathways are now being reinvestigated. For example, it appears quite possible, based on animal studies, that 22:6(n-3) in microalgae is formed by chain-shortening of 24:6(n-3) (Voss et al. 1991,

Buzzi et al. 1997, Sprecher et al. 1999), rather than elongation of 20:5(n-3) and further desaturation. However, there are no reports of this long-chain PUFA in microalgae to date. The high abundance of 18:5(n-3) in some dinoflagellates, prasinophytes, haptophytes and raphidophytes (see Bell et al. 1997 for references) may also be due to a similar chain-shortening mechanism from 20:5(n-3) as originally proposed by Joseph (1975). Chain-shortening of fatty acids by β-oxidation in *Phaeodactylum tricornutum* was confirmed in the ^{14}C-labelling studies of Moreno et al. (1979).

The fatty acid compositions of the triacylglycerols and different polar lipids in an alga can be very different. For example, Leblond and Chapman (2000) found that the phospholipid fractions in dinoflagellates contained the majority (over 75% in 12 of 16 strains) of the DHA present. In contrast, the highly unsaturated C_{18} fatty acids 18:4(n-3) and 18:5(n-3) were primarily recovered from a chloroplast-associated glycolipid fraction comprised of MGDG, DGDG and SQDG. These lipids have different physiological functions and thus their proportions (and that of their constituent fatty acids) can vary significantly in response to changes in light intensity, salinity and most importantly nutrient levels. In contrast to these results, Meireles et al. (2003) found that EPA in the haptophyte *Pavlova lutheri* was especially concentrated in MGDG (ca. 45%) and TAG (ca. 33%), whereas DHA was dispersed through the various classes, especially within TAG (ca. 27%), diphosphatidylglycerols (DPG, ca. 22%), and betaine lipids (21%).

Fatty acids from microalgae may be efficiently transferred to higher trophic levels such as fish larvae via intermediary zooplankton (Watanabe et al. 1983, Brett and Müller-Navarra 1997, Reitan et al. 1997). This process is often termed 'trophic upgrading'. The most used microalgae for boosting zooplankton with PUFA are those which contain high levels of EPA (e.g. *Nannochloropsis oculata*), DHA (e.g. *Isochrysis* sp. clone T-ISO), or both (e.g. *Pavlova lutheri*). Rotifers (*Brachionus plicatilis*) readily ingest *Pavlova* cells from which they can accumulate high concentrations of EPA, DHA and other PUFA within a few hours (Nichols et al. 1989). After just three hours of feeding on *Pavlova*, the distribution of fatty acids in the rotifer becomes almost indistinguishable from its algal food.

Effect of Culture Conditions on Lipid Compositions

Environmental conditions can lead to significant changes in the lipid composition of microalgae (Shifrin and Chisholm 1981). Sterol distributions are usually little affected, but environmental conditions can have a dramatic effect on fatty acid distributions (Dunstan et al. 1993). Other studies have demonstrated changes in fatty acid composition associated with light intensity (Thompson et al. 1990, 1993, Brown et al. 1993b), culture media (Ben-Amotz et al. 1985), temperature (James et al. 1989, Thompson et al. 1992) and pH (James et al. 1988, Guckert and Cooksey 1990), but many of the changes were species-specific. Such changes are a common response by microalgae to a particular environmental stress. In part this may simply be due to the changes in growth rate and physiological state rather than a specific response caused by environmental influences. Many studies have not corrected for changes in growth rate and thus the algae are not harvested at the same physiological state.

Growth stage at harvest in batch culture

Lipids of microalgae are often rich in polar lipids during the logarithmic phase growth, but many species accumulate triacylglycerol during the stationary phase when nitrogen is limiting (Dunstan et al. 1993). The fatty acid amount and composition of microalgae is dependent on both growth conditions and stage of harvest (Siron et al. 1989). For example, *Nannochloropsis oculata* contains more PUFA per cell in the logarithmic phase than in the stationary phase, whereas the reverse is true for *Pavlova lutheri* (Dunstan et al. 1993). In the diatom *Phaeodactylum tricornutum* the proportion of 20:5(n-3) first increased and then decreased with culture age, whereas 16:1(n-7) continued to increase (Siron et al. 1989).

A comprehensive study of effects of growth phase on the lipid content and composition of a dinoflagellate was carried out by Mansour et al. (2003). The lipid content per cell increased two fold from late logarithmic to linear growth phase and then decreased at the stationary phase. Changes in fatty acid content mirrored these changes, while the sterol content continued to increase with culture age. The largest changes occurred in the proportions of lipid classes. Triacylglycerols increased from an initial 8% (of total lipids) to 30% at the late stationary phase, with a concomitant decrease in the polar lipid fraction. The proportions of 16:0 and 22:6(n-3) increased while those of 18:5(n-3) and 20:5(n-3) decreased with increasing culture age.

Donato et al. (2003) observed significant changes in the amounts of fatty acids, sterols, carotenoids and α-tocopherol with culture age in the haptophyte *Diacronema vlkianum*. Sterol content increased up to the stationary phase and then declined in the decay phase, whereas α-tocopherol contents kept increasing perhaps reflecting the need for more anti-oxidants as the cells aged. The distribution of individual sterols only showed small changes. The highest content of PUFA was observed in the stationary phase cultures.

Semi-continuous and continuous culture

Semi-continuous culture is often used in aquaculture where a portion of the culture is used as feedstock and the volume is then topped up with fresh medium. Fábregas et al. (2001) examined the food value of the marine prasinophyte *Tetraselmis suecica* for *Artemia* when the alga was grown semi-continuously with renewal rates between 10% and 50%. At the low renewal rate the alga's biochemical composition was similar to that of the stationary-phase cultures commonly used in aquaculture and this produced poor growth and survival in the *Artemia* and low food-conversion efficiency compared to cultures maintained with a high renewal rate. The gross biochemical composition of the *Artemia* resembled that of the microalgal food except for total lipid content. Higher renewal rates resulted in higher lipid percentages in the microalga, but in *Artemia* the percentage of lipids decreased from 19% of the organic weight with a renewal rate of 10%, to 13% with a renewal rate of 50%. The proportion of protein in *Artemia* expanded with increasing renewal rates in the microalgal cultures from 45% to 65% of the organic weight, while the carbohydrate percentage decreased under the same conditions. These results emphasize the importance of controlling microalgal nutritional value for the success of aquaculture food chains in which filter feeders are involved.

If high rates of renewal are used (i.e. continuous culture), microalgae of a less variable biochemical composition can be obtained. For the diatom *Phaeodactylum tricornutum* grown in continuous culture for 1.5 to 7 d (López-Alonso et al. 2000), culture age had almost no influence on the fatty acid content (which remained around 11% of DW) and very little effect on the composition. The culture age had a greater impact on the proportions of lipid classes: TAG ranged from 43% to 69%, and galactolipids oscillated between 20% and 40%. In general, the content of polar lipids of the biomass decreased with culture age.

Effects of light intensity and light-dark cycle

Changes in fatty acid composition associated with light intensity have been demonstrated by Thompson et al. (1990, 1993) and Brown et al. (1993b). Logarithmic phase cultures of the diatom *Thalassiosira pseudonana* grown under a 12:12 h light-dark cycle (100 $\mu E\ m^{-2}s^{-1}$) contained a 25% greater proportion of 20:5(n-3) than similar cultures grown under continuous light (50 and 100 $\mu E\ m^{-2}s^{-1}$) (Brown et al. 1996). The content of saturated fats in microalgae can also be improved by culturing under high light conditions (Thompson et al. 1993).

Growth temperature

Reducing the growth temperature usually brings about an increase in the proportion of those fatty acids having a greater number of double bonds. Fatty acid compositions can respond rapidly to changes in temperature, with significant changes seen within two hours (Rousch et al. 2003). This is much faster than effects due to changes in other culturing factors such as nutrient status which may take several days to become significant. Okuyama et al. (1992) showed that a decrease in growth temperature from 5°C to 2°C was accompanied by a significant increase in levels of 18:5(n-3) and 18:4(n-3) in an un-named haptophyte. Intriguingly, the level of 22:6(n-3) changed little. These data led the authors to conclude that C_{18} PUFA with more than three double bonds, 18:5(n-3) in particular, serve as modulators of membrane fluidity.

The effects of temperature appear to be species-specific. For example, in the chlorophyte *Chlorella* the proportion of C_{18} PUFA increased with a decrease in temperature, whereas the proportion of 20:5(n-3) in the eustigmatophyte *Nannochloropsis* increased with temperature up to 25ºC and then declined greatly at 30ºC (James et al. 1989). Note that species of *Chlorella* contain very low contents of C_{20} and C_{22} PUFA, in contrast to the "marine *Chlorella*" used in the early days of aquaculture in Japan which is actually a eustigmatophyte (Maruyama et al. 1986).

Effect of nutrient regime

Changes in fatty acid composition due to the choice of culture media have been shown by Ben-Amotz et al. (1985). However, sterol distributions are usually fairly robust and distributional variations with changes in environmental conditions are usually small (Hallegraeff et al. 1991), except for a few noted exceptions (Piretti et al. 1997). Parrish et al (1999) showed that *Isochrysis galbana* grown in 85-lite cage culture turbidostats under conditions of nitrogen limitation had a significantly higher total lipid content than when grown under nutrient-replete conditions. This was due mainly to a doubling in the amount of less unsaturated triacylglycerol in the cells.

Increasing CO_2 availability can also enhance the production of lipids and fatty acids. Hu and Gao (2003) found that photoautotrophic conditions with enriched CO_2 (2800 µL CO_2 L^{-1}) and aeration gave the highest biomass yield, total lipid content (9% of DW), total fatty acids and polyunsaturated fatty acids (35% of total fatty acids) in the picoplankton *Nannochloropsis* sp.

Mixotrophic cultures gave a greater protein content but less carbohydrates. Adding sodium acetate (2 mM) decreased the amounts of the total fatty acids and EPA.

Nitrogen content is an important parameter, since in its absence proteins cannot be synthesized. Hence, López-Alonzo et al. (2000) found that saturated and monounsaturated fatty acids accumulated when the nitrogen concentration was decreased in continuous cultures of *P. tricornutum*. As N was lowered, TAG increased from 69% to 75%, neutral lipids from 6% to 8% and phospholipids from 73% to 79%. These changes were compensated for by a decrease in galactolipids from 21% to 12%. Increased fatty acid synthesis has been shown to occur when phosphorus is limited (Siron et al. 1989). When *Phaeodactylum tricornutum* was cultured in a P-deficient medium the composition of fatty acids was much like that observed in a senescent batch culture (Siron et al. 1989).

Lipid Oxidation and Toxic Fatty Acids

PUFA are particularly prone to oxidation and these oxidation products may have toxic effects (Hardwick et al. 1997). Autoxidation of fish oils (and by implication other marine oils) can also lead to polymer formation (Burkow and Henderson 1991). The stability of algal oils depends on their fatty acid composition, the physical and colloidal states of the lipids, the content of antioxidants, and the presence and activity of transition metals (Frankel et al. 2002). For example, the relatively high oxidative stability of an algal oil containing 42% DHA was completely lost after tocopherols and other antioxidants were removed (Frankel et al. 2002).

Many complex lipids are readily hydrolysed to yield free fatty acids and several studies have suggested that specific free fatty acids can be toxic (see below). The formation of free fatty acids due to autolysis on cell rupture seems to be a particular problem with diatoms (Berge et al. 1995). For this reason, it has been suggested that cells should be immediately placed in boiling water to deactivate lipolytic enzymes before lipid extraction (Budge and Parrish 1999).

Octadecapentaenoic acid [(all-*cis* $\Delta^{3,6,9,12,15}$-18:5; 18:5(n-3)] is an unusual fatty acid found in marine dinoflagellates, haptophytes, raphidophytes and prasinophytes (Joseph 1975; Dunstan et al. 1992; Bell et al. 1997; Parrish et al. 1998; Sellem et al. 2000), but only rarely is it found at higher trophic levels in the marine food web. Ghioni et al. (2001) found that 18:5(n-3) was readily metabolised by cell lines derived from turbot, gilthead sea bream and

Atlantic salmon to 18:4(n-3) *via* a 2-*trans*-18:5. This intermediate seems to be involved in the β-oxidation of both 18:5(n-3) and 18:4(n-3) and is thought to be generated by a Δ^3,Δ^2-enoyl-CoA-isomerase acting on 18:5(n-3).

The toxicity of the main PUFA in microalgae has been investigated using a range of bioassays including Microtox, diatom growth inhibition, and sea urchin gamete and embryo bioassays (Arzul et al.1998, Sellem et al. 2000). Long-chain PUFA with five or six double bonds are the most active. The 18:5(n-3) fatty acid from the toxic dinoflagellate *Gymnodinium* cf. *mikimotoi* delays or inhibits first cleavage of sea urchin (*Paracentrotus lividus*) eggs and provokes abnormalities in embryonic development (Sellem et al. 2000). This PUFA also causes toxic effects in the gills and intestine of the sea bass *Dicentrarchus labrax* leading to strong mucus production in the gills and alteration of ionocytes (Sola et al. 1999). Glycoglycerolipids containing 18:5(n-3) derived from microalgae can also be biologically active and potentially haemolytic (Parrish et al. 1998; Arzul et al. 1998). A mixture containing the C_{16} PUFA 16:2, 16:3 and 16:4 as well as 18:4 and 14:0 (inactive) from the diatom *Phaeodactylum tricornutum* inhibited growth of *Bacillus subtilis* and *Vibrio parahaeomolyticus* (Cooper et al. 1985).

Vitamins

Microalgae play an important role as the primary source of many essential vitamins that are passed up the food chain. In microalgae commonly used in aquaculture and grown under standard phototrophic conditions, ascorbic acid generally occurs in highest concentrations and shows the greatest variation (1 to 16 mg g^{-1} DW; Brown and Miller 1992). Other vitamins typically show a two- to four-fold difference between species (Table 12.3; Seguineau et al. 1996; Brown et al. 1999).

Limited information has been published on the effects of culture condition on vitamin content (Donato et al. 2003). Concentrations of α-tocopherol, riboflavin and thiamin can increase by up to two to three-fold during the stationary phase in phototrophic cultures (Brown and Farmer 1994, Seguineau et al. 1996, Brown et al. 1999). In another study, α-tocopherol in *Euglena gracilis* was increased from 400 μg g^{-1} (phototrophic cultures) to 1700 μg g^{-1} through light and nutrient optimization and including a heterotrophic culture phase (Ogbonna et al. 1999). Heterotrophic culture, together with strain mutagenesis has also been used to optimize extracellular production of ascorbic acid in *Chlorella pyrenoidosa* to concentrations between 1 to 2 g L^{-1} (Running et al. 1994).

Table 23.3 Range of vitamin content of microalgae (μg g^{-1} DW) used in aquaculture. Combined data from Brown et al. (1999), Seguineau et al. (1996) and Brown and Miller (1992). Requirements of shrimp from Conklin (1997) and marine fish from Tacon (1991) and NRC (1993). Retinol requirements of fish can be met through pro-vitamin A metabolites such as β–carotene.

Vitamin	*Concentration range (μg g^{-1})*	*Requirements of shrimp*	*Requirements of marine fish*
Ascorbic acid (C)	1,000 – 16,000	200	200
β–carotene	500 – 1,200		
niacin	110 – 470	40	150
α–tocopherol (E)	70 – 350	100	200
Thiamine (B_1)	30 – 110	60	20
Riboflavin (B_2)	25 – 50	25	20
pantothenic acid	14 – 38	75	50
Folates	7 – 24	10	5
Pyridoxine (B_6)	4 – 17	50	20
Cobalamin (B_{12})	1.7 – 7.4	0.2	0.02
Biotin	0.7 – 1.9	1	1
Retinol (A)	<0.25 – 2.2	1.6	1.9
Ergocalciferol plus Cholecalciferol (D_2, D_3)	<0.9	0.1	0.025

There have been some conflicting data published on the vitamin content of microalgae. While sometimes this may relate to strain and culture differences, differences in analytical methods undoubtedly have also contributed. For example, De Roeck-Holtzhauer et al. (1991) reported significantly higher contents of cobalamin (60 to 150 times), thiamine (7 to 20 times), pyridoxine (11 times), and α-tocopherol (6 to 10 times) compared to other studies (Seguineau et al. 1996; Brown et al. 1999). De Roeck-Holtzhauer et al. (1991) used HPLC with UV detection for analyzing cobalamin and pyridoxine (a relatively non-specific method), whereas other studies have used more specific microbiological assays. In another example, Hapette and Poulet (1990) found either acetonitrile, or metaphosphoric acid + acetic acid to be three to four times more efficient than methanol or ethanol for extracting ascorbate from plankton, and measured ascorbate concentrations were up to 10-times greater by HPLC with reverse-phase columns compared to using anion-exchange columns. These examples highlight the need to adopt harvesting, processing and storage conditions that minimize cell damage and vitamin loss (e.g. through oxidation), extraction procedures that maximize recovery and analytical methods that are specific and/or ensure interfering substances are removed prior to analysis.

Minerals

As previously stated, the mineral fraction can constitute between 10 to 40% of the DW of microalgae. The predominant elements include P, S, Ca, Na, K, Cl, Fe, Mg, Fe and Zn and for diatoms Si (Fabregas and Herrero 1986, Parsons et al. 1961) (Table 12.4). Species can differ significantly in their composition of these minerals and other trace metals; moreover their composition is affected by culture environment (Lee and Picard 1982). For example, *Synechococcus* sp. cells accumulated ions (Co, Zn, Ag, Sn, Hg, Pu and Am) in proportion to the concentration of the same ions in culture media (Fisher 1985).

Carotenoid and Chlorophyll Pigments

Microalgae contain a fascinating diversity of pigments which can often be used as chemotaxonomic characters in algal classification. This is discussed in more detail in chapter 2 (S.W. Jeffrey and Wright). Microalgal pigments are also important in nutritional studies as coloring agents for fish flesh and as antioxidants. Microalgae used in aquaculture usually contain 0.4 to 1.2 mg g^{-1} β-carotene (Seguineau et al. 1996), whereas under optimal conditions, *Dunaliella salina* may contain > 5% of its DW as β-carotene. *D. salina is* now grown on a very large scale in open ponds for commercial production of pigments (Borowitzka 1992). Similarly, under optimal conditions another chlorophyte *Haematococcus pluvialis* may contain up to 3% of its DW as astaxanthin (Lorenz and Cysewski 2000). There is also interest in the commercial production and extraction of other novel pigments from microalgae, including scytonemin from cyanobacteria as an ultraviolet sunscreen pigment (Proteau et al. 1993) and marennine from *Haslea ostrearia* as a natural colorant (Rossignol et al. 2000). Marennine gives oysters from northern France their characteristic 'blue-green' color (Turpin et al. 2001).

APPLICATIONS

Algal Production Systems for Aquaculture and Products

Typical indoor systems for microalgal mass culture include carboys (10 to 20 L), polythene bags (100 to 500 L) and tubs (1000 to 5000 L) (Donaldson

Table 12.4 Mineral element composition (mg g^{-1} DW) of selected microalgae. n.a. denotes not analyzed

	Tetraselmis suecica	*Isochrysis galbana*	*Dunaliella tertiolecta*	*Chlorella stigmatophora*	*Porphyridium cruentum*	*Pavlova lutheri*	*Skeletonema costatum*
Cl	37.2	50.8	24.2	24.5	n.a.	n.a.	n.a.
Ca	20.8	16.2	20.9	15.1	12.4	n.a.	n.a.
K	12.0	5.6	7.4	11.2	11.9	n.a.	n.a.
Na	10.4	7.2	9.2	9.8	11.3	n.a.	n.a.
S	n.a	n.a	n.a	n.a	14.10	n.a.	n.a.
Si	n.a	n.a	n.a	n.a	n.a	16.0	143.0
P	6.5	10.2	7.3	6.4	n.a.	30.0	17.0
Mg	7.8	11.5	6.3	7.8	6.3	n.a.	n.a.
Fe	1.0	3.6	2.0	1.8	6.6	n.a.	n.a.
Zn	1.50	0.60	0.30	0.30	3.7	n.a.	n.a.
Cu	0.65	0.20	0.07	0.11	0.08	n.a.	n.a.
Mn	0.047	0.040	0.057	0.040	0.47	n.a.	n.a.
Co	0.006	0.011	0.006	0.008	n.a.	n.a.	n.a.
Cr	n.a	n.a	n.a	n.a	0.01	n.a.	n.a.
Reference:	Fábregas and Herrero (1986)	Fábregas and Herrero (1986)	Fábregas and Herrero (1986)	Fábregas and Herrero (1986)	Rebolloso Fuentes et al. (2000)	Parsons et al. (1961)	Parsons et al. (1961)

1991). For larger volumes, outdoor tanks or ponds are used. Cultures are usually operated in batch or semi-continuous mode, though commercial systems are available that operate continuously (Seasalter Shellfish (Whitstable) Ltd., Reculver, Kent UK). Depending on their scale, hatcheries may produce between several hundred to tens of thousands of litres of microalgal culture daily. Seawater culture media enriched with nutrient mixture (e.g. f/2 and Walne media; Lavens and Sorgeloos 1996) can achieve cell densities ranging from 10^5 to 10^7 cells mL^{-1}. Using these standard systems production costs can range from US $50 to $200 kg^{-1} DW, which represents 20 to 50% of the operating costs of a hatchery (Coutteau and Sorgeloos 1992). The major factors contributing to production cost include labour, infrastructure, chemicals and lighting, and as there are clear economies of scale, costs become especially significant for small hatcheries. Consequently, there has been a lot of investigation for alternatives to production of fresh microalgae, and also more cost-efficient production systems.

Photobioreactors have been developed in various designs including enclosed tubular and flat plate systems operated either indoors or outdoors (Tredici and Materassi 1992, Chrismadha and Borowitzka 1994, Richmond 2000, Lee 2001). These can be considered as variations of the systems typically used in aquaculture, but with higher surface area to volume ratio (SA:VOL). Therefore, light is less likely to become limited so these systems have a high productivity and cell biomass at harvest and thus potentially can operate at a lower production cost. However these systems do have some disadvantages. Oxygen concentrations from photosynthesis can become elevated and give rise to photoinhibition, thus restricting productivity. Overheating of cultures can be a problem in outdoor systems because of the high SA:VOL. Due to their high cell densities the systems need turbulent flow to facilitate nutrient exchange and to prevent light-limitation, thereby making them unsuitable for fragile species. Despite the potential of photobioreactors most aquaculture strains have not been effectively cultured in such systems. Exceptions include *Nannochloropsis* spp. (Tredici and Materassi 1992) and *Skeletonema* spp. (Susan Blackburn et al., unpublished observations). For a more detailed description of photobioreactor systems, the reader is referred to reviews by Borowitzka (1999), Lee (2001) and Janssen et al. (2002).

Fermenter technology is well established for mass culturing bacteria and yeast, and there are some microalgae capable of heterotrophic growth. The advantages include a high-density and biomass production, and elimination of the need for light which is a major cost for phototrophic culture. This technology provides a greater degree of control and potentially better economy of production. Until recently, only a few microalgae of moderate usefulness for aquaculture, e.g. *Tetraselmis* spp. and *Chlorella* spp., have been grown effectively in heterotrophic systems (Chen and Johns 1995, Day and Tsavalos 1996). More recently, microalgae such as the dinoflagellate *Crypthecodinium cohnii* and diatom *Nitzschia* spp., or algal-like organisms known as thraustochytrids, have been grown commercially under heterotrophic conditions for aquaculture (Barclay et al. 1994, Harel et al. 2002). Their application, both as feeds for aquaculture and as a source of extracted products, is discussed further in later sections of this review. Production costs of between US $2 to 25 kg^{-1} DW microalgae have been projected for this technology (Gladue 1991); dried thraustochytrid preparations from the genus *Schizochytrium* are currently commercially available for between $25 to $50 kg^{-1} (Rotimac and AlgaMac 2000 from Aquafauna Biomarine Inc.; Docosa Gold from Sanders Brine Shrimp Co.).

Aquaculture Feeds and Nutrition

Microalgae are utilized as feedstocks in aquaculture for all growth stages of bivalve molluscs (e.g. oysters, scallops, clams and mussels), for the larval/early juvenile stages of crustaceans, abalone, and some fish species, and for live zooplankton used in aquaculture food chains. Over the last five decades, several hundred microalgae have been tested in aquaculture, but probably less than 20 have widespread use. Microalgae must possess a number of key attributes to be useful aquaculture species. They must be of a suitable size and shape for ingestion, e.g. from 2 to 20 µm for larval or juvenile filter feeders; 10 to 100 µm for grazers (Webb and Chu 1983, Jeffrey et al. 1992, Kawamura et al. 1998) and be readily digested. They must be amenable to mass culture with rapid growth rates, and also be stable in culture to any extremes of light, temperature, and nutrients as may occur within hatcheries. Further, they must have a good nutritional value, including an absence of toxins or heavy metals that might be transferred through the food chain.

Persoone and Claus (1980) identified *Isochrysis galbana*, *Isochrysis* sp. (T.ISO), *Pavlova lutheri*, *Tetraselmis suecica*, *Pseudoisochrysis paradoxa*, *Chaetoceros calcitrans* and *Skeletonema costatum* as successful microalgae for bivalve culture. Twenty years later, hatcheries are still using the same microalgae as feedstock with some additional species (Table 12.5). Mixtures of microalgae containing *Chaetoceros calcitrans*, *Isochrysis* sp. (T.ISO) and *Pavlova lutheri* are the most popular diets for larvae, early juvenile and

Table 12.5 Microalgae commonly used in aquaculture, either as individual diets or components of mixed diets. (++ denotes more popular than +)

	Bivalve molluscs	*Crustacean larvae*	*Juvenile abalone*	*Zooplankton (used for crustaceans and fish larvae)*
Isochrysis sp. (T.ISO)	++	+		++
Pavlova lutheri	++	+		++
Chaetoceros calcitrans	++	++		+
C. muelleri or *C. gracilis*	+	++		+
Thalassiosira pseudonana	+	+		
Skeletonema spp.	+	++		
Tetraselmis suecica	+	+		++
Rhodomonas spp.	+			
Pyramimonas spp.	+			
Navicula spp.	+	+	++	
Nitzschia spp.		+	++	
Cocconeis spp.			+	
Amphora spp.			+	
Nannochloropsis spp.				++

References: Brown et al. (1997), Reitan et al. (1997), Lee (1997), Kawamura et al. (1998), Wikfors and Ohno (2001), Cathy Johnston pers. comm. (CSIRO Collection of Living Microalgae).

broodstock (during hatchery conditioning) stages of bivalve molluscs (O'Connor and Heasman 1997, Brown and Robert 2002). Many of the microalgae successfully used for bivalves are also fed directly to crustaceans (especially shrimp) during the early larval stages, especially the diatoms *Chaetoceros* spp. and *Skeletonema* spp. Benthic diatoms such as *Navicula* spp. and *Nitzschia* spp. are commonly grown in mass-culture and then settled onto plastic plates as a diet for grazing juvenile abalone. *Isochrysis* sp. (T.ISO), *Pavlova lutheri*, *T. suecica* or *Nannochloropsis* spp. are commonly used to feed and enrich *Artemia* and rotifers, which are then fed on to later larval stages of fish and crustacean larvae.

The nutritional value of microalgae varies significantly between species, and it can change under different culture conditions (Enright et al. 1986a, Brown et al. 1997, Pernet et al. 2003). Microalgae with excellent nutritional properties, either as a single species or within a mixed diet, include *C. calcitrans*, *C. muelleri*, *P. lutheri*, *Isochrysis* sp. (T.ISO), *T. suecica*, *S. costatum* and *Thalassiosira pseudonana* (Enright et al. 1986b, Thompson et al. 1993, Brown et al. 1997). Factors that contribute to a microalga's nutritional value include its size and shape, digestibility (related to cell wall structure and composition), biochemical composition (e.g. nutrients, enzymes, toxins if present) and the requirements of the animal feeding on it. Early studies showed differences in the proximate (Parsons et al. 1961) and fatty acid compositions (Webb and Chu 1983) between microalgae, and many subsequent studies have attempted to correlate the nutritional value of microalgae with their biochemical profile. However, results from feeding experiments comparing microalgae differing in a specific nutrient are difficult to interpret because of the confounding effects of other microalgal nutrients or cell digestibility. Nevertheless, from examining a range of literature, including experiments where microalgae have been supplemented with compounded diets or emulsions, some general conclusions can be drawn (Knauer and Southgate 1999).

Despite large differences between microalgae, there is not always a correlation between their proximate composition and nutritional value, suggesting that other factors previously described are equally or more important. In studies where such factors are minimized, i.e. the use of similar species, or the same species grown under different conditions to reduce the effects of digestibility and trace nutrients, then high levels of carbohydrate produce the best growth for juvenile oysters (*Ostrea edulis*; Enright et al. 1986b) and larval scallops (*Patinopecten yessoensis*; Whyte et al. 1989) provided that PUFAs are also present in adequate proportions. Nevertheless, in another study assessing diatom species as part of a mixed diet, Knuckey et al. (2002) reported that the growth of juvenile Pacific oysters (*Crassostrea gigas*) was correlated with the amount of dietary protein.

The nutritional value of the algal protein component (i.e. protein plus free amino acids) is thought to be high if its essential amino-acid (EAA) composition is close to that of the feeding animal (Webb and Chu 1983). Ten amino acids are essential for maricultured animals: threonine, valine, methionine, isoleucine, leucine, phenylalanine, lysine, histidine, arginine and tryptophan, with proline also essential for molluscs (Webb and Chu 1983). A comparison of the content of EAAs from microalgae with, for example, the EAA content of the Pacific oyster showed a close correlation, suggesting that protein quality is not generally a factor contributing to the differences in nutritional value of microalgal species (Table 12.2). Studies with *Artemia* supplemented with microdiets show that betaine and the free amino acids alanine, glycine and arginine stimulate larval feeding in seabream (Koven et al. 2001). Similarly, microdiets supplemented with phospholipids, particularly phosphatidylcholine (PC), stimulated feeding activity leading to an increase of 45% in diet consumption. Dietary PC also appears to improve lipoprotein synthesis, resulting in enhanced transport of dietary lipids from the mucosa of the digestive tract to the body tissues of the fish (Koven et al. 2001).

Polysaccharide type, which in part is reflected in its constituent sugar composition, may contribute to differences in digestibility and hence nutritional value of microalgae. For example, *Chlorella* spp. are poorly digested by bivalve molluscs (Peirson 1983) and this is attributed to its thick cellulose cell wall. Parsons et al. (1961) proposed that a high content of glucose was related to a good nutritional value, though this has not been supported by subsequent observations (Webb and Chu 1983). Mannose-rich polymers are often associated with the siliceous frustules of diatoms (Hecky et al. 1973). Specifically, *Phaeodactylum tricornutum* has the highest proportion of mannose reported (46% of total sugars; Brown 1991) which could make it less digestible to bivalves (Epifanio et al. 1981) and explain why it has a low food value (Enright et al. 1986b). With respect to storage polysaccharides, it is likely that amylase and laminarase enzymes within bivalves would effectively break down the glucose and mannose-rich polysaccharides, i.e. glucans and mannans (Whyte 1987).

PUFAs derived from microalgae, i.e. docosahexaenoic acid (DHA), eicosapentaenoic acid (EPA) and arachidonic acid (AA) are essential for many marine larvae (Langdon and Waldock 1981, Sargent et al. 1997). A summary of the proportion of these important PUFAs in 46 strains of microalgae is shown in Fig. 12.1 (data from Volkman et al. 1989, 1991, 1993). The fatty acid content shows systematic differences according to the taxonomic group, although there were examples of significant differences between microalgae from the same class. While the importance of PUFAs is recognized, the quantitative requirements of larval or juvenile animals feeding directly on microalgae is not well established (Knauer and Southgate 1997). Thompson et al. (1993) found that the growth of Pacific oyster *C. gigas* larvae was not improved by feeding them microalgae containing more than 2% (total fatty acids) of DHA; moreover the percentage of dietary EPA was negatively correlated to larval growth. However, the authors found a positive correlation between the percentage composition of the short chain fatty acids 14:0 + 16:0 in microalgae, and larval growth rates. They reasoned that diets with higher percentages of the saturated fats were more beneficial for rapidly growing larvae because energy is released more efficiently from saturated fats than unsaturated fats.

Knauer and Southgate (1997) compared Pacific oyster (*Crassostrea gigas*) spat fed for 28 d on either *Dunaliella tertiolecta*, (which lacks fatty acids greater than C_{18}), or on a 80% ration of the alga and 20% gelatin-acacia microcapsules (GAM) containing varying amounts of either EPA or DHA or combinations of the two. The advantage of this technique is that it allows an accurate assessment of specific dietary components (Knauer and Southgate 1999). GAM containing either corn oil, corn oil containing up to 0.16% of EPA, 0.63% of DHA, or 0.32% of a EPA/DHA mixture did not improve shell length, DW, or ash-free DW (AFDW) of spat compared with spat fed *D. tertiolecta* alone. However, GAM containing 0.30 or 0.50% EPA resulted in spat with significantly higher AFDW than spat fed *D. tertiolecta* and there was a positive correlation between the level of EPA in GAM and AFDW of spat. The fatty acid profile of spat generally reflected that of the diet after 28 d, but the work also showed that unfed spat will selectively retain EPA and DHA.

Several studies have demonstrated the importance of lipids such as triacylglycerols as a source of energy in the early life stages of bivalves and the vital role played by the essential fatty acids (Coutteau et al. 1996). The quantity and quality of food is particularly important to the pre-spawning condition of adult bivalves. Endogenous reserves laid down in the eggs provide energy during embryogenesis and growth until larvae are able to feed independently; about 30% of the DW of eggs occurs as lipids, much of it present as neutral lipids. *Ostrea edulis* larvae benefit from a large initial lipid content and increased larval survival and successful metamorphosis in *Plactopectin magellanicus* is correlated with lipid content (Gallager et al. 1986, Delaunay et al. 1993). PUFA are thought to play a crucial role and thus the choice of algal species is important. Hendriks et al. (2003) showed that *Macoma balthica* kept on a broodstock diet supplemented with PUFAs, spawned a larger number of eggs per female and larger sized eggs than adults kept on a diet without PUFA supplementation. Once larvae start feeding their structural growth seems to be largely dependent on protein availability (Hendriks et al. 2003). Lipid emulsions can be used to supplement the lipid content of microalgae in the diet and bolster the content of essential fatty acids (Coutteau et al. 1996).

Despite a high variability in the vitamin content of microalgae (e.g. 16-fold difference in ascorbic acid; Brown and Miller 1992), to put this into context for aquaculture nutrition, data should be compared with the requirements of the consuming animal. Unfortunately, the requirements of larval or juvenile animals that feed directly on microalgae are, at best, poorly understood, but improved larval survival in the spiny lobster *Jasus edwardsii* has been demonstrated by feeding them on juvenile *Artemia* enriched with ascorbic acid (Smith et al. 2004). The requirements of juvenile or adults are far better known and, in the absence of information to the contrary, can serve as a guide for the larval animal. These data suggest that a carefully selected, mixed-algal diet should provide adequate concentrations of vitamins for aquaculture food chains (Table 12.3).

De novo synthesis of sterols has been shown in some bivalves, but it appears that even when present the rate of synthesis is too slow for optimum growth (Teshima et al. 1983, Knauer et al. 1999). Bivalves and other molluscs assimilate dietary sterols and therefore often contain quite complex distributions of sterols. Spat of the Pacific oyster *Crassostrea gigas* are unable to synthesize sterols, but they can bioconvert dietary sterols (Knauer et al. 1998). After six weeks feeding on either *Chaetoceros muelleri*, *Isochrysis* sp. or *Pavlova lutheri* the sterol composition of the spat resembled that of the diet, but not all sterols were assimilated with the same efficiency: 4-methyl sterols and dihydroxylated sterols were poorly assimilated (Knauer et al. 1999). Sterol contents decreased on starving, but the relative distributions remained similar apart from the concentration of cholesterol which remained high indicating the important role of this sterol in animal metabolism.

Crustaceans require cholesterol for growth and reproduction. They either obtain this directly from their diet or by bioconversion of alkylated sterols. Studies with the daphnid *Daphnia galeata* fed a cyanobacterium supplemented with 10 different sterols (Martin-Creuzburg and von Elert, 2004) have shown that Δ^5 sterols (sitosterol, stigmasterol and desmosterol) and Δ^7 sterols (cholesta-5,7-dien-3β-ol, ergosterol) meet the dietary requirements of this animal. However, cholest-7-en-3β-ol was less effective than cholesterol at supporting reproduction and somatic growth, while 5α-cholestanol and lanosterol did not support growth. The unusual Δ^4 sterol cholest-4-en-3β-ol adversely affected growth. These data suggest that *Daphnia* is unable to convert Δ^8-4,4,-dimethyl sterols such as lanosterol to cholesterol, but they are efficient at dealkylating C-24 sterols like other crustaceans (Ikekawa 1985, Teshima 1971).

Differences in the mineral composition of microalgae could potentially contribute to differences in the nutritional value, but to the best of our knowledge this has not been experimentally tested or validated. As microalgae can efficiently accumulate trace metals, microalgal-based food webs would represent a major route for the transfer of minerals in both natural marine ecosystems and intensive aquaculture systems. In one of the few studies reporting transfer of algal minerals to zooplankton, Lie et al. (1997) found a significant increase in manganese concentration in rotifers after feeding with *Isochrysis galbana*, though trends were not as marked with other minerals. Inclusion of dried algal biomass at < 5% within formulated feeds could fulfill the Cu and Mg requirement of freshwater fish (Fabregas and Herrero 1986).

Though much of the above discussion has focused on the nutritional role of microalgae as direct feeds in aquaculture, microalgae are also important for feeding or enriching zooplankton which in turn are fed to fish and other larvae. In addition to providing EAAs and energy, they contribute other key nutrients such as PUFAs, vitamins, pigments, sterols and minerals (see above) which are transferred through the food chain. For example, after 24 h rotifers fed on *Isochrysis* sp. (T.ISO) and *Nannochloropsis oculata* contained 2.5 and 1.7 mg ascorbic acid g^{-1} DW, respectively, whereas rotifers fed on baker's yeast (itself deficient in ascorbic acid) contained only 0.6 mg g^{-1} DW (Brown et al. 1998b). Similarly, *Artemia* become enriched in ascorbate by feeding on microalgae (Merchie et al. 1995). Little information is available on the transfer of other vitamins from microalgae through the food chain to fish larvae. Rønnestad et al. (1998) demonstrated that microalgal pigments transferred through to zooplankton may also contribute to nutritional value. For example, they found the major pigments in the copepod *Temora* were astaxanthin and lutein, whereas in *Artemia* it was canthaxanthin, and that profiles were related to dietary history. Copepods were superior feeds for halibut larvae and Rønnestad et al. (1998) ascribed this to the ability of the larvae to convert lutein and/or astaxanthin, but not canthaxanthin, into vitamin A. They recommended that *Artemia* should routinely be enriched with astaxanthin and lutein (a good source is the prasinophyte *Tetraselmis*) to improve their nutritional value.

A common observation during intensive fish and shrimp larval culture is that adding certain microalgae to culture tanks (i.e. 'green water') together with the zooplankton prey improves the performance of larvae (Tamaru et al. 1994, Reitan et al. 1997). The exact mode of action is unclear, but could include one or more of the following: (a) maintenance of the nutritional quality of the zooplankton (b) light attenuation (i.e. shading effects) having a positive effect on larvae, (c) an extra-cellular production of vitamins or other growth-promoting substances by algae, (d) a probiotic effect of the algae, (e) ingested microalgae may have triggered the digestion process or contributed to the establishment of an early gut flora in the larvae. Maintenance of NH_3 and O_2 balance has also been suggested, though this has not been experimentally proven (Tamaru et al. 1994). The most popular algae used for green water are *Nannochloropsis oculata* and *Tetraselmis suecica*. This approach may also be applied to extensive outdoor production systems by fertilizing ponds to stimulate the growth of natural assemblages of microalgae and zooplankton, as food for larvae introduced into the ponds.

Processed forms of microalgae have also been assessed as aquaculture feeds. One of the first was Algal 161 from CellSys, which was produced from *T. suecica* and cost US $180 kg^{-1}. The product had a moderate value as a diet component for molluscs (Laing et al. 1990), but it did not have a high-market penetration and is now unavailable. Algal pastes or concentrates have some potential as alternative diets as they can be used 'off-the-shelf', thus providing potential cost-efficiencies to hatcheries. Concentrates are prepared by centrifugation (≈ 500 to 1000-fold concentration) or flocculation (≈ 100 to 200-fold concentration). Concentrates prepared from diatoms appear the most promising with a shelf life of between two to eight weeks when stored ≤ 4°C. For hatcheries, concentrates can be prepared under two different scenarios, i.e. (a) by hatcheries on-site, as a back-up or as a means to store any overproduction of algae, or (b) production centralized at a large facility – with greater economies of scale – with concentrates dispatched to hatcheries upon request. Several companies are currently producing algal concentrates for commercial sale (e.g. Reed Mariculture; http://www.seafarm.com). Concentrates fed to the larvae and spat of Sydney rock oyster (Heasman et al. 2000) and Pacific oyster (McCausland et al. 1999, Brown and Robert 2002) were effective as partial diets (e.g. up to 80%) with growth rates similar to, or marginally inferior to, complete live diets.

Commercial products such as Docosa Gold and AlgaMac 2000 containing dried preparations of the thraustochytrid *Schizochytrium* sp. are now available for larval feeding (Barclay and Zeller 1996). These are useful when microalgal feeds do not provide the level of enrichment sought for the target larvae (Brown and McCausland 2000). *Schizochytrium* oils produce DHA to EPA ratios in zooplankton between 1 and 2, which is considered optimal for fish larval nutrition (Rodríguez et al. 1998).

Single Cell Oils

The term single cell oil (SCO) was coined more than 20 years ago by analogy to the term single cell protein to describe the oil or fats isolated from micro-organisms including microalgae (Materassi et al. 1980). Only a few species have been cultivated for SCOs since these need to meet a number of criteria including a high oil content and a high nutritional value with a high content of desirable fatty acids and an absence of toxins. The cells must be cultured economically under large-scale conditions and easily harvested (Borowitzka 1992, 1999). Heterotrophic culture in large fermenters is the method of choice (Boswell et al. 1992, Behrens and Kyle 1996, Certik and Shimizu 1999), although initial capital costs are high. Species of choice now include the thraustochytrid *Schizochytrium* sp. which is a good source of the PUFA 22:6(n-3) (Yaguchi et al. 1997, Gara et al. 1998). Unfortunately, few aquaculture species have been identified that can grow heterotrophically. *Tetraselmis* spp. are exceptions (Day and Tsavalos 1996), though these are generally recognised as having moderate food value, unless forming part of a mixed diet.

There has also been considerable interest in photobioreactor technology for photosynthetic autotrophs (Borowitzka 1992, 1999, Torzillo et al. 1993). Large-scale photobioreactors, either for indoor or outdoor production, have been assessed (Tredici and Materassi 1992; Chrismadha and Borowitzka 1994) using species such as *Nannochloropsis* spp. (Tredici and Materassi 1992) and *Skeletonema*. These are produced for a variety of markets, but the most common is as a dietary supplement and source of PUFA for inclusion in products such as infant milk formula and various foodstuffs (Behrens and Kyle 1996). Lee (2001) has suggested that the volumetric productivity and cost of production in these enclosed photobioreactors are no better than those achievable in open-pond cultures. The technical difficulty in sterilizing these photobioreactors has hindered their application for the production of high value pharmaceutical products. A comprehensive account of SCOs and their industrial applications can be found in the book by Kyle and Ratledge (1992).

Health foods

Certain microalgae and cyanobacteria have been used as a foodstuff for humans and domesticated animals for many centuries (Wikfors and Ohno 2001). The cyanobacterium *Spirulina* continues to be used in the daily diets of the indigenous populations of Africa and America (Mathew et al. 1995). Today, large amounts of cyanobacteria are commercially cultured for sale as a human dietary supplement (Allnutt 1996). These products are usually labelled as *Spirulina*, but the main ingredient is *Arthrospira platensis*. Cyanobacteria are a rich natural source of proteins, carotenoids, and other micronutrients such as antioxidants (phenolic acids, tocopherols and carotene; Miranda et al. 1998), but consumers need to be aware that some preparations from Asia are produced under poor sanitary conditions and hence can be contaminated.

Experimental studies in animal models have even demonstrated an inhibitory effect of *Spirulina* on oral carcinogenesis (Mathew et al. 1995). The original focus on *Spirulina* was mainly due to its adaptability to open pond growth, rather than to any intrinsic nutritional qualities. Indeed, like all cyanobacteria it lacks the longer-chain PUFA 20:5(n-3) and 22:6(n-3), but rather contains high contents of C_{18} PUFA. Similarly, much effort was expended to develop culture systems for the green alga *Chlorella* as a cheap source of protein, but it was only later discovered that *Chlorella* was refractory as a human food (reviewed by Wikfors and Ohno 2001). There are around 110 commercial producers of microalgae in the Asia-Pacific region (90% of which is in Asia), with annual production capacity of each operation ranging from 3 to 500 tonnes (Lee 1997). The commercially cultivated microalgae include *Chlorella, Spirulina, Dunaliella, Nannochloris, Nitzschia, Crypthecodinium, Schizochytrium, Tetraselmis, Skeletonema, Isochrysis* and *Chaetoceros*. Most of the commercially produced algal biomass is being marketed as health food, in the form of tablets and capsules. Microalgae and their extracts are also included in noodles, wine, beverages, breakfast cereals and cosmetics.

Unusual Lipid Constituents

Hydrocarbons (mainly henicosahexaene n-$C_{21:6}$ and squalene) occur in most species although they rarely constitute more than a few percent of the total lipids. The chlorophyte *Botryococcus braunii* is a notable exception since this contains high contents of unusual isoprenoid alkenes termed botryococcenes (Metzger and Casadevall 1983). *Botroyococcus* is one of the microalgae considered as a potential producer of petroleum substitutes (biodiesel; Baum 1994). Some diatoms including species from the genera *Haslea, Rhizosolenia, Pleurosigma* and *Navicula* contain C_{25} and C_{30} highly branched isoprenoid alkenes with 2-6 double bonds (Belt et al. 2001; Rowland et al. 2001a, b, Volkman et al. 1994, 1998). These alkenes occur widely in marine sediments and can be used as biomarkers for organic matter derived from diatoms. They have also been shown to be cytostatic to non-small cell lung cancer, but this has not led to a commercial product (Rowland et al. 2001a).

Long-chain (i.e. > C_{22}) acyclic lipids are not common in microalgae, but studies over the past decade have indicated a number of interesting exceptions. Volkman et al. (1992) identified C_{30}-C_{32} alcohols and diols in marine eustigmatophytes from the genus *Nannochloropsis*. At the time of their discovery, the function of such compounds was not known, but studies by

Gelin et al. (1999) have shown that these diols are building blocks for novel highly aliphatic biopolymers produced by these microalgae. Long-chain mid-chain hydroxy acids are also present in these microalgae and may be involved in the formation of these biopolymers (Gelin et al. 1997). Recent work has established that similar alkyl diols and hydroxyl fatty acids occur in some diatoms (Sinninghe Damsté et al. 2003).

Certain species of haptophytes including the genera *Emiliania*, *Gephyrocapsa*, *Isochrysis* and *Chrysotila* contain a suite of unusual very long-chain (C_{37}-C_{40}) unsaturated ketones termed alkenones (Volkman et al. 1980a, 1995, Conte et al. 1995, 1998, Rontani et al. 2004), as well as related alcohols, esters and alkenes (Conte et al. 1994, Rontani et al. 2004). Geochemists have shown considerable interest in these compounds since they are well preserved in sediments and the proportions of tri- and di-unsaturated C_{37} alkenones is strongly correlated with growth temperature (Brassell et al. 1986) so these compound distributions can be used as a paleothermometer. Some of the species used as aquaculture feeds (*Isochrysis*) contain alkenones, and a nutritional role has been suggested (Ben-Amotz and Fischler 1990), but this remains unproven. Rather, it appears that many animals are unable to effectively utilize these compounds and thus excrete them relatively unchanged (Volkman et al. 1980b, Knauer et al. 1999).

Other Bioactive Compounds

There has been increasing interest in tapping the genetic variation of microalgae and cyanobacteria as a source of new products, such as pharmaceuticals and pesticides (Borowitzka 1995). While this screening has identified many compounds with biological activity, commercial development is yet to occur, although the pace may quicken if the techniques of molecular biology for genetic manipulation are applied (Allnutt 1996). Ohta et al. (1998) tested 106 microalgae and cyanobacteria for anti-Herpes simplex virus (HSV-1) activity. Methanol extracts of *Dunaliella bioculata* and *D. primolecta* and the cyanobacteria *Lyngbya* sp. and *Lyngbya aerugineo-coerulea* all had inhibitory activity, but the green alga, *D. primolecta*, had the highest anti-HSV-1 activity (10 μg mL^{-1} of extract completely inhibited the cytopathic effect). The active substances were proposed to be pheophorbide-like compounds. Water-soluble (but not lipid-soluble) extracts of the marine microalgae *Chlorella stigmatophora* and *Phaeodactylum tricornutum* show significant anti-inflammatory, analgesic and free radical scavenging activity (Guzman et al. 2001).

Toxins initially isolated from fish or shellfish (e.g. paralytic shellfish poisoning - PSP; diarrhetic shellfish poisoning, DSP) are now known to originate from certain microalgae, especially dinoflagellates (Yasumoto and Satake 1998). These toxins have been used to investigate the structure and function of ion channels on cell membranes and to elucidate the mechanism of tumor promotion. Some cyanobacteria also contain similar toxins and a few diatoms such as *Pseudonitzschia* contain the amnesic toxin domoic acid (Douglas and Bates 1992, Bates 2000).

Other Applications

The future promise in microalgal biotechnology seems to be developing at the crossroads of molecular biology and novel aspects of cellular metabolism. (Allnutt 1996, Dunahay et al. 1996, Zaslavskaia et al. 2000, Stephanopoulos and Kelleher 2001). New products have also been developed for pharmaceutical and research applications. These include stable isotope biochemicals produced by algae in closed-system photobioreactors and as a source of extremely bright fluorescent pigments. Cryopreservation has also had a tremendous impact on the ability of strains to be maintained for long periods of time at low cost and maintenance while preserving genetic stability (Apt and Behrens 1999).

Specific microalgae can be used to improve water quality, help in the treatment of sewage and produce biomass. They can be used to produce hydrogen and biodiesel as alternative fuels (Wyman and Goodman 1993), fix N_2 for use as a biofertilizer, restore metal-damaged ecosystems, reducing CO_2 loads, and reclaim saline or alkaline infertile lands (Rai et al. 2000). Existing and potential uses of diatom biomass have been discussed by Lebeau and Robert (2003). One of the more unusual applications they mention is the use of silica from diatoms in the nanotechnology area.

Comparisons of Cultures with Microalgal Blooms

Biochemical data obtained from algal cultures seems to be a useful guide to the composition of the microalgae in natural environments even though environmental conditions may be very different. Environmental extremes may be infrequent in aquatic ecosystems, but they can occur when microalgae form large blooms that can strip the water of the major inorganic nutrients. The lipid compositional variations seen with batch cultures may provide a guide to the variations possible in nature and must be considered when comparing algal and sediment lipid data (Volkman et al. 1998).

Leblond et al. (2003) compared the fatty acid and sterol compositions of the harmful marine dinoflagellate, *Karenia brevis*, from a natural bloom with the same species in culture. A close correspondence was found between the natural and cultured samples. The fatty acid compositions in membrane phospholipids, chloroplast-associated glycolipids or storage triacylglycerols varied significantly. The glycolipid fraction contained abundant 18:5(n-3) while the phospholipid fraction contained small amounts of 28:8(n-3) and octacosaheptaenoic acid 28:7(n-6). The sterols occurred mainly in the non-esterified form and consisted of two main constituents.

Diatoms are readily ingested by zooplankton and have generally been considered a major source of food for marine food webs in productive regions. Recently, laboratory studies have demonstrated that the hatching success of copepods feeding on diatoms is dramatically impaired. Field data have now confirmed these observations (Miralto et al. 1999): in a diatom-dominated bloom only 12% of eggs hatched compared with 90% under post-bloom conditions. This effect has been linked to the presence of three tri-unsaturated C_{10} aldehydes (Miralto et al. 1999). Intriguingly, copepods can reproduce successfully on a diet of dinoflagellates that contain the potent neurotoxin saxitoxin and will accumulate the toxin making them toxic to fish (White 1981).

CONCLUSIONS

Microalgae still have a strong role to play in aquaculture as a preferred source of live and preserved feeds. The protein quality of all microalgae is high, but the sugar content and composition are variable, and in some instances may affect the nutritional value. The essential PUFAs 20:5(n-3) and 22:6(n-3) are key nutrients in animal nutrition, and most algae are rich in one or both of these acids. Chlorophytes, however, lack these acids and this contributes to their low food value. Microalgae are generally a rich source of vitamins, especially ascorbic acid, but species are variable in composition and in some instances this could contribute to differences in nutritional value in aquatic ecosystems. Because particular microalgae may be limiting in one or more of the key nutrients, mixed-algal diets provide a better balance and normally are used in mariculture.

The biochemical composition of microalgae can be manipulated readily by changing the growth conditions, but the effects vary from one species to another. Knowledge of how species respond to different environments is of practical use to aquaculturists, who may then grow the algae to optimize the level of specific nutrients needed by the animal. Microalgae are an important zooplankton food since important algal nutrients (fatty acids and vitamins) may be transferred to higher trophic levels via the zooplankton intermediates. Microalgae have been used in many other applications such as single cell oils and dietary supplements, but until their cost of production can be reduced this will tend to limit their use to speciality products.

ACKNOWLEDGEMENTS

We thank Jeannie-Marie LeRoi, Kelly Miller, Suzanne Norwood, Christine Farmer, Stephanie Barrett, Graeme Dunstan, Peter Mansour, Dion Frampton, Ian Jameson and Daniel Holdsworth for assistance with algal cultures and/or biochemical analyses. Dion Frampton and Ian Jameson provided helpful reviews of the manuscript. Our collaborators Dr S.W. Jeffrey and S.I. Blackburn are thanked for many valuable discussions on algal culture and access to cultures from the CSIRO Algal Culture Collection. Our early research was supported by Grants 86/81, 88/69, 90/63 and 91/59 from the Fisheries Research and Development Corporation, Grant A18831836 from the Australian Research Council and a Rural Credits Development Grant. We gratefully acknowledge funding from the CRC for Aquaculture and the Aquafin CRC for funding our studies.

REFERENCES

Allnutt, F.C.T. 1996. Cyanobacterial (bluegreen algal) biotechnology: past, present and future. J Sci. Ind. Res. India 55: 693-714.

Apt, K.E. and P.W. Behrens. 1999. Commercial developments in microalgal biotechnology. J. Phycol. 35: 215-226.

Arzul, G., P. Gentien and G. Bodennec. 1998. Potential toxicity of microalgal polyunsaturated fatty acids (PUFAs). pp. 53-62. *In* G. Baudimant, J. Guézennec, P. Roy, J.-F. Samain. [eds]. Marine Lipids. Proceedings of the Symposium Held in Brest, 19-20 November 1998. Actes de Colloques 27, Ifremer, Plouzane, France.

Barclay, W. and S. Zeller. 1996. Nutritional enhancement of n-3 and n-6 fatty acids in rotifers and *Artemia* nauplii by feeding spray-dried *Schizochytrium* sp. J. World Aqua. Soc. 27: 314-322.

Barclay, W.R., K.M. Meager and J.R. Abril. 1994. Heterotrophic production of long chain omega-3 fatty acids utilizing algae and algae-like microorganisms. J. Appl. Phycol. 6: 123-129.

Bates, S.S. 2000. Domoic-acid-producing diatoms: Another genus added! J. Phycol. 36: 978-983.

Baum, R. 1994. Microalgae are possible source of biodiesel fuel. Chem. Eng. News 72: 28-29.

Behrens, P.W. and D.J. Kyle. 1996. Microalgae as a source of fatty acids. J. Food Lipids 3: 259-272.

Bell, M.V., J.R. Dick and D.W. Pond. 1997. Octadecapentaenoic acid in a raphidophyte alga, *Heterosigma akashiwo*. Phytochemistry 45: 303-306.

Belt, S.T., G. Massé, W.G. Allard, J.-M. Robert and S.J. Rowland. 2001. C_{25} highly branched isoprenoid alkenes in planktonic diatoms of the *Pleurosigma* genus. Org. Geochem. 32: 1271-1275.

Ben-Amotz, A., T.G. Tornabene and W.H. Thomas. 1985. Chemical profile of selected species of microalgae with emphasis on lipids. J. Phycol. 21: 72–81.

Ben-Amotz, A. and R. Fischler. 1990. Long-chain polyunsaturated alkenones in *Isochrysis galbana* and their possible significance in the nutritional requirements of fish larvae. pp. 217-229. *In* H. Rosenthal and S. Sarig [eds.]. Research in Modern Aquaculture. Special Publication No. 11, European Aquaculture Society, Bredene, Belgium.

Bergen, B.J., J.G. Quinn and C.C. Parrish. 2000. Quality-assurance study of marine lipid-class determination using Chromarod/Iatroscan® thin-layer chromatography-flame ionization detector. Environ Toxicol Chem 19: 2189-2197.

Berge, J.P., J.-P. Gouygou, J.-P. Dubacq and P. Durand. 1995. Reassessment of lipid composition of the diatom *Skeletonema costatum*. Phytochemistry 39: 1017-1021.

Bidlingmeyer, B.A., S.A. Cohen and T.L. Tarvin. 1984. Rapid analysis of amino acids using pre-column derivitization. J. Chromatogr. 336: 93-104.

Blakeney, A.B., P.J. Harris, R.J. Henry and B.A. Stone. 1983. A simple and rapid preparation of alditol acetates for monosaccharide analysis. Carbohyd. Res. 113: 291-299.

Bligh, E.G. and W.J. Dyer. 1959. A rapid method of total lipid extraction and purification. Can. J. Biochem. Physiol. 37: 912-917.

Borowitzka, M. 1992. Algal biotechnology products and processes - matching science and economics. J. Appl. Phycol. 4: 267-279.

Borowitzka, M.A. 1995. Microalgae as sources of pharmaceuticals and other biologically-active compounds. J. Appl. Phycol. 7: 3-15.

Borowitzka, M.A. 1999. Commercial production of microalgae: Ponds, tanks, tubes and fermenters. J. Biotechnol. 70: 313-321.

Boswell, K.D.B., R. Gladue, B. Prima and D.J. Kyle. 1992. SCO production by fermentative microalgae. pp. 274-286. *In* D.J. Kyle and C. Ratledge [eds.]. Industrial Applications of Single Cell Oils. American Oil Chemists' Society, Champaign, USA.

Brassell, S.C., G. Eglinton, I.T. Marlowe, U. Pflaumann and M. Sarnthein. 1986. Molecular stratigraphy: a new tool for climatic assessment. Nature 320: 129-133.

Brett, M.T. and D.C. Müller-Navarra. 1997. The role of highly unsaturated fatty acids in aquatic food web processes. Freshw. Biol. 38: 483-499.

Brown, M.R. 1991. The amino acid and sugar composition of 16 species of microalgae used in mariculture. J. Exp. Mar. Biol. Ecol. 145: 79-99.

Brown,M.R. and S.W. Jeffrey. 1992. Biochemical composition of microalgae from the green algal classes Chlorophyceae and Prasinophyceae. 1. Amino acids, sugars and pigments. J. Exp. Mar. Biol. Ecol.161: 91-113.

Brown, M.R. and K.A. Miller. 1992. The ascorbic acid content of eleven species of microalgae used in mariculture. J. Appl. Phycology, 4: 205-215.

Brown, M.R. and C.A. Farmer. 1994. Riboflavin content of six species of microalgae used in mariculture. J. Appl. Phycol. 6: 61-65.

Brown, M.R. and S.W. Jeffrey. 1995. The amino acid and gross composition of marine diatoms useful for mariculture. J. Appl. Phycology, 7: 521-527.

Brown, M.R. and M.A. McCausland. 2000. Increasing the growth of juvenile Pacific oysters *Crassostrea gigas* by supplementary feeding with microalgal and dried diets. Aquacult. Res. 31: 671-682.

Brown, M. and R. Robert. 2002. Preparation and assessment of microalgal concentrates as feeds for larval and juvenile Pacific oyster (*Crassostrea gigas*). Aquaculture 207: 289-309

Brown, M.R., C.D. Garland, S.W. Jeffrey, I.D. Jameson and J.M. LeRoi. 1993a. The gross and amino acid compositions of batch and semi-continuous cultures of *Isochrysis* sp. (clone T.ISO), *Pavlova lutheri* and *Nannochloropsis oculata*. J. Appl. Phycol. 5: 285-296.

Brown, M.R. G.A. Dunstan, S.W. Jeffrey, J.K. Volkman, S.M. Barrett and J.M. LeRoi. 1993b. The influence of irradiance on the biochemical composition of the prymnesiophyte *Isochrysis* sp. (clone T-ISO). J. Phycol. 29: 601-612

Brown, M.R., G.A. Dunstan, K.A. Miller and S.A. Norwood. 1996. Effects of harvest stage and light on the biochemical composition of the diatom *Thalassiosira pseudonana*. J. Phycol. 32: 64-73.

Brown, M.R., S.W. Jeffrey, J.K. Volkman and G.A. Dunstan. 1997. Nutritional properties of microalgae for mariculture. Aquaculture 151: 315-331.

Brown, M.R., M.A. McCausland and K. Kowalski. 1998a. The nutritional value of four Australian microalgal strains fed to Pacific oyster *Crassostrea gigas* spat. Aquaculture 165: 281-293.

Brown, M.R., S. Skabo and B. Wilkinson. 1998b. The enrichment and retention of ascorbic acid in rotifers fed microalgal diets. Aquaculture Nutr. 4: 151-156.

Brown, M.R., M. Mular, I. Miller, C. Trenerry and C. Farmer. 1999. The vitamin content of microalgae used in aquaculture. J. Appl. Phycology 11: 247-255.

Budge, S.M. and C.C. Parrish. 1999. Lipid class and fatty acid composition of *Pseudo-nitzschia multiseries* and *Pseudo-nitzschia pungens* and effects of lipolytic enzyme deactivation. Phytochemistry 52: 561-566.

Burkow, I.C. and R.J. Henderson. 1991. Analysis of polymers from autoxidized marine oils by gel permeation HPLC using a light-scattering detector. Lipids 26: 227-231.

Buzzi, M., R.J. Henderson and J.R. Sargent. 1997. Biosynthesis of docosahexaenoic acid in trout hepatocytes proceeds via 24-carbon intermediates. Comp. Biochem. Physiol. 116B: 263-267.

Castell, J.D., D.E. Conklin, J.S. Craigie, S.P. Lall and K. Norman-Boudreau. 1986. Aquaculture nutrition. pp. 251–308. *In*: M. Bilo, H. Rosenthal and C. J. Sindermann [eds.]. Realism in Aquaculture: Achievements, Constraints, Perspectives, Proceedings of the World Conference on Aquaculture, Venice, Italy, 21-25 September, 1981. European Aquaculture Society, Bredene, Belgium.

Certik, M. and S. Shimizu. 1999. Production and application of single cell oils. Agro. Food Ind. Hi-Tech. 10: 26-32.

Chen, F. and M.R. Johns. 1995. A strategy for high cell density culture of the heterotrophic microalgae with inhibitory substrates. J. Appl. Phycol. 7: 43-46.

Chrismadha, T. and M.A. Borowitzka. 1994. Effect of cell density and irradiance on growth, proximate composition and eicosapentaenoic acid production of *Phaeodactylum tricornutum* grown in a tubular photobioreactor. J. Appl. Phycol. 6: 67-74.

Christie, W.W. 1982. Lipid analysis. Pergamon Press, Oxford, UK.

Christie, W.W. 1998. Some recent advances in the chromatographic analysis of lipids. Analusis 26: M34-M40.

Chu, F.E., J.L. Dupuy and K.L. Webb. 1982. Polysaccharide composition of five algal species used as food for larvae of the American oyster, *Crassostrea virginica*. Aquaculture 29: 241-252.

Clayton, J.R. Jr., Q. Dortch, S.S. Thoreson and S.I. Ahmed. 1988. Evaluation of methods for the separation and analysis of proteins and free amino acids in phytoplankton samples. J. Plankton Res. 10: 341-358.

Cohen, S.A., M. Meys and T.L. Tarvin. 1988. The Pico-Tag Method: A Manual of Advanced Techniques for Amino Acid Analysis. Waters Chromatography Division, Millipore Corporation, Milford, Massachusetts, USA.

Conklin, D.E. 1997. Vitamins. pp. 123–149. *In* L.R. D'Abramo, D.E. Conkin, and D.M. Akiyama [eds.]. Crustacean Nutrition, Advances in World Aquaculture. Vol. 6. World Aquaculture Society.

Conte, M.H., J.K. Volkman, and G. Eglinton. 1994. Lipid biomarkers of the Haptophyta. pp. 351-377. *In* J.C. Green and B.S.C. Leadbeater [eds]. The Haptophyte Algae. Systematics Association Special Volume No. 51. Clarendon Press, Oxford,. UK.

Conte, M.H., A. Thompson and G. Eglinton. 1995. Lipid biomarker diversity in the coccolithophorid *Emiliania huxleyi* (Prymnesiophyceae) and the related species *Gephyrocapsa oceanica*. J. Phycol. 31: 272-282.

Conte, M.H., A. Thompson, D. Lesley and R.P. Harris. 1998. Genetic and physiological influences on the alkenone/alkenoate versus growth temperature relationship in *Emiliania huxleyi* and *Gephyrocapsa oceanica*. Geochim. Cosmochim. Acta 62: 51-68.

Cooper, S.F., A. Battat, P. Marsot, M. Sylvestre and C. Laliberte. 1985. Identification of antibacterial fatty acids from *Phaeodactylum tricornutum* grown in dialysis culture. Microbios 42: 27-36.

Coutteau, P. and P. Sorgeloos. 1992. The use of algal substitutes and the requirement for live algae in the hatchery and nursery rearing of bivalve molluscs: an international survey. J. Shellfish Res. 11: 467-476.

Coutteau, P., J.D. Castell, R.G. Ackman and P. Sorgeloos. 1996. The use of lipid emulsions as carriers for essential fatty acids in bivalves: a test case with juvenile *Placopecten magellanicus*. J. Shellfish Res. 15: 259-264.

Daume, S., B.M. Long and P. Crouch. 2003. Changes in amino acid content of an algal feed species (*Navicula* sp.) and their effect on growth and survival of juvenile abalone (*Haliotis rubra*). J. Appl. Phycol. 15: 201-207.

Day, J.G. and A.J. Tsavalos. 1996. An investigation of the heterotrophic culture of the green alga *Tetraselmis*. J. Appl. Phycol. 8: 73-77.

Delaunay, F., J. Marty, J. Moal and J.-F. Samain. 1993. The effect of monospecific algal diets on growth and fatty acid composition of *Pecten maximus* (L.) larvae. J. Exp. Mar. Biol. Ecol. 173: 163-179.

De Roeck-Holtzhauer, Y., I. Quere and C. Claire. 1991. Vitamin analysis of five planktonic microalgae and one macroalga. J. Appl. Phycol. 3: 259-264.

Derrien, A., L.J.M. Coiffard, C. Coiffard, and Y. De Roeck-Holtzhauer. 1998. Free amino acid analysis of five microalgae. J. Appl. Phycol.10: 131-134.

Dobson, G., W.W. Christie and B. Nikolovada-Myanova. 1995. Silver ion chromatography of lipids and fatty acids. J. Chromatogr. B-Biomed. Appl. 671: 197-222.

Donaldson, J. 1991. Commercial production of microalgae at Coast Oyster Company. pp 229-236. *In* W. Fulks and K.L. Main [eds.]. Rotifer and Microalgae Culture Systems.

Proceedings of a US-Asia Workshop, Honolulu, Hawaii, January 28-31, 1991. The Oceanic Institute, Hawaii, USA.

Donato, M., M.H. Vilela and N.M. Bandarra. 2003. Fatty acids, sterols, α-tocopherol and total carotenoids composition of *Diacronema vlkianum*. J. Food Lipids 10: 267-276.

Douglas, D.J. and S.S. Bates. 1992. Production of domoic acid, a neurotoxic amino acid, by an axenic culture of the marine diatom *Nitzschia pungens f- multiseries* Hasle. Can. J. Fish. Aquat. Sci. 49: 85-90.

Dubois, M., K.A. Gillies, J.K. Hamilton, P.A. Rebers and F. Smith. 1956. Colorimetric method for the determination of sugars and related substances. Anal. Chem. 28: 350-356.

Dunahay, T.G., E.E. Jarvis, S.S. Dais and P.G. Roessler. 1996. Manipulation of microalgal lipid production using genetic engineering. Appl. Biochem. Biotechnol. 57/58, 223-231.

Dunstan, G.A., J.K. Volkman, S.W. Jeffrey and S.M. Barrett. 1992. Biochemical composition of microalgae from the green algal classes Chlorophyceae and Prasinophyceae. 2. Lipid classes and fatty acids. J. Exp. Mar. Biol. Ecol. 161: 115–134.

Dunstan, G.A., J.K. Volkman, S.M. Barrett, and C.D. Garland. 1993. Changes in the lipid composition and maximisation of the polyunsaturated fatty acid content of three microalgae grown in mass culture. J. Appl. Phycol. 5: 71-83.

Dunstan, G.A., J.K. Volkman, S.M. Barrett, J.M. Leroi and S.W. Jeffrey. 1994. Essential polyunsaturated fatty acids from fourteen species of diatom (Bacillariophyceae). Phytochemistry 35: 155-161.

Eichenberger, W. 1993. Betaine lipids in lower plants: distribution of DGTS, DGTA and phospholipids, and the intracellular localization and site of biosynthesis of DGTS. Plant Physiol. Biochem. 31: 213-221.

Eichenberger, W. and C. Gribi. 1997. Lipids of *Pavlova lutheri*: Cellular site and metabolic role of DGCC. Phytochemistry 45: 1561-1567.

Eichenberger, W., H. Gfeller, P. Grey, C. Gribi, and R.J. Henderson. 1996. Gas chromatographic mass spectrometric identification of betaine lipids in *Chroomonas salina*. Phytochemistry 42: 967-972.

Enright, C.T., G.F. Newkirk, J.S. Craigie and J.D. Castell. 1986a. Growth of juvenile *Ostrea edulis* L. fed *Chaetoceros calcitrans* Schütt of varied chemical composition. J Exp. Mar. Biol. Ecol. 96: 15–26.

Enright, C.T., G.F. Newkirk, J.S. Craigie and J.D. Castell. 1986b. Evaluation of phytoplankton as diets for juvenile *Ostrea edulis* L. J Exp. Mar. Biol. Ecol. 96: 1-13.

Epifanio, C.E., C.C. Valenti and C.L. Turk. 1981. A comparison of *Phaeodactylum tricornutum* and *Thalassiosira pseudonana* as foods for the oyster, *Crassostrea virginica*. Aquaculture 23: 347-353.

Fábregas, J. and C. Herrero. 1986. Marine microalgae as a potential source of minerals in fish diets. Aquaculture 51: 237-243.

Fábregas, J., A. Otero, A. Dominguez and M. Patino. 2001. Growth rate of the microalga *Tetraselmis suecica* changes the biochemical composition of *Artemia* species. Mar. Biotechnol. 3: 256-263.

Falk, M., D.G. Smith, J. McLachlan and A.G. McInnes. 1966. Studies on chitan [β-(1-4)-linked 2-acetamido-2-deoxy-D-glucan] fibres of the diatom *Thalassiosira fluviatilis* Hustedt. Can. J. Chem. 44: 2269-2281

Fisher, N.S. 1985. Accumulation of metals by marine picoplankton. Mar. Biol. 87: 137-142.

Flynn, K.J. 1990. Composition of intracellular and extracellular pools of amino acids, and amino acid utilization of microalgae of different sizes. J. Exp. Mar. Biol. Ecol. 139: 151-166.

Frankel, E.N., T. Satue-Gracia, A.S. Meyer and J.B. German. 2002. Oxidative stability of fish and algae oils containing long chain polyunsaturated fatty acids in bulk and in oil-in-water emulsions. J. Agric. Food Chem. 50: 2094-2099.

Gallager, S.M., R. Mann and G.C. Sasaki. 1986. Lipids as an index of growth and viability in three species of bivalve larvae. Aquaculture 56: 81-103.

Gara, B., R.J. Shields and L. McEvoy. 1998. Feeding strategies to achieve correct metamorphosis of Atlantic halibut, *Hippoglossus hippoglossus* L., using enriched *Artemia*. Aquaculture Res. 29: 935-948.

Geider, R.J. and J. La Roche. 2002. Redfield revisited: variability of C:N:P in marine microalgae and its biochemical basis. Eur. J. Phycol. 37: 1-17.

Gelin, F., I. Boogers, A.A.M. Noordeloos, J.S. Sinninghe Damsté, R. Riegman and J.W. de Leeuw. 1997. Resistant biomacromolecules in marine microalgae of the classes Eustigmatophyceae and Chlorophyceae: Geochemical implications. Org. Geochem. 26: 659-675.

Gelin, F., J.K. Volkman, C. Largeau, S. Derenne, J.S. Sinninghe Damsté and J.W. de Leeuw. 1999. Distribution of aliphatic, nonhydrolyzable biopolymers in marine microalgae. Org. Geochem. 30: 147-159.

Geresh, S. and S. Arad. 1991. The extracellular polysaccharides of the red microalgae: chemistry and rheology. Biores. Technol. 38: 350-356.

Ghioni, C. A.E.A. Porter, I.H. Sadler, D.R. Tocher and J.R. Sargent. 2001. Cultured fish cells metabolize octadecapentaenoic acid (all-*cis* Δ3,6,9,12,15-18:5) to octadecatetraenoic acid (all-*cis* Δ6,9,12,15-18:4) *via* its 2-*trans* intermediate (*trans* Δ2, all-*cis* Δ6,9,12,15-18:5). Lipids 36: 145-152.

Gladue, R. 1991. Heterotrophic microalgae production: Potential for application to aquaculture feeds. pp. 275-286. *In* W. Fulks and K.L. Main [eds.]. Rotifer and Microalgae Culture Systems. Proceedings of a US-Asia Workshop, Honolulu, Hawaii, January 28-31, 1991 The Oceanic Institute, Hawaii, USA.

Guckert, J.B. and K.E. Cooksey. 1990. Triglyceride accumulation and fatty acid profile changes in *Chlorella* (Chlorophyta) during high pH-induced cell cycle inhibition. J. Phycol. 26: 72-79.

Guillard, R.R.L. 1975. Culture of phytoplankton for feeding marine invertebrates. pp. 29-60. *In* W.L. Smith and M.H. Chanley [eds]. Culture of Marine Invertebrate Animals. Plenum Press, New York, USA.

Guillard, R.R.L. and J.H. Ryther. 1962. Studies of marine planktonic diatoms. I. *Cyclotella nana* Hustedt, and *Detonula confervacea* (Cleve) Gran. Can. J. Microbiol. 8: 229-239.

Guzman, S., A. Gato and J.M. Calleja. 2001. Antiinflammatory, analgesic and free radical scavenging activities of the marine microalgae *Chlorella stigmatophora* and *Phaeodactylum tricornutum*. Phytother. Res. 15: 224-230.

Hallegraeff, G.M., P.D. Nichols, J.K. Volkman, S.I. Blackburn and D.A. Everitt. 1991. Pigments, fatty acids, and sterols of the toxic dinoflagellate *Gymnodinium catenatum*. J. Phycol. 27: 591-599.

Handa, N. and K. Yanagi. 1969. Studies on water extractable carbohydrates of the particulate matter from the northwest Pacific Ocean. Mar. Biol. 4: 197-207.

Hapette, A.M. and S.A. Poulet. 1990. Variation of vitamin C in some common species of marine plankton. Mar. Ecol. Prog. Ser. 64: 69-79.

Hardwick, S.J., K.L.H. Carpenter, N.S. Law, C. Vanderveen, C.E. Marchant, R. Hird and M.J. Mitchinson. 1997. Toxicity of polyunsaturated fatty acid esters for human monocyte-macrophages: the anomalous behaviour of cholesteryl linolenate. Free Radical Res. 26: 351-362.

Harel, M., W. Koven, I. Lein, Y. Bar, P. Behrens, J. Stubblefield, Y. Zohar and A.R. Place. 2002. Advanced DHA, EPA and ArA enrichment materials for marine aquaculture using single cell heterotrophs. Aquaculture 213: 347-362.

Harrison, P.J., P.A. Thompson and G.S. Calderwood. 1990. Effects of nutrient and light limitation on the biochemical composition of phytoplankton. J. Appl. Phycol. 2: 45–56.

Heasman, M., J. Diemar, W. O'Connor, T. Sushames and L. Foulkes. 2000. Development of extended shelf-life micro-algae concentrate diets harvested by centrifugation for bivalve molluscs – a summary. Aquacult. Res. 31: 637-659.

Hecky, R.E., K. Mopper, P. Kilham and E.T. Degens. 1973. The amino acid and sugar composition of diatom cell-walls. Mar. Biol. 19: 323-331.

Hendriks, I.E., L.A. van Duren and P.M.J. Herman. 2003. Effect of dietary polyunsaturated fatty acids on reproductive output and larval growth of bivalves. J. Exp. Mar. Biol. Ecol. 296: 199-213.

Hu, H.H. and K.S. Gao. 2003. Optimization of growth and fatty acid composition of a unicellular marine picoplankton, *Nannochloropsis* sp., with enriched carbon sources. Biotechnol. Lett. 25: 421-425.

Ikekawa, N. 1985. Structures, biosynthesis and function of sterols in invertebrates. pp. 199-230. In *Sterols and Bile Acids*, H. Danielsson and S. Sjovall ed. Elsevier/North Holland Biomedical, Amsterdam, the Netherlands.

James, C.M., A.M. Al-Khars and P. Chorbani. 1988. pH dependent growth of *Chlorella* in a continuous culture system. J. World Agric. Soc. 19: 27-35.

James, C.M., S. Al-Hinty and A.E. Salman. 1989. Growth and ω3 fatty acid and amino acid composition of microalgae under different temperature regimes. Aquaculture 77: 337-357.

Janssen, M., J. Tramper, L.R. Mur and R.H. Wijffels. 2002. Enclosed outdoor photobioreactors: Light regime, photosynthetic efficiency, scale-up, and future prospects. Biotechnol. Engineer. 81: 193-210.

Jeffrey, S., J.-M. LeRoi and M.R. Brown. 1992. Characteristics of microalgal species for Australian mariculture. pp. 164-173. *In* G.L. Allan and W. Dall [eds.]. Proceedings of the National Aquaculture Workshops, Pt. Stephens, NSW Australia, April 1991.

Jeffrey, S.W., H.S. MacTavish, W.C. Dunlap, M. Vesk and K. Groenewoud. 1999. Occurrence of UVA- and UVB-absorbing compounds in 152 species (206 strains) of marine microalgae. Mar. Ecol. Prog. Ser. 189: 35-51.

Joseph, J.D. 1975. Identification of 3,6,9,12,15-octadecapentaenoic acid in laboratory cultured photosynthetic dinoflagellates. Lipids 10: 395-403.

Kato, M., M. Sakai, K. Adachi, H. Ikemoto and H. Sano. 1996. Distribution of betaine lipids in marine algae. Phytochemistry 42: 1341-1345.

Kawamura, T., R.D. Roberts and C.M. Nicholson. 1998. Factors affecting the food value of diatom strains for post-larval abalone *Haliotis iris*. Aquaculture 160: 81-88.

Knauer, J. and P.C. Southgate. 1997. Growth and fatty acid composition of Pacific oyster (*Crassostrea gigas*) spat fed a microalga and microcapsules containing varying amounts of eicosapentaenoic and docosahexaenoic acid. J. Shellfish Res. 16: 447-453.

Knauer, J. and P.C. Southgate. 1999. A review of the nutritional requirements of bivalves and the development of alternative and artificial diets for bivalve aquaculture. Rev. Fish. Sci. 7: 241-280.

Knauer, J., R.G. Kerr, D. Lindley and P. Southgate. 1998. Sterol metabolism of Pacific oyster (*Crassostrea gigas*) spat. Comp. Biochem. Physiol. 118B: 81-84.

Knauer, J., S.M. Barrett, J.K. Volkman and P.C. Southgate. 1999. Assimilation of dietary phytosterols by Pacific oyster *Crassostrea gigas* spat. Aqua. Nutr. 5: 257-266.

Knuckey R.M., M.R. Brown, S.M. Barrett and G.M. Hallegraeff. 2002. Isolation of new nanoplanktonic diatom strains and their evaluation as diets for the juvenile Pacific oyster (*Crassostrea gigas*). Aquaculture 211: 253-274.

Koven, W., S. Kolkovski, E. Hadas, K. Gamsiz and A. Tandler. 2001. Advances in the development of microdiets for gilthead seabream, *Sparus aurata*: a review. Aquaculture 194: 107-121.

Kyle, D.J. and C. Ratledge [eds.]. 1992. Industrial Applications of Single Cell Oils. American Oil Chemists' Society, Champaign.

Laing, I., A.R. Child and A. Janke. 1990. Nutritional value of dried algae diets for larvae of manila clam (*Tapes philippinarum*). J. Mar. Biol. Assn. U.K. 70: 1-12.

Langdon, C.J. and M.J. Waldock. 1981. The effect of algal and artificial diets on the growth and fatty acid composition of *Crassostrea gigas* spat. J. Mar. Biol. Assn. U.K. 61: 431-448.

Lavens, P. and P. Sorgeloos. 1996. Manual on the production and use of live food for aquaculture. FAO Fisheries Technical Paper 361, Food and Agriculture Organization of the United Nations, Rome, Italy.

Lazarus, W. 1973. Purification of plant extracts for ion-exchange chromatography of free amino acids. J. Chromatogr. 87: 169-178.

Lebeau, T. and J.M. Robert. 2003. Diatom cultivation and biotechnologically relevant products. Part II: Current and putative products. Appl. Microb. Biotechnol. 60: 624-632.

Leblond, J.D. and P.J. Chapman. 2000. Lipid class distribution of highly unsaturated long chain fatty acids in marine dinoflagellates. J. Phycol. 36: 1103-1108.

Leblond, J.D., and T.J. Evans and P.J. Chapman. 2003. The biochemistry of dinoflagellate lipids, with particular reference to the fatty acid and sterol composition of a *Karenia brevis* bloom. Phycologia 42: 324-331.

Lee, B.H. and G.A. Picard. 1982. Chemical analysis of unicellular algal biomass from synthetic medium and sewage effluent. Can. Inst. Food. Sci. Technol. J. 15: 58–64.

Lee, Y.K. 1997. Commercial production of microalgae in the Asia-Pacific rim. J. Appl. Phycol. 9: 403-411.

Lee, Y.K. 2001. Microalgal mass culture systems and methods: their limitation and potential. J. Appl. Phycol. 13: 307-315.

Lie, Ø., H. Haaland, G.I. Hemre, A. Maage, E. Lied, G. Rosenlund, K. Sandnes and Y. Olsen. 1997. Nutritional composition of rotifers following a change in diet from yeast and emulsified oil to microalgae. Aquacult. Nutrition 5: 427-438.

López Alonso, D., E.H. Belarbi, J.M. Fernandez-Sevilla, J. Rodriguez-Ruiz and E. Molina Grima. 2000. Acyl lipid composition variation related to culture age and nitrogen concentration in continuous culture of the microalga *Phaeodactylum tricornutum*. Phytochemistry 54: 461-471.

Lorenz, R.T. and G.R. Cysewski. 2000. Commercial potential for *Haematococcus* microalgae as a natural source of astaxanthin. Trends Biotechnol. 18: 160-167.

Mansour, M.P., J.K. Volkman, A.E. Jackson and S.I. Blackburn. 1999. The fatty acid and sterol composition of five marine dinoflagellates. J. Phycol. 35: 710-720.

Mansour, M.P., J.K. Volkman and S.I. Blackburn. 2003. The effect of growth phase on the lipid class, fatty acid and sterol composition in the marine dinoflagellate, *Gymnodinium* sp. in batch culture. Phytochemistry 63: 145-153.

Martin-Creuzburg, D. and E. von Elert. 2004. Impact of 10 dietary sterols on growth and reproduction of *Daphnia galeata*. J. Chem. Ecol. 30: 483-500.

Maruyama, I., T. Nakamura, T. Matsubayashi, Y. Ando and T. Maeda. 1986. Identification of the alga known as "marine *Chlorella*" as a member of the Eustigmatophyceae. Jap. J. Phycol. 34: 319-325.

Materassi, R., C. Paoletti, W. Balloni and G. Florenzano. 1980. Some considerations on the production of lipid substances by microalgae and cyanobacteria. pp. 619-626. *In* G. Shelef and C. J. Soeder [eds.]. Algae Biomass. Elsevier, Amsterdam., the Netherlands.

Mathew, B., R. Sankaranarayanan, P.P. Nair, C. Varghese, T. Somanathan, B.P. Amma, N.S. Amma and M.K. Nair. 1995. Evaluation of chemoprevention of oral cancer with *Spirulina fusiformis*. Nutr. Cancer-Int. J. 24: 197-202.

McCausland, M.A., M.R. Brown, S.M. Barrett, J.A. Diemar and M.P. Heasman. 1999. Evaluation of live and pasted microalgae as supplementary food for juvenile Pacific oysters (*Crassostrea gigas*). Aquaculture 174: 323-342.

Meireles, L.A., A.C. Guedes and F.X. Malcata. 2003. Lipid class composition of the microalga *Pavlova lutheri*: eicosapentaenoic and docosahexaenoic acids. J. Agric. Food Chem. 51: 2237-2241.

Merchie, G., P. Lavens, P. Dhert, M. Dehasque, H. Nelis, A. De Leenheer and P. Sorgeloos. 1995. Variation of ascorbic acid content in different live food organisms. Aquaculture 134: 325-337.

Metzger, P. and E. Casadevall. 1983. Structure de trois nouveaux "botryococcenes" synthetises par une souch de *Botryococcus braunii* cultivee en laboratoire. Tetrahedron Lett. 24: 4013-4016.

Miralto, A., G. Barone, G. Romano, S.A. Poulet, A. Ianora, G.L. Russo, I. Buttino, G. Mazzarella, M. Laabir, M. Cabrini and M.G. Giacobbe 1999. The insidious effect of diatoms on copepod reproduction. Nature 402: 173-176.

Miranda, M.S., R.G. Cintra, S.B.M. Barros and J. Mancini. 1998. Antioxidant activity of the microalga *Spirulina maxima*. Brazilian J. Med. Biol. Res. 31: 1075-1079.

Molina Grima, E., A. Robles Medina, A. Giménez Giménez, J.A. Sánchez Pérez, F. García Camacho and J.L. García Sanchez. 1994. Comparison between extraction of lipids and fatty acids from microalgal biomass. J. Amer. Oil Chem. Soc. 71: 955-959.

Moreno, V.J., J.E.A. de Moreno and R.R. Brenner. 1979. Biosynthesis of unsaturated fatty acids in the diatom *Phaeodactylum tricornutum*. Lipids 14:15-19.

Mykelstad, S. 1974. Production of carbohydrates by marine planktonic diatoms, I. comparison of nine different species in culture. J. Exp. Mar. Biol. Ecol. 15: 261-274.

Newman, S.J., W.C. Dunlap, S. Nicol and D. Ritz. (2000). Antarctic krill (*Euphausia superba*) acquire a UV-absorbing mycosporine-like amino acid from dietary algae. J. Exp. Mar. Biol. Ecol. 255: 93-110.

Nichols, P.D., D.G. Holdsworth, J.K. Volkman, M. Daintith and S. Allanson. 1989. High incorporation of essential fatty acids by the rotifer *Brachionus plicatilis* fed on the prymnesiophyte alga *Pavlova lutheri*. Aust. J. Mar. Freshw. Res. 40: 645-655.

Nordbäck, J., E. Lundberg and W.W. Christie. 1998. Separation of lipid classes from marine particulate material by HPLC on a polyvinyl alcohol-bonded stationary phase using dual-channel evaporative light-scattering detection. Mar. Chem. 66: 165-175.

NRC (National Research Council). 1993. Nutrient requirements of fish. National Academy Press, Washington DC., USA.

O'Connor, W.A., M.P. Heasman. 1997. Diet and feeding regimens for larval doughboy scallops, *Mimachlamys asperrima*. Aquaculture 158: 289-303.

Ogbonna, J.C., S. Tomiyama and H. Tanaka. 1999. Production of α-tocopherol by sequential heterotrophic-photoautotrophic cultivation of *Euglena gracilis*. J. Biotechnol. 70: 213-221.

Ohta, S., F. Ono, Y. Shiomi, T. Nakao, O. Aozasa, T. Nagate, K. Kitamura, S. Yamaguchi, M. Nishi and H. Miyata. 1998. Anti-Herpes Simplex virus substances produced by the marine green alga, *Dunaliella primolecta*. J. Appl. Phycol. 10: 349-355.

Okuyama, H., N. Morita and K. Kogame. 1992. Occurrence of octadecapentaenoic acid in lipids of a cold stenothermic alga, prymnesiophyte strain B. J. Phycol. 28: 465-472.

Parrish, C.C., G. Bodennec and P. Gentien. 1992. Separation of polyunsaturated and saturated lipids from marine phytoplankton on silica gel-coated Chromarods. J. Chromatogr. 607: 97-104.

Parrish, C.C., G. Bodennec, and P. Gentien. 1998. Haemolytic glycoglycerolipids from *Gymnodinium* species. Phytochemistry 47: 783-787.

Parrish, C.C., J.S. Wells, Z. Yang and P. Dabinett. 1999. Growth and lipid composition of scallop juveniles, *Placopecten magellanicus*, fed the flagellate *Isochrysis galbana* with varying lipid composition and the diatom *Chaetoceros muelleri*. Mar. Biol. 133, 461-471.

Parsons, T.R., K. Stephens and J.D.H. Strickland. 1961. On the chemical composition of eleven species of marine phytoplankters. J. Fish. Res. Bd. Canada 18: 1001–1016.

Peirson, W.M. 1983. Utilization of eight algal species by the Bay scallop, *Argopecten irradians concentricus* (Say). J. Exp. Mar. Biol. Ecol. 68: 1-11.

Pernet, F., R. Tremblay, E. Demers and M. Roussy. 2003. Variation of lipid class and fatty acid composition of *Chaetoceros muelleri* and *Isochrysis* sp. grown in a semicontinuous system. Aquaculture 221: 393-406.

Persoone, G. and C. Claus. 1980. Mass culture of algae: a bottleneck in the nursery culturing of molluscs. pp. 265-285. *In* G.S.C.J. Shoeder [ed.]. Algae Biomass. Elsevier, Amsterdam, the Netherlands.

Piretti, M.V., G. Pagliuca, L. Boni, R. Pistocchi, M. Diamante and T. Gazzotti. 1997. Investigation of 4-methyl sterols from cultured dinoflagellate algal strains. J. Phycol. 33: 61-67.

Proteau, P.J., W.H. Gerwick, F. Garciapichel and R.W. Castenholtz. 1993. The structure of scytomenin, an ultraviolet sunscreen pigment from the sheaths of cyanobacteria. Experientia 49: 825-829.

Rai, L.C., H.D. Kumar, F.H. Mohn and C.J. Soeder. 2000. Services of algae to the environment. J. Microbiol. Biotechnol. 10: 119-136.

Rebolloso Fuentes, M.M., G.G. Acién Fernández., J.A. Sánchez Pérez and J.L. Guil Guerrero. 2000. Biomass nutrient profiles of the microalga *Porphyridium cruentum*. Food Chem. 70: 345-353.

Reitan, K.I., J.R. Rainuzzo, G. Øie and Y. Olsen. 1997. A review of the nutritional effects of algae in marine fish larvae. Aquaculture 155: 207-221.

Renaud, S.M., L.V. Thinh and D.L. Parry. 1999. The gross composition and fatty acid composition of 18 species of tropical Australian microalgae for possible use in mariculture. Aquaculture 170: 147-159.

Richmond, A. 2000. Microalgal biotechnology at the turn of the millennium: a personal view. J. Appl. Phycol. 12: 441-451.

Riegman, R., W. Stolte, A.A.M. Noordeloos and D. Slezak. 2000. Nutrient uptake, and alkaline phosphate (EC 3:1:3:1) activity of *Emiliania huxleyi* (Prymnesiophyceae) during growth under N and P limitation in continuous cultures. J. Phycol. 36: 87-96.

Rodríguez, C., J.A. Pérez, P. Badía, M.S. Izquierdo, H. Fernández-Palacios and A. Lorenzo Hernández. 1998. The n–3 highly unsaturated fatty acid requirements of gilthead seabream (*Sparus aurata* L.) larvae when using an appropriate DHA/EPA ratio in the diet. Aquaculture 169: 9-23.

Rønnestad, I., S. Helland and Ø Lie. 1998. Feeding *Artemia* to larvae of Atlantic halibut (*Hippoglossus hippoglossus* L.) results in lower larval vitamin A content compared with feeding copepods. Aquaculture 165: 159-164.

Rontani J.-F., B. Beker and J.K. Volkman. 2004. Long-chain alkenones and related compounds in the benthic haptophyte *Chrysotila lamellosa* Anand HAP 17. Phytochemistry 65: 117-126.

Rossignol, N., P. Jaouen, J.M. Robert and F. Quéméneur. 2000. Production of exocellular pigment by the marine diatom *Haslea ostrearia* in a photobioreactor equipped with immersed ultrafiltration membranes. Bioresource Technol. 73: 197-200.

Rousch, J.M., S.E. Bingham and M.R. Sommerfeld. 2003. Changes in fatty acid profiles of thermo-intolerant and thermo-tolerant marine diatoms during temperature stress. J. Exp. Mar. Biol. Ecol. 295: 145-156.

Rowland, S.J., S.T. Belt, E.J. Wraige, G. Massé, C. Roussakis and J.-M. Robert. 2001a. Effects of temperature on polyunsaturation in cytostatic lipids of *Haslea ostrearia*. Phytochemistry 56: 597-602.

Rowland, S.J., W.G. Allard, S.T. Belt, G. Massé, J.-M. Robert, S. Blackburn, D. Frampton, A.T. Revill and J.K. Volkman. 2001b. Factors influencing the distributions of polyunsaturated terpenoids in the diatom, *Rhizosolenia setigera*. Phytochemistry 58: 717-728.

Running, J.A., R.J. Huss and P.T. Olson. 1994. Heterotrophic production of ascorbic acid by microalgae. J. Appl. Phycol. 6: 99-104.

Rupérez, F.J., D. Martín, E. Herrera and C. Barbas. 2001. Chromatographic analysis of α-tocopherol and related compounds in various matrices. J. Chromatogr. A 935: 45-69.

Sargent, J.R., L.A. McEvoy and J.G. Bell. 1997. Requirements, presentation and sources of polyunsaturated fatty acids in marine fish larval feeds. Aquaculture 155: 117-127.

Seguineau, C., A. Laschi-Loquerie, J. Moal and J.F. Samain. 1996. Vitamin requirements in great scallop larvae. Aqua. Int. 4: 315–324.

Sellem, F., D. Pesando, G. Bodennec, A. El Abed and J.P. Girard. 2000. Toxic effects of *Gymnodinium* cf. *mikimotoi* unsaturated fatty acids to gametes and embryos of the sea urchin *Paracentrotus lividus*. Water Res. 34: 550-556.

Shifrin, N.S. and S.W. Chisholm. 1981. Phytoplankton lipids: interspecific differences and effects of nitrate, silicate and light-dark cycles. J. Phycol. 17: 374-384.

Shiran, D., I. Khozin, Y.M. Heimer and Z. Cohen. 1996. Biosynthesis of eicosapentaenoic acid in the microalga *Porphyridium cruentum*.1: The use of externally supplied fatty acids. Lipids 31: 1277-1282.

Sieburth, J.M., M.D. Keller, P.W. Johnson and S.M. Myklestad. 1999. Widespread occurrence of the oceanic ultraplankter, *Prasinococcus capsulatus* (Prasinophyceae), the diagnostic 'Golgi-decapore complex' and the newly described polysaccharide 'capsulan'. J. Phycol. 35: 1032-1043.

Silversand, C. and C. Haux. 1997. Improved high-performance liquid chromatographic method for the separation and quantification of lipid classes: application to fish lipids. J. Chromatogr. B 703: 7-14.

Singh, S., S. Arad and A. Richmond. 2000. Extracellular polysaccharide production in outdoor mass cultures of *Porphyridium* sp. in flat plate glass reactors. J. Appl. Phycol. 12: 269-275.

Singh, S.C., R.P. Sinha and D.P. Hader. 2002. Role of lipids and fatty acids in stress tolerance in cyanobacteria. Acta Protozool. 41: 297-308.

Sinninghe Damsté, J.S., S. Rampen, W.I.C. Rijpstra, B. Abbas, G. Muyzer and S. Schouten. 2003. A diatomaceous origin for long-chain diols and mid-chain hydroxy methyl alkanoates widely occurring in Quaternary marine sediments: Indicators for high-nutrient conditions. Geochim. Cosmochim. Acta 67: 1339-1348.

Siron, R., G. Giusti and B. Berland. 1989. Changes in the fatty acid composition of *Phaeodactylum tricornutum* and *Dunaliella tertiolecta* during growth and under phosphorus deficiency. Mar. Ecol. Prog. Ser. 55: 95-100.

Smedes, F. and T.K. Askland. 1999. Revisiting the development of the Bligh and Dyer total lipid determination method. Mar. Poll. Bull. 38: 193-201.

Smith, G.G., M.R. Brown and A.J. Ritar. 2004. Feeding juvenile *Artemia* enriched with ascorbic acid improves larval survival in the spiny lobster *Jasus edwardsii*. Aquac. Nutr. 10: 105-112.

Sola, F., A. Masoni, B. Fossat, J. Porthe-Nibelle, P. Gentien and G. Bodennec. 1999. Toxicity of fatty acid 18:5n3 from *Gymnodinium* cf. *mikimotoi*: I. Morphological and biochemical aspects on *Dicentrarchus labrax* gills and intestine. J. Appl. Toxicol. 19: 279-284.

Sprecher, H., Q. Chen and F.Q. Yin. 1999. Regulation of the biosynthesis of 22:5n-6 and 22:6n-3: a complex intracellular process. Lipids 34: S153-S156.

Stephanopoulos, G. and J. Kelleher. 2001. Biochemistry - How to make a superior cell. Science 292: 2024-2025.

Tacon, A.G.J. 1991. Vitamin nutrition in shrimp and fish. pp. 10-41. *In* D.M. Akiyama and R.K.H. Tan [eds.]. Proceedings of the Aquaculture Feed Processing and Nutrition Workshop, Thailand and Indonesia, September 1991, American Soybean Association, Singapore.

Tamaru, C.S., R. Murashige and C.S. Lee. 1994. The paradox of using background phytoplankton during the larval culture of striped mullet, *Mugil cephalus* L. Aquaculture 119: 167-174.

Tartarotti, B., G. Baffico, P. Temporetti and H.E. Zagarese. 2004. Mycosporine-like amino acids in planktonic organisms living under different UV exposure conditions in Patagonian lakes. J. Plankt. Res. 26: 753-762.

Teshima, S. 1971. Bioconversion of β–sitosterol and 24-methylcholesterol to cholesterol in marine crustacea. Comp. Biochem. Physiol. 39B: 815-822.

Teshima, S.-I. and A. Kanazawa, and H. Sasada. 1983. Nutritional value of dietary cholesterol and other sterols to larval shrimp, *Penaeus japonica* Bate. Aquaculture 31: 159-167.

Thomas, W.H., D.L.R. Seibert, M. Alden, A. Neori and P. Eldridge. 1984. Yields, photosynthetic efficiencies and proximate composition of dense marine microalgal cultures. III. *Isochrysis* sp. and *Monallantus salina* experiments and comparative conclusions. Biomass 5: 299-316.

Thompson, P.A., P.J. Harrison and J.N.C. Whyte. 1990. Influence of irradiance on the fatty acid composition of phytoplankton. J. Phycol. 26: 278-288.

Thompson, P.A., M.-X. Guo and P.J. Harrison. 1992. Effects of variation in temperature. I. On the biochemical composition of eight species of marine phytoplankton. J. Phycol. 28: 481-488.

Thompson, P.A., and M.-X. Guo and P.J. Harrison. 1993. The influence of irradiance on the biochemical composition of three phytoplankton species and their nutritional value for larvae of the Pacific oyster (*Crassostrea gigas*). Mar. Biol. 117: 259-268.

Tocher, D.R. and J.R. Sargent. 1984. Analysis of lipids and fatty acids in ripe roes of some northwest European marine fish. Lipids 19: 492-499.

Torzillo, G., P. Carlozzi, B. Pushparaj, E. Montaini and R. Materassi. 1993. A 2-plane tubular photobioreactor for outdoor culture of *Spirulina*. Biotechnol. Bioeng. 42: 891-898.

Tredici, M.R. and R. Materassi. 1992. From open ponds to vertical alveolar panels: The Italian experience in the development of reactors for the mass cultivation of phototrophic microorganisms. J. Appl. Phycol. 4: 221-231.

Turpin, V., J.M. Robert, P. Goulletquer, G. Masse and P. Rosa. 2001. Oyster greening by outdoor mass culture of the diatom *Haslea ostrearia* Simonsen in enriched seawater. Aqua. Res. 32: 801-809.

Vesk, M., S.W. Jeffrey and G.M. Hallegraeff. 1990. Golden-brown algae: Prymnesiophyceae and Chrysophyta. pp. 96-114. *In* M.N. Clayton and R.J. King [eds.]. Biology of Marine Plants. Longman Cheshire, Melbourne, Australia.

Volkman, J.K., 1986. A review of sterol markers for marine and terrigenous organic matter. Org. Geochem. 9: 83-99.

Volkman, J.K. and P.D. Nichols. 1991. Applications of thin layer chromatography-flame ionization detection to the analysis of lipids and pollutants in marine and environmental samples. J. Planar Chromatogr. 4: 19-26.

Volkman, J.K., G. Eglinton, E.D.S. Corner and J.R. Sargent. 1980a. Novel unsaturated straight-chain C_{37}-C_{39} methyl and ethyl ketones in marine sediments and a coccolithophorid *Emiliania huxleyi*. pp. 219-227. In A.G. Douglas and J.R. Maxwell [eds.] Advances in Organic Geochemistry. 1979. Pergamon Press, Oxford, UK.

Volkman J.K., E.D.S. Corner and G. Eglinton. 1980b. Transformations of biolipids in the marine food web and underlying sediments. pp. 185-197. *In* Colloques Internationaux du CNRS No. 293. Biogéochimie de la matière organique à l'interface eau-sédiment marin. CNRS Paris, France.

Volkman J.K., S.W. Jeffrey, P.D. Nichols, G.I. Rogers and C.D. Garland. 1989. Fatty acid and lipid composition of 10 species of microalgae used in mariculture. J. Exp. Mar. Biol. Ecol. 128: 219-240.

Volkman, J.K., G.A. Dunstan, S.W. Jeffrey and P.S. Kearney. 1991. Fatty acids from microalgae of the genus *Pavlova*. Phytochemistry 30: 1855-1859.

Volkman, J.K., S.M. Barrett, G.A. Dunstan and S.W. Jeffrey. 1992. C_{30}-C_{32} alkyl diols and unsaturated alcohols in microalgae of the class Eustigmatophyceae. Org. Geochem. 18: 131-138.

Volkman, J.K., M.R. Brown, G.A. Dunstan and S.W. Jeffrey. 1993. The biochemical composition of marine microalgae from the class Eustigmatophyceae. J. Phycol. 29: 69-78.

Volkman, J.K., S.M. Barrett and G.A. Dunstan. 1994. C_{25} and C_{30} highly branched isoprenoid alkenes in laboratory cultures of two marine diatoms. Org. Geochem. 21: 407-413.

Volkman, J.K., S.M. Barrett, S.I. Blackburn and E.L. Sikes. 1995. Alkenones in *Gephyrocapsa oceanica* - implications for studies of paleoclimate. Geochim. Cosmochim. Acta 59: 513-520.

Volkman, J.K., S.M. Barrett, S.I. Blackburn, M.P. Mansour, E.L. Sikes and F. Gelin. 1998. Microalgal biomarkers: A review of recent research developments. Org. Geochem. 29: 1163-1179.

Voss, A., M. Reinhart, S. Sankarappa and H. Sprecher. 1991. The metabolism of 7,10,13,16,19-docosapentaenoic acid to 4,7,10,13,16,19-docosahexaenoic acid in rat liver is independent of a 4-desaturase. J. Biol. Chem. 266: 19995-20000.

Watanabe, T., C. Kitajima and S. Fujita. 1983. Nutritional values of live organisms used in Japan for mass propagation of fish: a review. Aquaculture 34: 115–143.

Webb, K.L. and F.E. Chu. 1983. Phytoplankton as a food source for bivalve larvae. pp. 272–291. *In* G.D. Pruder, C.J. Langdon and D.E. Conklin [eds.]. Proceedings of the Second International Conference on Aquaculture Nutrition: Biochemical and Physiological Approaches to Shellfish Nutrition, Louisiana State University, Baton Rouge, LA., USA.

White, A. 1981. Marine zooplankton can accumulate and retain dinoflagellate toxins and cause fish kills. Limnol. Oceanogr. 26: 103-109.

Whyte, J.N.C. 1987. Biochemical composition and energy content of six species of phytoplankton used in mariculture of bivalves. Aquaculture 60: 231–241.

Whyte, J.N.C., N. Bourne and C.A. Hodgson. 1989. Influence of algal diets on biochemical composition and energy reserves in *Patinopecten yessoensis* (Jay) larvae. Aquaculture 78: 333-347.

Wikfors, G.H., G.W. Patterson, P. Ghosh, R.A. Lewin, B.C. Smith and J.H. Alix. 1996. Growth of post-set oysters, *Crassostrea virginica,* on high-lipid strains of algal flagellates *Tetraselmis* spp. Aquaculture 143: 411-419.

Wikfors, G.H. and M. Ohno. 2001. Impact of algal research in aquaculture. J. Phycol. 37: 968-974.

Wyman, C.E. and B.J. Goodman. 1993. Biotechnology for production of fuels, chemicals, and materials from biomass. Appl. Biochem. Biotechnol. 39: 41-59.

Yaguchi, T., S. Tanaka, T. Yokochi, T. Nakahara and T. Higashihara. 1997. Production of high yields of docosahexaenoic acid by *Schizchytrium* sp. strain SR21. J. Amer. Oil Chem. Soc. 74: 1431-1434.

Yasumoto,T. and M. Satake. 1998. Bioactive compounds from marine microalgae. Chimia 52: 63-68.

Zaslavskaia, L.A., J.C. Lippmeier, P.G. Kroth, A.R. Grossman and K.E. Apt. 2000. Transformation of the diatom *Phaeodactylum tricornutum* (Bacillariophyceae) with a variety of selectable marker and reporter genes. J. Phycol. 36: 379-386.